Studies in Big Data

Volume 39

Series editor

Janusz Kacprzyk, Polish Academy of Sciences, Warsaw, Poland
e-mail: kacprzyk@ibspan.waw.pl

The series “Studies in Big Data” (SBD) publishes new developments and advances in the various areas of Big Data- quickly and with a high quality. The intent is to cover the theory, research, development, and applications of Big Data, as embedded in the fields of engineering, computer science, physics, economics and life sciences. The books of the series refer to the analysis and understanding of large, complex, and/or distributed data sets generated from recent digital sources coming from sensors or other physical instruments as well as simulations, crowd sourcing, social networks or other internet transactions, such as emails or video click streams and other. The series contains monographs, lecture notes and edited volumes in Big Data spanning the areas of computational intelligence incl. neural networks, evolutionary computation, soft computing, fuzzy systems, as well as artificial intelligence, data mining, modern statistics and Operations research, as well as self-organizing systems. Of particular value to both the contributors and the readership are the short publication timeframe and the world-wide distribution, which enable both wide and rapid dissemination of research output.

More information about this series at http://www.springer.com/series/11970

Bhabani Shankar Prasad Mishra
Himansu Das · Satchidananda Dehuri
Alok Kumar Jagadev
Editors

Cloud Computing for Optimization: Foundations, Applications, and Challenges

Editors
Bhabani Shankar Prasad Mishra
School of Computer Engineering
KIIT University
Bhubaneswar, Odisha
India

Himansu Das
School of Computer Engineering
KIIT University
Bhubaneswar, Odisha
India

Satchidananda Dehuri
Department of Information
and Communication Technology
Fakir Mohan University
Balasore, Odisha
India

Alok Kumar Jagadev
School of Computer Engineering
KIIT University
Bhubaneswar, Odisha
India

ISSN 2197-6503 ISSN 2197-6511 (electronic)
Studies in Big Data
ISBN 978-3-030-08832-3 ISBN 978-3-319-73676-1 (eBook)
https://doi.org/10.1007/978-3-319-73676-1

Printed on acid-free paper

This Springer imprint is published by the registered company Springer International Publishing AG part of Springer Nature
The registered company address is: Gewerbestrasse 11, 6330 Cham, Switzerland

Bhabani Shankar Prasad Mishra dedicates this work to his parents: Gouri Prasad Mishra and Swarnalata Kar, wife: Dr. Subhashree Mishra and kids: Punyesh Mishra and Anwesh Mishra.

Himansu Das dedicates this work to his wife Swagatika Das for her love and encouragement and also to his parents—Jogendra Das and Suprava Das, for their endless support and guidance.

Satchidananda Dehuri dedicates this work to his wife: Dr. Lopamudra Pradhan, and kids: Rishna Dehuri and Khushyansei Dehuri, also his mother: Kuntala Dehuri, who has always been there for him.

Alok Kumar Jagadev dedicates this work to his wife and kids.

Preface

A computing utility has been a dream of computer scientists, engineers, and industry luminaries for several decades. With a utility model of computing, an application can start small and grow to be big enough overnight. This democratization of computing means that any application has the potential to scale. Hence, an emerging area in the name of cloud computing has become a significant technology trend in current era. It refers to applications and services that run on a distributed network using virtualized resources and accessed by common internet protocols and networking standards. It is distinguished by the notion that resources are virtual and limitless and that details of the physical systems on which software runs are abstracted from the user. Moreover, cost saving, access to greater computing resources, high availability, and scalability are the key features of cloud which attracted people. Cloud provides subscription-based access to infrastructure (resources, storage), platforms, and applications. It provides services in the form of IaaS (Infrastructure as a Service), PaaS (Platform as a Service), and SaaS (Software as a Service).

The purpose of this volume entitled "*Cloud Computing for Optimization: Foundations, Applications, and Challenges*" is to make the interested readers/researchers about the practice of using a network of remote servers hosted on the internet to store, manage, and process data, rather than local server or a personal computer while solving highly complex nonlinear optimization problem. In addition, this volume also magnetizes and sensitizes the readers and researchers in the area of cloud computing by presenting the recent advances in the fields of cloud, and also the tools and techniques.

To achieve the objectives, this book includes sixteen chapters contributed by promising authors.

In Chap. 1, Nayak et al. have highlighted a detail survey on the applicability of nature-inspired algorithms in various cloud computing problems. Additionally, some future research directions of cloud computing and other applications areas are also discussed. Nowadays, many organizations are using cloud computing successfully in their domain of interest, and thereby popularity is growing; so because of this, there has been a significant increase in the consumption of resource by different data centers. Hence, urgent attention is required to develop optimization techniques for saving resource consumption without compromising the performance. These solutions would not only help in reducing the excessive resource allocation but would also reduce the costs without much compromise on SLA violations thereby benefitting the cloud service providers.

In Chap. 2, authors discuss the optimization of resource allocation so as to provide cost benefits to the cloud service users and providers. Radhakrishnan and Saravanan in Chap. 3 illustrate the resource allocation in cloud IaaS. How to optimize the VM instances allocation strategy using the novel ANN model has been presented. Further, several issues in implementing the resource allocation are also discussed. Cloud federation has become a consolidated paradigm in which set of cooperative service providers share their unused computing resources with other members of federation to gain some extra revenue.

Chapter 4 gives emphasis on different approaches for cloud federation formation based on game theory and also highlights the importance of trust (soft security) in federated cloud environment. Different models for cloud federation formation using coalition game and the role of a cloud service broker in cloud federation are also presented in this chapter. The major components of resource management systems are resource provisioning and scheduling; in Chap. 5, author discusses the essential perceptions behind the cloud resource provisioning strategies. Then, the author has proposed QoS parameters based resource provisioning strategies for workflow applications in cloud computing environment.

Ritesh in Chap. 6 presents consolidation in cloud environment using optimization techniques. Author has highlighted that in cloud computing, moving large size VM from one data center to other data center over wide area network is challenging task. In Chap. 7, Rao et al. describe different issues and the performances over the virtual machine migration in cloud computing environment. Specifically, authors make the reader to learn about the architectural design of working and storage structures of a key virtualization technology, VMware.

In Chap. 8, Dash et al. present a survey on the various frameworks to develop SLA-based security metrics in addition to different security attributes and possible threats in cloud. Along the line in Chap. 9, to maintain security and privacy at cloud system, Sengupta presents a dimension reduction based intrusion detection system on a cloud server. Deshpande et al. in Chap. 10 have discussed methods and technologies that form the digital guardians of our connected world. In addition, it adapts a case study based approach to understand the current scenario and best practices with respect to cloud security. Cook et al. in Chap. 11 pursue two main works: i) analyze the different components of cloud computing and IoT and ii) present security and privacy problems that these systems face. Developing cloud-based IDS that can capture suspicious activity or threats and prevent attacks and data leakage from both the inside and outside the cloud is the topic of interest in Chap. 13.

In Chap. 12, Chakrabarty et al. have proposed a hybrid model of IoT infrastructure to overcome some of the issues of existing infrastructure. This model will be able to transfer data reliably and systematically with low latency, less bandwidth, heterogeneity, and maintaining the Quality of Service (QoS) befittingly. In Chap. 14, Barik et al. discuss the concept of edge-assisted cloud computing and its relation to Fog-of-things. Further, they have also proposed application-specific architectures GeoFog and Fog2Fog that are flexible and user orientated. In Chap. 15, Limbasiya and Das present a secure smart vehicle cloud computing system for smart cities which is useful to identify the vehicle user in establishing a communication session to share a remarkable information. In Chap. 16, Sahoo et al. have presented various techniques related to cloud-based transcoding system including video transcoding architecture and performance metrics to quantify cloud transcoding system.

Topics presented in each chapter of this book are unique to this book and are based on unpublished work of contributed authors. In editing this book, we attempted to bring into the discussion all the new trends, experiments, and products that have made cloud computing such a dynamic area. We believe the book is ready to serve as a reference for larger audience such as system architects, practitioners, developers, and researchers.

Bhubaneswar, Odisha, India — Bhabani Shankar Prasad Mishra
Bhubaneswar, Odisha, India — Himansu Das
Balasore, Odisha, India — Satchidananda Dehuri
Bhubaneswar, Odisha, India — Alok Kumar Jagadev

Acknowledgements

The making of this edited book was like a journey that we had undertaken for several months. We wish to express our heartfelt gratitude to our families, friends, colleagues, and well-wishers for their constant support throughout this journey. We express our gratitude to all the chapter contributors, who allowed us to quote their remarks and work in this book.

We thank Santwana Sagnika for helping us in the process of compilation of this edited volume.

We wish to acknowledge and appreciate Mrs. Varsha Prabakaran, Project Co-ordinator, Book Production of Springer and her entire team of associates who proficiently guided us through the entire process of publication.

Finally, we offer our gratitude and prayer to the Almighty for giving us wisdom and guidance throughout our lives.

Contents

Chapter 1
Nature Inspired Optimizations in Cloud Computing: Applications and Challenges

Janmenjoy Nayak, Bighnaraj Naik, A. K Jena, Rabindra K. Barik and Himansu Das

Abstract Cloud computing is an emerging area of research and is useful for all level of users from end users to top business companies. There are several research areas of cloud computing including load balancing, cost management, workflow scheduling etc., which has been the current research interest of researchers. To deal with such problems, some conventional methods are developed, which are not so effective. Since, last decade the use of nature inspired optimization in cloud computing is a major area of concern. In this chapter, a detailed (yet brief) survey report on the applicability of nature inspired algorithms in various cloud computing problems is highlighted. The chapter aims at providing a detailed knowledge about nature inspired optimization algorithms and their use in the above mentioned problems of cloud computing. Some future research directions of cloud computing and other application areas are also discussed.

Keywords Cloud computing · Nature inspired optimization · Load balancing
Work flow scheduling · Cost optimization

J. Nayak (✉)
Department of Computer Science and Engineering, Sri Sivani College of Engineering, Srikakulam AP-532402, India
e-mail: mailforjnayak@gmail.com

B. Naik
Department of Computer Applications, Veer Surendra Sai University of Technology, Burla, Sambalpur 768018, Odisha, India
e-mail: mailtobnaik@gmail.com

A. K. Jena · H. Das
School of Computer Engineering, KIIT University, Bhubaneswar, Odisha, India
e-mail: ajay.bbs.in@gmail.com

H. Das
e-mail: das.himansu2007@gmail.com

R. K. Barik
School of Computer Applications, KIIT University, Bhubaneswar, Odisha, India
e-mail: rabindra.mnnit@gmail.com

B. S. P. Mishra et al. (eds.), *Cloud Computing for Optimization: Foundations, Applications, and Challenges*, Studies in Big Data 39,
https://doi.org/10.1007/978-3-319-73676-1_1

1.1 Introduction

With the satisfaction of some constraints, choosing the best solution among the available solution for solving a problem is optimization. Every optimization problem is either minimization or maximization depending on the nature of the problem. In our real lives also, optimization is there at almost everywhere. While solving engineering problems, the main objective of any form of optimization is to make a healthy balance in between exploration and exploitation. The key elements of any optimization are constraints (obstacles), design variable and objective function (heart of optimization). There are various types of optimization algorithms depending on the number of objective functions, types of objective functions, types of constraints, variable types and nature of optimization. Based on these criteria the optimization may be single objective/multiobjective/multobjective with pareto optimal solutions, local/global, smooth/non smooth, stochastic/deterministic, continuous/discrete, constrained/unconstrained etc. The type of optimization may be varied, but for solving any complex engineering problem it is very difficult to choose the exact optimization method, which is suitable for that particular problem. With the successive growth of science and technology, the real life optimization problems are becoming more complex in nature. The earlier developed traditional optimization algorithms fail to explicate the exact and real solution of the nonlinear and non differential problems in large search space. The basic limitations to these algorithms are that they suffer from early convergence, use of complicated stochastic functions and higher order derivatives in solving the equations. During last few decades, some popular optimization algorithms have already proved their effectiveness in resolving different real life problems. In 1992, John Holland and Goldberg (1989) developed the most popular evolutionary algorithm called Genetic algorithm (GA). In comparison to the gradient search based methods, this algorithm performs better at local optima and having very less chance to trap at local minima positions. Then in 1995, Kennedy and Eberhart developed a stochastic swarm inspired technique called PSO. It is one of the popular stochastic and heuristic based search method on till date. Several variants of such category physical, chemical and nature based algorithms are introduced during the last decade. Although they have been used to resolve variety of complex problems, but still they suffers from some major issues such as convergence criteria, when they are being single handedly applied. This is due to the extensive use of controlling parameters such as population size, environmental conditions, no. of iterations etc. Therefore, such variations have been developed by integrating some modifications in the parameters or leading any form of hybridization of the algorithms to explore their capability to resolve complex problems. As, any major change in the parameter selection may change the functional aspects of the whole algorithm, so hybridization is not the exact solution to solve these complex problems.

The applicability of nature inspired optimization algorithms have been a recent interest among the researchers of various research community. The main reason behind the success rate of nature inspired and swarm based algorithms are having the capability to solve the NP-hard problems. To resolve the real life problems, some of the earlier developed optimization techniques fails and the solutions of many real life problems have been obtained by heat and trail methods. So, this is the basis for the researchers to focus towards the development of some competitive optimization algorithms, which are efficient to resolve the complex problems. We all are surrounded by a beautiful scenario called nature and now-a-days most of the algorithms are natured inspired algorithms, which are developed with the concepts of nature. Nature is one of the amazing creations of God and has always been a great source of inspiration for all of us. Moreover, some biological systems are also responsible for the development of novel optimization algorithms. So, most of the nature inspired algorithms are based on biological processes or the behavior of some of the nature's creation (animals or insects). Among the biological inspired algorithms, swarm based algorithms draw a special attention due to their larger applicability in various applications. Such algorithms are inspired by collective nature or behavior of some swarms such as bees, birds, frogs, fishes etc. Various examples of swarm based algorithms may be PSO [1], CS [2], Bat inspired algorithm (BA) [3], Bacteria foraging algorithm (BFA) [4], ABC optimization [5], BCO [6], Wolf Search (WS) [7], Cat Swarm Optimization (CSO) [8], Firefly Algorithm (FA) [9], Monkey Search Algorithm (MSA) [10] etc. There are also some algorithms such as Atmosphere cloud model [11], Biogeography based algorithm [12], Brain storm optimization [13], Differential evolution algorithm [14], Japanese tree frogs calling [15], Flower pollination algorithm [16], Great salmon run [17], Group search optimizer [18], Human-Inspired Algorithm [19], Invasive weed optimization [20], Paddy Field Algorithm [21], Queen-bee evolution [22], Termite colony optimization [23] etc. which are nature inspired, bio inspired but not swarm based. The concept behind all the swarm based algorithms is the collective behavior and coordination among the swarm elements in an open environment. The advantage of using swarm based algorithms relies on information sharing between numerous agents, for which self organization, co-evolution and learning throughout the iterations possibly helps to provide high efficiency [24]. In fact, each of the individual swarm behaves and acts in a collective manner so that, the processes such as foraging, reproduction, task allocation among themselves etc. is easier. Based on the locally obtained information among each other, the necessary decisions are taken in a decentralized manner.

Since last two decades, the use of internet has become very popular among all levels of users starting from business to automation industry at some cheaper price. More storage, access and processing of data with a vast utility makes the internet more successful, which in turn responsible for the evolvement of a new era called cloud computing. Cloud computing is basically dealt with hosting and delivering the services over the network. Cloud services are more popular and demanding due to their flexibility towards the use of resources as per user's choice. Today, most of the major IT industries are effectively using the cloud services to fulfill their requirements for reshaping their business. Although the research of cloud computing is on hype, but some of its past issues are not yet resolved and some new issues are arising. Issues like provision of automated service, migration of virtual machines, server consolidation, effective management of energy, traffic management etc. needs urgent attention. However, some problems such as task scheduling, load balancing, job/flow scheduling, resource optimization, resource allocation etc. are being solved by various researchers with the use of optimization algorithms. Most of them have used the nature inspired optimization techniques for efficiently solving those problems. In this chapter, a detailed survey has been elucidated on the use of nature inspired optimization algorithms in different problems of cloud computing with the observations, analytical discussions and suggestions for better improvement in the use of resources.

The remainder of this chapter is organized as follows. Section 1.2 provides the overview of cloud computing and its architecture. Section 1.3 describes various research challenges of cloud computing. Section 1.4 discuses on analysis of broad areas of research on cloud computing using different optimization techniques, Sect. 1.5 focuses on future research challenges on cloud computing and finally Sect. 1.6 concludes the chapter.

1.2 Cloud Computing: An Overview

Now-a-day's cloud computing is an emerging technology used for hosting and delivering cloud services over the internet throughout the globe and provides solution for deploying on demand services to satisfy the requirement of service provider and the end users based on service level agreement (SLA). It is used both in commercial and scientific applications. Cloud computing allow users to access large amount of computing power services in virtualized environment. It provides on-demand, dynamic, scalable, flexible services to the end users on pay-as-you-use basis. Cloud providers [25] such as Google, Amazon, Microsoft etc. established their data centers for hosting cloud computing applications throughout the globe. These services [28] are called Infrastructure as a service (IaaS), Platform as a service (PaaS), and Software as a Service (SaaS). This cloud services are provided for different functionalities such as computing services, network services, storage services etc. to the end user by employing Service level Aggrements(SLA).

Many researchers have defined the cloud computing with their own aspects. Buyya et al. [25] have explained cloud computing as "It is a distributed and parallel computing environment which consist of a huge collection of virtualized and inter-connected computing resources that are dynamically presented and provisioned as one or more unified computing resources based on SLA".

The National Institute of Standards and Technology (NIST) [29] defined the cloud computing as "it is a pay-per-use model for enabling convenient, on-demand computing resource access to a common pool of computing resources that can be dynamically released and provisioned with service provider effort or negligible management effort".

1.2.1 Layered Architecture of Cloud Computing

The cloud computing services are broadly classified based on the abstraction level of ability and service provided by the service provider. This abstraction level can also represented as layered architecture in cloud environment. The layered architecture [27] of cloud computing can be represented by four layers such as data center layer, platform layer, infrastructure layer and application layer.

1.2.1.1 Datacenter Layer

This layer is accountable for managing physical resources such as servers, switches, routers, power supply, and cooling system etc. in the data center of the cloud environment. All the resources are available and managed in data centers in a large numbers to provide services to the end user. The data center consists of large number of physical servers, connected through high speed devices such as router and switches.

1.2.1.2 Infrastructure Layer

It is a virtualization layer where physical resources are partitioned into set of virtual resources through different virtualization technologies such as Xen, KVM and VMware. This layer is the core of the cloud environment where cloud resources are dynamically provisioned using different virtualization technologies.

1.2.1.3 Platform Layer

This layer consists of application software and operating system. The objective of this layer is to deploy applications on directly on the virtual machines.

1.2.1.4 Application Layer

This layer consists of different actual cloud services which are used by cloud users. These applications provide services to the end user as per their requirements.

1.2.2 Cloud Business Model

Cloud computing environment deploys service oriented business model to satisfy the customer requirements. It means physical resources are delivered as services on demand basis to the end user. This section focuses on services provided by the cloud providers and are classified into three types according to the abstraction level and service level.

1.2.2.1 Infrastructure as a Service (IaaS):

IaaS provides on demand physical resources such as CPU, storage, memory, etc. in terms of virtualized resources like virtual machines. Each virtual machine has its own computing capability to do certain operations as per the user requirements. Cloud infrastructure employs on-demand service provisioning of servers in terms of virtual machines. Example of IaaS providers are EC2, GoGrid etc.

1.2.2.2 Platform as a Service (PaaS):

It provides an environment on which the developer will create and deploy applications. This offers high level of abstraction to make the cloud environment more easily accessible by the programmer. Example of PaaS providers are Google AppEngine, Microsoft Azure etc.

1.2.2.3 Software as a Service (SaaS):

It provides on demand services or applications over Internet to the end users. This model of delivering of services is called Software as a Service (SaaS). It will eliminate the overhead of software maintenance and simplifies the development process of end users. Example of PaaS providers are Rackspace, salesfource.com etc.

1.2.3 Types of Cloud

There are different types of clouds available in cloud environment.

1.2.3.1 Public Cloud

The public clouds are operated and owned by the organizations that provide cloud services to the customers as pay-as-you-use model. In public cloud, users can easily access the services and computing resources throughout the globe publicly. All the users share the same computing resources with limited design, security protections, and availability resources. All the services are managed and controlled by the cloud providers. The services are delivered to the customer through I internet from a third party service provider. The vendors of public cloud providers are Google,

Amazon, Salesforce, Microsoft, etc. The infrastructure of the public cloud provider is Rackspace, Amazon, Gogrid, Terramark etc.

1.2.3.2 Private Cloud

The private clouds are exclusively allow services to be accessible for a single enterprise only. As it is private in nature, it is more secured than public cloud. In private cloud, only authorized organization can access and use the computing resources and services. Private clouds are hosted within its own data center or externally with a cloud provider which provides a more standardized process of its privacy and scalability. This private cloud may be advisable for organizations that does not favor a public cloud due to common allotment of physical resources. These services are provided and managed by the organization itself. The vendors of the private cloud providers are Vmware, IBM, Oracle, HP etc. The infrastructure of the private cloud providers are Vmware, Eucalyptus, IBM, etc.

1.2.3.3 Hybrid Cloud

This Cloud is a hybridization of both private and public cloud models. In hybrid cloud, critical and secure services are provided by the private cloud and other services are provided by the public clouds. Hybrid cloud provides on-demand, dynamically provisioned resources to the end users. It also combines the services provided from public and private Clouds. Example of hybrid cloud is ERP in private cloud and emails on public cloud.

1.2.3.4 Community Cloud

The community cloud distributes and manages infrastructure between numerous organizations between specific communities by internally, externally or by a third party and hosted externally or internally. These types of cloud is specifically designed for specific community of people to serve the end-users requirements.

1.2.4 Research Challenges in Cloud Environments

Most of the works of grid computing [26, 30–33] concepts such as virtualization, resource allocation, scheduling, service discovery, scalability and on-demand service are ported to cloud computing. Now, the biggest challenges of these concepts are task scheduling [35], resource allocation [37], and energy efficient scheduling [34], etc. in large scale environments [36]. Here this section focuses on aforesaid concepts and its current research directions.

1.2.4.1 Task Scheduling

Scheduling of task [35] with existing resources is one of the emerging research direction for research community. There are two types of applications used in cloud like independent tasks called Bag-of-tasks or interconnected independent tasks called workflows. Workflow scheduling can be either deterministic or non-deterministic. Deterministic means execution path can be determined in advance by a directed acyclic graph (DAG) but in non-deterministic algorithm execution path is determined dynamically. The workflow scheduling is a NP-Complete problem which deals with different factors such as dynamicity, heterogeneity, elasticity, quality of services, analysis of large volume of data etc. In this scheduling, it is difficult to find the global optimum solution. In cloud computing environment makespan, cost, elasticity, and energy consumption etc. are most important factor to determine the quality of services. In commercial cloud, applications are categorized into single service oriented or workflow oriented. One of the most important problem in cloud is task-to-resource mapping. This problem has three folds: selection of virtual machines, determination of best resource provisioning algorithm for virtual machines, scheduling of task on virtual machines.

1.2.4.2 Resource Allocation

The cloud computing environment provides infinite number of computing resources to the cloud users so that they can dynamically increase or decrease their resource uses according to their demands. In resource allocation model having two basic objectives as cloud provider wants to maximize their revenue by achieving high resource utilization while cloud users wants to minimize their expenses while meeting their requirements. The main objective of the resource allocation [37] in a cloud environment is to dynamically allocate VM resources among the users on demand basis without exceeding their resources capabilities and expense prices. When a cloud provider allocates resource to multiple users, they proportionally occupy cloud provider capacity and increase the expense. When a cloud provider allocates resource to multiple users, they proportionally occupy cloud provider capacity and increase the expense. The goal is to assign each user to a require resource in order to maximize the resource utilization and minimize the total expense.

1.2.4.3 Cloud Federation

One of the major challenges in cloud computing environment is scalability. When one cloud is not enough to provide services to the users or demand exceeds the maximum capacity of the single cloud provider, then to achieve user level satisfaction, we go for cloud federation [38]. To achieve scalability and user level satisfaction cloud federation is required. The basic objective of cloud federation is to provide dynamic-on-demand users request to achieve quality of services. The cloud federation consists of several cloud providers joined by themselves by mutual collaboration among themselves or service level agreement. The cloud providers in the federation having excess resource capacity can share their resources with other members of federation.

1.2.4.4 VM Consolidation

The VM consolidation [39] is used to improve the dynamic utilization of physical resources to reduce energy consumption by dynamically reallocating VMs using live migration technologies. VM consolidation tries to pack the active VMs in the minimum number of physical servers with the goal of energy saving and maximization of server resource utilization. Cloud computing architectures take advantage of virtualization technologies to implement the VM consolidation concept. Virtualization is the most emerging technology that reduces energy consumption of datacenters by detaching the virtual machines from the physical servers and allowing them to be positioned on the physical servers where energy consumption can be improved. Cloud providers try to reduce their operational costs of the data centers by increasing the number of customers by increasing the number of VMs but within limited number of physical servers, and also sinking power consumptions of the data center. VMs should be distributed among the minimum number of physical servers such a way that the utilization of each physical server is also maximized. As a result, consolidation provides the higher efficiencies with less number of machines those are switched on which leads to the less energy consumption in the data centers.

1.2.4.5 Energy Efficient Scheduling

Scheduling in cloud computing environment is a problem of assigning tasks to a particular machine to complete their work based on Service Level Agreements. Energy aware task scheduling is a process of assigning tasks to a machine in such a way that minimum energy is used. This minimization can be achieved by implementing the problem in various scheduling algorithms in-order to get the best result.

In cloud computing the service consumers can access the required resources without being present physically in the working location based on pay which means they pay for the amount of resources they use by signing SLA with the cloud service provider. Now-a-days, energy costs increases as the data center resource management cost are increased rapidly while maintaining high service level performance. In datacenters, the quantity of energy consumption by a server is varied dynamically depending on its current workloads. The power consumption of a computing server is specified by its number of processors are currently working and that power consumption of that processor is estimated mainly by its CPU utilization. In data centers carbon dioxides emissions are increasing day by day due to the high energy consumptions and massive carbon footprints are incurred due to huge amount of electricity consumed for the power supply and cooling the several servers those are hosted in those data centers.

The main objective is to minimize energy consumption in cloud data centers by performing energy-aware task scheduling [34]. Minimizing the energy utilization of cloud data centers is a challenging issue because of its large dynamic computing applications requirements. So, there is a need of suitable green cloud computing environment that not only minimize energy consumption in cloud data centers, but also reduces the operational costs in terms of energy consumption without violating SLA policy in cloud data centers.

1.3 Emerging Research Areas of Cloud Computing

There are several earlier and newly evolved research areas of cloud computing. Every time the researchers are developing some highly efficient techniques to handle the real life problems of cloud scenario. This section highlights with some important research areas in which the use of nature inspired optimization algorithms can be realized.

1.3.1 Energy Optimization

In any cloud scenario, energy optimization is a challenging task as the data centers must have to utilize the resources in an efficient manner. In order to manage the power consumption, it is essential to find out the list of power providers and the way to manage the usage. In reality, energy is being used for maintaining the working status of any computing facilities. So, the inefficient utilization may be helpful to originate the waste of power. Most of the researchers use the dynamic reduction technique in the number of clusters to handle such problem. Also, through virtualization technique the target may be achieved by VM migration. Energy optimization can be achieved in various ways. First, the operation of different physical machines may be optimized by automatically changing the frequency and voltage through the dynamic voltage and frequency scaling technique. The main intend of such technique is to optimize the power and alleviate the heat generated from processors. Second, the method of power capping is used in the data centers, which helps to budget the power at system level. Besides that, this technique is useful for individual allocation of power to each server in the configured cluster, which is called power shifting. In this technique, preference is given to systems having high priority for more power rather than the systems having low priority. Another important recent development for saving the power is C-states systems [40], which are outfitted with processors. By using this technique, the fallow components of idle systems are turned off to better utilize the power. Although, this technique is good for long time use systems like personal laptops, but it has the limitation like deep sleep. Deep sleep is a mode, where the

Table 1.1 Literatures of nature inspired algorithms in energy optimization

Developed strategy	Optimization algorithm	Purpose	Year	Reference
–	Multi objective PSO	Identification of power models of enterprise servers	2014	[41]
Smart VM overprovision	Mixed integer linear programming	Solving the energy efficiency problem	2017	[42]
–	Convex predictive optimization	Energy efficient resource utilization	2017	[43]
Dynamic backlight scaling optimization	–	Minimization of energy consumption	2014	[44]
–	Generic optimization	Cost optimality	2014	[45]
–	Improved PSO	Optimization of energy in virtual data centers	2013	[46]
–	SLA based optimization	Minimization of energy cost	2012	[47]
–	Immune Clonal optimization	Energy efficiency	2014	[48]
–	PSO	Virtual machine allocation	2014	[49]
ACO-VM placement	ACO	Effective use of physical resources	2014	[50]

processors cannot move forward by sticking at some point and requires long time to activate. Moreover, system's kernel failure is also another problem associated with deep sleep. Apart from these techniques, some other techniques are developed for suitable optimization of energy and are listed in Table 1.1.

1.3.2 Load Balancing

In any cloud scenario, the total execution time of the tasks must be reduced for better performance. Virtual machines (VM) are known as the processing units of clouds. For business purpose, execution of tasks by VM is faster and run parallel. As a result, problems occur in scheduling of tasks within the existing resources. All the resources must be fully utilized by the scheduler for efficient scheduling. One or multiple VMs are assigned to more than one task for parallel completion. In such scenario, the scheduler must be ensured that all the VMs are assigned to equal tasks and the load is equivalent for all; rather some VMs have more load and some are in idle state. So, all the loads must be distributed/assigned equally to all VMs and scheduler is responsible for that. The main objective of load balancing algorithms is to ensure the maximum use of available resources along with the improvement in response time of client's offered application. Also, the algorithms are useful for rapid execution of applications during the variability of workload in runtime [51]. Finding in both homogeneous and heterogeneous environments, load balancing algorithms are of two types such as: static and dynamic. If the variation is low at node, then static method is applicable. If the load will vary time to time like cloud environments, then these methods are not suitable. So, dynamic algorithms are more preferable than static methods with an extra overhead of cost factor. Specifically in heterogeneous environments, dynamic algorithms are more preferable. For dynamic load balancing, a number of nature inspired based methods are proposed and remain successful. Many authors have studied the literatures of load balancing algorithms in various aspects (Table 1.2).

Babu and Krishna [59] have proposed a novel dynamic load balancing technique based on honey bee mating optimization algorithm. Their main aim is to achieve maximum throughput as well as maintaining suitable balance in loads across the VMs. They claim that, their method is also useful for minimization in waiting time of the tasks while those wait in the queue. With the combination of ant colony optimization, Nishant et al. [60] has developed one method for the effective distribution of loads among the nodes. The developed method is helpful for both under loaded and over loaded nodes in efficient load distribution. Apart from these some other load balancing methods based on nature inspired optimization are developed and are listed in Table 1.3.

Table 1.2 Literatures survey articles of load balancing algorithms

Authors	Main Focus Area of Survey	Year	Reference
Ghomi et al.	Hadoop Map Reduce load balancing category Natural Phenomena-based load balancing category Agent-based load balancing category General load balancing category Application-oriented category Network-aware category Workflow specific category	2017	[52]
Milani and Navimipour	Various Parameters Advantages & Disadvantages	2016	[53]
Mesbahi and Rahmani	Design metrics with pros and cons	2016	[54]
Kanakala et al.	Metrics such as throughput, speed, complexity	2015	[55, 56]
Ivanisenko and Radivilova	Distributed environment Efficiency factors	2015	[57]
Farrag and Mahmoud	Intelligent cloud computing algorithms	2015	[58]

Table 1.3 Load balancing techniques based on nature inspired optimizations

Algorithm	Objective	Advantages	Reference
GA	Minimization self adaptation of load distribution	Cost effective	[61]
PSO	Minimization of execution time	Best solution for VM migration	[62]
GA	Minimization of make span	Improved performance than traditional hill climbing methods	[63]
ABC	Minimization of make span	Better efficiency than other scheduling	[64]
Improved PSO	Minimization of response time	Avoidance of local optima and better than GA	[65]
ACO	Minimization of execution time	Better efficiency in make span	[66]
PSO and Cuckoo Search	Minimization of make span	Better performance than traditional methods	[67]
GA	Minimization of VMs	Minimization of make span	[68]
PSO	Better resource allocation	Efficiency in allocating VMs	[69]
PSO	Load balancing with VM	Efficient VM migration	[70]
GA	Efficient task scheduling	Better resource utilization	[71]
ACO	Efficient Load distribution	Better efficiency	[72]
ACO	Load balancing at data center	Efficiency in performance	[73]
ACO	Balancing total load	Minimization in make span	[74]
PSO	Task scheduling between overloaded and under loaded VMs	Better resource utilization	[75]
BCO	Efficient load balancing in VMs	Better utilization and performance	[76]
Compare and Balance	Load balancing in server	Efficiency in response time	[77]
PSO	System load balancing	Reduce down time	[78]
Multi Particle Swarm Optimization	Minimization of task overhead	Better resource utilization	[79]
ACO and PSO	Scheduling of VMs	Better performance than only PSO and ACO	[80]
Orthogonal Taguchi Based-Cat Swarm Optimization	Minimization of execution time	Better system utilization	[81]

1.3.3 Task Scheduling

Since its inception, cloud computing has drastically reduced the financial and maintenance cost for various application deployment. Due to the property of high scalability, clients are not bothering about any resource or respective revenue loss [82, 83]. By VMs, several systems connected over the internet can easily use the resources at any of the remote place systems. One of the major objective of cloud computing is to increase or generate the revenue as much as possible at both sides i.e. cloud provider and clients. Task scheduling has been a important aspect in cloud computing, as ineffective task scheduling may lead to huge revenue loss, degradation in the performance etc. So, the scheduling algorithms must be effective one to efficiently handle the problems such as use of response time, resource utilization, make span, cost of data communication etc. Zhu and Liu [84] have developed one multidimensional based cloud computing framework with genetic algorithm for efficient task scheduling. They considered the metrics such as completion time of tasks and economic needs of the clients for improvement in efficiency in task scheduling. As compared to traditional genetic algorithms, their performance result is very good in terms of efficiency. Performance comparison with energy aware task scheduling and two different optimization techniques like GA and CRO is conducted by Wu et al. [85]. From their simulation results, they claim for better utilization of make span in case of CRO algorithm as compared to GA. In a hetero generous environment, Tao et al. [86] have developed one model for handling the task scheduling problem with GA and case library, Pareto solution based hybrid method. Both make span and energy consumption are taken into consideration and after the successful simulation study, they are able to minimize the both with optimized resource utilization. As compared to other algorithms, they claim that their hybrid method performs well for solving the task scheduling problem in a heterogeneous environment. Li et al. [87] have proposed one ant colony algorithm based task scheduling algorithm for minimizing the make span for a given set of tasks. Their proposed method is able to handle all types of task scheduling problem irrespective to the size and also, outperforms some other method like FCFS. Apart from these works, some other nature inspired methods [88–95] are developed for handing the task scheduling problem in the cloud environment.

1.3.4 Work Flow Scheduling

Proper scheduling of work flow is very much necessary for effectively managing the inter-dependent tasks with efficient resource utilization. Work flow is nothing but the completion of number of activities required to complete a task. There may be various components of workflow like data, programs, reports, sequencing etc. The structure to work with such components is any process, data aggregation, data segmentation and distribution/redistribution. As compared to job scheduling, work flow scheduling is more popular due to its efficiency in determining the optimal solution in an efficient way for dealing with complex applications by considering different constraints among the tasks. Most of the researchers use directed acyclic graph to represent the workflow scheduling. It is very important to consider the computation cost and completion time while scheduling the work flow in any cloud scenario. If the work flow is to be computed manually by any IT staff, then some proper background knowledge is required for right execution [96]. For the cloud environment, it is essential to meet the optimization criteria of work flow scheduling to accomplish the proficient work flow management system. With the existing computing resources, the work flow management system identifies, control and executes the work flows. The execution order of work flows are handled by computational logic representation. For the problems like process optimization, well management of process, system integration, re-scheduling, improvement in maintainability etc., the work flow management system may be developed [97]. For effective cost optimization, a number of nature inspired based or soft computing based approaches (table 4) are developed in the cloud environment. Both the single objective and multi objective algorithms are able to tackle the cost optimization problem. The optimization criterias may be makespan, availiability, budget allocation, service level agreement, reliability control, waiting time etc.

1.4 Analysis and Discussion

Cloud computing is a vast area of research having multiple subfields. Various research areas within the vicinity of cloud computing are discussed in the previous section and some new areas such as resource scheduling, software deployment, network infrastructure optimization, web service composition etc. are also at the blooming stage. Also, some other close related areas of cloud computing like grid computing, virtualization, utility computing and smart computing are the interest of different researchers. Among the other related areas, task scheduling, load balancing, energy optimization, workflow scheduling [98–110] are the most frequently applied and solved research issues. Based on the keyword search in Google scholar, it is found that most of the problems related to load balancing and task scheduling are solved during the last decade. An analysis on the number of published research article is conducted and may be realized in Fig. 1.1.

In the above discussed three areas, the use of nature inspired algorithms is also frequent for all applications. It is found that compared to other algorithms the use of PSO & ACO is more in all applications. For example, PSO has been frequently used in solving the load balancing problem in any homogeneous or heterogeneous cloud environments as compared to other algorithms. At the same time after PSO, ACO has been applied for effective optimization of problems like task scheduling and work flow scheduling. However, some other algorithms like cat swarm optimization, bee colony optimization, honey bee optimization, chemical reaction optimization, genetic algorithm etc. are also used to solve various problems of cloud computing. A typical comparative analysis may be realized in Figs. 1.2, 1.3, 1.4 and 1.5.

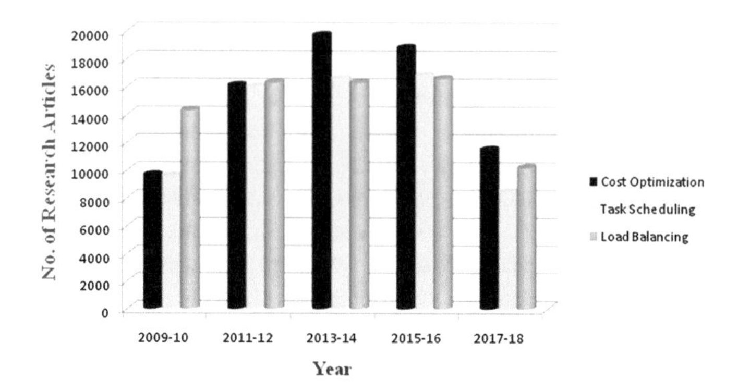

Fig. 1.1 Number of research articles published in major areas of cloud computing

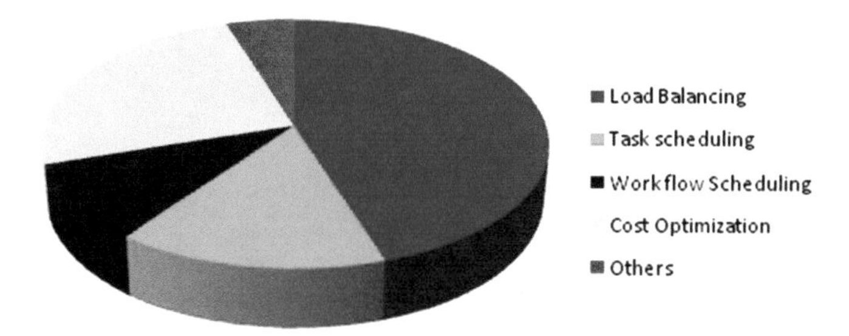

Fig. 1.2 Applications of PSO in various research areas of cloud computing

Fig. 1.3 Applications of ACO in various research areas of cloud computing

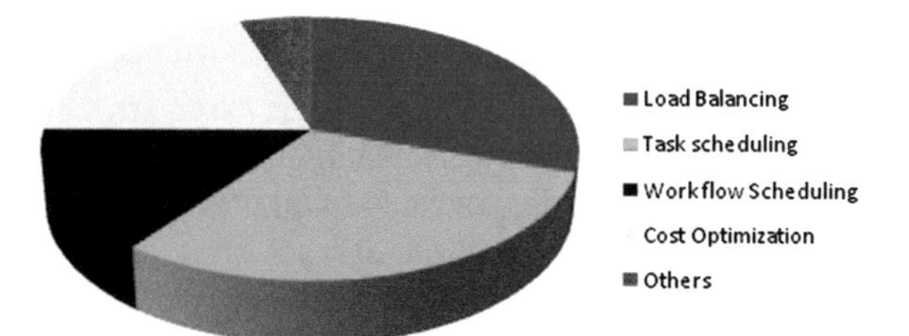

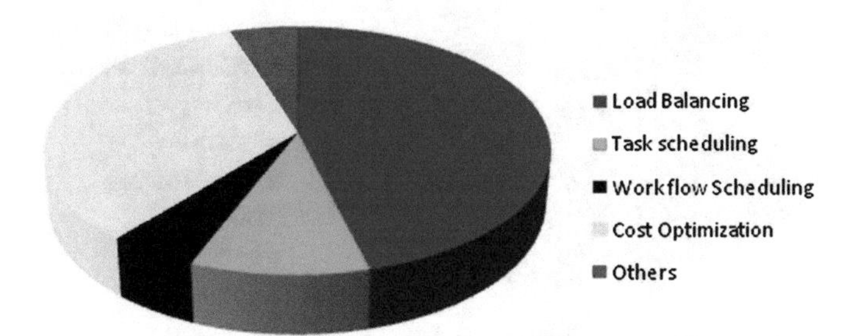

Fig. 1.4 Applications of GA in various research areas of cloud computing

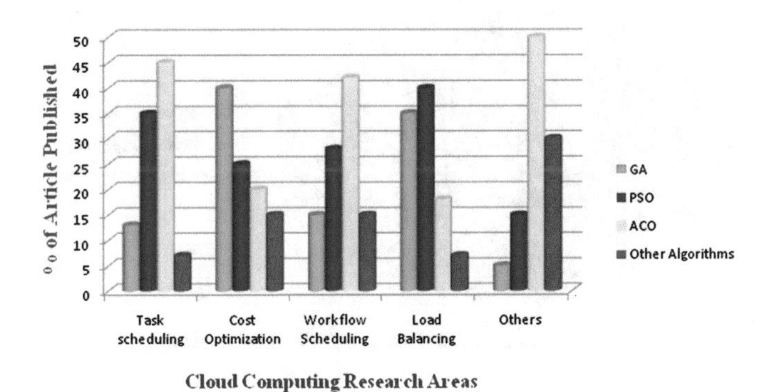

Fig. 1.5 Comparative analysis of algorithms in various research areas of cloud computing

In Figs. 1.2 and 1.3, the application of PSO and ACO algorithms are visualized in different research areas of cloud computing. The PSO algorithm is mostly applied in the area of load balancing with its various advanced versions such as improved PSO, modified PSO, Pareto based PSO, fuzzy PSO etc. But, for solving the scheduling problem in cloud, ACO based techniques are mostly proposed as compared to others. In Fig. 1.4, the case of ever popular algorithm called GA has been analyzed and found mostly in the problem of load balancing. Also, the success rate of GA in cost optimization is more as compared to others. Moreover, it is observed that other than a single optimization, now-a-days more focus is paid in hybrid algorithms for better results.

In Fig. 1.5, an overall comparison is made in between all algorithms in consideration to major applications of cloud computing. It can be analyzed that, other than GA, PSO and ACO, some other nature inspired algorithms are frequently used in cloud environment.

Figure 1.6 analyzes the development of different load balancing techniques and it may be concluded that, the ratio of development using nature inspired algorithms is

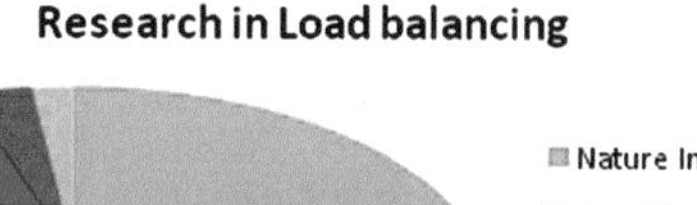
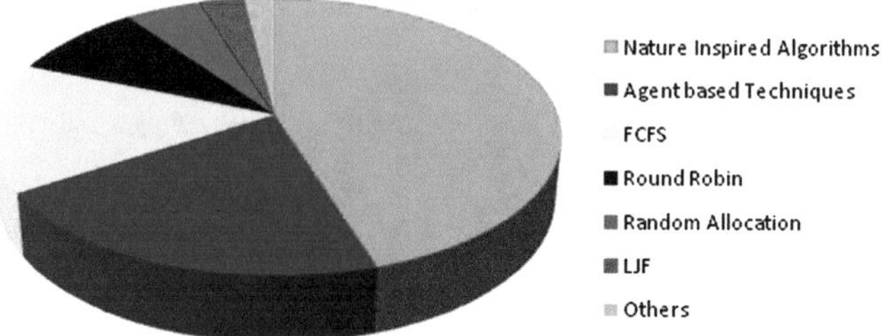

Fig. 1.6 Analysis of various load balancing techniques

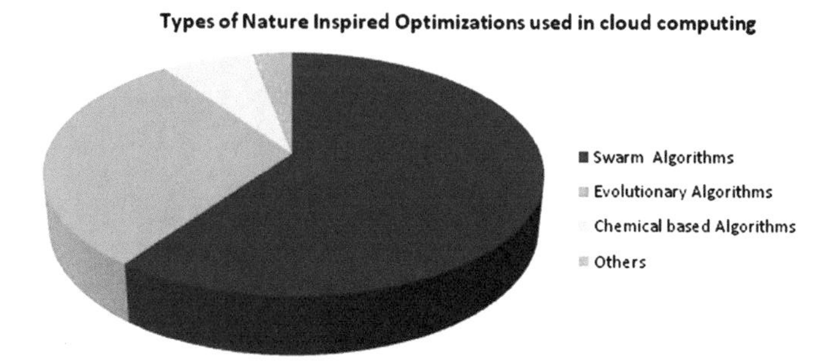

Fig. 1.7 Types of nature inspired algorithms in solving cloud computing problems

more than other methods i.e. agent based techniques, FCFS based methods, round robin, random allocation, LJF and others.

So, the applicability of nature inspired optimization algorithms is quite important in all respect of cloud computing scenario. In both static and dynamic environments, these algorithms can cope with the structure of the problem and produces effective solution with marginal errors for further improvements. Moreover, a brief analysis is demonstrated by distinguishing the type of nature inspired algorithms for cloud computing in Fig. 1.7. From the literatures, it is conveyed that most of the cloud computing research are based on swarm based algorithms other than evolutionary based, physical and chemical based optimizations. The reason behind this, is swarm based algorithms are capable of handling nonlinearity, producing more chance at global optima, less error rate, good convergence and less complexity. The algorithms such as cuckoo search, firefly optimizations are multiobjective type techniques and have less chance to trap at local optima solutions. On the other side, in evolutionary based optimizations e.g. GA, due to crossover and mutation operations, the complexity is more. Also, evolutionary algorithms are stochastic in nature means random variables are used in these techniques and after every run, new different solution will be generated. In swarm based algorithms, there is no any central controller to control the behaviour of swarm members in the population and only through some behavioral aspects, they communicate with each other. Also, swarm based algorithms are more scalable than evolutionary algorithms. Other than these two, some other techniques such as CRO [111, 112], biogeography based optimization [113], big bang big crunch algorithm [114], intelligent water drop algorithm [115], simulated annealing [116] etc. are also applied to solve different issues of cloud computing.

1.5 Further Research Challenges

Undoubtly, cloud computing is an important emerging area of research and it is attractive to all level of users starting from one end user to large business or software industry peoples. Despite this fact, it is found still there are some issues to be addressed in different domains. Some of the research challenges with future directions are discussed by Zhang et al. [117]. For any type of digital service delivery, resources are to be used effectively. The service of cloud computing is totally based on internet and for delivering any service, the users have to fully rely on internet connection. None of the any cloud service provider can assure for full phase service, as they may fall at any moment. Although, demanding a service level agreement is an optional way, but still some more ways to be find out for effective solution. Another key factor of cloud computing is effective way of scheduling of resources. Always it is the prime responsibility of service provider to apportion and de-apportion of resources with the satisfaction of service level objectives and decreasing the cost. But achieving this goal is not an easy task and for massive demands, it becomes very difficult to map to QoS level. Security and privacy issues are always been an important deal to cloud computing. Any of the cloud infrastructures may not fully ensure for complete secure communication. However, some of the risk based approaches are proposed to deal such challenges and further research is required to reach at complete assurance level. Balancing loads across the data center is another important area of concern. Some research has been developed for dealing with above problem with virtual machine migration and upto certain extent those are successful. For eg., Xen [118] has proposed one live based virtual machine migration technique with a very short downtime in a range of 10ms to 1s. For maximum utilization of resource and minimum consumption of energy, most of the developers use the server consolidation approach. To deal with the optimality mode of server consolidation, many heuristic based approaches [119–121] are developed. Due to server consolidation, the performance of applications must not be hampered and there must be a quick response for any congestion(if occurs). Apart from these issues, some other important research challenges such as effective management of energy, data security with more confidentiality, avoiding the network traffic by analyzing the demands, dealing with storage and capability to maintain consistency etc. are evolving day by day and advance research is needed to tackle all these problems.

1.6 Conclusion

Since last two decades, nature-inspired optimizations are quite popular for their capability to produce promising solutions for diversified applications. However, it does not sense that there is no need to focus urgent attention as they are in infancy stage. In this chapter, the applicability of nature inspired optimization algorithms are surveyed for different research areas of cloud computing. Applications such as load balancing, task scheduling, workflow scheduling, cost optimization etc. have remained the main focus for this survey. Also, the type and nature of various nature inspired algorithms are analysed in various perspectives. It is found that, there has been a frequent use of PSO and ACO for solving almost all types of problems in cloud computing. Other than some of the conventional methods of cloud computing, nature inspired algorithms are robust, scalable and effective to use.

However, realizing the reality, it can be concluded that still there is a long way to go for such algorithms. These algorithms are quite efficient in producing the optimal solutions, but some significant gap among the theory and practice may still found. Research challenges like meticulous mathematical analysis, analysis of convergence to get optimality condition, suitable tradeoff between exploration and exploitation, accurate tuning of algorithmic parameters etc. are yet to be solved. Apart from these, some recently developed nature inspired algorithms such as multi verse optimization, lion optimization, whale optimization, dragonfly optimization, virus colony search optimization, elephant herding optimization, social spider optimization, social emotional optimization, moth search algorithm, intelligent water drop algorithm, krill herd algorithm, wind driven optimization, kidney inspired algorithm, bird swarm algorithm, ant lion optimization, salp swarm algorithm, flower pollination algorithm, grey wolf optimization, intrusive tumor growth optimization etc. are yet to be applied and may be the future concern.

Appendix

GA: Genetic algorithm

ABC: Artificial bee colony

PSO: Particle swarm optimization

DE: Differential evolution

ACO: Ant colony optimization

CRO: Chemical reaction optimization

BCO: Bee colony optimization

CSO: Cat swarm optimization

FCFS: First come fist serve

LJF: Longest job first

References

1. J. Kennedy, Russell Eberhart, Particle swarm optimization. Proc. IEEE Int. Conf. Neural Networks **4**, 1942–1948 (1995)
2. X.S. Yang, S. Deb, Cuckoo search via Lvy flights, in *World Congress on Nature and Biologically Inspired Computing (NaBIC 2009)* (IEEE, 2009), pp. 210–214
3. X.S. Yang, A new meta-heuristic bat-inspired algorithm, Nature Inspired Cooperative Strategies for Optimization (NICSO 2010), p. 6574
4. K.M. Passino, Biomimicry of bacterial foraging for distributed optimization and control. IEEE Control Syst. **22**(3), 52–67 (2002)
5. D. Karaboga, B. Basturk, A powerful and efficient algorithm for numerical function optimization: artificial bee colony (ABC) algorithm. J. Global Optim. **39**(3), 459–471 (2007)
6. D. Teodorovi, M. DellOrco, Bee colony optimizationa cooperative learning approach to complex transportation problems, in *Advanced OR and AI Methods in Transportation: Proceedings of 16th MiniEURO Conference and 10th Meeting of EWGT (13–16 September 2005) Poznan: Publishing House of the Polish Operational and System Research* (2005), pp. 51–60
7. R. Tang, S. Fong, X.S. Yang, S. Deb, Wolf search algorithm with ephemeral memory, in *IEEE Seventh International Conference on Digital Information Management (ICDIM)* (2012), pp. 165–172
8. S.C. Chu, P.W. Tsai, J.S. Pan, Cat swarm optimization, in *PRICAI 2006: Trends in artificial intelligence* (Springer Berlin Heidelberg, 2006), pp. 854–858
9. X.S. Yang, Firefly algorithm, stochastic test functions and design optimization. Int. J. Bio-Inspired Comput. **2**(2), 78–84 (2010)
10. A. Mucherino, O. Seref, Monkey search: a novel metaheuristic search for global optimization. Data Min. Syst. Anal. Optim. Biomed. **953**(1), 162–173, AIP Publishing (2007)
11. G.W. Yan, Z.J. Hao, A novel optimization algorithm based on atmosphere clouds model. Int. J. Comput. Intell. Appl. **1**(2 (01)) (2013)
12. D. Simon, Biogeography-based optimization. IEEE Trans. Evolut. Comput. **12**(6), 702–713 (2008)
13. Y. Shi, An optimization algorithm based on brainstorming process, *Emerging Research on Swarm Intelligence and Algorithm Optimization* (2014)
14. R. Storn, K. Price, Differential evolution a simple and efficient heuristic for global optimization over continuous spaces. J. Global Optim. **11**(4), 341–359 (1997)
15. H. Hernndez, C. Blum, Distributed graph coloring: an approach based on the calling behavior of Japanese tree frogs. Swarm Intell. **6**(2), 117–150 (2012)
16. X.S. Yang, Flower pollination algorithm for global optimization, in *Unconventional computation and natural computation* (Springer Berlin Heidelberg, 2012), pp. 240–249
17. A. Mozaffari, A. Fathi, S. Behzadipour, The great salmon run: a novel bio-inspired algorithm for artificial system design and optimization. Int. J. Bio-Inspired Comput. **4**(5), 286–301 (2012)
18. S. He, Q.H. Wu, J.R. Saunders, Group search optimizer: an optimization algorithm inspired by animal searching behavior. IEEE Trans. Evolut. Comput. **13**(5), 973–990 (2009)

19. L.M. Zhang, C. Dahlmann, Y. Zhang, Human-inspired algorithms for continuous function optimization, in *IEEE International Conference on Intelligent Computing and Intelligent Systems, ICIS 2009*, Vol. 1 (2009), pp. 318–321
20. A.R. Mehrabian, C. Lucas, A novel numerical optimization algorithm inspired from weed colonization. Ecolo. Inf. **1**(4), 355–366 (2006)
21. U. Premaratne, J. Samarabandu, T. Sidhu, A new biologically inspired optimization algorithm, in *International Conference on Industrial and Information Systems (ICIIS)* (IEEE, 2009), pp. 279–284
22. S.H. Jung, Queen-bee evolution for genetic algorithms. Electron. Lett. **39**(6) (2003)
23. R. Hedayatzadeh, F.A. Salmassi, M. Keshtgari, R. Akbari, K. Ziarati, Termite colony optimization: A novel approach for optimizing continuous problems, in *18th Iranian Conference on In Electrical Engineering (ICEE)* (IEEE, 2010), pp. 553–558
24. I. FisterJr, X.S. Yang, I. Fister, J. Brest, D. Fister, A brief review of nature-inspired algorithms for optimization (2013). arXiv preprint arXiv:1307.4186
25. R. Buyya, C.S. Yeo, S. Venugopal, J. Broberg, I. Brandic, Cloud computing and emerging IT platforms: Vision, hype, and reality for delivering computing as the 5th utility. Future Gener. Comput. Syst. **25**(6), 599–616 (2009)
26. I. Foster, Y. Zhao, I. Raicu, S. Lu. Cloud computing and grid computing 360-degree compared. in *Grid Computing Environments Workshop, IEEE, GCE'08* (2008), pp. 1–10
27. Q. Zhang, L. Cheng, R. Boutaba, Cloud computing: state-of-the-art and research challenges. J. Internet Serv. Appl. **1**(1), 7–18 (2010)
28. M. Armbrust, A. Fox, R. Griffith, A.D. Joseph, R. Katz, A. Konwinski, G. Lee et al., A view of cloud computing. Commun. ACM **53**(4), 50–58 (2010)
29. P. Mell, T. Grance, The NIST definition of cloud computing. National Institute of Standards and Technology, Information Technology Laboratory, Technical Report Version 15, (2009)
30. H. Das, G.S. Panda, B. Muduli, P.K. Rath. The complex network analysis of power grid: a case study of the West Bengal power network. in *Intelligent Computing, Networking, and Informatics* (Springer, New Delhi, 2014), pp. 17–29
31. Das, Himansu, S.K. Mishra, D.S. Roy, The topological structure of the Odisha power grid: a complex network analysis. IJMCA **1**(1), 012–016 (2013)
32. Das, Himansu, D.S. Roy, A grid computing service for power system monitoring. Int. J. Comput. Appl. **62**(20) (2013)
33. H. Das, A.K. Jena, P.K. Rath, B. Muduli, S.R. Das. Grid computing-based performance analysis of power system: a graph theoretic approach. in *Intelligent Computing, Communication and Devices* (Springer, New Delhi, 2015), pp. 259–266
34. I. Kar, R.N Ramakant Parida, H. Das, Energy aware scheduling using genetic algorithm in cloud data centers. in *IEEE International Conference on Electrical, Electronics, and Optimization Techniques (ICEEOT)* (2016), pp. 3545–3550
35. Sarkhel, Preeta, H. Das, L.K. Vashishtha. Task-Scheduling Algorithms in Cloud Environment. in *Computational Intelligence in Data Mining* (Springer, Singapore, 2017), pp. 553–562
36. C.R. Panigrahi, M. Tiwary, B. Pati, H. Das, Big Data and Cyber Foraging: Future Scope and Challenges. in *Techniques and Environments for Big Data Analysis* (Springer International Publishing, 2016), pp. 75–100
37. X. Xin, H. Yu, A game theory approach to fair and efficient resource allocation in cloud computing. Math. Probl. Eng. **2014**, 1–14 (2014)
38. L. Mashayekhy, M.M. Nejad, D. Grosu, Cloud federations in the sky: Formation game and mechanism. IEEE Trans. Cloud Comput. **3**(1), 14–27 (2015)
39. A. Corradi, M. Fanelli, L. Foschini, VM consolidation: A real case based on openstack cloud. Future Gener. Comput. Syst. **32**, 118–127 (2014)
40. S. Nedevschi, L. Popa, G. Iannaccone, S. Ratnasamy, D. Wetherall, Reducing network energy consumption via sleeping and rate-adaptation, in *NSDI*, Vol. 8 (2008), p. 323336
41. P. Arroba et al., Server power modeling for run-time energy optimization of cloud computing facilities. Energy Procedia **62**, 401–410 (2014)

42. M.A. Sharkh, A. Shami, An evergreen cloud: Optimizing energy efficiency in heterogeneous cloud computing architectures. Veh. Commun. **9**(11), 199–210 (2017)
43. D.-M. Bui et al., Energy efficiency for cloud computing system based on predictive optimization. J. Parallel Distrib. Comput. **102**, 103–114 (2017)
44. Chun-Han Lin, Pi-Cheng Hsiu, Cheng-Kang Hsieh, Dynamic backlight scaling optimization: a cloud-based energy-saving service for mobile streaming applications. IEEE Trans. Comput. **63**(2), 335–348 (2014)
45. M. Ferrara, E. Fabrizio, J. Virgone, M. Filippi, A simulation-based optimization method for cost-optimal analysis of nearly zero energy buildings. Energy Build. **84**, 442–457 (2014)
46. S. Wang, Z. Liu, Z. Zheng, Q. Sun, F. Yang, Particle swarm optimization for energy-aware virtual machine placement optimization in virtualized data centers, in *IEEE International Conference on Parallel and Distributed Systems (ICPADS)* (2013), pp. 102–109
47. H. Goudarzi, M. Ghasemazar, M. Pedram, *IEEE 12th IEEE/ACM International Symposium on SLA-based optimization of power and migration cost in cloud computing* (Cloud and Grid Computing (CCGrid), In Cluster, 2012), pp. 172–179
48. W. Shu, W. Wang, Y. Wang, A novel energy-efficient resource allocation algorithm based on immune clonal optimization for green cloud computing. EURASIP J. Wirel. Commun. Netw. **1** (2014)
49. A.P. Xiong, C.-X. Xu, Energy efficient multiresource allocation of virtual machine based on PSO in cloud data center. Math. Probl. Eng. (2014)
50. X.F. Liu, Z.H. Zhan, K.J. Du, W.N. Chen, Energy aware virtual machine placement scheduling in cloud computing based on ant colony optimization approach. in *Proceedings of the 2014 ACM Annual Conference on Genetic and Evolutionary Computation* (2014), pp. 41–48
51. D.L. Eager, E.D. Lazowska, J. Zahorjan, Adaptive load sharing in homogeneous 647 distributed systems. IEEE Trans. Softw. Eng. **12**(5), 662675
52. E.J. Ghomi, A.M. Rahmani, N.N. Qader, Load-balancing Algorithms in Cloud Computing: A Survey. J. Netw. Comput. Appl. **88**, 50–71 (2017)
53. A.S. Milani, N.J. Navimipour, Load balancing mechanisms and techniques in the cloud environments: systematic literature review and future trends. J. Netw. Comput. Appl. **71**, 8698 (2016)
54. M. Mesbahi, A.M. Rahmani, Load balancing in cloud computing: a state of the art survey. Int. J. Mod. Educ. Comput. Sci. **8**(3) (2016)
55. V.R.T. Kanakala, V.K. Reddy, Performance analysis of load balancing techniques in cloud computing environment. TELKOMNIKA Indones. J. Electr. Eng. **13**(3), 568573 (2015a)
56. V.R.T. Kanakala, V.K. Reddy, Performance analysis of load balancing techniques in cloud computing environment. TELKOMNIKA Indones. J. Electr. Eng. **13**(3), 568573 (2015b)
57. I.N. Ivanisenko, T.A. Radivilova, Survey of major load balancing algorithms in distributed system. in *IEEE Information Technologies in Innovation Business Conference (ITIB)* (2015), pp. 89–92
58. A.A.S. Farrag, S.A. Mahmoud, Intelligent Cloud Algorithms for Load Balancing problems: A Survey. IEEE in *Proceedings of the Seventh International Conference on Intelligent Computing and Information Systems (ICICIS 'J 5)* (2015), pp. 210–216
59. L.D. Dhinesh Babu, P. Venkata Krishna, Honey bee behavior inspired load balancing of tasks in cloud computing environments. Appl. Soft Comput. **13**(5), 2292–2303 (2013)
60. K. Nishant, P. Sharma, V. Krishna, C. Gupta, K.P. Singh, R. Rastogi, Load balancing of nodes in cloud using ant colony optimization. in *IEEE UKSim 14th International Conference on Computer Modelling and Simulation (UKSim)* (2012), pp. 3–8
61. J. Gu, J. Hu, T. Zhao, G. Sun, A new resource scheduling strategy based on genetic algorithm in cloud computing environment. JCP J. Comput. **7**(1), 42–52 (2012)
62. F. Ramezani, J. Lu, F.K. Hussain, Task-based system load balancing in cloud computing using particle swarm optimization. Int. J. Parallel Progm. Int. J. Parallel Program. **42**(5), 739–754 (2013)
63. K. Dasgupta, B. Mandai, P. Dutta, J.K. Mandai, S. Dam, A Genetic Algorithm (GA) based load balancing strategy for cloud computing. Int. Conf. Comput. Intell. Model. Tech. Appl. **10**, 340–347 (2013)

64. B.K. Ruekaew, W. Kimpan, Virtual machine scheduling management on cloud computing using artificial bee colony, in *Proceedings of the International MultiConference of Engineers and Computer Scientists*, vol. 1 (2014)
65. H. Yuan, C. Li, M. Du, Optimal virtual machine resources scheduling based on improved particle swarm optimization in cloud computing. JSW J. Softw. **9**(3), 705–708 (2014)
66. M.A. Tawfeek, A. EI-Sisi, A.E. Keshk, F.A. Torkey, Cloud task scheduling based on ant colony optimization, in *2013 8th International Conference on Computer Engineering & Systems (ICCES)*, vol. 12, no. 2 (2015), pp. 64–69
67. A. Al - maamari, F.A. Omara, Task scheduling using hybrid algorithm in cloud computing environments. IOSR J. Comput. Eng. **17**(3), 96–106 (2015)
68. S. Dam, G. Mandal, K. Dasgupta, P. Dutta, Genetic algorithm and gravitational emulation based hybrid load balancing strategy in cloud computing. in *IEEE Third International Conference on 2015 Computer, Communication, Control and Information Technology (C3it)* (2015), pp. 1–7
69. T.S. Ashwin, S.G. Domanal, R.M. Guddeti, A Novel Bio-Inspired Load Balancing Of Virtual Machines In Cloud Environment. in *IEEE International Conference On Cloud Computing In Emerging Markets (Ccem)* (2014), pp. 1–4
70. S. Aslanzadeh, Z. Chaczko, Load balancing optimization in cloud computing: applying endocrine-particle swarm optimization. in *2015 IEEE International Conference On Electro/Information Technology (Eit)* (2015), pp. 165–169
71. T. Wang, Z. Liu, Y. Chen, Y. Xu, X. Dai, Load balancing task scheduling based on genetic algorithm in cloud computing, in *IEEE 12th International Conference On Dependable. Autonomic And Secure Computing (Dasc)* (2014), pp. 146–152
72. Z. Zhang, X. Zhang, A load balancing mechanism based on ant colony and complex network theory in open cloud computing federation. in *2010 IEEE 2nd International Conference On 2010 May 30 IEEE Industrial Mechatronics And Automation (Icima)*, Vol. 2, pp. 240–243
73. E. Gupta, V. Deshpande, A technique based on ant colony optimization for load balancing in cloud data center. in *2014 IEEE International Conference On Information Technology (Icit)* (2014), pp. 12–17
74. R. Kaur, N. Ghumman, Hybrid improved max min ant algorithm for load balancing in cloud. in *International Conference On Communication, Computing and Systems* (IEEE Icccs2014)
75. K. Pan, J. Chen, Load balancing in cloud computing environment based on an improved particle swarm optimization. in *2015 6th IEEE International Conference On Software Engineering And Service Science (Icsess)* (2015), pp. 595–598
76. K.R. Babu, A.A. Joy, P. Samuel, Load balancing of tasks in cloud computing environment based on bee colony algorithm. in *2015 IEEE Fifth International Conference On Advances In Computing And Communications (Icacc)* (2015), pp. 89–93
77. R. Achar, P.S. Thilagam, N. Soans, P.V. Vikyath, S. Rao, A.M. Vijeth, Load balancing in cloud based on live migration of virtual machines. in *2013 Annual IEEE India Conference (Indicon)* (2013), pp. 1–5
78. F. Ramezani, J. Lu, F.K. Hussain, Task-based system load balancing in cloud computing using particle swarm optimization. Int. J. Parallel Program. **42**(5), 739–754 (2014)
79. S. Mohanty, P.K. Patra, S. Mohapatra, M. Ray, MPSO: A Novel Meta-Heuristics for Load Balancing in Cloud Computing. Int. J. Appl. Evolut. Comput. (IJAEC) **8**(1), 1–25 (2017)
80. K.M. Cho, P.W. Tsai, C.W. Tsai, C.S. Yang, A hybrid meta-heuristic algorithm for VM scheduling with load balancing in cloud computing. Neural Comput. Appl. **26**(6), 1297–1309 (2015)
81. D. Gabi, A.S. Ismail, A. Zainal, Z. Zakaria, Solving task scheduling problem in cloud computing environment using orthogonal Taguchi-Cat algorithm. Int. J. Electr. Comput. Eng. (IJECE) **7**(3) (2017)
82. M.G. Avram, Advantages and challenges of adopting cloud computing from an enterprise perspective. Procedia Technol. **12**, 529534 (2014)
83. M.D. Assuno, A. Costanzo, R. Buyya, A cost-benefit analysis of using cloud computing to extend the capacity of clusters. Clust. Comput. **13**(3), 335347 (2010)

84. Y. Zhu, P. Liu, Multi-dimensional constrained cloud computing task scheduling mechanism based on genetic algorithm. Int. J. Online Eng. **9**, 1518 (2013)
85. L. Wu, Y.J. Wang, C.K. Yan, Performance comparison of energy-aware task scheduling with GA and CRO algorithms in cloud environment, in *Applied Mechanics and Materials* (2014), p. 204208
86. F. Taoa, L.Z. Ying Fengb, T.W. Liaoc, CLPS-GA: A case library and Pareto solution-based hybrid genetic algorithm for energy-aware cloud service scheduling. Appl. Soft Comput. **19**, 264279 (2014)
87. K. Li, G. Xu, G. Zhao, Y. Dong, D. Wang, Cloud task scheduling based on load balancing ant colony optimization. in *2011 IEEE Sixth Annual Chinagrid Conference (ChinaGrid)* (2011), pp. 3–9
88. S. Xue et al., An ACO-LB algorithm for task scheduling in the cloud environment. J. Softw. **9**(2), 466473 (2014)
89. S. Xue, J. Zhang, X. Xu, An improved algorithm based on ACO for cloud service PDTs scheduling. Adv. Inf. Sci. Serv. Sci. **4**(18), 340348 (2012)
90. S. Kaur, A. Verma, An efficient approach to genetic algorithm for task scheduling in cloud computing environment. Int. J. Inf. Technol. Comput. Sci. **10**, 7479 (2012)
91. Z. Yang et al., Optimized task scheduling and resource allocation in cloud computing using PSO based fitness function. Inf. Technol. J. **12**(23), 70907095 (2013)
92. S. Zhan, H. Huo, Improved PSO-based task scheduling algorithm in cloud computing. J. Inf. Comput. Sci. **9**(13), 38213829 (2012)
93. K. Kaur, N. Kaur, K. Kaur, *A Novel Context and Load-Aware Family Genetic Algorithm Based Task Scheduling in Cloud Computing, Data Engineering and Intelligent Computing* (Springer, Singapore, 2018), pp. 521–531
94. H.B. Alla, S.B. Alla, A. Ezzati, A. Mouhsen, A novel architecture with dynamic queues based on fuzzy logic and particle swarm optimization algorithm for task scheduling in cloud computing. in *Advances in Ubiquitous Networking 2* (Springer, Singapore, 2017) pp. 205–217
95. K.R. Kumari, P. Sengottuvelan, J. Shanthini, A hybrid approach of genetic algorithm and multi objective PSO task scheduling in cloud computing. Asian J. Res. Soc. Sci. Humanit. **7**(3), 1260–1271 (2017)
96. W.-J. Wang, Y.-S. Chang, W.-T. Lo, Y.K. Lee, Adaptive scheduling for parallel tasks with QoS satisfaction for hybrid cloud environments. J. Supercomput. **66**(2), 129 (2013)
97. E.N. Alkhanak, S.P. Lee, R. Rezaei, R.M. Parizi, Cost optimization approaches for scientific workflow scheduling in cloud and grid computing: A review, classifications, and open issues. J. Syst. Softw. **113**, 1–26 (2016)
98. W.N. Chen, J. Zhang, A set-based discrete PSO for cloud workflow scheduling with user-defined QoS constraints. in *2012 IEEE International Conference on Systems, Man, and Cybernetics (SMC)* (2012), pp. 773–778
99. Z. Wu, Z. Ni, L. Gu, X. Liu, A revised discrete particle swarm optimization for cloud workflow scheduling. in *2010 IEEE International Conference on Computational Intelligence and Security (CIS)* (2010), pp. 184–188
100. S. Pandey, L. Wu, S.M. Guru, R. Buyya, A particle swarm optimization-based heuristic for scheduling workflow applications in cloud computing environments. in *2010 24th IEEE international conference on Advanced information networking and applications (AINA)* (2010), pp. 400–407
101. Z. Wu, X. Liu, Z. Ni, D. Yuan, Y. Yang, A market-oriented hierarchical scheduling strategy in cloud workflow systems. J. Supercomput. 1–38 (2013)
102. C. Wei-Neng, J. Zhang, An ant colony optimization approach to a grid workflow scheduling problem with various QoS requirements. in *IEEE Transactions on Systems, Man, and Cybernetics, Part C (Applications and Reviews)*, Vol. 39, No. 1 (2009), pp. 29–43
103. H. Liu, D. Xu, H.K. Miao, Ant colony optimization based service flow scheduling with various QoS requirements in cloud computing. in *IEEE 2011 First ACIS International Symposium on Software and Network Engineering (SSNE)*, pp. 53–58

104. J. Yu, R. Buyya, Scheduling scientific workflow applications with deadline and budget constraints using genetic algorithms. Sci. Program. **14**, 217–230 (2006)
105. A.K. Talukder, M. Kirley, R. Buyya, Multiobjective differential evolution for scheduling workflow applications on global grids. Concurr. Comput. Pract. Exp. **21**(13), 1742–1756 (2009)
106. S. Pandey, L. Wu, S.M. Guru, R. Buyya, A particle swarm optimization-based heuristic for scheduling workflow applications in cloud computing environments. in *IEEE 2010 24th IEEE international conference on Advanced information networking and applications (AINA)* (2010), pp. 400–407
107. H.B. Alla, S.B. Alla, A. Ezzati, A. Mouhsen, A novel architecture with dynamic queues based on fuzzy logic and particle swarm optimization algorithm for task scheduling in cloud computing. in *Advances in Ubiquitous Networking 2* (Springer, Singapore, 2017), pp. 205–217
108. B. Kumar, M. Kalra, P. Singh, Discrete binary cat swarm optimization for scheduling workflow applications in cloud systems. in *3rd International Conference on Computational Intelligence and Communication Technology (CICT)* (IEEE, 2017), pp. 1–6
109. S. Prathibha, B. Latha, G. Suamthi, Particle swarm optimization based workflow scheduling for medical applications in cloud. Biomed. Res. **1**(1) (2017)
110. D. Gabi, A.S. Ismail, A. Zainal, Z. Zakaria, Solving Task Scheduling Problem in Cloud Computing Environment Using Orthogonal Taguchi-Cat Algorithm. Int. J. Electr. Comput. Eng. (IJECE) **7**(3) (2017)
111. Y. Xu, K. Li, L. He, L. Zhang, K. Li, A hybrid chemical reaction optimization scheme for task scheduling on heterogeneous computing systems. IEEE Trans. parallel Distribut. Syst. **26**(12), 3208–3222 (2015)
112. Y. Jiang, Z. Shao, Y. Guo, A DAG scheduling scheme on heterogeneous computing systems using tuple-based chemical reaction optimization. Sci. World J. (2014)
113. S.S. Kim, J.H. Byeon, H. Yu, H. Liu, Biogeography-based optimization for optimal job scheduling in cloud computing. Appl. Math. Comput. **247**, 266–280 (2014)
114. V. Kumari, M. Kalra, S. Singh, Independent task scheduling in cloud environment using Big Bang-Big Crunch approach. in *2nd International Conference on Recent Advances in Engineering and Computational Sciences (RAECS)* (IEEE, 2015), pp. 1–4
115. S. Selvarani, G. Sadhasivam, An intelligent water drop algorithm for optimizing task scheduling in grid environment. Int. Arab J. Inf. Technol. **13**(6) (2016)
116. G. Guo-Ning, H. Ting-Lei, Genetic simulated annealing algorithm for task scheduling based on cloud computing environment, in *Proceedings of International Conference on Intelligent Computing and Integrated Systems* (2010), pp. 60–63
117. Q. Zhang, L. Cheng, R. Boutaba, Cloud computing: state-of-the-art and research challenges. J. Internet Serv. Appl. **1**(1), 7–18 (2010)
118. XenSource Inc, Xen, www.xensource.com
119. F. Farahnakian, A. Ashraf, T. Pahikkala, P. Liljeberg, J. Plosila, I. Porres, H. Tenhunen, Using ant colony system to consolidate VMs for green cloud computing. IEEE Trans. Serv. Comput. **8**(2), 187–198 (2015)
120. S.E. Dashti, A.M. Rahmani, Dynamic VMs placement for energy efficiency by PSO in cloud computing. J. Exp. Theor. Artif. Intell. **28**, 97–112 (2016)
121. F. Farahnakian, A. Ashraf, P. Liljeberg, T. Pahikkala, J. Plosila, I. Porres, H. Tenhunen, Energy-aware dynamic VM consolidation in cloud data centers using ant colony system. in *Cloud Computing (CLOUD)* (2014), pp. 104–111

Chapter 2
Resource Allocation in Cloud Computing Using Optimization Techniques

Gopal Kirshna Shyam and Ila Chandrakar

2.1 Introduction

The aim of cloud computing is to provide utility based IT services by interconnecting a huge number of computers through a real-time communication network such as the Internet. Since many organizations are using cloud computing which are working in various fields, its popularity is growing. So, because of this popularity, there has been a significant increase in the consumption of resources by different data centres which are using cloud applications [1–4]. Hence, there is a need to discuss optimization techniques and solutions which will save resource consumption but there will not be much compromise on the performance. These solutions would not only help in reducing the excessive resource allocation, but would also reduce the costs without much compromise on SLA violations, thereby benefitting the Cloud service providers. In this chapter, we discuss on the optimization of resource allocation so as to provide cost benefits to the Cloud service users and Cloud service providers.

2.2 Resource Allocation Using Meta-Heuristic Optimization Techniques

Cloud computing offers various resource allocation services like computation, storage etc. in a virtualized environment [5]. The virtual machine in Cloud allocates the job and schedules it efficiently. The key issues in using cloud is task schedul-

G. K. Shyam (✉) · I. Chandrakar
School of Computing and Information Technology, REVA University,
Bengaluru 560 064, India
e-mail: gopalkrishna@revainstitution.org

I. Chandrakar
e-mail: ilaprithvi@gmail.com

B. S. P. Mishra et al. (eds.), *Cloud Computing for Optimization: Foundations, Applications, and Challenges*, Studies in Big Data 39,
https://doi.org/10.1007/978-3-319-73676-1_2

ing and resources utilization. Scheduling allocates different types of jobs using the existing resources. Scheduling is decided based on the feedback of the Quality of Services (QoS), which handles the different tasks in the job allocation [6]. Therefore, in order to schedule the tasks, numerous heuristic techniques exist such as Particle Swarm Optimization (PSO), Genetic Algorithms (GA), Ant Colony Optimization Algorithms (ACO), Artificial Bee Colony Algorithms (ABC), is used to solve the task scheduling and resource problems. This section shall discuss algorithm to solve task scheduling and resource allocation problem in Cloud computing. The algorithms aim at providing better efficient scheduling mechanism which increases the performance and efficiency of the system by minimizing the execution time (makespan), execution cost, deadline etc.

2.2.1 Particle Swarm Optimization (PSO)

PSO is a swarm based meta-heuristic algorithm simulating the nature such as a flock of insects, bird's gesture or schooling of fish to discover the optimal solutions. Algorithm 1 provides PSO algorithm aimed at reducing the cost function. It is an universal optimization algorithm, where the optimized results for multi-dimensional searches can be made to appear as a point or surface. The fitness values examine the particles. In PSO, swarm is considered as population and participants like insects, birds or fishes generated by random velocities and situations are considered as particles [1]. The algorithm is easy to implement and it contains only few parameters for modification [2].

To minimize the usage of energy in Cloud data center, energy efficient virtual machine allocation algorithm is suggested through the PSO technique and the energy efficient multi-resource allocation model. This algorithm can escape dropping into local optima, which is very commonly found in traditional algorithms [3].

Another technique known as Position Balanced Parallel Particle Swarm Optimization (PBPPSO) algorithm is given for allocation of resources in IaaS Cloud [7]. It discovers the resource optimizations for the group of jobs with less make span and lesser cost.

2.2.2 Genetic Algorithms (GA)

GA is a probabilistic optimization algorithm imitating the progression of natural evolution. The biological evolution process in chromosomes is the idea behind GA, which is survival of the fittest. An advantage of Genetic Algorithm (GA) is to resolve the problem of resource allocation and recommend a new model to enhance the result of the decision making process.

An objective of GA algorithm, that we have discussed here is discovering trade-off solutions between completion time of tasks and system energy consumption. It is

Algorithm 1 Pseudo-code for PSO

Inputs: size of the problem, size of the population;
Output: Optimized solution;
Description:

```
population = θ
s_g_optimal = θ
for (i = 1 to size_of_population) do
S_velocity = Random_Velocity();
S_position = RandomPosition(size_of_population);
s_g_optimal = s_position ;
if Cost(S_s_best) <= Cost(S_g_best)
S_g_optimal = (S_s_best);
end
end
while: StopCondition () do
for each p ε population do
S_velocity = UpdateVelocity(S_velocity, S_s_best, S_s_best);
S_position = UpdatePosition(S_position, S_velocity);
if Cost(S_position) <= Cost(S_s_best)
s_s_best = s_position;
if Cost(S_s_best) < = Cost(S_g_best)
s_g_optimal = S_g_optimal;
end
end
end
end
return P_g_best
```

applicable for Cloud computing data centers that perform joint allocation of network and computational resources. The algorithm allows to explore solutions space and to search for the optimal solution in an efficient manner. It is scalable as well as energy efficient and is based on a model developed to capture specifics of the data center network topology and device power consumption.

2.2.3 Ant Colony Optimization Algorithms (ACO)

An ant colony framework for adaptive resource allocation in Cloud computing environments is done by hosting the applications with specified QoS necessities as a response time and throughput. It minimizes the usage of energy of data center resources by making an allowance for the dynamic workloads of servers by means of various ant agents. The issue of VM placement is expressed as a multi-objective combinatorial optimization problem, whose core goal is to optimize concurrently total resource wastage and energy consumption.

Algorithm 2 Pseudo-code for Ant Colony optimization

Inputs: An instance P of a CO problem model P=S, f, ω;
Output: P_best;
Description:

```
InitializePheromoneValues (T);
P_best = null;
while (terminations conditions not met) do
S_iter = θ;
for ( j = 1 to n ) do
S = ConstructSolution (T)
if (s is a valid solution) then
P_g_best = (P_p_best)
s = LocalSearch = s optional;
if (f(s) < f_best or (P_best=null) ) then P_best = S ;
s_iter = s_iter ∪ s
end
end
ApplyPheromoneUpdate (T, S_iter, P_best );
end
return P_best;
```

2.2.4 *Cooperative Game Based Resource Allocation*

The work in Ref. [8] proposed a scheme for bargaining cooperation based resource by making use of game theory approach to achieve energy consumption, cost and resource utilization requirements. The scheme is evaluated based on parameters like bargaining steps, job submission, job execution and expected payoffs.

To achieve the above mentioned requirement, a cooperation based resource bargaining game is shown in Fig. 2.1. The sequence of bargaining when combined gives rise to four cases. The cases are with respect to the following: (i) resources are available and jobs are more, (ii) resources are available and jobs are less, (iii) resources are not available and jobs are more, and (iii) resources are not available and jobs are less. The payoff matrix is constructed for cooperative game such that the parties involved get computational credits and none of them lose to another by a huge margin. The equilibrium is obtained by maximizing the payoffs for each of the cases so that: (i) SU can submit, and (ii) SP can execute as many jobs as possible.

2.2.5 *Description of Cooperative Game*

Here we are considering that SU and SP are competing fairly to acquire resources i.e. they are cooperating each other since resources are limited in the CDC. The three factors to be taken into consideration for game are set of players SUs and SPs, strategies of each players and utility of each player.

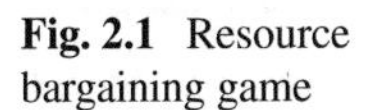

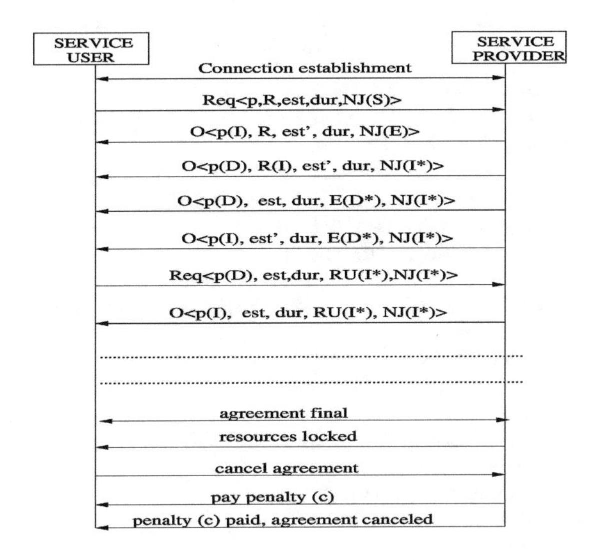

Fig. 2.1 Resource bargaining game

In Fig. 2.1, SU starts resource bargaining game by stating the price that it can pay (p), required resource (R), expected start time of the job (est), duration of execution (dur), number of jobs to be submitted (NJ(S)). The SP initially bargains with increasing price (P(I)), a different expected start time of the job (est'), and states the number of jobs it can execute (NJ(E)). In the subsequent steps, it provides attractive offers to SUs, with decrease cost (p(D)) and increased number of jobs (NJ(I)). It also offers to increase the price if numbers of jobs are not increased (NJ(I*)). The offers also considers cost if decreased energy consumption (E(D*)) and increased resource utilization (RU(I*)) are taken into account. The SP offers to lock the resources when SP commits, but penalizes them when they cancel the order after committing. These bargaining steps are limited for cooperative bargaining while for non-cooperative game (since SUs and SPs are selfish and look to fulfill their own needs), the bargaining steps increases drastically.

A payoff matrix algorithm is discussed in Algorithm 3.

2.2.6 Nash Equilibrium in a Cooperative Game

To identify Nash equilibria on a payoff matrix, rule goes as follows: if, in the payoff pair of the cell, the first payoff number is the maximum of the column of the cell and if the second number is the maximum of the row of the cell - then the cell shows a Nash equilibrium. This rule can be applied to a 3x3 matrix shown in Table 2.1. Nomenclature used in Table 2.1 are as follows: LCJ (lesser cost job_submission), ICJ (increased cost job_submission), LCJE (lesser cost job execution), ICJE (increased cost job execution), DEJ (decreased energy job_submission), IEJ (increased energy

Algorithm 3 Payoffs Algorithm

Inputs: priority of SU and SP;
Output: number of job submissions by SU and successful offers by SP;
Description:

IF1 (p(D) or p(I) is prioritized by SP and SU)
find job submissions by SU
find successful job offers by SP
end **IF1**
IF2 (p(D) or p(I) prioritized by SP **OR** (p(D) or p(I) is prioritized by SP **AND** RU(D) or RU(I) is prioritized by SU)
find job submissions by SU
find successful job offers by SP
end **IF2**
IF3 (E(D) or E(I) is prioritized by SP **AND** p(D) or p(I) is prioritized by SU) **AND** SU prioritize either RU(D) or RU(I))
find job submissions by SU
find successful job offers by SP
end **IF3**
IF4 (RU(D) or RU(I) is prioritized by SP **AND** SU prioritize either p(D) or p(I)) **OR** (SP and SU prioritize either RU(D) or RU(I))**OR** (SP prioritize either RU(D) or RU(I) **AND** SU prioritize either E(D) or E(I))
find job submissions by SU
find successful job offers by SP
end **IF4**

Table 2.1 Payoff matrix for different strategies of players

		Service Providers		
		Cost	Energy	Resource Utility
Service Users	Cost	LCJE (ICJE) LCJ (ICJ)	DEJE (IEJE) DEJ (IEJ)	LRUJE (HRUJ) LRUJ (HRUJ)
	Energy	LCJE (ICJE) DEJ (IEJ)	DEJE (IEJE) DEJ (IEJE)	LRUJE (HRUJE) DEJ (IEJ)
	Resource Utility	LCJE (ICJE) LRUJ (HRUJE)	DEJE (IEJE) LRUJ (HRUJE	LRUJE (HRUJE) LRUJE (HRUJE)

job_submission), DEJE (decreased energy job execution), IEJE (increased energy job execution), LRUJ (lesser resource utilized jobs), HRUJ (higher resource utilized jobs), LRUJE (lesser resource utilized job execution), HRUJE (higher resource utilized job execution). So, the actual mechanics of finding equilibrium cells is: searching for the maximum of a column and then checking whether the second member of the pair is the maximum of the row. After meeting these conditions, the cell represents a Nash equilibrium.

2.2.7 *Non-cooperation Based Resource Bargaining*

The sequence diagram shown in Fig. 2.1, is applicable to non-cooperative bargaining when the number of bargaining steps are increased to a large value. Each of the entities, namely SU and SP become selfish and fail to arrive at common goals. Payoff matrix constructed in such case reveals that there is a large difference between job submissions and job executions. Nash equilibrium is obtained when SU decides not to accept services (from SP) above a certain expected value, and SP decides not to offer any services (to SU) below a certain expected value. Expected values are payoffs in terms of cost, energy consumption and resource utilization.

2.2.8 *Non-cooperative Game Based Resource Allocation*

In Cloud computing, the resources are normally available as VMs that appear as traditional servers. The consideration taken by us is that a data centre of Cloud has a large number of requests, and each request requires for multidimensional resource like CPU, memory, disk, network bandwidths etc. and this will be taken care by a VM. Analogously, each PM has a capacity across each of these dimensions.

Recently, game theory are used to solve different resource competition problems in Cloud computing. Kwok et al. [9] studied a non-cooperative game in a hierarchical Grid computing, in which the machines are selfishness. Kong et al. [10] studied the mechanism design on stochastic virtual resource allocation, in which the value function of a VM is a private information. Auction based [11] methods are such that resources are held by incentive providers and the users act strategically to maximize their utilities.

2.2.9 *Resource Allocation Using Uncertainty Principle of Game Theory*

Resource management is a significant issue in Cloud computing, as the on-demand resource allocation needs to be offered to cloud consumers. The different approaches on resource management in Cloud computing are:

1. Game Theoretic Resources Management in Cloud.
   Nowadays, game theory is used to solve different resource allocation problems in Cloud computing. A study on non-cooperative games for the load balancing and

virtual machine placement problem is discussed [12]. They focus on the existence of Nash equilibrium and care little about the solution for an optimal allocation strategy. A method for the distributed resource allocation problem in federated Cloud which studies both noncooperative and cooperative games [13]. It shows that the cooperative allocation game has a stronger motivation for providers to contribute resources.

2. Fair Resources Allocation.
   One of the big challenge for resource allocation in Clouds is the coordination of resource sharing. Fair allocation for hadoop is used in many works which divides resources as fixed-size partitions, or slots [14]. Another fair technique is the max-min fairness, which tries to maximize the minimum resource each user received. This approach can be enhanced by providing a weighted max-min fairness model to support some policies which considers different factors, like priority, reservation, and deadline [15]. Some approaches have recently been proposed to quantify fairness [16].
   But, most of them considers the fairness of single type resource allocation problem. The fair allocation problem for multiple types of resources allocation has been studied in [17]. In this work, they present fairness approach for dominant resource, which considers the problem by computing the dominant share of each user. [18] extends the DRF approach by leveraging a technical framework and studied the indivisibilities allocation. They proved that this mechanism satisfies three properties of fairness. But their work also has some drawbacks because do not consider resource wastage. Our work makes use of DRF approach to measure the fairness of resource allocation, as well as exploiting a way to improve the resource utilization for greater optimization.
3. Efficient Resource Allocation.
   For a Cloud having number of heterogeneous physical servers in data centre, achieving efficient resource consumption is another interesting direction of resource allocation [19, 20]. A method of the resource allocation for a heterogeneous mix of workloads and presentation of a system to manage data centre for increasing the resource consumption of servers has been discussed [21]. Maximization of resource utilization and optimal execution efficiency has been proposed by a novel scheme DOPS [22]. A study of the efficiency of resource allocation of MapReduce Clouds is discussed [23]. They propose a spatiotemporal trade-off technique to scale MapReduce clusters dynamically to improve energy consumption while simultaneously improving performance.

Reference [24] shows a resource aware multiple job scheduling technique for MapReduce that improves resource utilization while meeting completion time goals. Reference [25] presents a resource allocation system that uses user-assigned priorities to offer different service levels and adjusts resource allocations dynamically to fit the requirements.

Although some existing researches study on the trade-off between fairness and efficiency [26], some of them consider the case for multiple types of resources allocation in Cloud. Developing a framework to address the fairness efficiency trade-off with multiple resources is discussed [27]. The limitation of this approach is that the characterization of the parameters is not done which otherwise, could have ensured the fairness properties being satisfied.

2.3 Game Theory Based Concurrent and Cooperative Negotiation of Resource Cost in Cloud Computing Environment Using Agents

Agent technology is proclaimed to be a flexible promising solution for network resource management and QoS (Quality of Service) control. Cloud computing using such technology will be helpful in designing mechanisms for: (i) analyzing consumer job's resource requirements, (ii) reducing service negotiation time by providing cooperative environment in case of consumers and providers having potential conflicting interests, where consumers prefer availing best resources at minimum cost for job execution and providers wish efficient utilization of resources with a maximum profit, and (iii) ensuring successful delivery of services. This section presents an agent-based scheme for the entities namely, consumers, brokers and service providers, where each entity comprises of an agency with set of agents and database. The scheme works in following steps: (i) broker agency communicates with resource provider agency to receive the updates of resource availability and cost, (ii) consumer agency upon receiving the job resource requirements communicates to broker agency concurrently for execution of several jobs, (iii) broker agency plays a cooperative game concurrently with consumer agencies for resource cost negotiation and payoffs, and (iv) broker agency attempts to provide the best services to the consumer agencies by communicating to resource provider agency. Some of the advantages of concurrent and cooperative resource cost negotiation scheme are as follows:

- Helping consumer agent determine the best framework for each individual need based on a number of factors. This includes proenhanced visioning assistance and budget guidance, as well as identifying selection and integration of disparate services across multiple brokers.
- Cost-effective resources and infrastructure advantages, including the ability to negotiate.
- Time efficient resource cost analysis and negotiation for different combinations.
- Enhanced security, allows organizations to develop a customized solution which balances security and the cost benefits.
- Assurance that there will be no disruption on upgradation, repair, and maintenance activities.
- Cooperative negotiations leading to long term relations between consumers, brokers and service providers.

The computational grid design with a distributed resource management scheme is introduced in [5], that uses rational agents for managing the servers sharing reliable resources. The automated multiparty negotiation algorithm used by these agents are motivated by economic value (e.g., revenue) of the strategies employed. In this case, the cooperation is formed to share the resource to process the computing tasks from multiple computing grids, such that the cost is minimized. The concept of virtual organization in computational grid is introduced in [28]. The agent-based service composition scheme [29, 30] decentralizes decision making, limited knowledge, and considers rationality of agents for cooperation.

Reference [31] presents a simple game theory approach to propose a cooperative resource management in Cloud computing systems. It involves simple negotiation process, wherein service users and service providers negotiate to achieve Nash equilibrium. Reference [32] discusses the support for single composite service request, considering a centralized approach. Coalition formation among agents has also been discussed. The coalitions solely emerge through interactions, but number of interactions are not defined. Message handling is very complicated because of a lot of dependencies among the agents. The work given in [33] better fits in for Grid computing scenario. It considers off-line, clairoyant scheduling with no preemption on time-sharing processors. Further, it presents the interest of collaboration between independent parties.

Research carried out in [34] presents the first work on truthful online auction design in Cloud computing where users with heterogeneous demands come and leave on the fly. The nature and dynamics of user's demand in Cloud markets necessitates designing online mechanisms [35]. But, such online mechanisms make no discussion about future demands. To bolster many-to-many consumer-to-Cloud negotiations, [29] devises a novel interaction protocol and a novel negotiation strategy that is characterized by adaptive concession rate (ACR) and minimally sufficient concession (MSC). Game theory has been applied to study different issues in utility computing (e.g., grid and pervasive computing) [14, 15]. For example, in [14], a scheduling algorithm to guarantee QoS constraints in an open Cloud computing framework is modeled as a game. The assignment of tasks to the available resources is obtained from Nash equilibrium.

The agreement settlement among Internet service providers (ISP) is studied using cooperative game theory [16]. The Shapley value is applied for the fair profit sharing which motivates any selfish ISP to apply the routing and connection management to achieve Nash equilibrium among all ISPs.

We discuss cooperative resource cost negotiation scheme expediting flexible pricing of Cloud services that adopts many-to-many negotiation model for negotiating resource cost between consumers and providers. Since a Cloud service may be dynamically composed using multiple types of Cloud resources, external resource brokers can help in bargaining multiple types of Cloud resources with multiple groups of Cloud providers offering several types of computational resources. The resource cost negotiation environment for the proposed scheme is shown in Fig. 2.2. It consists of resource consumer agency (RCAg), resource broker agency (RBAg), and resource provider agency (RPAg) and ICSP (IaaS Cloud service provider). RBAg interacts with RPAg to obtain information about resources in ICSP. RPAg is bounded to ICSP. RCAg submits job to RBAg, and a resource cost negotiation takes place between these agencies.

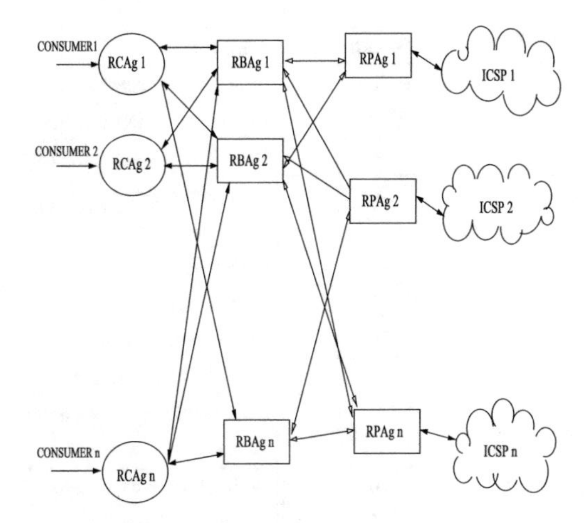

Fig. 2.2 Resource cost negotiation environment

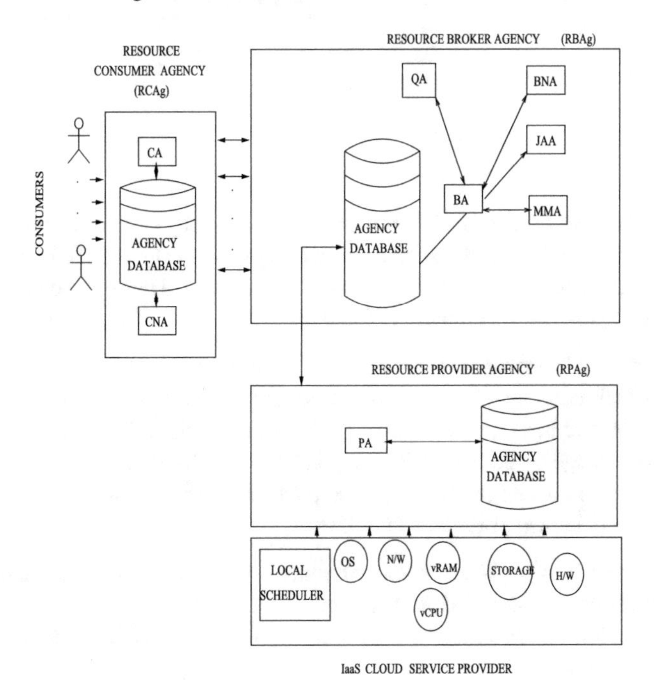

Fig. 2.3 Agencies framework

2.3.1 Agencies

The agencies of resource consumer, resource broker and resource provider have one or more static agents and database which are shown in Fig. 2.3. They are discussed as follows. Table 2.2 lists set of notations used in description of the agencies framework.

Table 2.2 List of notations

Notations	Meaning
RCAg	Resource consumer agency
RBAg	Resource broker agency
RPAg	Resource provider agency
CA	Consumer agent
BA	Broker agent
QA	Query agent
PA	Provider agent
MMA	Matchmaker agent
JAA	Job Allocation agent
CNA	Consumer negotiation agent
BNA	Broker negotiation agent
ICSP	IaaS Cloud service provider
CNG	Cooperative negotiation game
M	Mappers
R	Reducers
VM	Virtual machines

Resource consumer agency (RCAg): It involves agency database and two agents, namely, consumer agent (CA) and consumer negotiation agent (CNA). RCAg functions are designed to act on behalf of resource consumers. The significance of agency database and agents are presented as follows.

Agency database: It provides detailed information about several brokers in the Cloud market and maintains history of consumer(s) and broker(s) interactions. The agents use the database to pass messages among themselves and to know the frequency of application(s) usage by consumer(s). The database facilitates in deciding the offered price for resource cost negotiation and the payoffs. Further, consumer's services are updated to the database at regular intervals.

Consumer agent (CA): The functions of agent are: (i) receiving and mapping consumer requirements (workload) to available RBAg, (ii) selecting the best RBAg (in terms of the quality and cost), and (iii) receiving and handling the virtualized service from RBAg to consumers.

The workload received by CA is a set of several jobs. Jobs can require VMs from several combinations, such as small, medium and large as shown in Table 2.3. The specifications of small, medium, and large VMs are same as in Amazon EC2. Algorithm 4 illustrates concurrent analysis of user's job resource requirements by RBAg from several RCAg(s). It proceeds in two phases, a distributed 'map' operation followed by a distributed 'reduce' operation. At each phase a configurable number of M 'mapper' processors and R 'reducer' processors are assigned to work on the problem, by CA and BA respectively. In the map phase each mapper reads

Table 2.3 Summary of the possible combinations ($3^4 = 81$) for resource types (S: Small VM, M: Medium VM, L:Large VM)

Combinations	*CPU*	*Memory*	*Storage*	*Network*
1	S	S	S	S
2	S	S	S	M
3	S	S	M	L
4	S	L	M	L
5	M	L	M	L
6	L	L	M	L
7	L	M	M	L
–	–	–	–	–
81	L	L	L	L

approximately 1/Mth of the input (in this case combination of resources), from the consumers. Each mapper then performs a 'map' operation to compute frequencies of resource requests. In other words, the map task consists of emitting a user-resource combinations pair for each RCAg, represented by $(u_i, [c1 \ldots cn]) \rightarrow [(c_i, u_k)]$. These frequencies are sorted (from higher to lower value), and written to the local file system of the mapper. At the next phase reducers are each assigned a subset of combinations. The reduce step groups the pairs by several combinations and creates an index entry for each combinations, represented by $[(c_i, u_k)] \rightarrow (c_i, [u_{i1} \ldots u_{im}])$.

Algorithm 4 Requested Resource Combinations count

Input: Resource request by RCAg(s)
Output: The mapper emits an intermediate user-resource pair for each combinations;
The reducer sums up all counts for each resource

```
1: class Mapper
2:    method Map(userid u; combination c)
3:    for all u requesting combination c do
4:       return (u;c)
5: class Reducer
6:    method Reduce(combination c; counts [n1; n2; : : :])
7:    sum=0
8:    for all count n ε counts [n1; n2; : : :] do     sum=sum + n     return (combination c; count sum)
```

Consumer negotiation agent (CNA): It negotiates resource cost with BNA of RBAg. The strategy of CNA negotiation is discussed next.

CNA Negotiation Strategy: Initially, CNA discovers the state of set of resources that RBAg provides, which helps in applying appropriate negotiation strategies. For example, if a CNA knows that the resource competition is low, it may lower its offer price. The offer price of CNA is a function of (R, est, dl, t) of requesting resources R at time t with earliest start time est and latest start time dl with following considerations. Firstly, the deadline: the CNA lowers the offer price when the deadline approaches. Secondly, the cost of resource R: intuitively, a CNA has to offer more for a resource with a higher cost. Thirdly, the demand to supply ratio of a resource R: higher the ratio, higher will be the offered price for the resource.

Resource Broker Agency (RBAg): RBAg consists of an agency database, a query agent (QA), a broker agent (BA), a broker negotiation agent (BNA), a matchmaker agent (MMA), and a job allocation agent (JAA). Their responsibilities are explained as follows.

Agency database: It maintains the information about several RCAgs which use its services, and a history of consumer(s) and broker(s) interactions. The agents use this database to pass messages among themselves. The database aids in deciding the offered price for resource cost negotiation and the payoffs. The updates regarding broker's services are added to the database at regular time intervals. Further, agency database of an RBAg contains information about availability of resources from multiple RPAgs, which facilitates in providing virtualized service(s) to RCAg.

Query agent (QA): It responds to the queries of CA and the responses are updated in agency database instantly.

Broker agent (BA): It controls and coordinates activities of an RBAg. This agent triggers JAA, MMA, and BNA. It manages the availability and usability of resources. BA runs on several machines, manages all the jobs, and decides whether to provision new VM from reserved resources or send request to ICSP for extra resources. The BA performs following tasks: (i) predicting future incoming workloads based on history of most frequent / least frequent accessed applications from agency database, (ii) provisioning necessary VMs in advance from ICSP whenever required, (iii) allocating jobs to VMs, (iv) releasing idle VMs (which do not have jobs running on them), and (v) dynamically allocating VMs for un-allocated jobs through PA (of RPAg).

Broker negotiation agent (BNA): It performs resource price negotiations. Initial payoffs are decided for consumers based on the adopted strategy. Successive payoffs are decided on the basis of cooperative/non-cooperative nature of consumers involved.

Job allocation agent (JAA): JAA is assigned a job by BA, which is decomposed into parallelizable subjobs as presented in [36]. JAA requests MMA to discover corresponding resources for job scheduling.

Matchmaker agent (MMA): MMA discovers and recommends resources for job scheduling as follows: (i) MMA interacts with BA to obtain the statistical status information of each available resource, (ii) resources are sorted in a descending order of their reliability, and (iii) MMA chooses resource with higher reliability value based on less failure history of VMs.

All virtual resources (reserved or dynamically allocated) are provisioned from ICSPs (via RPAg). This helps to pre-configure all the necessary VMs images to run consumer jobs. All the incoming jobs are enqueued. Algorithm 5 presents the resource proposals sent and received by CA. Algorithm 6 presents role of BA in processing jobs. Owing to space constraints, the algorithms are made self explanatory.

BNA Negotiation Strategy: The BNA adopts a very simple negotiation strategy in which it accepts almost all CA offers (within its computational complexity) hoping that in future the profit (and the rewards) may increase through cooperative strategies.

All the offers are sorted by decreasing revenue and offers are greedily picked in the order, starting with the first offer, until no offers remain. The BNA decides the schedule (the start time and end time of providing resources) specified in the offer before forwarding it to JAA. If BNA can fulfill all the agreements offered from CNA, it sends an acceptance message to CNA. If BNA cannot fulfill the agreement(s), it sends a decommitment message(s) to the CNA.

Resource provider agency (RPAg): It comprises of agency database and a provider agent (PA). Their responsibilities are discussed further.

Agency database: It provides information about the resource provisioning services of the ICSP to which it is connected. It stores information about broker(s) requests and ICSPs response, and also whether the resources are reserved well in advance by the CA/BA or not. The agent uses this database to predict and allocate appropriate VMs for incoming workloads. The updates regarding ICSP's services are added to the database regularly.

Provider agent (PA): It can be seen as the owner of a virtual organization, wherein some RBAg are registered as members. PA offers and encapsulates Cloud resources as well as decides cost for resource availability from ICSPs. If the resources are reserved well in advance by BAs, it ensures seamless execution of consumer jobs. However, if the resource(s) are not reserved, it dynamically allocates VMs to incoming workloads of BAs.

Algorithm 5 Resource proposals by CA

Inputs: set of jobs with CA;
Output: Virtualized service to consumers;
Description:
1: Collect consumer's job requirements
2: for job (i = 1 to n) do
3:     Request (and Receive) proposal for i_{th} job from BAs
4:     Send Finalized proposal to a BA
5: end for
6: Initiate resource cost negotiation game between CNA and BNA
7: Decide payoffs through negotiating agents, discussed in section III.C
8: Integrate outputs as a virtualized service to consumers
9: Throw error, if any

Algorithm 6 Job processing by BA

Inputs: Finalized proposal from CAs;
Output: Instantiation of a proposal processing by (BA or PA)
Description:
1: if (Reserved resource for i_{th} job in BA) then
2: Send i_{th} job to ICSP for execution
3: else
4: send i_{th} job resource requirements to PA
5: process execution of i_{th} job dynamically

2.3.2 Cooperative Resource Cost Negotiation Game

We proceed to model the RCAg-RBAg resource cost negotiation as a cooperative game. The consumer's budget to amass resources is limited. Also, the resources vital for an application to run are limited in the data center (DC) of an ICSP acquired by RBAg. Hence cooperative resource cost negotiation between RCAg and RBAg is needed for achieving better payoffs. The RBAg offers services and receives compensation from the CA. The cooperative negotiation game considered consists of three factors: (i) set of players, i.e., CNA from RCAg and BNA from RBAg, (ii) strategies for each player, and (iii) payoff for each player.

Let RCAg's total resource cost (that it can pay to RBAg) be referred to as the offered price of cooperating and deceiving CNA (negotiating on behalf of RCAg) denoted by OP_{CNA} and $ncOP_{CNA}$, respectively, where $ncOP_{CNA} < OP_{CNA}$. Also, let the RBAg's total cost be specified as the offered price of cooperating and deceiving BNA (negotiating on behalf of RBAg) denoted by OP_{BNA} and $ncOP_{BNA}$, respectively, where $ncOP_{BNA} < OP_{BNA}$. If RBAg decides not to allow expected quality of services (refer to expected quality of service as Q), it saves cost but it risks the violation of SLAs which may result in even greater costs in terms of penalty. This encourages RBAg to play a cooperative resource cost game.

The RCAg's experience (on services received from RBAg) could be quantified in terms of satisfaction, where a service delivered on expected lines, i.e., of quality Q, results in satisfaction $s(Q)$. On the other hand, if the service delivered is not on the expected lines, i.e., of quality Q', the satisfaction is $s(Q')$, where $s(Q') < s(Q)$.

The interactions between CNA and BNA could be recursive in resource cost negotiation. In such relationships, the players do not seek the immediate maximization of payoffs but instead the long-run optimal solution. Such situations are modeled in game theory by repeated game models.

Let us consider the RCAg-RBAg interaction model as an infinite horizon repeated game, since the RCAg may keep requesting RBAg for the particular service, but the number of such requests are not known well in advance.

A repeated game makes it possible for the players to condition their negotiation moves on the complete previous history of the various stages, by employing strategies that define appropriate actions for each round. In our scheme, the harsh trigger strategy is used by the CNA, such that if the CNA is not satisfied in one of the stages (i.e., the RBAg does not provide the service as promised in SLAs), the CNA may punish BNA by leaving the relationship forever in the next stage (e.g., stop interacting with the specific RBAg for subsequent requests of the particular service). Given such a strategy, the BNA has stronger incentives to cooperate and provide the service promised, since it faces the threat of losing its consumers from RCAg. Another popular strategy used in this scheme is for a player to mimic the actions of his opponent, giving him the incentive to play cooperatively. This way the player will be rewarded with a similar replicating behaviour. We refer to this as a tit-for-tat strategy.

We employ the harsh strategy as a viable strategy for the CNA and the tit-for-tat strategy as a feasible strategy for the BNA. In addition, we define two more probable strategies for the CNA and one more desirable strategy for the BNA. For the CNA we define: (a) the deceive-and-exit strategy, and (b) the deceive-and-joinback strategy, and for the BNA we define the deceive-and-join-back strategy. The deceive-and-exit strategy is defined so as to allow the CNA to employ non-cooperative behaviour. With this strategy, the CNA leaves after deceiving from cooperation, i.e., does not continue interaction with the particular BNA, in order to avoid any punishment for deceiving. The deceive-and-join-back strategy is defined to capture a case where the BNA deceives and joins back in the subsequent rounds. It is a cooperative strategy that provides the opportunity to the BNA to join back and cooperate with CNA, without penalising it heavily. The discount factor is taken into account while employing the strategies.

2.3.3 *Payoffs Calculation*

We present several cases for payoff calculations.

- **Case 1: CNA playing the harsh strategy**. BNA could either play the tit-for-tat strategy, i.e., cooperate in the current round, or play the deceive-and-join-back strategy. If the BNA cooperates, then: $CPV_c^{BNA} = \frac{|OP_{CNA} - OP_{BNA}|}{1-df}$. If the BNA deceives, then: $CPV_{nc}^{BNA} = OP_{CNA} - ncOP_{BNA} + \frac{df.0}{1-df}$. For the BNA to be motivated for cooperation, CPV in case of cooperation must be greater than CPV in case of non-cooperation. Thus, $CPV_c^{BNA} > CPV_{nc}^{CNA} = \frac{OP_{CNA} - OP_{BNA}}{1-df} > OP_{CNA} - ncOP_{BNA} + \frac{df.0}{1-df}$. If the CNA plays harsh strategy, the BNA is motivated to cooperate, when $df > \frac{OP_{BNA} - ncOP_{BNA}}{OP_{CNA} - ncOP_{BNA}}$.
- **Case 2: CNA playing the deceive-and-exit strategy**. Considering BNAs possible strategies, it could either cooperate or deceive in the current round. If BNA cooperates then: $CPV_c^{BNA} = ncOP_{CNA} - OP_{BNA} + \frac{df.0}{1-df}$. If BNA deceives then: $CPV_{nc}^{BNA} = NCOP_{CNA} - ncOP_{BNA} + \frac{df.0}{1-df}$. If CNA plays deceive-and-exit strategy, BNA is not motivated to cooperate since $CPV_{nc}^{BNA} > CPV_c^{BNA}$.

Table 2.4 Limits of cooperation by CNA and BNA

	Cooperation strategies	Punishment strategies
BNA	$(df)_c^{BNA} > \frac{OP_{BNA}-ncOP_{BNA}}{OP_{CNA}-ncOP_{BNA}}$	$(df)_{pun}^{BNA} > \frac{OP_{BNA}-ncOP_{BNA}}{OP_{CNA}-OP_{BNA}}$
CNA	$(df)_c^{CNA} > \frac{OP_{CNA}-ncOP_{CNA}}{OP_{BNA}-ncOP_{CNA}}$	$(df)_{pun}^{CNA} > \frac{OP_{CNA}-ncOP_{CNA}}{OP_{BNA}-ncOP_{BNA}}$

- **Case 3: BNA playing tit-for-tat strategy**. If CNA cooperates then: $CPV_c^{SU} = \frac{|OP_{SP}-OP_{SU}|}{1-df}$. $CPV_c^{CNA} = \frac{|OP_{BNA}-OP_{CNA}|}{1-df}$. If the CNA deceives then: $CPV_{nc}^{CNA} = OP_{BNA} - ncOP_{CNA} + \frac{df.0}{1-df}$. For the CNA to be motivated for cooperation, CPV in case of cooperation must be preferable than CPV in case of non-cooperation. Thus, $CPV_c^{CNA} > CPV_{nc}^{CNA} = \frac{OP_{BNA}-OP_{CNA}}{1-df} > OP_{BNA} - ncOP_{CNA} + \frac{df.0}{1-df}$. If the BNA cooperates, the CNA is motivated to cooperate when: $df > \frac{OP_{CNA}-ncOP_{CNA}}{OP_{BNA}-ncOP_{BNA}}$.
- **Case 4: BNA playing the deceive-and-joinback strategy**. Considering CNAs possible strategies, it could either cooperate or deceive in the current round. If CNA cooperates then: $CPV_c^{CNA} = s(Q') - OP_{CNA} + \frac{df.0}{1-df}$. If CNA deceives then: $CPV_{nc}^{CNA} = s(Q') - ncOP_{CNA} + \frac{df.0}{1-df}$. If BNA plays deceive-and-joinback strategy, BNA is not motivated to cooperate since $CPV_{nc}^{CNA} > CPV_c^{CNA}$.

The cooperation limits for each player (BNA and CNA), with several strategies used by an opponent, are summarized in Table 2.4.

2.3.4 Agents Interaction

The interaction of agents in game is discussed. Agents in RCAg, RBAg, and RPAg cooperatively work together to discover suitable VMs with optimal cost for service execution of jobs. The sequence of operations in the proposed scheme are as follows:

1. CA sends job execution request (along with job requirements) to MMA of RBAg.
2. BA communicates to JAA of RBAg about job requirements.
3. JAA informs MMA to recommend specified number of suitable VMs.
4. MMA searches logs and discovers suitable VMs that meet job requirements (logs hold resource history containing details of each used VMs).
5. MMA sorts them in descending order of required resource.
6. MMA recommends expected number of VMs to JAA.
7. JAA informs BA to decide optimal cost of VMs.
8. BA triggers BNA to play a cooperative negotiation game with CNAs to negotiate resource cost.
9. CNA and BNA play cooperative resource cost negotiation game.

10. BNA returns cost of each VMs (satisfiable to both negotiating agents) to JAA.
11. JAA schedules jobs to RPAg, and receives result from them after execution.
12. JAA sends job results to CAs.

2.4 Simulation

We have used JADE simulator to test the operational effectiveness of the work. This section describes the simulation model, performance parameters and simulation procedure.

2.4.1 Simulation Model

The simulated Cloud environment consists of RCAg, RBAg, RPAg and ICSPs. The behavior of ICSPs considered is similar to that of Amazon EC2, and Microsoft Azure, discussed in [37]. The values of simulation parameters are presented in Table 2.5. The

Table 2.5 Simulation parameters

Symbol	Values
RCAg	[50–100] nodes
RBAg	[250–500] nodes
RPAg	[250–750] nodes
ICSP	4 nodes
Mappers	10 nodes
Reducers	5 nodes
No. of resource types per BA	[10–25] applications
Quantity of resources per BA	[10, 50] VMs
Unit cost of a resource	[10, 100]$
No. of resource types per CA	[1, 10] VMs
VMs per ICSPs	[25, 400] nodes
Processor's speed	[2, 4] Mghz
Processor's memory	[4, 8] GB
Jobs	[50, 500] applications
Jobs Arrival rate	[5–100] jobs/milliseconds
Payoffs	[0–90]$
Jdeadline	[60, 150] sec./job
Max_negotiations	[2,10] iterations
Consumers	[10, 250]
Buffered resource	[100, 200] VMs

Table 2.6 Payoffs for different simulation sets

Strategies	Simulation set 1	Simulation set 2	Simulation set 3
User leaves	(0, 0)	(0, 0)	(0, 0)
One defects, one cooperates	(4, 1)	(100, 1)	(90, 10)
Both defect	(2, 2)	(40, 40)	(50, 50)
Both cooperate	(3, 3)	(60, 60)	(60, 60)

topology and setting of individual clusters are derived from the grid test-bed DAS-2. The experimental workload (tasks stream) is generated by using Lublin-Feitelson model, which is derived from the workload logs of real supercomputers.

The implementation of the CNA-BNA interactive negotiation game is based on a publicly available Matlab implementation of the Iterated Prisoner's Dilemma Game [38], which has been extended to include all existing and proposed strategies examined in Algorithm 3. We use simple numbers as payoffs to get scores for different strategy combinations. These numbers follow the relationships of the payoffs, as described in general case in the repeated game models. The payoffs for the different simulation sets are summarized in Table 2.6.

2.4.2 Simulation Procedure

Simulation procedure is as follows.

1. Deploy RCAg, RBAg and RPAg to create the Cloud environment
2. Concurrent job submission is done by RBAg for resource analysis
3. Employ MMA to find the eligible VMs and then recommend required number of VMs
4. Employ BNA to play concurrent negotiation game with CNAs for negotiating resource price
5. Employ JAA to schedule user jobs to RPAg, collect results from RPAg and send-back to CAs
6. Compute performance parameters.

2.5 Results

This section presents the results of the proposed scheme obtained during the simulation.

Analysis of waiting time, negotiation time, and execution time: To examine the effects of resource requirements on response time, we enlarge the workload's

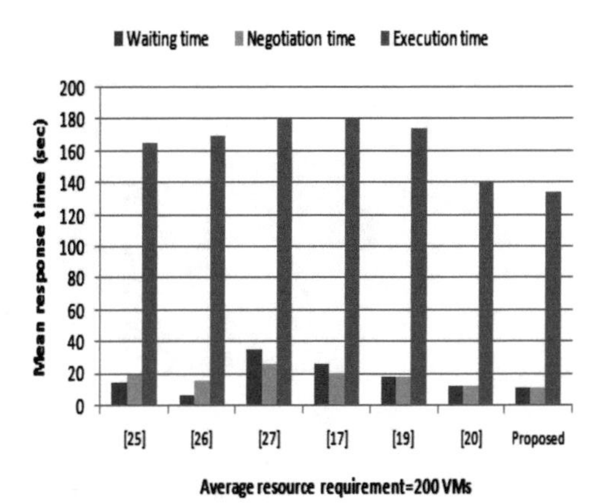

Fig. 2.4 Response time versus resource requirement, when VMs = 200

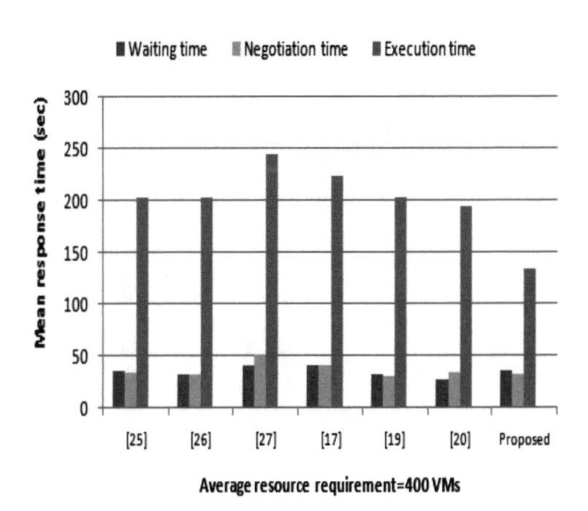

Fig. 2.5 Response time vs resource requirement, when VMs = 400

resource requirement from 200 to 400 VMs. The results are shown in Figs. 2.4 and 2.5 respectively. Comparing with auction model, [35] is effective in reducing communication-related costs. However, our simulation results indicate that, its negotiation time increases significantly when resource requirement increases to higher levels. By examining the logs, we found that re-negotiation occurs more frequently than before, that is, [35] can not efficiently finish the trading for all resource requirements. Based on the above simulation results, we can say that the proposed method is effective in reducing the negotiation time.

Analysis of payoffs: Figures 2.6 and 2.7 illustrate payoff analysis for CNA and BNA respectively. The most profitable strategy for CNA and BNA is Tit-for-Tat strategy for all payoffs except for the payoff received from the combination with the CNA's deceive-and-exit strategy. However, the difference between the payoffs received by the BNA from playing either the Tit-for-Tat strategy or the deceive-and-join-back strategy, in combination with the CNA deceive-and-exit strategy, is negligible. Also, the combination of the two most profitable strategies in the same game profile offers the highest cumulative payoffs to both players.

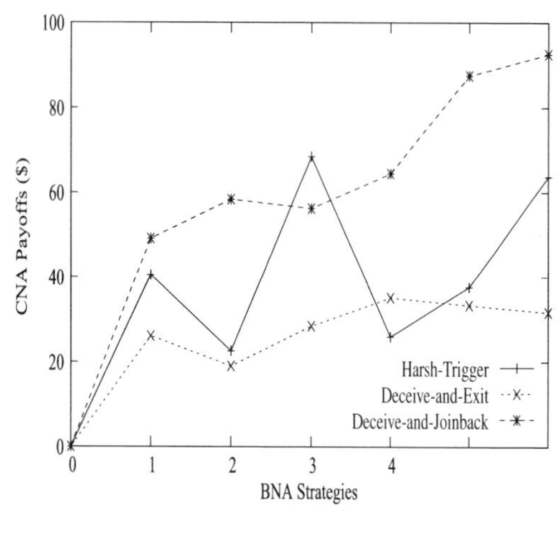

Fig. 2.6 CNA payoffs versus BNA strategies

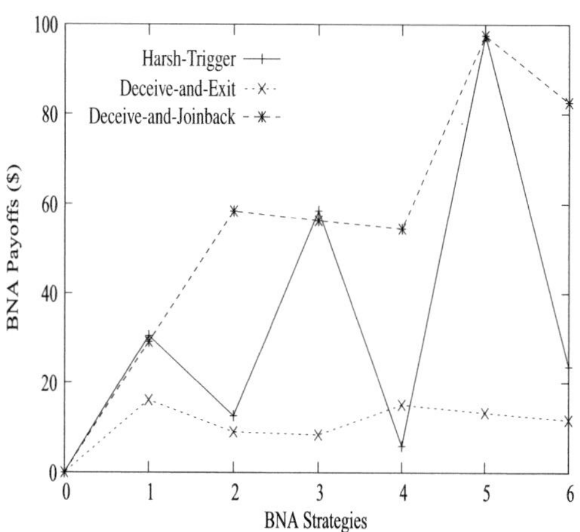

Fig. 2.7 BNA payoffs versus BNA strategies

2.6 Summary

Cloud computing is, no doubt, an emerging trend in ICT domain. Resource allocation is an important aspect for improving the performance of large-scale Cloud applications. Moreover, performance evaluation of different workload models catering to different applications is the need of the hour. Resource allocation techniques and algorithms in Cloud environment under different configurations and requirements are needed. Obviously, there is still lack of tools for enabling developers to compare different resource allocation strategies regarding both computing servers and user workloads. To fill this gap in tools for evaluation and modeling of Cloud environments and applications, this chapter discusses different optimization techniques for resource allocation in Cloud computing environment.

References

1. J. Kennedy, Particle swarm optimization, in *Encyclopedia of Machine Learning* (Springer, US, 2010), pp. 760–766
2. Y. Shi, R. Eberhart, A modified particle swarm optimizer, in *IEEE International Conference on Evolutionary Computation Proceedings of World Congress on Computational Intelligence* (Anchorage, AK, 1998), pp. 69–73
3. X. An-Ping, X. Chun-Xiang, Energy efficient multiresource allocation of virtual machine based on PSO in Cloud data center. Math. Probl. Eng. 8–15 (2014)
4. S.E. Dashti, A.M. Rahmani. Dynamic VMs placement for energy efficiency by PSO in Cloud computing. J. Exp. Theor. Artif. Intell. 1–16 (2015)
5. A.S. Banu, W. Helen, Scheduling deadline constrained task in hybrid IaaS cloud using cuckoo driven particle swarm optimization. Indian J. Sci. Tech. **8**(16), 6 (2015)
6. Y. Qiu, P. Marbach, Bandwidth allocation in ad hoc networks: a price-based approach. Proc. IEEE INFOCOM **2**(3), 797–807 (2013)
7. R.S. Mohana, A position balanced parallel particle swarm optimization method for resource allocation in cloud. Indian J. Sci. Tech. **8**(S3), 182–8 (2015)
8. P. Ghosh, K. Basu, S.K. Das, A game theory based pricing strategy to support single/multiclass job allocation schemes for bandwidth-constrained distributed computing system. IEEE Trans. Parallel Distrib. Syst. **18**(4), 289–306 (2010)
9. Y. Kwok, K. Hwang, S. Song, Selfish grids: game theoretic modeling and NAS/PAS benchmark evaluation. IEEE Trans. Parallel Distrib. Syst. **18**(5), 621–636 (2007)
10. Z. Kong, C. Xu, M. Guo, Mechanism design for stochastic virtual resource allocation in non-cooperative Cloud systems, in *Proceedings of 2011 IEEE International Conference on Cloud Computing, Cloud* (2011), pp. 614–621
11. U. Kant, D. Grosu, Auction-based resource allocation protocols in grids, in *Proceedings of 16th International Conference on Parallel and Distributed Computing and Systems, ICPDCS* (2004), pp. 20–27
12. S. Caton, O. Rana, Towards autonomic management for cloud services based upon volunteered resources. Concurr. Comput. Pract. Experi. **24**(9), 992–1014 (2012)
13. J. Espadas, A. Molina, G. Jimnez, M. Molina, D. Concha, A tenant-based resource allocation model for scaling software-as-a-service applications over cloud computing infrastructures. Future Gener. Comput. Syst. **29**(1), 273–286 (2013)
14. J. Bi, Z. Zhu, R. Tian, Q. Wang, Dynamic provisioning modeling for virtualized multi-tier applications in Cloud data center, in *Proceedings of the 3rd IEEE International Conference on Cloud Computing (Cloud '10)* (2010), pp. 370–377
15. D.C. Vanderster, N.J. Dimopoulos, R. Parra-Hernandez, R.J. Sobie, Resource allocation on computational grids using a utility model and the knapsack problem. Future Gener. Comput. Syst. **25**(1), 35–50 (2009)
16. D. Ye, J. Chen, Non-cooperative games on multidimensional resource allocation. Future Gener. Comput. Syst. **29**(6), 1345–1352 (2013)
17. M. Hassan, B. Song, E.N. Huh, Game-based distributed resource allocation in horizontal dynamic Cloud federation plat- form, in *Algorithms and Architectures for Parallel Processing*. Lecture Notes in Computer Science (Springer, Berlin, 2011), pp. 194–205
18. Scheduling in Hadoop (2012), https://www.Cloudera.com/blog/tag/scheduling
19. C.A. Waldspurger, Lottery and Stride Scheduling: Flexible Proportional-Share Resource Management, Massachusetts Institute of Technology (1995)
20. T. Lan, D. Kao, M. Chiang, A. Sabharwal, An axiomatic theory of fairness in network resource allocation, in *Proceedings of the Annual Joint Conference of the IEEE Computer and Communications Societies* (2010), pp. 1–9
21. D.C. Parkes, A.D. Procaccia, N. Shah, Beyond dominant resource fairness: extensions, limitations, and indivisibilities, in *Proceedings of the 13th ACM Conference on Electronic Commerce* (Valencia, Spain, 2012), pp. 808–825

22. X. Wang, X. Liu, L. Fan, X. Jia, A decentralized virtual machine migration approach of data centers for cloud computing. Math. Prob. Eng. Article ID 878542, 10 (2013)
23. D.C. Erdil, Autonomic cloud resource sharing for inter cloud federations. Future Gener. Comput. Syst. **29**(7), 1700–1708 (2013)
24. M. Steinder, I. Whalley, D. Carrera, I. Gaweda, D. Chess, Server virtualization in autonomic management of heterogeneous workloads, in *Proceedings of the 10th IFIP/IEEE International Symposium on Integrated Network Management* (2007), pp. 139–148
25. S. Di, C.L. Wang, Dynamic optimization of multiattribute resource allocation in self-organizing clouds. IEEE Trans. Parallel Distrib. Syst. **24**(3), 464–478 (2013)
26. M. Cardosa, A. Singh, H. Pucha, A. Chandra, Exploiting spatio-temporal tradeoffs for energy-aware MapReduce in the cloud. IEEE Trans. Comput. **61**(12), 1737–1751 (2012)
27. T. Sandholm, K. Lai, MapReduce optimization using regulated dynamic prioritization, in *Proceedings of the 11th International Joint Conference on Measurement and Modeling of Computer Systems* (Seattle, Wash, USA, 2009), pp. 299–310
28. A. Ghodsi, M. Zaharia, B. Hindman, A. Konwinski, S. Shenker, I. Stoica, Dominant resource fairness: fair allocation of multiple resource types, in *Proceedings of the 8th USENIX Conference on Networked Systems Design and Implementation* (Boston, Mass, USA, 2011), pp. 24–28
29. K.M. Sim, Agent-based cloud computing. Trans. Serv. Comput. IEEE **5**(4), 564–577 (2012)
30. K.M. Sim, Complex and concurrent negotiations for multiple interrelated e-markets. Trans. Syst. Man Cybern. IEEE **43**(1), 230–245 (2013)
31. G.K. Shyam, S.S. Manvi, Co-operation based game theoretic approach for resource bargaining in cloud computing environment, in *International Conference on Advances in Computing, Communications and Informatics (ICACCI)* (2015), pp. 374–380
32. I. Uller, R. Kowalczyk, P. Braun, Towards agent-based coalition formation for service composition, in *Proceedings of the IEEE/WIC/ACM International Conference on Intelligent Agent Technology (IAT)* (2006), pp. 73–80
33. F. Pascual, K. Rzadca, D. Trystram, Cooperation in multi-organization scheduling, in *Proceedings of International Euro-Par Conference* (2007)
34. Hong Zhang, et al., A framework for truthful online auctions in cloud computing with heterogeneous user demands, in *Proceedings of International Conference on Computer Communications* (IEEE, Turin, Italy, 2013), pp. 1510–1518
35. Lena Mashayekhy, et al., An online mechanism for resource allocation and pricing in clouds. Trans. Comput. IEEE. https://doi.org/10.1109/TC.2015.2444843
36. Tony T. Tran et al., Decomposition methods for the parallel machine scheduling problem with setups. J. Comput. Springer **28**(1), 83–95 (2015)
37. IaaS providers, http://www.tomsitpro.com/articles/iaas-providers,1-1560.html. Accessed 24 March 2016
38. G. Taylor, Iterated Prisoner's Dilemma in MATLAB: Archive for the "Game Theory", Category (2007), https://maths.straylight.co.uk/archives/category/game-theory

Chapter 3
Energy Aware Resource Allocation Model for IaaS Optimization

A. Radhakrishnan and K. Saravanan

Abstract This chapter illustrates the resource allocation in cloud IaaS. We detail how to optimize the VM instances allocation strategy using the novel ANN model. This chapter narrates the functionality and workflow of the system using the NFRLP and EARA algorithms. Further, several issues in implementing the resource allocation are also detailed. This chapter illustrates how the artificial neural network and genetic algorithm techniques are used in IaaS frame work to efficiently allocate the resources for VMs.

Keywords IaaS optimization · Energy efficiency · Resource allocation · ANN

3.1 Introduction

Cloud computing is a new computing paradigm, which offers computing resources for solving complex IT related problems in fast and low cost under pay per usage basis through internet. These features attract the industries and individuals to load their computation and data in cloud environment that significantly reduces infrastructure setup and maintenance cost. Cloud provides its services in three fundamental models namely SaaS, PaaS and IaaS. Among the three service models, IaaS is called as foundation of cloud computing because of its resource provisioning capability to all other cloud service models. Since all cloud services are depends on IaaS for resources, resource allocation is an important activity in IaaS to provide better cloud service. In IaaS service model, the entire computing environment is provided as a resource to service requests in the form of Virtual Machines (VMs). The VMs are created as

---

A. Radhakrishnan (✉)
Department of Information Technology, Anna University College of Engineering, Nagercoil 629004, Tamilnadu, India
e-mail: radhakrishnan.a@auttvl.ac.in

K. Saravanan
Department of Computer Science and Engineering, Anna University Regional Campus, Tirunelveli 627007, Tamilnadu, India
e-mail: saravanan.krishnan@auttvl.ac.in

B. S. P. Mishra et al. (eds.), *Cloud Computing for Optimization: Foundations, Applications, and Challenges*, Studies in Big Data 39,
https://doi.org/10.1007/978-3-319-73676-1_3

machine instances in cloud data centers servers, which could work as a dedicated computer system to consumers. The cloud data centers are located in geographically different locations with high configuration servers and interconnected together to share their workload.

In cloud environment, both consumer and providers are much interested parties to augment their profits, particularly the providers are much concern about maintaining trust and Service Level Agreements (SLA) in front of loyal consumers. The violation of SLA brings down the trust worthiness about cloud provider [1]. For instance, one of the prime factors for SLA violation is resource failure during service consumption. The efficient and strategic resource allocation in IaaS greatly supports the provider to deliver failure free resource to consumers and handle more number of service requests without any hitch. It also facilitates the consumers to complete their task on time with low rental cost. This chapter discusses the optimized resource allocation model in IaaS aimed to fulfill the expectation of both consumer and provider as well as elevate the cloud service in high altitude. In the novel approach, the resource allocation in IaaS is takes place based on the energy level of resources in present as well as near future. It helps the provider to provide reliable and trustworthy resources to service requests. The energy aware resource allocation model utilizes genetically weight optimized Artificial Neural Network (ANN) to predict the near future energy level of resources in cloud data centers. Typically, the initial weight factors in ANN are assigned randomly during learning, whereas the projected model possess genetic algorithm to optimize the initial weights of ANN. The genetically optimized weight enriches the ANN to improve its prediction accuracy. As per the novel strategy, the resources present and near future energy level are taken in to account for allocating a resource to service request. The resources found to be heavily loaded in near future never to be targeted for allocation. This feature supports the resource manager in cloud to distribute the workloads in balanced manner as well as minimize the migration of VMs that occurs due to hefty workload. The minimization of VMs migration automatically improves the performance of data centers.

In cloud data centers, workload of each physical server is most reliable substance to measure their energy level. In the energy based resource allocation model, the ANN is associated with each computing servers that could predicts its near future workload based on their recent past workloads. This powerful ANN extends great support to the resource allocator to identify the reliable resource for service requests. The novel model encompasses the algorithms of Near Future Resource Load Prediction (NFRLP) and Energy Aware Resource Allocation (EARA), which implements the strategic approach of resource allocation for IaaS service requests. It certainly allays the fear of cloud consumers about the failure of resources in mid of the computation and makes the cloud service as trustable and reliable service.

Cloud computing is spreading from private enterprises to public sectors as it meets the agility expected by both sectors. The reason behind the rapid adaptation of cloud computing is its features of service delivery mechanism such as ubiquitous, scalable and elastic. A typical cloud computing framework consists of four major areas that is depicted in Fig. 3.1.

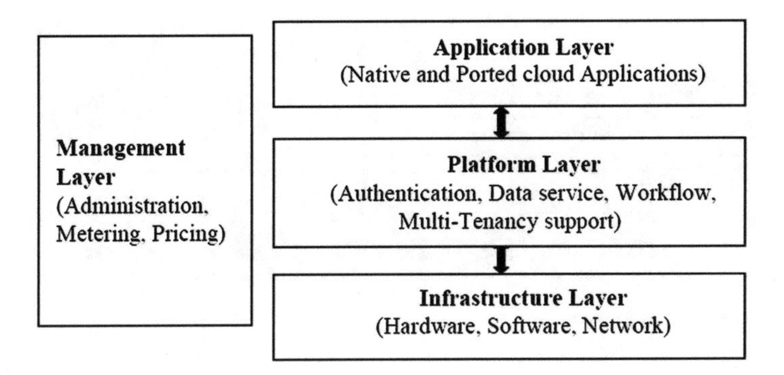

Fig. 3.1 Typical cloud computing framework

The application layer in the frame work consists of two different type of application specifically native cloud application that is developed to be working in cloud platform. Another application is ported in cloud, which were originally developed in licensed software and ported into cloud environment. The middle layer of the cloud framework is called platform layer that offers common framework model for building application. The major services of this layer are data service, workflow, authorization/authentication and multitenancy support.

The bottom layer is infrastructure layer that contains the most basic information technology resources with service capability. It delivers the whole infrastructure requirements such as hardware, software and other associated resources for the development and execution of cloud applications. The cloud framework also contains management layer to monitor and manage the cloud environment activities. The core management activities are analysis and administration across the layers, metering, pricing and billing capabilities.

In the cloud computing framework, the bottom most infrastructure layer is playing a vital support to the top layers by providing required IT resources. Since all layers are completely depends on infrastructure layer for its service, infrastructure service is called as foundation of cloud computing. Any refinement and revamp in infrastructure layer activities reflects in entire cloud computing service so that this chapter mainly focuses on infrastructure service especially revamping its prime responsibility of resource provisioning. The introduction about the core functionalities of infrastructure layer and the methodologies used for revamping its resource provisioning is as follows.

3.1.1 Introduction About IaaS Service Framework

The service architecture of IaaS is shown in Fig. 3.2, which consist of four important compartments namely cloud computing system with virtualization, cloud storage system, content delivery network and IT infrastructure.

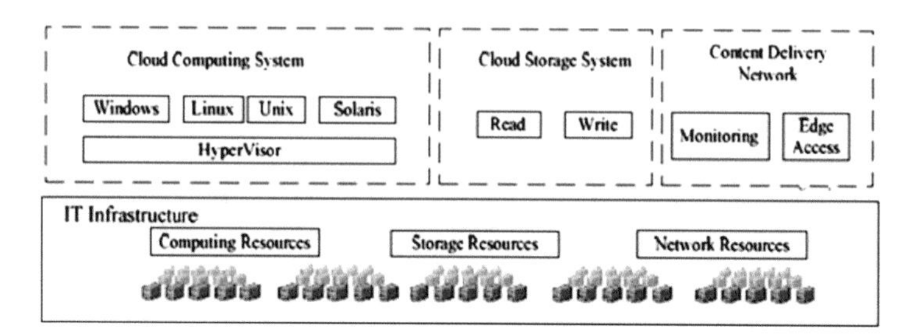

Fig. 3.2 IaaS service framework

The IaaS platform utilizes virtualization as their prime technique that provides illusion on hardware. This enables portability of high level functions and sharing or aggregation of physical resources. Virtualization can be applied to all aspects of computing systems such as storage, memory, processor, and networks. This feature allow multiple guest OS to run simultaneously on a single system so that a system in cloud environment is able to handle multiple service requests. The popular virtualization techniques currently used in IaaS framework is VMware, XEN and KVM.

The virtualization technique supports IaaS model to deliver VMs as a resource to service requests. The VMs are created on cloud data center servers as machine instances that could act as an independent computer system. A hypervisor or virtual machine monitor is a software that could create and run VMs on physical server. A machine a hypervisor runs one or more VMs is called host machine, each VMs are called guest machines. The VMs can be classified into different types such as small, medium and high according to their capability in terms of processor, memory and bandwidth. Allocating reliable resource to VMs is a challenging and responsible task of IaaS framework because it avoids VMs failure and frequent migration. The strategic allocation of resources to VMs leads cloud service as a trustworthy and reliable service to all walks of cloud consumers. This chapter illustrates how the artificial neural network and genetic algorithm techniques are used in IaaS frame work to efficiently allocate the resources for VMs.

3.1.2 Overview of Artificial Neural Network and Genetic Algorithm

The ANN is a computation system inspired by the structure, processing methodology and learning ability of a biological brain. The ANN is mainly used for prediction process through its learning capability. The important component in this system is the structure of information processing. It is build by large number of highly interconnected processing elements known as neurons, which are working together to solve any specific problem. ANN attempts to imitate the way a human brain works.

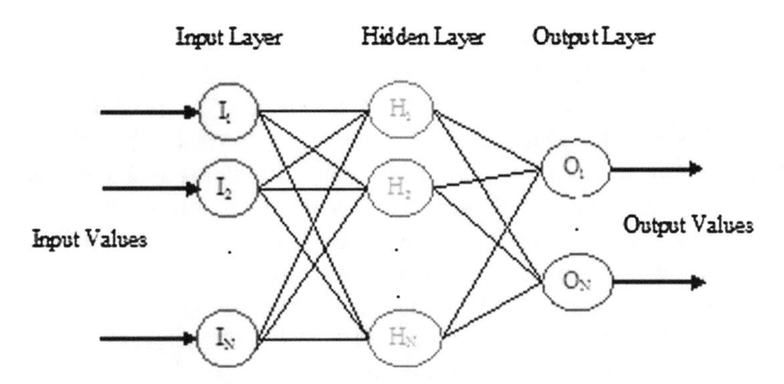

Fig. 3.3 Generic structure of artificial neural network

The ANN architecture consists of three core layers, which are input layer, output layer and hidden layer of neurons. The generic architecture of ANN is depicted in the Fig. 3.3.

In ANN architecture, the layers are well connected with each others. The input layer units are connected to the hidden layer units, and the hidden layer units are connected to the output layer units. The input layer represents the information which is fed into the ANN as input. The hidden neurons functions are identified by the activities of the input neurons and the connecting weights between input and hidden unit. Similarly, the role of the output unit depends on the activity of the neurons in the hidden unit and the weights assigned between the hidden and output unit. In ANN, the hidden neurons are free to construct their own representation of the input. Each node in ANN has the same node function. The node function is a composite of the weighted sum and a differential nonlinear activation function. The nonlinear activation function is any one of the following namely, logistic, hyperbolic tangent, and identity. They were used to transform the incoming signal into an output signal.

The learning process in ANN is a result of modification of the network weights with the learning algorithms at some extent. The main objective of learning is to find a set of weight matrices that would applied to the network to map any input to correct output. The popular learning paradigms are namely supervised, unsupervised and reinforcement learning. There are several classes of ANN, which are classified according to their learning mechanisms. The arrangements of neurons into layers and the pattern of connecting within and in-between layers are generally called as the types of neural networks. The most fundamental classification of ANN falls into three categories namely single layer feed forward network, multi layer feed forward network and recurrent network.

The term Genetic algorithm (GA) was initially introduced as a probabilistic optimization algorithm. The GA is a family of computational model influenced by natural progression. The original idea came from the biological evolution process in chromosomes. The GA exploits the idea of the survival of fittest where best solutions are recombined with each other to form better solutions.

There are three basic processes in GA, which are selection, crossover and mutation. Selection is a process for choosing a pair of chromosomes which are fitted for survival. The crossover operation performs the exchange of genes between two individual chromosomes for reproducing. Mutation process refers to arbitrarily changing the chromosomes. In GA, set of individual number named as population. Each individual denotes the life form of chromosome. There is a fitness function to determines how fit each individual for survival and select the fitted individuals from the population to reproduce. The two selected chromosomes performs crossover and split again, followed by mutation that produce two new individuals. The process is repeated until best individual is good enough. The term fitness in GA measures the goodness of a chromosome, how well the chromosome fits to solve the problem.

In the energy aware resource allocation approach for VMs, the ANN mechanism utilizes GA for optimizing its initial weights matrices. This kind of weight optimization increases the accuracy of ANN prediction. The subsequent section illustrates the existing resource allocation methodologies in IaaS framework followed by the functionality and utilization of energy aware resources allocation technique using ANN with GA.

3.2 Existing Resource Allocation Methodologies in IaaS Framework

Resource allocation is the process of assigning cloud resources to the required cloud applications; inefficient allocation will lead to resource contention, scarcity of resources and resource fragmentation. The complexity of finding an optimal resource allocation is high in huge system such as data centers and grid. Since the resource availability and demand are dynamic and scalable, various methods for resource allocation in cloud should be followed. The researchers found different type of resource allocation strategies in cloud, but the prime focus of all are, cloud must provide QoS guaranteed computing environment to the end user.

The resource allocation strategies are based on policy, execution time of applications, utility function, Gossip protocol, hardware resource dependency, auction and SLA. Kuo and Kuan [2] has proposed the most fit-processor policy for resource allocation. This policy assigns the job to the cluster, which produces a leftover processor distribution. The fit policy requires a search process to determine the target cluster, which leads to high time complexity. The work done by Shin and Akkan [3] creates decentralized user and virtual resource management for IaaS through new layer called domain layer. Based on role based access control, the virtualized resources are allocated to users through domain layer. Li et al. [4] has proposed task execution time and preemptable scheduling for resource allocation. It avoids resource contention and increases the resource utilization by using different modes of renting. The major problem of this approach is that estimating the execution time of a job. The advanced reservation based resource allocation strategy for distributed environment is proposed by Melendez and Majumdar [5]. It is based on any scheduling criteria for assigning a job to resource in heterogeneous environment subjected to advanced reservation of resource. The work done by Kumar et al. [6], have assigned a virtual machine to a job based on the cost and speed of available VMs in IaaS. It allows the user to select the VMs according to their budget level.

Zhen et al. [7] have proposed stochastic approximation method to model and analyze the QoS performance of virtual resources. This approach enforces the VMs to report their credentials truthfully, based on that the virtual resources would be allocated to the requests. Wuhib and Stadler [8] have suggested a Gossip-based protocol for resource allocation in cloud. This protocol implements a distributed cloud resources allocation scheme, which assigns the applications that have time dependent memory demands and dynamically increases a global utility function. The work given by Niyato et al. [9] has proposed Gossip based co-operative VMs with VMs allocation. In which, cost is reduced by sharing the available resources to the VMs. The co-operative formations of organization are based on network gaming approach so that none of the organization will be deviated. The work proposed by Nguyen et al. [10] considers the utility function to assess the application fulfillment in resource allocation. The utility function computes the workload of enterprise online applications and CPU-intensive application then resources could be allocated to applications accordingly. Weisong et al. [11] has proposed job optimization scheduler for resource allocation. The jobs are classified by scheduler based on their hardware resource dependency such as CPU, memory, network bandwidth and I/O. Afterward, the scheduler detect the parallel jobs of different category, based on the category resources are allocated. Xiaoyi et al. [12] have analyzed adaptive resource co-allocation approach based on CPU consumption amount. Gradient climbing approach is used to determine the actual CPU share that each VM occupies on a system and its optimizing level. It mainly focuses on CPU and memory resources for co-allocation and does not consider the dynamic nature of resource request.

The auction based resource allocation strategy is proposed by Wei et al. [13]. This methodology is based on seal based biding, where the cloud provider collects the entire sealed bid from users and determine the price. The resource is distributed to the users based on highest bidder. This strategy does not enlarge the profit due to its truth telling property under severe constraints. The main objective of this resource allocation strategy is to increase the profits of both cloud consumer and provider by balancing the resource supply and demand according to market level. Xindong et al. [14] has introduced market based resource allocation strategy using equilibrium theory. It determines the number of fractions can be used in one VM and can adjust dynamically according to the varied resource requirement of workload. The proposed market economy system balances the resource supply and demand in the market system. Uthaya and Saravanan [15] defined Kalman filter prediction model for predicting resource demand in both reservation and on-demand pricing.

The nature of the application is also considered for resource allocation in cloud. Tram and John [16] has designed a virtual infrastructure allocation strategy for workflow based application, where resources are allocated based on the workflow representation of the application. The application logic is interpreted and exploited to produce execution schedule estimation. Xiong et al. [17] has designed a database replicas allocation strategy in cloud, in which the resource allocation module separates into two levels. The first level optimally divides the resources among the process and the second level expands database replicas based on learned predicted model. In cloud, SLA must be ensured during service to sustain the trust over cloud computing. Hence, various resource allocation strategies are suggested by the prime consideration of SLA. The work done by Linlin et al. [18] has focused on SLA driven user based QoS parameters to maximize the profit of SaaS provider, which map the customer request into infrastructure parameters and policies. This approach maximizes the cost by optimizing the resource allocation within a VM. Ma et al. [19] has proposed a framework for resource management for SaaS provider to effectively control the SLA of their users. It mainly concentrates on SaaS provider profit and considerably reduces the resource wastage and SLA violation.

The aforementioned various resource allocation strategies are the results of intense analysis by researchers in cloud domain. All the resource allocation strategies are designed to consider one or more prime parameters such as policy, SLA, auction, execution time, hardware dependency and utility function. The main objectives of all allocation strategies are providing failure free and optimized resource allocation for cloud consumers, through which the cloud service should be made as affordable and trustworthy service to all kind of consumers. The highlighted resource allocation strategy of this chapter is illustrated in the remaining sections.

3.3 The IaaS Cloud Architecture with Energy Aware Resource Allocation Model

The IaaS architecture for cloud computing service is depicted in the following Fig. 3.4, which encompass various functional components along with the energy aware resource allocation methodology for handling service requests. The cloud data center consists of number of high configuration servers to handle the requests. It requires a mechanism to equally distribute the load across the data center servers in order to meet QoS expectation of consumers from different locations. The elastic service of IaaS leads very hard to implement load balancing due to the reasons of (i) The provider must deliver enough resources to hold peak time resource usage, (ii) The service should grow depends on the development of resource consumption, (iii) The service provider should be able to predict the growth of resource consumption well in advance and strategic planning must be made to meet the requests. These complexities are one of the major hindrances for the growth of IaaS service. The energy aware resource allocation strategy lends support to address these issues effectively.

The computing hosts in the cloud hub is called as data center servers, which are the foundation of cloud as it renders all kind of resources to cloud services. In the above architecture, the genetically weight optimized ANN is attached with each computing hosts to predicts their near future availability when it required [20, 21]. It helps the cloud resource allocator to provide reliable computing resources to consumers.

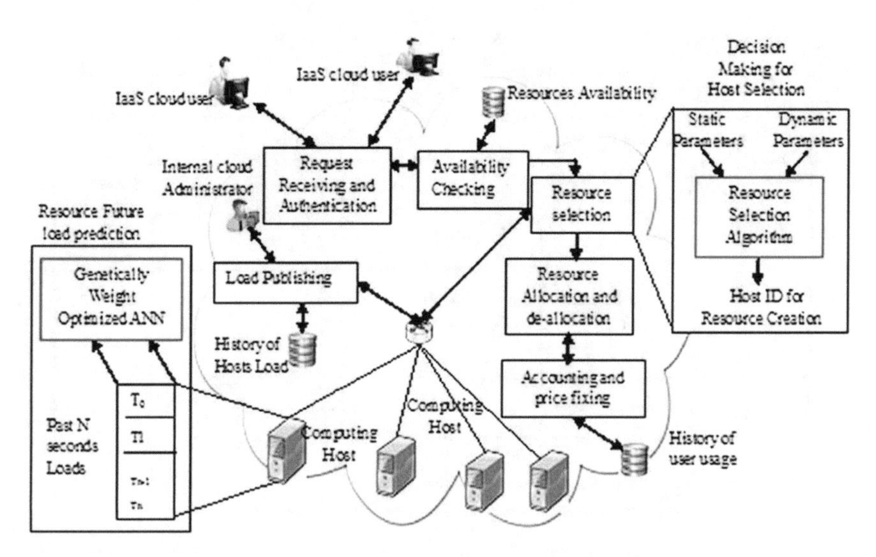

Fig. 3.4 The IaaS cloud architecture with resource allocation model

3.3.1 Architectural Components and Its Functionalities

The architecture contains a variety of functional components and their functionalities are well defined in sequences (Fig. 3.5) to handle the service request. The major architectural components as well as their functionalities are described as follows.

Request Receiving and Authentication Unit: This unit receives the IaaS service request from the users. The user authentication is examined and authorized based on cloud security strategy which is either third party or internal authentication. If the authentication of the user is valid, then the request is forwarded to resource availability checking unit.

Resource Availability Checking Unit: It keeps the availability details of resources in the cloud hub. The authenticated requests are evaluated by this unit. If the hub is able to provide resource for a request then it triggers resource selection unit, otherwise request is readdressed to some other cloud hub. This unit is updated by resource allocation and deallocation unit.

Resource Selection Unit: This unit selects a suitable host for resource allocation with the support of activating genetically weight optimized ANN attached in all computing hosts to predict their near future load. Finally, it chooses a suitable host whose present and near future load is adequate compared with all other computing nodes. The ANN of a host is activated if its current load exceeds mid of its capacity during selection.

Resource Load Prediction Process: It is fabricated by genetically weight optimized trained ANN that is attached with all computing hosts. It takes recent past loads of computing host from queue data structure to predict its near future load. The prediction process is initiated by resource selection unit. The predicted near

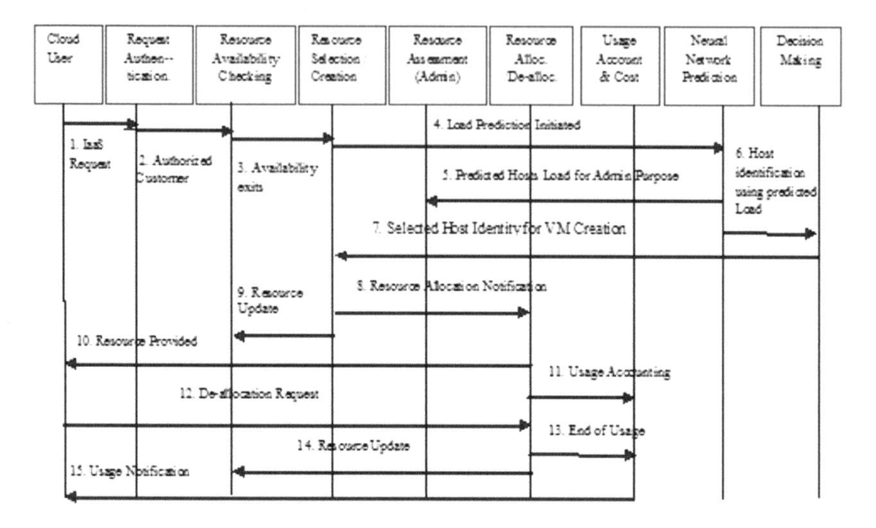

Fig. 3.5 Sequence diagram of architecture activities

future load is reported to internal cloud admin and resource selection unit for further process.

Resource Allocation and Reallocation Unit: It allocates the requested resource to aspirants based on the resource identity obtained from resource selection unit. This activity reflects in accounting unit for metering the service as well as resource availability unit for availability update. It also takes care about reallocation of resources after usage is over.

Accounting and Price Fixing Unit: This unit is responsible for accounting the resource usage details such as how long the resource was consumed. The usage charges also calculated based on the metering and pricing policies of the provider. The invoice of the payment voucher is sent to the cloud user to enable payment.

3.3.2 Workflow of Architecture

1. The IaaS requester log on by valid authentication and submit their request details.
2. The requested resources availability is checked in cloud hub database.
3. If availability exits then assess the current load of working hosts as well as its near future load if necessary by ANN for host identification.
4. The current and predicted load of each host is brought to the attention of hub administrator as well as resource selection unit for VMs creation.
5. Resource selection unit calls decision making algorithm for identifying suitable host by providing dynamic and static parameters such as resources predicted load, requested type of resource and user category details.

6. Decision making process is done by EARA algorithms that identifies targeted physical host for virtual machine creation with requested resource capability and notifies the same to resource selection unit.
7. The created virtual machine is allocated to customer, the update is made on resource availability database and accounting unit.
8. Customer received requested resource as virtual machine instance, after usage relinquish request send to the cloud.
9. Revoke allocated virtual machine from customer and notify to accounting unit and resource availability units.
10. The cost of resource usage is informed to customer to make payment.

3.3.3 The Genetically Weight Optimized ANN Model for Load Prediction

The ANN model of near future workload prediction for cloud data center host is depicted in Fig. 3.6. The ANN must be trained with substantial amount of data because the trained model only directs the network to behave in response to the external stimulus or initiate activities on its own. The training and testing dataset are collected from the load traces of grid workstation node, which is available in the URL http://people.cs.uchicago.edu/~lyang/load/, herein mystere10000.data file is selected for training and testing. The 200 samples of host load were sequentially taken to form experimental dataset.

The ANN model is build with 5 input nodes, 5 hidden nodes and one output node. The hosts recent past five seconds loads from t_n to t_{n-4} are given as input to predict the load at t_{n+1} s. The recent past loads are kept in queue data structure associated with each physical host in cloud data center. In ANN prediction model, the initial weights between the units are optimized by genetic algorithm. Typically the initial weights are given as random. For optimizing initial weights of ANN, optimization starts at the multiple random points of initial weights, the weights are encoded into chromosome format of symbol strings. Each chromosome is evaluated by fitness function based on mean square error value in ANN training. Once this has been done, all the chromosomes are operated by the selection, crossover and mutation of

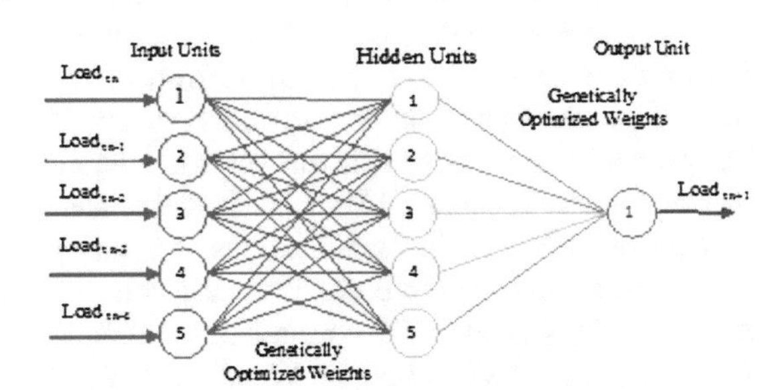

Fig. 3.6 ANN model for hosts near future load prediction

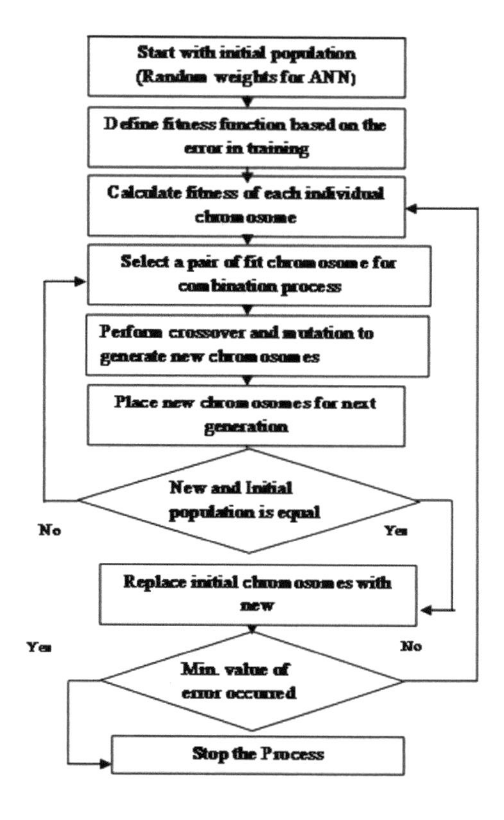

Fig. 3.7 Flowchart of ANN weight optimization by GA

GA. The chromosomes which correspond to the best interconnecting weights are progressively evolved. At last, the weights optimized with GA are given as initial weights for ANN training. The procedure for the initial weight optimization in ANN using GA is depicted in Fig. 3.7 as a flowchart.

3.3.4 *Categorization of Resource Requests and Consumers*

As per the service agreements, the customer raises the service request to the provider. An IaaS provider might have multiple cloud hubs in geographically different locations and can transfer the service request from one cloud hub to other hub if necessary. In the IaaS architecture, the customers can specify the required infrastructure accordingly the VMs are created with respect to the requirement specification of customers on cloud hosts. The created VMs are named related to their capability that is helpful to identify the VMs. The classifications of VMs are specified in Table 3.1.

The cloud consumers are classified based on their usage as well as SLA agreement executed with provider. The consumers are allotted to a grade based on their earlier usage or contract with provider. When request arrives for VMs, the corresponding

Table 3.1 Resource classification

Sl. no	Customer request specification	Virtual machine category
1	Requested resource is more than 2.5 GHz CPU and/or RAM size is more than 3 GB	High
2	Requested CPU capability is in-between 1.5 and 2.5 GHz and/or in-between 1.5 and 3 GB RAM size	Standard
3	Requested CPU power is less than 1.5 GHz and/or RAM size is less than 1.5 GB	Small

Table 3.2 Categorization of consumer

Sl. no	Type of customer	Grade
1	Consumers with contract	A
2	More than hundred hours usage	B
3	Less than hundred hours usage	C

grade of the requested customer is obtained from cloud database. The taxonomy of cloud customer is denoted in Table 3.2.

The host identification process to launch the requested VMs utilizes consumer grade and virtual machine category as key parameters for decision making. The preference is given to loyal and long term customers during resource allocation. The providers are much careful in avoiding SLA violation and providing trustworthy service to their consumers.

3.3.5 Energy Aware Resource Allocation Algorithm in IaaS Cloud

The main objective of the proposed algorithms Energy Aware Resource Allocation (EARA) and Near Future Resource Load Prediction (NFRLP) are to efficiently handle IaaS service requests in terms of trust and cost worthy. The algorithm EARA is developed to receive the required resource specification from consumers and then choose an appropriate physical host in data center to create a virtual machine with required resource capability. The NFRLP extends support to EARA, which classifies the hosts of cloud hub into three categories such as heavy, medium and low according to their current and predicted load level in near future. The heavily loaded hosts are never considered for resource allocation [22]. Also, the algorithm EARA accords preference to loyal customers in resource allocation, which are presented in Figs. 3.8 and 3.9 respectively.

The algorithm EARA consists of two sun functions, NFRLP (N) and Host-Class (Load, Host-Id). The sub function NFRLP (N) is used to predict the near future load

```
Algorithm 1: Energy Aware Resource Allocation (EARA)
EARA (VM-Type, Cus-ID, N)
        Input : VM-Type, Cus-ID, N
        Output: Host-ID {
        Array: Heavy [n][2], Medium [n][2], Low [n][2]
        Integer: Load; Character: Cus-Grade
1.      If ( Cus-ID is in blocked list of service) {
2.              Service request is dropped, break; }
3.      Else {
                /*The customer gets any one of the grade among A, B and C. */
4.      Cus-Grade = Obtain customer grade from cloud database through Cus-ID;
5.      loop {
6.              For each Physical host in cloud hub {
7.              If Current load is less than its average load then load = current load
8.              Else Load = Call NFRLP (N)
9.              Call Host-Class (Load) } }
10.     End Loop
11.     SORT Arrays Low, Medium and Heavy according to load value;
        /* checking whether cloud hosts are fully loaded */
12.     If (Low and Medium Array are null && VM- Type == 'Heavy' or 'Standard')
13.     { Forward request to another cloud hub;  break; }
14.     Else {
15.             Switch (VM-Type)
        Case 'High':
16.             IF (Low Array != null) {
17.             Return Host-ID of first host in Low Array;
18.             break } Else
19.             Return Host-ID of first host in Medium Array; break;
20.     Case 'Standard':
21.             If ( Cus-Grade== 'A' or 'B') {
22.             If ( Low Array != null) {
23.             Return Host-ID of first host in Low Array;
24.             break; } Else
25.             Return Host-ID of less load host in Medium Array ; } break;
26.     Case 'Small':
27.             If (Low and Medium Array are null) {
28.             Return Host-ID of lowest load host in Heavy Array;  break; }
29.             If ( Cus-Grade == 'A') {
30.             Return Host-ID of first host in Low Array;
31.             break; } Else
32.             Return Host-ID of lowest load host in Medium Array;
33.             break;   } } }
```

Fig. 3.8 Pseudo code of Energy Aware Resource Allocation algorithm (EARA)

of each physical host. The parameter N specifies how long past load are taken into account for prediction process. The function Host-Class (Load, Host-Id) performs the categorization of hosts according to their load specified by parameter load. The physical hosts in the cloud hub are having its identity that is represented by parameter Host-Id. The pseudo code of the algorithm NFRLP is given in Fig. 3.9.

The proposed algorithms are developed to identify the right resources for IaaS requests after ensuring their energy level in present as well as near future. To reduce the overhead, the ANN mechanism for load prediction is activated when the current load of resources are above their average level. It gives the new dimension in resource allocation strategy in cloud environment. The pseudo codes of the algorithms are converted into JAVA code while implementing cloud environment.

```
• Sub Function NFLP(N)
      Algorithm 2: Finding near future load of physical hosts.
      NFLP (N) {
                /* Queue is attached with each physical host to monitor its past loads */
  1.  Fetch past N seconds load of host from queue data structure attached in host.
  2.  The obtained loads are applied to trained genetically weight optimized neural networks to
      predict its near future load;
  3.  Return the predicted load to main function;
  4.  }

• Sub Function Host-Class(Load)
      Algorithm 3: Classification of hosts according to predicted future load.
      Host-Class (Load) {
      /* categorized hosts according to their load value */
  1.  If (Load value is less than 50%) {
  2.   The Host-ID and Load are move to array Low;
  3.   break; }
  4.  Else {
  5.  If (Load value is between 50% and 75%) {
  6.   The Host-ID and Load are move to array Medium;
  7.   break; }
  8.   Else
  9.  The Host-ID and Load are allotted to array Heavy;
  10.  Return }}
```

Fig. 3.9 Pseudo code of near future load prediction algorithm

3.4 Implementation Results of Proposed Methodology in Cloud Environment

The implementation of the IaaS architecture with energy aware resource allocation strategy is done as two parts. The first part contains the implementation and performance evaluation of genetically weight optimized various ANNs and the second part contains the implementation and performance comparison of novel resource allocation strategy in IaaS cloud environment.

3.4.1 Experimental Setup and Prediction Performance Results of ANN

The training of ANN is required sizeable amount of data, the dataset for training and testing is collected from load traces of grid workstation node that can be found in the URL in http://people.cs.uchicago.edu/~lyang/load/. The 200 samples of host load were taken successively to run the experimental setup. The transformation of identified dataset for neural network training is specified in Table 3.3.

The training model of ANN possesses 5 input nodes, 5 hidden nodes and one output node. The identified 200 load samples are further split into training, validation and testing purposes. The experiment is conducted in MATLAB environment, the implemented MATLAB program is converted into JAVA code using MATLAB Builder JA tool for the evaluation in cloud environment. The prediction performance is evaluated by parameters such as CPU time taken for training that is used to measure efficiency, and Mean Absolute Error (MAE) that is used to measure accuracy.

Table 3.3 ANN training dataset model

Input value1	Up to	Input value m	Output value
Load (1)	..	Load (m)	Load (m+1)
Load (2)	..	Load (m+1)	Load (m+2)
..	..	..	..
Load (t−2)	..	Load (t−m−1)	Load (t−1)
Load (t−1)	..	Load (t−m)	Load (t)

Table 3.4 Prediction performance result comparisons

ANN models	Parameters					
	Standard way of optimized weight			Genetically optimized weight		
	MAE	R-Square	CPU time (s)	MAE	R-Square	CPU time (s)
BPNN	0.037	0.949	8	0.01	0.962	6
ELNN	0.023	0.971	6	0.006	0.984	3
JNN	0.002	0.995	5	0.0005	0.996	2

The R-Square value is used to measure the successful fitting between targets and predicted. The R-Square value closer to one indicates a better fit of prediction.

The prediction performance comparison between standard weight and genetically weight optimized three neural networks namely Back Propagation Neural Network (BPNN), Elman Neural Network (ELNN) and Jordon Neural Network (JNN) is shown in Table 3.4. In standard weight optimization, JNN possess high accuracy of prediction among three neural networks. The R-Square value of JNN is near to one and the time taken by the CPU is 5 s to predict the given load samples, which is less compared with the other two neural networks.

The evaluation parameters in Table 3.4 shows that the genetic optimization of weight increases the performance of all neural networks compared with standard weight optimization. The performance of ELNN and JNN are very close in prediction but comparing the CPU time taken, JNN took only two seconds to predict the samples. The time taken is drastically decreased by genetically optimized weights. Hence, the cloud architecture for IaaS service management affixes genetically weight optimized JNN for its computing host near future load prediction. The reasons behind the selection of JNN among other neural network are its prediction accuracy and time taken for prediction.

3.4.2 *Experimental Setup and Performance Results of Energy Aware Resource Allocation Strategy in Cloud Environment*

Cloud architecture setup is made in CloudAnalyst toolkit that provides simulation environment for IaaS cloud computing service. It also facilitates to customize the functionality of core IaaS system components such as data centers, VMs creation, resource allocation policies and request making pattern. CloudAnalyst is another simulation tool from cloudbus.org for modeling and analysis of the large scale cloud data center environment. This simulation is used to measure the statistical measurement of response time of simulated applications, usage patterns of applications, time taken by data centers for servicing user requests and cost of the application. The CloudAnalyst simulation environment splits the data centers in the earth into six regions. In experimental setup, three data centers and seven user hubs are configured. Each data center and user base is having unique identifier and located in different regions. The position of user bases (UB) and data centers (DC) in simulation environment is shown in following Fig. 3.10 where the user bases which are distance from data centers are getting network latency compared with user bases near to data centers.

In simulation screen, the data centers are not exists in the region of user bases 1, 3 and 7 therefore the requests of these user bases are sent to data centers in another regions. It increases the network latency that reflects in response time of consumer application in these user bases. The three data centers are made to serve the consumer requests irrespective of regions. The physical hosts in data centers are configured with Linux OS, X86 architecture, 8 GB RAM, 1terabyte storage, 4 cores and 1 G.Bit/s

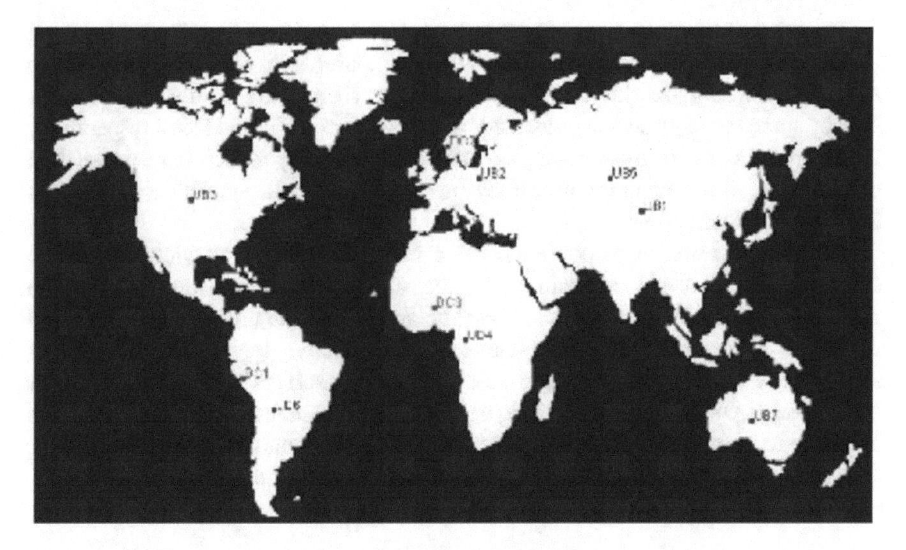

Fig. 3.10 User bases and data centers in simulation environment

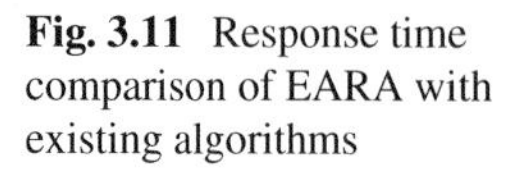

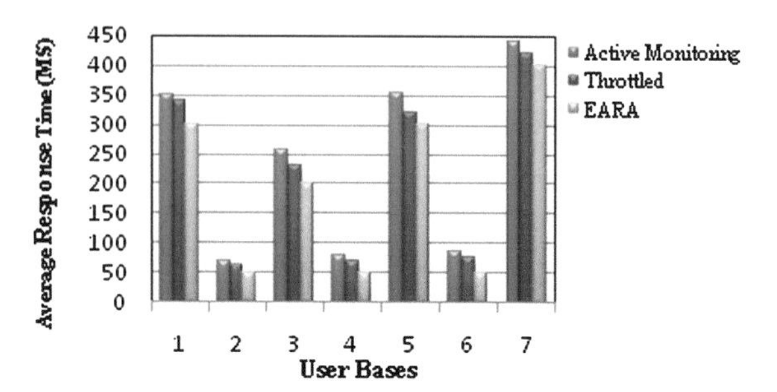

Fig. 3.11 Response time comparison of EARA with existing algorithms

Band width. The simulation duration is set as sixty hours and made hundred requests per hour from each user hubs. The requests are equally mixed up to acquire different category of VMs.

The performance of the proposed algorithm EARA is evaluated against the existing resource allocation algorithms Throttled and Active Monitoring. The simulation is first run with existing algorithms with the aforesaid set up. After that, the EARA is executed with the same simulation setup, where the VMs are created as per the request specification of customers in reliable physical host that is predicted by genetically weight optimized ANN. The CloudAnalyst simulation tool helps to evaluate the performance of the proposed technique in terms of response time of user hubs and data centers processing time to serve the requests. The response time of user request is measured by the time taken between request initiation and completion by provider data center. The average response time of user hubs using the existing resource algorithms Throttled and Active Monitoring as well as the novel algorithm EARA is depicted in Fig. 3.11.

The average response time of user hubs in Fig. 3.11 shows that all user hubs get less response time in algorithm EARA when compared with existing Throttled and Active Monitoring algorithms. The user hubs 1, 3, 5 and 7 are having high network latency since the requests are processed by data centers DC-1 and DC-2 respectively which are far away from request bases. The network latency reflects in response time of user bases. The remaining user hubs are close to the data centers hence they get low network latency.

The minimization of response time is a major concern for service provider to sustain profit, trust and maintain SLA. The reduction of response time helps the consumer to complete their task in less time and reduce the rental cost of the hired resources from cloud provider. In data center perspective, the EARA algorithm is significantly reducing processing time of data center servers because of their near future load also consider for assigning tasks. In experimental set up three data centers are made to cater the need of user bases. Table 3.5 represents the average processing time of data centers using EARA algorithm and existing algorithms.

Table 3.5 Data centers processing time comparisons

Data center	Active monitoring	Throttled	EARA
DC-1	14.62	12.17	10.11
DC-2	16.11	13.98	12.01
DC-3	8.72	7.64	6.19

Table 3.5 shows that EARA considerably reduces the processing time of data centers compared with other two algorithms. The data center DC-3 had very less processing time since it handles only the request of user base 4 due to service proximity based request routing. The remaining user bases 3, 6 and 7 are handled by data center DC-1 and user bases 1, 2 and 5 are handled by data center DC-2. The diminution of data centers processing time minimizes the internal cost of the service provider. The overall performance of the energy aware resource allocation methodology EARA is proved that it can effectively optimize the infrastructure requests in IaaS cloud environment.

3.5 Advantages of Energy Aware Resource Allocation Methodology

The energy aware resource allocation strategy in IaaS cloud optimization renders number of support to the both: cloud consumer and provider in various dimension. The major activities of cloud which are benefited by the novel methodology are listed below.

- Reliable resource provisioning to cloud consumers
- Reduce the rental cost of resources by completing the tasks on time
- Increase the number of request handling at provider site
- Support to maintain Service Level Agreements
- Support to sustain trust about cloud provider
- Load balancing among cloud data center servers
- Alleviate the fear of SMEs by providing failure free resources
- Reduce the migration of VMs among data centre servers.

The enhancement on above activities makes cloud computing is more trustable and reliable service to all kind of cloud consumers. It also supports to green world initiative by conserving the energy of cloud data centers, which reduces the carbon emission of data centers.

3.6 Conclusion

The cloud computing offers many IT related service options to large and small IT business companies and individuals. It opens up the world of computing to a broad range of uses and increases the ease of use by giving access through internet. The cloud services are basically classified into three major categories namely SaaS, PaaS and IaaS. These services greatly cut off consumers infrastructure setup cost and maintenance cost by loading their computation and data in cloud environment. Among all cloud service, IaaS is called as foundation of cloud service because resources for all service models are obtained from IaaS. The performance of all the services are highly depends on IaaS. In spite of lot of benefits provided by cloud, the cloud consumers especially the Small and Medium Enterprises (SMEs) are worried to utilize the cloud service due to the standing issues of trust and getting reliable resource for their computation. The SMEs are playing vital role for the key success of any technological development, currently SMEs are stay away from cloud business. Since IaaS cloud service can allay these kinds of issues, this projected methodology was concentrated on enriching IaaS service internal activities to mitigate the standing issue of reliable resource provisioning and maintaining trust.

The IaaS cloud service provides VMs as resource to consumers. The provided VMs can act as dedicated computer systems to customers to cater their needs. The VMs are created on physical host of cloud provider data centers. The reliable host selection for launching VMs as per the requirements of consumers is a challenging task for service provider. The deficient selection of hosts for VMs creation leads VMs failure in middle of the computation or deprived the performance of VMs due to the heavy workload of physical host. In the energy aware resource allocation approach, the physical hosts selection for VMs creation is based on their current and near future availability. The availability of computing host is measured by its workload. The load prediction process is fabricated by genetically weight optimized ANN. The ANN load prediction model is trained and validated by the real load samples of grid workstation node. The future load prediction of a computing host is based on its recent past load history. The load prediction mechanism is embedded in the algorithm EARA, which facilitates the VMs creation process to identify the reliable physical host for VMs creation. The reliable host selection improves the performance of VMs and avoids VMs failure during service. The performance of EARA is validated in CloudAnalyst tool. The results demonstrates that the response time of consumer applications and processing time of provider data centers are decreased immensely by using EARA compared with existing methodologies in cloud environment.

References

1. K. Saravanan, M. Rajaram, An exploratory study of cloud service level agreements - state of the art review. KSII Trans. Internet Inform. Syst. **9**(3), 843–871 (2015). https://doi.org/10.3837/tiis.2015.03.001. ISSN 1976-7277.IF0.561

2. H. Kuo-Chan, L. Kuan-Po, Processor allocation policies for reducing resource fragmentation in multi cluster grid and cloud environments (IEEE, 2010), pp. 971–976
3. D. Shin, H. Akkan, Domain-based virtualized resource management in cloud computing, in *2010 6th International Conference on Collaborative Computing: Networking, Applications and Work-sharing (CollaborateCom)* (IEEE, 2010), pp. 1–6
4. J. Li, M. Qiu, J.W. Niu, Y. Chen, Z. Ming, Adaptive resource allocation for preemptable jobs in cloud systems, in *2010 10th International Conference on Intelligent Systems Design and Applications (ISDA)* (IEEE, 2010), pp. 31–36
5. J.O. Melendez, S. Majumdar, Matchmaking with limited knowledge of resources on clouds and grids, in *2010 International Symposium on Performance Evaluation of Computer and Telecommunication Systems (SPECTS)* Ottawa, ON (2010), pp. 102–110
6. K. Kumar, J. Feng, Y. Nimmagadda, Y.H. Lu, Resource allocation for real-time tasks using cloud computing, in *2011 Proceedings of 20th International Conference on Computer Communications and Networks (ICCCN)* (IEEE, 2011), pp. 1–7
7. K. Zhen, Z.X. Cheng, G. Minyi, Mechanism design for stochastic virtual resource allocation in non cooperative cloud systems, in *IEEE 4th International Conference on Cloud Computing* (2011), pp. 614–621
8. F. Wuhib, R. Stadler, Distributed monitoring and resource management for large cloud environments, in *2011 IFIP/IEEE International Symposium on Integrated Network Management (IM)* (IEEE, 2011), pp. 970–975
9. D. Niyato, K. Zhu, P. Wang, Cooperative virtual machine management for multi-organization cloud computing environment, in *Proceedings of the 5th International ICST Conference on Performance Evaluation Methodologies and Tools* (ICST (Institute for Computer Sciences, Social-Informatics and Telecommunications Engineering), 2011), pp. 528–537
10. H. Nguyen Van, F. Dang Tran, J.M. Menaud, Autonomic virtual resource management for service hosting platforms, in *Proceedings of the 2009 ICSE Workshop on Software Engineering Challenges of Cloud Computing* (IEEE Computer Society, 2009), pp. 1–8
11. H. Weisong, T. Chao, L. Xiaowei, Q. Hongwei, Z. Li, L. Huaming, Multiple job optimization in MapReduce for heterogeneous workloads, in *IEEE 6th International Conference on Semantics, Knowledge and Grids* (2010), pp. 35–140
12. L. Xiaoyi, L. Jian, Z. Li, X. Zhiwei, Vega ling cloud: a resource single leasing point system to support heterogeneous application modes on shared infrastructure (IEEE, 2011), pp. 99–106
13. L. Wei-Yu, L. GuanYu, L. Hung-Yu, Dynamic auction mechanism for cloud resource allocation, in *IEEE/ACM 10th International Conference on Cluster, Cloud and Grid Computing* (2010), pp. 591–592
14. Y. Xindong, X. Xianghua, W. Jian, Y. Dongjin, RAS-M: Resource allocation strategy based on market mechanism in cloud computing (IEEE, 2009), pp. 256–263
15. M. Uthayabanu, K. Saravanan, Optimizing the cost for resource subscription policy in IaaS cloud. Int. J. Eng. Trends Technol. (IJETT), Seventh Sense Res. Group **6**(5), 296 (2014)
16. T.H. Tram, M. John, Virtual resource allocations distribution on a cloud infrastructure (IEEE, 2010), pp. 612–617
17. P. Xiong, Y. Chi, S. Zhu, H.J. Moon, C. Pu, H. Hacigm, Intelligent management of virtualized resources for database systems in cloud environment, in *2011 IEEE 27th International Conference on Data Engineering (ICDE)* (IEEE, 2011), pp. 87–98
18. W. Linlin, K.G. Saurabh, R. Buyya, SLA–based resource allocation for SaaS provides in cloud computing environments. IEEE. 195–204 (2011)
19. R.T. Ma, D.M. Chiu, J.C. Lui, V. Misra, D. Rubenstein, On resource management for cloud users: a generalized kelly mechanism approach. Electr. Eng. Tech. Rep. (2010)
20. A. Radhakrishnan, V. Kavitha, Energy conservation in cloud data centers by minimizing virtual machines migration through artificial neural network. Comput. Springer–Verlag Wien, **98**, 1185–1202 (2016)
21. A. Radhakrishnan, V. Kavitha, Proficient decision making on virtual machine allocation in cloud environment. Int. Arab. J. Inf. Technol. **14** (2017)
22. S. Vinothina, R. Sridaran, G. Padmavathi, A survey on resource allocation strategies in cloud. Int. J. Adv. Comput. Sci. Appl. **3**, 98–104 (2012)

Chapter 4
A Game Theoretic Model for Cloud Federation

Benay Kumar Ray, Sunirmal Khatua and Sarbani Roy

Abstract Cloud federation has become a consolidated paradigm in which set of cooperative service providers share their unused computing resources with other members of the federation to gain some extra revenue. Due to increase in consciousness about cloud computing, demand for cloud services among cloud users have increased, thus making it hard for any single cloud service provider to cope up with cloud users demands and satisfying the promised quality of service. Hence, the cloud federation overcomes the limitation of each cloud service provider for maintaining their individual cloud resources. This chapter emphasizes on different approaches for cloud federation formation based on game theory and also highlights the importance of trust (soft security) in federated cloud environment. Different models for cloud federation formation using coalition game and the role of a cloud service broker in cloud federation are presented in this chapter.

4.1 Introduction

In recent years, Cloud computing has emerged as a new computing paradigm, where different cloud services like Infrastructure as a Service (IaaS), Software as a Service (SaaS), Platform as a Service (PaaS) are provision on-demand. The on demand provisioning of computing resources has attracted many big organization and led to rapid increase in cloud market due to their cost benefit. Moreover, with the increase in consciousness and growth in the cloud, requirement for computing resources has increased in such a way that, single service provider available resources become

B. K. Ray (✉) · S. Roy
Department of CSE, Jadavpur University, Kolkata, India
e-mail: benayray@gmail.com

S. Roy
e-mail: sarbani.roy@cse.jdvu.ac.in

S. Khatua
Department of CSE, University of Calcutta, Kolkata, India
e-mail: skhatuacomp@caluniv.ac.in

B. S. P. Mishra et al. (eds.), *Cloud Computing for Optimization: Foundations, Applications, and Challenges*, Studies in Big Data 39,
https://doi.org/10.1007/978-3-319-73676-1_4

insufficient to dealt with cloud users resources requests. Therefore, not able to deliver cloud services with committed QoS. This necessitates service providers to reshape their business plan in order to deliver uninterrupted cloud service to cloud users by increasing their resource scaling capabilities and availability (uptime) of resources. Hence federated cloud offers a practical platform to service provider for delivering uninterrupted cloud service to cloud users.

Cloud federation is a paradigm where the cloud environments of two or more service providers can collaborate and share their unused computing resources with other member service providers to obtain some extra revenue. In federation user applications need to be provider independent so that they can be easily migrated across multiple cloud service providers within the federation. In addition to this, the security, privacy and independence of the members of federation should be preserved to motivate cloud providers to take part in federation. Cloud federation provides substantial advantages to service providers and cloud users. Cloud federation enables service providers to earn some extra revenue by sharing their idle or underutilized computing resources. Second, cloud federation also allow service providers to increase their geographic space and allow to overcome unexpected rise in resources (virtual machine) request without having to invest in new infrastructure. Third, cloud users can avoid vendor lockin scenarios if their associated cloud service providers support more than one federations standards.

Cloud broker plays an important role in forming the cloud federation. A cloud broker is a middleware that provides cloud services to cloud users but may not provide any of its own computing resources. Cloud federation introduces new avenues of research based on the assistance it receives from the cloud broker, such as:

- Formation and management of the cloud federation.
- determining the number of computing resources, each cloud service provider should contribute within the federation and how the profits can be shared among the member of federation.
- Standardizing and monitoring the QoS and promised SLA among the cloud service providers within the federation.
- Providing a single unified view to the applications and cloud users.
- Formation and management of Trust of the cloud service providers for ensuring security of sensitive data and computation within the federation.

Forming cloud federation among different cloud service providers and sharing revenue among them are often complicated acts; so cloud broker helps different cloud service providers to work in federation.

Recently, popularity of game theory has considerably increased in the research field of cloud computing. This chapter will focus on a broker based cloud federation framework using game theory. We present the formation of cloud federation as a cooperative (coalition) game where different service providers collaborate to form a federation to cope up with fluctuation of users resources demands. The framework calculate the profit of every member service providers of federation based on their contribution of resources in federation and help them to gain highest profit while being part of federation. The cloud broker determines the individual satisfaction level of each CSP in a federation and based on that the existing federations are merged or split. It produces a stable federation partition where not a single service providers get better incentives in a different federation. Cloud broker is responsible for managing cloud federation, allocation of resources to cloud service users, service level monitoring etc. If these service providers can collaborate to form a federation then the available computing resources of these service providers can be combined to maximize their profit by using idle resources and support more cloud users. One of the main problems for service providers to take part in federation is the absence of trust between different service providers taking part in federation. Moreover, to guarantee the committed QoS and security of critical and sensitive data of cloud users, it is necessary to evaluate the trust of service providers and then form federation. The framework maintains the trust of each CSPs and the cloud broker selects trusted cloud service providers based on their individual QoS parameters like performance, scalability, availability. Cloud broker also combined services of more than one cloud service providers and provides an combined framework to the service providers or cloud users. This framework provides immense advantage to both cloud users and as well as service providers as cloud broker provides a single entry point to multiple heterogeneous cloud environments. Our focus is on laying emphasis on different approaches for cloud federation formation based on game theory and also highlighting the importance of trust (soft security) in federated environment.

The chapter is organized as follows. The Sect. 4.2 presents the role of cloud broker in cloud ecosystem. Section 4.3 provides brief overview of cooperative game theory. A game theoretical model for cloud federation formation is discussed in Sect. 4.4 followed by discussion of different framework based on coalition game theory. Section 4.5 discusses about the importance of trust in cloud federation. Section 4.6 concludes this chapter.

4.2 Role of a Cloud Broker in the Cloud Ecosystem

A cloud broker is a middleware that provides cloud services to cloud users but may not provide any of its own computing resources. Due to increase in cloud computing demands, the need for some expert to provide the optimal cloud offerings for enterprise business and technical requirements is also increasing. The provisioning task becomes more complicated in a heterogeneous cloud environment. Cloud service broker plays an important role by leveraging specialize expertise to provision cloud services in such a heterogeneous environment. In order to deliver reliable services, cloud service broker should be trusted and should have sound knowledge of the available cloud market. Trusted cloud service broker will make cloud service more secure to select and manage complicated cloud services, in federated cloud environments. A cloud service broker help cloud users by clearly defining technical and business needs while cautiously assessing the security policies, infrastructure capabilities and unique differentiating features provided by every service providers. Thus cloud broker provides a brokerage services to the cloud user's. The important characteristics of cloud service brokers are:

- It is a middleware between cloud service providers and cloud users.
- It does not provide cloud service of its own.
- Manages relationships between cloud service providers and cloud users.
- Provide single platform to deal with heterogeneous cloud providers.
- Cloud service broker provision resources optimally and add some value to interaction.

Increase in consciousness and growth in the cloud market, resulted in increase of variety of heterogeneous cloud services, thus increasing the need for specialized expertise (cloud service broker). As the demands of cloud service increases and number of service provider and infrastructure increases, the complexity service offered by cloud market also increases. Due to increase in complexity of cloud market, users have to manage many different heterogeneous services in terms of cloud interfaces, type of instances and price schema. For example, virtual machine instances are characterized, based on their configuration (number of core, memory, storage, compute unit) and each virtual machine provided by different service providers are of different quality as it these service providers provides service with different quality. In this scenario, the problem for cloud user was to select, best cloud service provider, who can deliver good quality service at low price. The cloud service broker's help cloud users to save their time by analyzing best negotiated services from different service providers and availing the cloud users with information about the best quality services at negotiated price. After analyzing best negotiated service, the broker provides the cloud users with a short list of selected service providers.

Based on above discussion, cloud brokering procedure will be essential to overcome variety of heterogeneous cloud services, for example, choosing computing resources for task at negotiated price, management of Service Level Agreement, monitoring of service for SLA violation. Ray et al. [1] in their work, proposed (a) broker based cloud service provider selection architecture and (b) game theory based SLA negotiation framework is proposed. The objective of cloud service broker was to determine a most suitable cloud provider, who can deliver good quality service at low price to cloud user and negotiating SLA on behalf of both cloud providers and cloud users and determine optimal value for price and quality for both cloud providers and cloud users. Their service broker architecture is given in Fig. 4.1.

In the proposed architecture, as shown in Fig. 4.1 [1], the service broker is considered as third party between cloud provider and cloud user. Their service broker assists cloud user to select the most suitable provider who will provide cloud service based on negotiation on parameters like price and quality. The working principle of the architecture is describe as follows, (a) Client submit task and SLA template to Resource Request Requirement (RRR), (b) Forward resource request to Request Analyzer (RA), (c) RA check for similar task in history, (d) If task matched then its corresponding resource details are forward to Resource selection module otherwise task information is forward to Service Provider Broker (SPB) module, (e) SPB submit the details of matched resource with task to Execution Time Analyzer module, (f) Execution Time Analyzer module, after analyzing time t for task on different selected resource, submitted to RA module, (g) RA forward cloud user SLA template to SLA

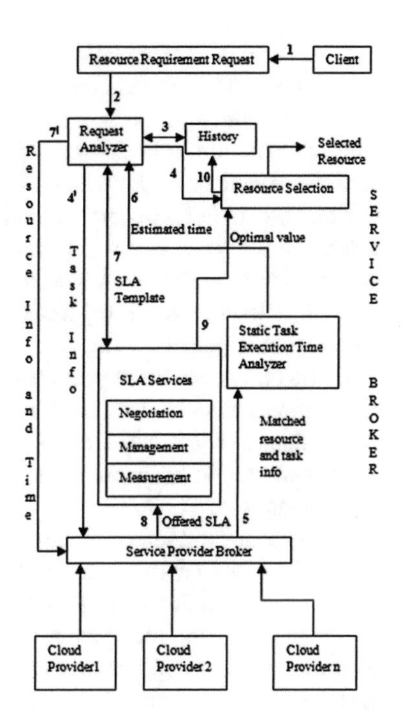

Fig. 4.1 Cloud broker architecture

negotiation module and ask initial offered value (price and quality) to execute task for estimated time t from SPB module (consist of different service providers), (h) best values for chargeable price and quality of service of service providers are submitted to resource selection module and (i) History module is updated.

Cloud service broker can deliver different categories of services. Cloud service broker are categorized based on functionality provided by them. As stated by National Institute of Standards and Technology [NIST], (2011) cloud service brokers are of three different categories and are differentiated based on their functionality. They are discussed below:

Service Aggregation: service broker unites many heterogeneous cloud services into one or more new services. The cloud service broker offer data integration and guarantee secure movement of critical data among users and service providers.

Service Intermediation: Cloud service broker upgrade a cloud service by enhancing some particular capacity and deliver value-added services to cloud users. The capability enhancement includes access and identity management, security enhancements, etc.

Service Arbitrage: Service arbitrage and service aggregation are almost similar, but the only difference is, that the combined cloud services not stable. Service arbitrage means a cloud service broker is flexible to pick cloud services from multitude of cloud services.

Cloud service broker plays an major part in forming the federation. A cloud broker aggregates and integrates cloud service providers and managed their services from multiple providers through a single entry point. Cloud brokering establishes relationships with multiple cloud service providers. Forming cloud federation between multiple private and public clouds and sharing revenue are often complicated acts; so cloud broker helps different cloud service providers to work in federations. Cloud federation introduces new avenues of research based on the assistance it receives from the cloud service broker. Some of them are, (a) Formation and management of cloud at federation level, (b) Management of resources at cloud provider level and (c) Standardizing and monitoring the QoS and promised SLA.

Cloud broker can make profit from aggregating all types of services from multiple cloud providers and delivering that service to users and thus easily include their own value-added services into the overall solution. Users of cloud broker gain substantial amount of benefit by continuously outsourcing their IT needs while cloud broker able to manage the complexity, cost and risk of using cloud services. Currently many cloud users, beyond selecting the cloud service providers, are looking for trusted third party (cloud broker) for monitoring, managing of cloud services provided by cloud providers. In addition to strategic and technical challenges, commonly associated with migration of application workload to other cloud, the cloud providers also need support over post-deployment, so that they may maximize their return from their investment. For instance, it is very essential to obtain suitable service broker such that it can guide a user to understand the technical and business aspects of cloud.

4.3 Overview of Cooperative Game Theory

Game theory is the branch of mathematics which dealt with the study of strategic decision making. Specifically, determines strategies for dealing with situations where the outcome of one players choice of action depends on the actions of other players. The cloud federation formation framework is modeled based on problem of a cooperative game. In cooperative game, a group of players can take a set of joint actions. These groups of rational players are referred to as coalition and enforce a cooperative behavior. The outcome of a cooperative game will be specified by which coalitions forms, and the combine action that group takes. Cooperative game theory examines condition where set of players can cooperate to create value by joining coalitions.

In coalition game two important components are the players and the coalition value. Let currently set of rational players be indicated by $\xi = [1, m]$ who interact with each other to form cooperative group (coalition) and try to improve their payoff in the coalition game. In a coalition game coalition $\mathbb{F} \subseteq \xi$ represents group of player cooperative with each other and act as a single entity in a given game. On the other hand the coalition value u(payoff), is defined as the utility of a coalition in a cooperative game. Again, based on the considered definition of coalition value, different properties of a coalition game can be defined. Therefore a coalition game can be defined by the pair (ξ, u). The mathematical model of the Coalition Game is given by:

Definition: A coalition $\mathbb{F}$ in any subset of ξ. The set of all coalitions is denoted by 2^{ξ}. A coalition game is mapping $u : 2^{\xi} \rightarrow \mathbb{R}$ such that $u(\emptyset) = 0$. For any coalition $\mathbb{F} \subseteq \xi$, the number $u(\mathbb{F})$ is called the worth of $\mathbb{F}$.

In general, a coalition value can be of three different types:

Characteristic form: The characteristic form is the most popular form of coalition value in game theory and it was first introduced by Neumann [2]. The value of coalition in characteristic form depends on each players of coalition, i.e., the value or utility of any coalition $\mathbb{F} \subseteq \xi$ is not relied on the coalitions formed between the players which are not the part of $\mathbb{F}$.

Partition form: Any coalition $\mathbb{F} \subseteq \xi$ are in partition form if utility of coalition game is dependent on the coalitions formed by the members in $\xi \backslash \mathbb{F}$ and as well as members of $\mathbb{F}$. The concept of the partition form was introduced by Thrall and Lucas [3].

Graph form: In coalition game, the value of game may be strongly affected if members of coalition may communicate through pairwise links in a graph. Therefore in such coalition game, the value of game is considered in graph form because a different game value can be determined for each different graph structure. The characteristic form and the partition form are not suitable to determine the value of a coalition $\mathbb{F}$, as they are not dependent on the connection between the members of coalition. The concept of modeling interconnection graph within coalition game was first introduced by Myerson [4].

In any coalition game, the value of any coalition denotes the total utility obtained by a coalition. The payoff of a player, denotes the total utility obtained by an individual player. Based on how payoff is divided among the members of coalition, the coalition game can either be with non-transferable utility or with transferable utility. In transferable utility game the total utility obtained by any coalition $\mathbb{F}$ can be shared in any way among the players of coalition $\mathbb{F}$. For an example, in transferable utility game, if the value represents an amount of money, it can be distributed in any way among the coalition members. Based on property of transferable utility, the total utility obtained can be distributed in any order among the players of coalition (using some fairness rule). The amount of utility that an each players of coalition obtained from the sharing of total utility $u(\mathbb{F})$ comprise the players payoff and is represented by y^j. The vector $y \in \mathbb{R}^{\mathbb{F}}$ with each element y^j being the payoff of players $j \in \mathbb{F}$ constitutes a payoff allocation.

Coalition game with transferable utility is a popular and accepted method, there exist a number of circumstances, where the value of coalition cannot be assigned a single real number or there exists a rigid restriction on the utility division. These types of coalitional games are known as game with non-transferable utility. The concept was known using basis of non-cooperative strategic games according to Aumann and Peleg [5]. In a game of non-transferable utility, the payoff received by each player of coalition $\mathbb{F}$ depends on the joint action taken by the players of that coalition. In game theory literature most aspect of coalition game theory are studied in characteristic form with non-transferable utility or transferable utility. For these types of coalition game, different game properties and solution concepts are defined. Some of the properties of a game in characteristic form with transferable utility are provided below:

- **Superadditive**: The subset of two disjoint coalition will have an incentive to cooperate only when profit obtained in case of cooperation are greater than the profit obtained alone according to Drechsel [6].

$$\mathbb{F}_1 \cap \mathbb{F}_2 = \varnothing \Rightarrow u(\mathbb{F}_1 \cup \mathbb{F}_2) \geq u(\mathbb{F}_1) + u(\mathbb{F}_2)$$

- **Monotone**: A cooperative game (ξ, u) in characteristic form is monotone if for all $\mathbb{F}_1, \mathbb{F}_2 \in 2^{\xi}$.

$$\mathbb{F}_1 \subseteq \mathbb{F}_2 = u(\mathbb{F}_1) \leq u(\mathbb{F}_2)$$

- **Symmetric**: A coalition game (ξ, u) is symmetry if the coalition value $u(\mathbb{F}_1)$ only based on the number of players in the coalitions $\mathbb{F}_1$ according to Gilles [7]. Hence there is some function $f : \xi \rightarrow R$ such that $u(\mathbb{F}_1) = f(|\mathbb{F}|)$ for all $\mathbb{F}_1$ subset ξ.
- **Constant-sum**: A cooperative game (ξ, u) is constant sum if for every coalition $\mathbb{F}_1$ subset ξ is

$$u(\mathbb{F}_1) + u(\xi \setminus \mathbb{F}_1) = u(\xi)$$

- **Simple**: A cooperative game (ξ, u) is simple if, for each coalition $\mathbb{F}_1$ subset of ξ we have either $u(\mathbb{F}_1) = 0$ or $u(\mathbb{F}_1) = 1$.

4.3.1 Classification of Coalitional Game Theory

Based on the properties, coalitional game can be mapped into three different groups. The three different types of coalition game are (i) Canonical coalitional games, (ii) Coalition formation games and (iv) Coalitional graph games.

4.3.1.1 Canonical Coalitional Games

Canonical coalitional games refer to the group of most popular type of coalition game, which is well studied in cooperative game theory. These groups of coalition game are widely used, thoroughly formalized, well understood and its solution concepts are clear. To classify the game in canonical form, the game must satisfy the following properties:

1. The value of the game must be in characteristic form or the value may be mapped to this form through some assumptions. The characteristic form will either be of the transferable utility form or non-transferable utility form.

2. The game follows a superadditive property. Increasing the number of participating players in the coalition will not decrease its value, i.e. no group of players will worsen off while joining the coalition. That is, cooperation among members of coalition will always be beneficial.
3 Proper study is needed to divide utility and gain among the members of coalition in a fair manner, so that grand coalition (coalition of all the players) can be stabilized.

The motive of canonical game is to examine the stability of grand coalition under certain payoff allocation rule. Hence assessing the total profit that the grand coalition can obtained and fair payoff division is also another important key objective of canonical game. In game theory, various methods and broad range of concepts are available for solving canonical game. Almost all the solution concepts, which are used for solving canonical game, satisfy all the properties and key objective of canonical game. Thus the solution concepts which are used for finding solution of canonical game are (i) The core, (ii) the Shapley value and (iii) the nucleolus.

The Core

The existing game theoretical literature considers the core as the most important concept for solving a canonical coalitional game. In canonical coalitional game the core defines the set of payoff allocations which assured that not a single player in a group will have an incentive to join new coalition $\mathbb{F} \subseteq \xi$ while leaving the grand coalition ξ. Therefore the main idea of the core in canonical games is to distribute the payoff to players that will stabilize the grand coalition. It is to be noted that in canonical game, the grand coalition will help to generate maximum payoff due to superadditivity. Though mathematically, the core definition is different for both transferable utility and non- transferable utility game, the important concept of core is applied to both. The core of a canonical coalitional game (ξ, u) in a characteristic form with transferable utility game is defined as follows:

$$core_{Tu} = \{y : \sum_{i \in \xi} y_j = u(\xi)\} \ and \ \{\sum_{i = \mathbb{F}} y_j \geq u(\mathbb{F}) \forall \mathbb{F} \subseteq \xi\}$$

Therefore, the core denotes the set of payoff allocation where none of the coalition $\mathbb{F} \subseteq \xi$ has benefit to leave grand coalition and form other coalition $\mathbb{F}$ and reject the allocated payoff to the players. The concept of core guarantee that deviation from grand coalition does not occur due to the reason that in the core any allocated payoff y, guarantees at least $u(\mathbb{F})$ amount of utility for every single coalition $\mathbb{F} \subset \xi$. Note that, for an NTU game, an analogous definition of the core can be used according to Saad et al. [8].

At any instance, in canonical coalition game, solution of grand coalition will be stable and optimal, only if obtained payoff allocation lies in the core. This highlights the important of core in solution concept. However, the core of a coalitional game suffers from several drawbacks, such as (a) In canonical coalition game if the core is empty, there does not exist payoff allocation that can stabilize grand coalition,

(b) If there exists a payoff allocation that can stabilize the grand coalition, then the core will be a vast set and providing a fair payoff allocation from this set will be a complicated task. So to avoid this type of problem, there exists many classes of canonical coalitional games where the core is non-empty and based on this class of game respective properties may be derived. More detailed explanation can be found in [8].

The Shapley Value

To deal with the drawback of the core, the Shapley value is used as a solution concept for canonical coalition game. Shapley value deals with the concept of value. Shapley provides a unique approach to assign unique payoff vector (value) to each game in coalition form given by (ξ, u). Let $\phi_j(u)$ denote the payoff provided to player j by the Shapley value ϕ with characteristic function u. Therefore, $\phi(u) = (\phi_1(u), \phi_1(u), \ldots, \phi_n(u))$ is the value function that is assigned to each possible characteristic function of a n-person game. According to Shapley, there exist a unique mapping, Shapley value $\phi(u)$ from the space of all coalition games to $\mathbb{R}^{\xi}$.

Alternative definition of Shapley value considers the order of each player joining in grand coalition. In coalition game, if a player randomly joins the grand coalition, the allocated payoff (calculated based on Shapley value) to player $j \in \xi$ is said to be the expected marginal contribution to player j. Thus based on this interpretation for any canonical coalition game with transferable utility game (ξ, u), the payoff $\phi_j(u)$ assigned to each player $j \in \xi$ of Shapley value $\phi(u)$ can be given as:

$$\phi_j(u) = \sum_{\mathbb{F} \subseteq \xi \setminus \{j\}} \frac{|\mathbb{F}|!(\xi - |\mathbb{F}| - 1)!}{\xi!}[u(\mathbb{F} \cup \{j\}) - u(\mathbb{F})]$$

From above equation, $u(\mathbb{F} \cup \{j\}) - u(\mathbb{F})$ denotes the marginal distribution of each player $j \in \mathbb{F}$. The weight before $u(\mathbb{F} \cup \{j\}) - u(\mathbb{F})$ denotes the probability of player j facing the $\mathbb{F}$, while joining in random order. Therefore from $\frac{|\mathbb{F}|!(\xi - |\mathbb{F}| - 1)!}{\xi!}$ it can be noted that in coalition $\mathbb{F}$ the players can be positioned at the start of ordering in $|\mathbb{F}|!$ ways and the remaining players can be positioned at the end of an ordering in $(\xi - |\mathbb{F}| - 1)$ ways. The probability of occurrence of such order is $\frac{|\mathbb{F}|!(\xi - |\mathbb{F}| - 1)!}{\xi!}$ and the finally calculated value $\phi_j(u)$ denotes the expected marginal contribution of player j. The Shapley value is mostly used, but the problem of using this solution concept is that the complexity of finding the solution by Shapley value will increase as the numbers of players increases.

The Nucleolus

Other than core and Shapley value, solution concept used for n-person canonical coalition game is nucleolus. The solution concept of nucleolus was introduced by Schmeidler [9]. Nucleolus as a solution concept in cooperative canonical coalitional game (ξ, u), provides a payoff allocation that reduces the dissatisfaction of players in coalition from obtained allocation. Thus nucleolus help to find imputation y, that minimize the maximum dissatisfaction. For a given coalition $\mathbb{F}$, the measure of dissatisfaction of an imputation y for $\mathbb{F}$ is defined as the excess and is given by:

$$e(y, \mathbb{F}) = u(\mathbb{F}) - \sum_{j \in \mathbb{F}} y_j$$

It measures the excess by which coalition $\mathbb{F}$ falls short of its potential in the imputation y. Thus, this is one of the main reasons behind the concept of nucleolus.

4.3.1.2 Coalition Formation Games (CFG)

In many cooperative scenario, cooperation among the members may accompanied with an extra hidden cost, and hence limit from getting advantage of this cooperation. Thus formation of grand coalition may not be always guaranteed. In this type of scenario, using of canonical coalitional games for modeling the cooperative behavior between the players is not preferred. In CFG cost of cooperation and network structure play a vital role. Some important characteristics of a coalition formation game are discussed below:

1. Coalition formation game is non-superadditive in nature and grand coalition formation is not guaranteed. This due to the reason that, though coalition formation brings profit to its members, the profits of member are limited to the cost of coalition formation.
2. The coalition formation game can be in characteristic form or in partition form.
3. The CFG important aim is to study the coalitional network structure. From coalitional structure, the question which can be answered are (j) what is the size of optimal coalition, (ii) which coalition will form and (iii) how structures characteristics, can be assess and many other.

In contrast to canonical game finding solution to coalition formation game is much more difficult. This is because in many coalition game problems, coalition formation bear an extra cost for an information exchange process and negotiation process, thus reducing the overall profit obtained from coalition formation. In coalition formation game, finding solution in presence of coalition structure by using previous solution concept is not easy and need significant changes in their definition. But even after significant change in definition, finding solution of game is not straightforward and can be complicated. Therefore it will be quite challenging to find optimal coalition structure and characterizing their formation. In case of coalition formation game no formal and unified solution concepts exist unlike canonical coalition game. Therefore in many research works, dealing with coalition formation game, the solution concepts which are defined are specific to the problem.

In this type of coalition game, the most important part is to determine the coalition partition that is appropriate to the respective game. The respective coalition structure of the game in transferable utility, helps to maximize the total payoff of the game and in non-transferable utility game, the coalition structure is found with

Pareto optimal payoff distribution. For finding this type of solution a centralized approach is used and such approaches are mostly NP-complete as has been shown [10] and [11]. The problem is NP-complete because, finding optimal coalition partition (which provide highest payoff) required continuous iteration over all coalition partitions of player in ξ. If players in ξ increases then number of coalition partition in set ξ also increases exponentially [12]. In order to avoid complexity of centralized approach, in many practical application it was observed that the process of coalition formation takes place in a distributed manner. In distributed process, players can make their own decision to join or leave current coalition. Hence the demand for distributed approach and complexity of the centralized approach has requirement for huge development in the coalition formation. The objective of this requirement is to find low complexity centralized approach and distributed algorithm for coalition formation. The different approaches used for distributed coalition formation are (i) Markov chain-based methods, (ii) heuristic methods, (iii) set theory based methods and (iv) method that use negotiation technique or bargaining theory from economics. Again it is noted that, there are two approaches for coalition formation, fully reversible and partially reversible [13]. In coalition formation game, with fully reversible approach, players can leave or join particular coalition with no limitation whereas in partially reversible approach, players are not allowed to leave coalition after coalition formation. Though fully reversible approach is flexible and practical, forming such an approach is complicated. In case of partial reversible approach, it is easy to construct but their practicality is limited, because players are not allowed to leave from current coalition and join new coalition thus breaking the agreement. Depending on the type of application, most favorable approach can be selected.

4.3.1.3 Coalitional Graph Games (CGG)

Overview of two types of cooperative coalition game (a) canonical game and (b) coalition formation game are discussed above. The utility or value of coalition for these games is not depended on the internal connection of players within the coalition. But in many coalition games, the impact of interconnection or communication between the players can be reflected on the characteristic function and outcome of the game. The interconnection between the players can be captured by graph. This graph helps to determine connectivity of the players among each other in coalition. Therefore some of the properties that differentiate CGG from others are discussed below:

1. The cooperative coalitional graph game can be of type transferable utility or non-transferable utility and the value of the game may depend on internal interconnection between players within the coalition.
2. The internal structures of player within the coalition have great impact on outcome and characteristics of the game.

The main objective of CGG are given below:

1. Study the properties like stability, efficiency, etc. of the formed directed or undirected network graph. If network graph is provided for some coalition game, then analyzing properties like efficiency and stability is the only goal of the game.
2. Coalitional graph games are a suitable tool for the games where hierarchy governs the interactions between the players.

In [4] Myerson et al. first proposed the idea of coalitional graph game, through the graph function form for transferable utility games. In their work, Myerson et al. tried to define fair solution for the game, starting from canonical coalitional game with transferable utility and in presence of undirected graph (interconnects the players) in the game. Afterwards these fair solution proposed by Myerson et al. was defined as Myerson value. Myerson et al. work, motivated the future work significantly, and many new approaches was developed for coalition graph game and one of them is network formation games. Network formation game is a mixture of two game, CGG and non-cooperative games. The main aim of Network formation game is to study the interactions between the players in a group that wants to form a graph. The more details literature can be found in [4, 8].

4.4 Coalition Game in Cloud Federation Formation

Coalition games have huge application within cloud federation. For instance, in cloud federation cooperation always required between different cloud service providers. Cooperation helps cloud service providers to maximize the use of idle resource and increase their profit. This section presents three different models of cloud federation formation. Ray et al. [14] in their work proposed a cloud federation framework based on satisfaction level. Mashayekhy et al. have also model cloud federation formation problem using coalitional game theory. Their model let the cloud service providers to decide of its own, to form a federation and obtaining the highest gross profit [15]. The author of [16] have present trust based cloud federation formation model based on hedonic coalition game theory. The different models are discussed as follows:

4.4.1 Cloud Federation Formation Based on Satisfaction Level

The objective of the work by Ray et al. in [14] was to find best federation for individual cloud service provider. The criterion for selecting most suitable federation is, on the basis of two important Quality of Service (QoS) attributes price and availability. Availability a is stated as the uptime of the cloud service. Price p on the other hand is the amount paid by cloud user for using the cloud service. It is assumed that cloud service provider with high availability will have higher price. Thus relation between

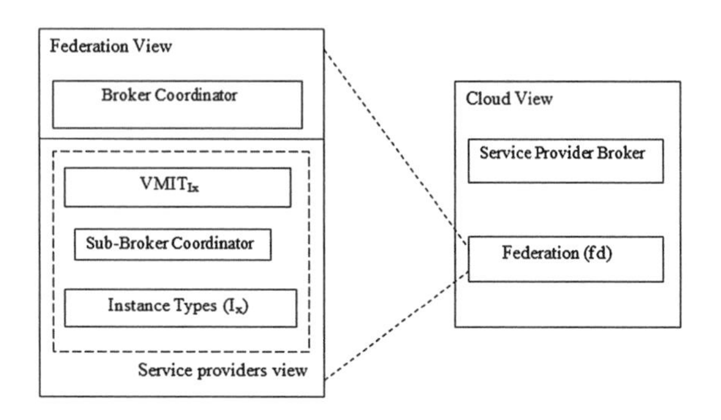

Fig. 4.2 Broker based cloud federation architecture

availability and price is given by $p_i = \rho * a_i$ where $a = \{a_1, \ldots, a_n\}$ denotes the availability, $p = \{p_1, \ldots, p_n\}$ denotes the price of ξ number of players and ρ is considered as the proportionality constant and its value represent the price of cloud service provider with availability 1. It has been assumed that cloud service provider with same availability will have same price.

Cloud brokers are identified as the key component for federation. The role and responsibility of the service broker is to find the best federation from a set of federations. Ray et al. [14] describes, broker based cloud federation formation framework. The cloud federation framework are divided into two module (a) Federation View and (b) Cloud View (see Fig. 4.2). The detail discussion of each component of individual module is discussed below.

Cloud View

- **Service Provider Broker (SPB)**: manage a set of service providers and cloud federation.
- **Federation**: Federation are formed among different service providers. Each federation consist its own broker coordinator and heterogeneous set instance of different cloud service providers.

Federation View

- **Broker Coordinator (BC)**: handles all the resource requests by user on behalf of federation and provide each and every update to SPB.
- **Virtual machine instance type** ($VMIT_{I_x}$): retain information of homogeneous types of instances of different cloud service provider of particular federation where $I = \{I_a, \ldots, I_y\}$ denotes different instance types.
- **Sub Broker Coordinator (SBC)**: keep details of all instances in $VMIT_{I_x}$. It also handles all the request received from BC.

A coalition game with transferable utility is used to form federation among different service providers. Here a set of rational player is denoted by ξ and u is the value of the game, i.e., the total profit achieved by member of cloud service providers in a federation. Here players are referred as cloud service providers. Broker coordinator manage the cooperation among the members of cloud service providers in federation, based on optimal arrangement order $\alpha = \{E_j^{SP^i} | j = [1, z] \, and \, SP^i \in \xi\}$. Therefore cloud users requests are processed based on arrangement order of cloud service providers in α. Again, availability of $VMIT_{I_x}$ (service of resource) of cloud service providers in federation is achieved by cooperation among them. When, unavailability of each cloud service providers will have effect on overall availability of federation. Thus based on conditional probability, probability of availability of $VMIT_{I_x}$ in federation fd_j is given by:

$$A_{fd_j}^{VMIT_x}(\alpha) = 1 - \prod_{i=1}^{\xi} u_{I_x}^{SP^i}$$

Each instance type $I_x^{SP^i}$ are differentiated according to amount of compute unit $Cu_{I_x}^{SP^i}$ amount of storage $m_{I_x}^{SP^i}$, amount of memory $s_{I_x}^{SP^i}$ and specific number of cores $Co_{I_x}^{SP^i}$. The cost incurred by instance type $I_x^{SP^i}$ of cloud service providers are given by:

$$c_{I_x}^{SP^i} = Co_{I_x}^{SP^i} \cdot pc + Cu_{I_x}^{SP^i} \cdot pcu + m_{I_x}^{SP^i} \cdot p_m + s_{I_x}^{SP^i} \cdot p_s$$

In these equation pc denotes the cost of single core, pcu is the cost of each compute unit, p_m is the cost of one GB of memory and p_s is the cost of one GB of storage. The cost of federation is formulated based on the availability of each $VMIT_{I_x}$ in federation. Therefore the total cost of federation fd_j will be define as the sum of cost of all instances $I_x^{SP^i}$ of all the service providers in federation and is given by:

$$C_{fd_j}(\alpha) = \sum_{I_x \in I} \sum_{i=1}^{\xi} \prod_{k=1}^{i-1} u_{I_x}^{SP^k} \cdot a_{I_x}^{SP^i} \cdot c_{I_x}^{SP^i} \tag{4.1}$$

In this equation $a_{I_x}^{SP^i}$ represent the availability of cloud providers SP^i of I_x types of instance. The chargeable price for each instance depends on the availability provided by federation. Hence that is defined as follows:

$$P_{fd_j}(\alpha) = \sum_{I_x \in I} \rho_{I_x}^{SP^i}(\alpha) \cdot C_{fd_j}(\alpha) \tag{4.2}$$

The characteristic function (payoff) of federation is model based on total profit obtained by group of cloud service provider in federation. That is, payoff $u_\alpha(fd_j)$ is the difference between the revenue $P_{fd_j}(\alpha)$ received by cloud users and cost $C_{fd_j}(\alpha)$ incurred by cloud federation. Process of Shapley value is used for fair allocation of payoff among the members of federation in this coalition game. The payoff of cloud federation is given by:

$$u_\alpha(fd_j) = P_{fd_j}(\alpha) - C_{fd_j}(\alpha) \tag{4.3}$$

To form federation they introduce the concept of satisfaction level. The different measured satisfaction levels are (i) $sat^{SP^i}(fd_j)$ denotes the satisfaction level of SP^i in federation fd_j, (ii) $sat^{SP^i}(I(SP^i))$ denotes the satisfaction level of single service provider in federation or satisfaction level of service provider without being in any federation and (iii) $sat(fd_j)$ denotes the satisfaction level of federation. The value of satisfaction level $sat^{SP^i}(fd_j)$ and $sat^{SP^i}(I(SP^i))$ are calculated based on cloud providers two QoS parameters, availability and profit and satisfaction level $sat(fd_j)$ is determined based on payoff value and availability of federation. The satisfaction level are model based on idea describe in [14].

The Coalition (federation) formation algorithm follows a distributed approach. It is assumed that at any instant of time any cloud service providers may merge new federation or split from existing federation. At each time period SPB checks for available of each type of instances for every federation. If available resource (virtual machine) is below certain threshold value, then BC of those federation are informed by SPB. BC on being informed it request to join with new federation or cloud service providers. Based on satisfaction level, some set of preference rules are defined over possible federation (coalition) partition. The preference rule are (i) a cloud service providers merge with new federation only if, its satisfaction level maximizes during formation of federation otherwise service provider will leave the federation (ii) federation will collaborate with other providers, if its satisfaction level increases in newly formed federation.

4.4.2 Merge and Split Based Cloud Federation Formation

Mashayekhy et al. [15] proposed the model for the problem of cloud federation formation based on hedonic coalition game. According to their proposed model every single service provider form federation to serve users' virtual machine requests. Their model maximize the overall profit obtained by federation by delivering required request to cloud users. Further their proposed model consider group of service providers and group of cloud users that submits their requirements about different types of virtual machine instances. Out of this group of service providers a subset of this group will be selected as a federation to serve the required virtual machine request of users.

A hedonic game with additional properties of fairness and stability has been chosen as a model for the proposed game, with well-defined preference relations. Further their proposed model enables the formation of a federation yielding the highest total profit. Merge and split operations are performed on cloud federations in order to form a federation serving the user with his requested resources. According to Mashayekhy et al. every service providers obtain its discrete profit based on its market share. Their cloud federation model provides a stable federation structure, i.e., not single service providers, will get extra benefit to join with new federation or split from an existing federation to join with new federation.

4.4.2.1 Proposed Model

Their system model is comprised of (a) group of service providers, (b) a cloud broker, and (c) cloud users. In their model broker is considered as the trusted third party entrusted with 1. receiving requests, 2. executing CFFM, 3. receiving payment, and 4. profit division. A set of cloud providers serve resources requests of cloud users. Each VM instance are characterized by their total cores, size of memory, and storage. A specific capacity is reserved by a cloud provider for private users, and the balance virtual machine load is contributed to the federation according to current load conditions.

A user request is consists of the number instances of different virtual machine types. The broker receives the request, and charges a user based on the virtual machine instances types apportioned to him, who finally pays an amount that is free of the service provider providing the virtual machine instances.

4.4.2.2 IP-CFPM

The problem of federation formation with the objective of profit maximization is formulated as an integer program, called IP-CFPM. The total profit is determined by the weighted sum of the profits per VM instance, the weights being the number of allocated virtual machine instances of different types, which are the decision variables here and are constrained to be integers. The constraints ensure that the resources (cores, memory and storage) delivered by a service providers taking part in federation formation process are less than their respective total amounts of resources. A separate constraint ensures that every single provider at least provide one virtual machine in federation.

4.4.2.3 Game Formulation

The cloud federation game is a coalitional game that models federations among different service providers. A cloud federation game is defined based on the concept of coalitional game (I, u) with transferable utility. Every single service provider in I (set of service providers) is considered to be player of the game, and u is the utility function, of the federation $F \subset I$. In this game, profit is considered as the characteristic function i.e., if federation is formed then profit is considered as the utility of the game otherwise the utility is 0 if no federation can be formed. Their cloud federation game satisfies two main properties (a) *Fairness* means a fair division of the profit achieved by a federation by dividing the total profit among the members of federation and (b) *Stability* means a federation should be stable, providing no incentives to the participating cloud providers to leave the federation.

4.4.2.4 Profit Division

The system employs a fair profit distribution among the member of federation based on the share of the service providers. The author have used normalized Banzhaf value for distribution of profit among the members of federation. The normalized Banzhaf value ensures that the member of federation which contributes more resources in federation obtain higher profit. The main idea of Banzhaf value technique is to divide the payoff for the grand federation that takes into consideration the power of the players and represents the average marginal contribution of provider C_i over all possible federations containing C_i. Although computing the Banzhaf value for a game is NP-hard, some of the possible federations are checked by iteratively applying the merge and split rules and their values are estimated. These checked values are used to define an estimated Banzhaf value.

4.4.2.5 Cloud Federation Formation Framework

The salient features of their proposed model are:

1. Players are partitioned into disjoint sets.
2. The formed federation partition consist of the single federation to deliver the service to the corresponding requests of the users.
3. The remaining service providers are free to take part in federation formation with other service providers to service the requests of the users, without affecting the decision of the other service providers participating in the corresponding federation.
4. The preference relation for the hedonic game is based on the characteristic function, which is in turn based on profit.
5. A CSP prefers a federation over another if the characteristic function for that federation with respect to that CSP higher than that for the other, i.e. if it yields more profit.
6. The merge comparison relations and split comparison relations are defined for service provider.
7. The merge comparison compares the merged federation to its two constituents and is valid only if all CSPs prefer the merged federation to the constituents.
8. The split comparison compares two disjoint federations to their merged form and is valid if any CSP prefers the constituents to the merged form.
9. Two rules - Merge Rule and Split Rule are defined based on these comparisons.
10. According to merge rule two federations will merge to form a new federation if total profit of new federation increases.
11. According to the split rule a federation will splits if there is present atleast one subfederation that achieves identical profit with respect to its constituent service providers.
12. A split may lead to decrease in the profit of the other sub-federations.

4.4.2.6 Algorithm CFFM

The CFFM mechanism is carry out by a broker and uses merge and split rules discussed previously. The CFFM algorithm takes users request as input. From an initial federation structure consisting of singleton federations a final federation is obtained by repeatedly solving IP-CFPM and applying the merge and split rules in two procedures - MergeFederations() and SplitFederation(). The proposed algorithm converges to a federation partition consisting of only single federation and produces an individually stable federation. CFFM avoids performing an extensive search on the set of federation by employing the merge and split procedures.

4.4.3 Trust Based Federation

The trust-based model of cloud federation formation, called DEBT (Discovery, Establishment, and Bootstrapping Trust, Wahab et al. [16]) aims to address the problem of credibility assignment to different cloud service providers in the scenario where it is uncertain that they will adhere to their commitments detailed in the clauses of official contracts such as SLAs. The malicious agents (here, other providers) may collude to give false information regarding the trustworthiness of a provider. Their purpose is to artificially raise or lower a provider's face value. Also, in the situation where there is no past interaction between two providers, it is important to assign initial trust values. This is called trust bootstrapping. The DEBT framework provides a stepwise guide for services to build trust relationships starting from discovering services and collecting feedback, down to bootstrapping new services and aggregating feedback in a collusion-resistant manner. Here trust is modeled as a private relationship between each pair of services rather than a global absolute value that can be referenced by all services. Based on DEBT it is possible to design a trust-based hedonic coalitional game with nontransferable utility that aims to find the coalition partition that minimizes the number of malicious members.

4.4.3.1 System Model

Since services may be either truthful or collusive in judging the other services, each pair of services $(\mathbb{F}_i, \mathbb{F}_j)$ has a belief in credibility, $Cr(\mathbb{F}_i \rightarrow \mathbb{F}_j) = n, Cr(\mathbb{F}_j \rightarrow \mathbb{F}_i) = m$, also called a *credibility score*, that represents each services accuracy level in judging the other services, where n and m are two decimal numbers. Each service $\mathbb{F}_i$ develops a measure of belief in the trustworthiness of another provider $\mathbb{F}_j$, referred to as $belief_{\mathbb{F}_i}^{\mathbb{F}_j}(T)$ and a belief in its maliciousness, denoted by $belief_{\mathbb{F}_i}^{\mathbb{F}_j}(M)$ before forming a coalition. These belief measures are based on a satisfaction metric which is the ratio of the number of satisfactory experiences to the total number of experiences. Thus, these measures depend on past history between the two services.

The community formation is designed as a hedonic coalition game, which results in a coalition structure or partition. It is a set of non-empty coalitions that partitions the existing set of services into mutually exclusive and exhaustive subsets. The utility of service $\mathbb{F}_i$ in a certain coalition C_k is obtained by summing up $\mathbb{F}_i$'s beliefs in trustworthiness in C_k's members. This utility is given by

$$U_{\mathbb{F}_i}(C_k) = \sum_{\mathbb{F}_j \in C_k} belief_{\mathbb{F}_i}^{\mathbb{F}_j}(T)$$

4.4.3.2 Attack Model

There are two situations when attacks are likely: (1) during trust establishment (2) during and after communities formation. During trust establishment, collusion attacks may take place between services to mislead the results. During and after communities formation, passive malicious services may misbehave to save their resources and gain illegal advantage over the other services. Two major attack types are considered

- **Collusion Attacks**: Such attacks occur when several malignant services collaborate together to give misleading judgments either to increase the trust score of some services (i.e., a promoting attack) or to decrease the trust score of some other services (i.e., a slandering attack).
- **Passive Attacks**: Such attacks occur when passive malicious services cheat about their available resources and/or QoS potential during communities formation in order to increase their chances of being grouped into powerful communities. After communities are formed, these malicious services would renege on their agreements with both services and clients by benefiting from the other services resources (e.g., physical computing infrastructure) and refraining from sharing their own resources to dedicate them for their own workload.

4.4.4 DEBT Framework

DEBT (Discovery, Establishment, and Bootstrapping Trust) framework presents a step by step guide for services to build trust relationships beginning from discovering services and collecting feedback, down to bootstrapping new services and aggregating feedback in a collusion-resistant manner. The brief description is provided below.

4.4.4.1 Service Discovery

DEBT incorporates a discovery algorithm that allows services to inquire about each other from their direct neighbors is proposed to collect judgments for the purpose of establishing trust. The algorithm capitalizes on the concept of tagging in online social networks (e.g., Facebook, LinkedIn) to explore the direct neighbors of a certain service. The basic idea is to keep tagging or nominating intermediate services until identifying all the reachable direct neighbors of the intended service. The algorithm is a variation of the Breadth-First Search (BFS) strategy in graph theory.

4.4.4.2 Trust Establishment

During the trust establishment process, services may collude with each other and provide dishonest judgments, generating false trust results. Moreover, these services usually tend to refrain from revealing their opinions due to a lack of incentives for doing so, which leads to meaningless or biased computations of the aggregate trust value.

In recognition of these problems, the DEBT system includes an aggregation model for the collected judgments that is able to overcome the collusion attacks even when attackers are the majority and an incentive model for the services to motivate them to participate in the trust establishment process. An aggregation technique based on the Dempster-Shafer theory of evidence which takes into account the existence of colluding services is used. The performance of the aggregation technique depends heavily on the *credibility scores* assigned to the services. The proposed aggregation technique overcomes the collusion attacks even when attackers are the majority if and only if (1) the credibility values are between 0 and 1, and (2) the credibility scores of the trustworthy raters are higher than those of colluding ones. Thus, the authenticity of this metric is essential to the trust establishment process and should be protected. Therefore, the credibility metric should be updated continuously to ensure that truthful services always hold higher credibility scores than those of colluding ones.

4.4.4.3 Trust Bootstrapping

Trust bootstrapping is assessing trust, i.e. allocating initial trust values for newly deployed Web services, in the absence of historical information about the past behavior of newcomers. The bootstrapping mechanism is based on the concept of endorsement in online social networks that is resilient to white-washing. Each service maintains a dataset that records its previous interactions with several services having different functional and non-functional specifications. Whenever a request from service i to bootstrap a new service j is received, the services that are interested in the request train a decision tree classifier on their datasets to predict an initial trust value for j. The decision tree classifier analyzes services training dataset that contains properties and specifications for some existing services (e.g., providers name, deployment country, etc.) and learns the patterns of the data by pairing each set of inputs with the expected output (e.g., the judgment on the services).

4.4.5 *Trust-Based Hedonic Coalitional Game*

4.4.5.1 Game Formulation

In the proposed game the players are the services that seek to form multi-cloud communities. The objective is to form trusted communities wherein the number of malicious members is minimal. The proposed coalitional game is a *non-cohesive* game and utility of the game is non-transferable. It is hedonic in nature, with a preference relation defined as follows:

$$P_{\mathbb{F}_i}(C) = \begin{cases} -\infty, & \text{if } a \in C \text{ and } belief^a_{\mathbb{F}_i}(T) < belief^a_{\mathbb{F}_i}(M), \\ 0, & \text{if } C \in h_{\mathbb{F}_i}(t), \\ U_{\mathbb{F}_i}(C), & \text{otherwise} \end{cases} \quad (4.4)$$

where $h_{\mathbb{F}_i}$ (t) represents the history set of service $\mathbb{F}_i$ at time t, consisting of all the coalitions it has joined and left at any time $t' < t$.

4.4.5.2 Hedonic Coalition Formation Algorithm

The system relies on a distributed hedonic coalition formation algorithm that enables services to make decisions about which coalitions to join so that the number of malicious services is minimized in the final coalition structure. The algorithm takes as input an initial partition of services at a certain time t and outputs the final coalition partition. The algorithm converges to a final coalition partition consisting of a number of disjoint coalitions which is Nash-stable as well as individually stable.

4.5 Role of Trust in Forming Cloud Federation

In above sections, following are discussed: (1) the importance of formation of cloud federation by different cloud service providers (2) different technique based on game theory for cloud federation formation (3) role of broker in cloud federation formation. But above of all these topics discussion about importance of trust or soft security in cloud federation formation is important. Formation of cloud federation by cloud service providers has various advantages, but though has various challenges to overcome like allocation of optimum resource, interoperablity service issue, migration of resource, establishing trust between cloud service providers in federation. Absence of trust between cloud service providers in federation, establish the major problems for cloud service providers in adoption of federation. To guarantee committed QoS and security of critical data of cloud users, it will be necessary to assess and build trust between multitudes of cloud providers. Again in federation, image of instances or partial data objects are migrated from one cloud service provider to other and hence there is matter of concern for privacy and security of cloud user's. Thus in order to guarantee, privacy and security of data of cloud user on new cloud provider's platform, it is necessary to identify trusted service provider. Thus evaluating trust of each participating cloud service providers in federation is identified as essential condition to participate in Cloud federation according to Kanwal et al. [17]. Trust is an important issue for cloud federation. The issues which need to be taken care off are as follows:

1. Defining specification of security policy for cloud federation.
2. Defining techniques to manage and implement security policies for cloud federation.
3. Some security measure must be present to guarantee the security of the information. The considered security measures are (a) Encryption of data integrity, (b) access controls, (c) confidentiality of the data and (d) availability of data.
4. Determining reliable cloud service supplied by different providers.

The trustworthiness of cloud service providers can be determined based on their reputation and reliability. The reputation can be defined based on the opinions of cloud users, cloud broker and other cloud service providers, whereas reliability of particular cloud service providers is measured through observation obtained by the cloud user while using the service according to Zant et al. [18]. In cloud federation every security system may depend on the trust of different cloud service providers in federation [19]. A very basic security problem faced by cloud service providers in federation is the service which is delivered to cloud user is trusted or reliable. Thus to tackle this problem there is a need to develop trust based security mechanism in cloud federation. In case of soft security one service provider in federation may not trust any other member of same federation initially but if some security mechanism is used to select cloud service providers in federation then every member of federation may trust on each other as trustworthiness of each cloud service providers, have already been verified. If suppose, there are no security mechanism available for selecting service providers then no service provider will have belief on service delivered by other cloud service providers in federation, because there is no trust between them. Therefore if some security mechanism is provided then these service providers will surely trust each other in federation, thus increasing overall Quality of delivered services. The trust of cloud service provider can be obtain based on defined set of rules in security system.

Though considerable amount of work over the last few years have been done on trust evaluation and establishment in cloud computing, the issue of trust evaluation in cloud federation is important challenge to overcome. Some of the works done on this respect are discussed below:

Mashayekhy et al. [20] in their work, have proposed data protection framework for cloud federation. Their framework considers the data privacy and security restrictions as well as minimizes the cost of outsource computing resources to the service providers which are member of federation. They have proposed two conflict graph for restricting privacy of data (1) one graph denotes the conflicts among virtual machine and (2) second graph represents the conflicts among virtual machine. An integer program is used to formulate the problem of data protection in cloud federation. An extensive experimental analysis is performed and obtained result is compared with the optimal result.

Kanwal et al. in [17] have also proposed the security model to guarantee the security of critical data of cloud users. Thus proposed protocol will help each cloud service providers to determine the trustworthiness of each cloud providers before taking part in federation to share their resources in reliable and trusted way. The main objective the author was to establish two-way trust between home and foreign cloud service providers. They have formulated trust value based on service level agreement of provider and feedback obtained from registered cloud users. Based on this two parameters final value of trust is formulated and this trust score is considered as the overall level of trustworthiness of each cloud service providers. On getting trust score, the credentials of trust are interchanged between home and foreign cloud service providers. The credentials are (a) aggregated trust value, (b) service level agreement of service providers and (c) their Level of Trust.

In [21] Messina et al. dealt with problem of measurement of trust of cloud service providers in cloud federation environment. They have proposed their model as an extension of RRAF model [22]. The RRAF model is used widely with the advancement of multi-dimensional approach. This new approach of RRAF model solve many reliability issue, which originally exists in cloud computing. Their trust model identify trusted service provider based on (a) measuring of reputation, is obtained based on recommendation made by other cloud service provider and user and (b) evaluation of services delivered in the past. This technique helps cloud users to make a fine grained evaluation of cloud services, which are delivered in the past by service providers.

4.6 Conclusion

With the advent of cloud computing technology, it has become utmost importance to extract benefits out of idle or underutilized resources maintained by different cloud service providers to yield more revenue. This requirement has led to the birth of the concept of cloud federation. This chapter deals with how the concept of cooperative game theory can be applied in forming the cloud federation. It also discusses different framework of cloud federation formation based on coalition game theory where different cloud service providers form a cooperative group based on some preference rule and examines the benefit of a cloud service broker in cloud federation. Establishing trust between cloud service providers is necessary, so this chapter presents some brief theoretical idea about the importance of trust in cloud federation and discusses some existing problem and its solution.

References

1. B.K. Ray, S. Khatua, S. Roy, Negotiation based service brokering using game theory, Applications and Innovations in Mobile Computing (AIMoC) (Kolkata, 2014), pp. 1–8
2. L.J. Neumann, O. Morgenstern, *Theory of Games and Economic Behavior* (Princeton University Press, Princeton, 1947)
3. R.M. Thrall, W.F. Lucas, Nperson games in partition function form. Naval Research Logistics Quarterly **10**(1), 281–298 (1963)
4. R.B. Myerson, Graphs and cooperation in games. Math. Oper. Res. **2**(3), 225–229 (1977)
5. R.J. Aumann, B. Peleg, Von Neumann-Morgenstern solutions to cooperative games without side payments. Bull. Am. Math. Soc. **66**(3), 173–179 (1960)
6. J. Drechsel, Selected Topics in Cooperative Game Theory, *Cooperative Lot Sizin Games in Supply Chains* (Springer, Berlin, 2010), pp. 5–39
7. R.P. Gilles, *The Cooperative Game Theory of Networks and Hierarchies*, vol. 44 (Springer Science and Business Media, Berlin, 2010)
8. W. Saad, Z. Han, M. Debbah, A. Hjorungnes, T. Basar, Coalitional game theory for communication networks. Signal Process. Mag. IEEE **26**(5), 77–97 (2009)
9. D. Schmeidler, The nucleolus of a characteristic function game. SIAM J. Appl. Math. **17**(6), 1163–1170 (1969)
10. K.R. Apt, A. Witzel, A generic approach to coalition formation. Int. Game Theory Rev. **11**(03), 347–367 (2009)
11. T. Arnold, U. Schwalbe, Dynamic coalition formation and the core. J. Econ. Behav. Organ. **49**(3), 363–380 (2002)
12. T. Sandholm, K. Larson, M. Andersson, O. Shehory, F. Tohm, Coalition structure generation with worst case guarantees. Artif. Intell. **111**(1), 209–238 (1999)
13. D. Ray, *A Game-Theoretic Perspective on Coalition Formation* (Oxford University Press, Oxford, 2007)
14. B.K. Ray, S. Khatua, S. Roy, Cloud Federation Formation Using Coalitional Game Theory, *Distributed Computing and Internet Technology* (Springer International Publishing, Berlin, 2015), pp. 345–350
15. L. Mashayekhy, D. Grosu, A coalitional game-based mechanism for forming cloud federations, in *Proceedings of the 2012 IEEE/ACM Fifth International Conference on Utility and Cloud Computing* (IEEE Computer Society, 2012), pp. 223–227
16. O.A. Wahab, J. Bentahar, H. Otrok, A. Mourad, Towards trustworthy multi-cloud services communities: a trust based hedonic coalitional game, IEEE Transactions on Services Computing (2016)
17. A. Kanwal, R. Masood, M.A. Shibli, Evaluation and establishment of trust in cloud federation, in *Proceedings of the 8th International Conference on Ubiquitous Information Management and Communication* (ACM, 2014), p. 12
18. B.E. Zant, N.E. Zant, N.E. Kadhi, M. Gagnaire, Security of cloud federation, in *2013 International Conference on Cloud Computing and Big Data (CloudCom-Asia)* (IEEE, 2013), pp. 335–339
19. J. Ma, M.A. Orgun, Trust management and trust theory revision. IEEE Trans. Syst. Man Cybern. Part A: Syst. Hum. **36**(3), 451–460 (2006)
20. L. Mashayekhy, M.M. Nejad, D. Grosu, A framework for data protection in cloud federations. in *2014 43rd International Conference on Parallel Processing (ICPP)*, (IEEE, 2014), pp. 283–290
21. F. Messina, G. Pappalardo, D. Rosaci, C. Santoro, G.M. Sarn, A trust model for competitive cloud federations. Complex, Intelligent, and Software Intensive Systems (CISIS) (2014), pp. 469–474
22. S. Garruzzo, D. Rosaci, The roles of reliability and reputation in competitive multi agent systems. in *On the Move to Meaningful Internet Systems: OTM 2010* (Springer, Berlin, Heidelberg, 2010), pp. 326–339

Chapter 5
Resource Provisioning Strategy for Scientific Workflows in Cloud Computing Environment

Rajni Aron

Abstract Cloud computing has emerged as a computing paradigm to solve large-scale problems. The main intent of Cloud computing is to provide inexpensive computing resources on a pay-as-you-go basis, which is promptly gaining momentum as a substitute for traditional information technology (IT)-based organizations. Therefore, the increased utilization of Clouds makes successful execution of scientific applications a vital research area. As more and more users have started to store and process their real-time data in Cloud environments, resource provisioning and scheduling of huge Data processing jobs becomes a key element of consideration for efficient execution of scientific applications. The base of any real-time system is a resource, and to manage the resources to handle workflow applications in Cloud computing environment is a very tedious task. An inefficient resource management system can have a direct negative effect on performance and cost and indirect effect on functionality of the system. Indeed, some functions provided by the system may become too expensive or may be avoided due to poor performance. Thus, Cloud computing faces the challenge of resource management, especially with respect to choosing resource provisioning strategies and suitable algorithms for particular applications. The major components of resource management systems are resource provisioning and scheduling. If any system is able to fulfill the requirements of these two components, the execution of scientific workflow applications will become much easier. This chapter discusses the fundamental concepts supporting Cloud computing and resource management system terms and the relationship between them. It reflects the essential perceptions behind the Cloud resource provisioning strategies. The chapter also identifies requirements based on user's applications associated with handling real-time data. A model for resource provisioning based on user's requirements to maximize efficiency and analysis of scientific workflows is also proposed. QoS parameter (s) based resource provisioning strategy has been proposed for workflow applications in cloud computing environment. Validation of resource provisioning strategies is presented in this book chapter.

R. Aron (✉)
Lovely Professional University, Phagwara, India
e-mail: rajni3387@gmail.com

B. S. P. Mishra et al. (eds.), *Cloud Computing for Optimization: Foundations, Applications, and Challenges*, Studies in Big Data 39,
https://doi.org/10.1007/978-3-319-73676-1_5

5.1 Introduction

The concept of Cloud computing gives us an illusion of availability of infinite computing resources on demand. The emerging trends in this field have up to much extent eliminated advance commitments related to provisioning by cloud users [30]. The smaller companies in this field now can start small and increase h/w resources as and when the need arises. It provides the users ability to pay-per-use of computing resources even on a short-term basis. Any application here comes with three requirements: computation, storage and network. Resources here are virtualized in order to give the illusion of infinite capacity as well as to hide the implementation of how they are multiplexed and shared [15].

The basis of distinguishing between various utility computing offerings basically lies on the level of abstraction presented to the programmer and the level of management of resources. Whenever a demand for a service varies with time or the demand is unknown in advance, utility computing is preferred to run a private cloud. There are mainly five critical obstacles to the growth of cloud computing such as

1. Bottlenecks in Data Transfer,
2. Performance Unpredictability
3. Scalable Storage,
4. Occurrence of Bugs in Large Distribution Systems and
5. Quick Scaling On Demand.

If we concentrate on the second obstacle i.e. Performance Unpredictability, it can be resolved through:

i. Improved VM support
ii. Provisioning of VMs

Taking view of the above observations and analyzing the current public utility computing like Amazon Web Service, Google App Engine and Microsoft Azure, resource provisioning and scheduling has become important research area in a cloud computing environment [29].

Workflow and bag of tasks are the two main categories of applications usually considered for execution in a cloud environment. Bag of tasks consists of a number of independent tasks whereas workflow is a combination of several mutually dependent tasks. Bag of tasks having negligible inter-task dependencies are almost expected to get a predictable performance under some suitably chosen provisioning policy. The picture is not exactly same in case of workflows driven mostly by inter-task dependency. Vast use of Clouds by scientific workflows has come up with many challenging issues related to execution performance. Execution performance of scientific workflows get influenced by task dependency which might lead to massive inter-task communication involving heavy data traffic.

The traditional way to evaluate the performance of a workflow system is to record and compare the execution time. Our work which focuses mainly on workflows and being well aware with inter task dependency has taken count of the total data movement too. In a cloud computing environment having a limited bandwidth guided by the Internet, reduction in the total data movement leads to the reduced execution time too.

In [17], the cost is only computed according to the BTUs. However, data and communications represent other costs, such as storage and communication, that should be taken into account. In a cloud computing environment providing a limited bandwidth based on the Internet, reduction in the total data movement leads to the reduced execution time correspondingly [29]. Furthermore, it leads to the reduced cost of data transfer. We focused on resource capacity in terms of bandwidth for execution of scientific applications. We proposed a resource provisioning model and provisioning strategy considering data transfer and data storage impact on VM provisioning strategies. The main contribution of the work are thus:

- To show need resource provisioning strategies to handle large volume data jobs.
- To proposed model for resource provisioning in Cloud Computing Environment.
- QoS parameter based resource provisioning strategy for scientific workflows which have more data dependencies.

The motivation of our work stems from the challenges in managing and an efficient utilization of the resources in cloud computing environment. In real life situations, there are many constraints including (i) satisfying user's requirements in terms of QoS parameters (ii) minimizing consumption of resources for workflow application by choosing appropriate strategy and (iii) minimizing the time for execution of an application by meeting the deadline requirements which will be submitted by the cloud user. The main intent of this work is to provision the resources along with achieving the practical constraints mentioned above. This paper presents resource provisioning model and resource provisioning policy. By implementing the model itself, actual provisioning and scheduling of resources can be well anticipated.

The rest of paper is organized as follows. Relationship between cloud computing and resource management systems along with resource provisioning's requirements is presented in Sect. 5.2. Section 5.3 discusses related work in the area of resource provisioning policies and communication effect on data intensive workflows. Cloud resource provisioning model is proposed in Sect. 5.4. Section 5.5 explains our assumptions and the problem we address in this paper. In Sect. 5.6, we proposed PCH based hyper-heuristic resource provisioning strategies. We present the existing different strategies, which are evaluated in Sect. 5.7. The experimental set up is described in Sect. 5.8 and the results are further discussed in same section. Section 5.9 concludes with our paper and future work plans.

5.2 Relationship Between Cloud Computing and Resource Management Systems

Cloud computing is an elaborate and complex distributed computing paradigm providing an easy to use, on-demand network access to a shared pool of configurable computing resources involving networks, servers, storage, applications and services that can be dynamically provisioned and released with least management effort as well as service providers interaction. This cloud model can be visualized as an amalgamation of five essential characteristics and three service models. Cloud Computing services offerings can broadly be put into three major categories: Infrastructure-as-a-Service (IaaS), Platform-as-a-Service (PaaS) and Software-as-a-Service (SaaS). Cloud Computing is primarily concerned about sharing of the resources. Any application in this paradigm comes with three requirements: computation, storage and network. Resources here are virtualized giving us an illusion of infinite capacity as well as hiding the implementation of their multiplexing and sharing [3]. In Cloud computing, huge volumes of data is distributed over dynamic and geographically remote sites probably all over the world. In case of an application running on a site not holding the required data, reliable and efficient data management facilities are required to take care of the required movements across various machine [8]. The term resource management system is used for managing large number of resources having resource provisioning policies and scheduling procedure. Cloud resource management systems as a research area has captivated the attention of many research organizations, academia and industries world-wide. Efficient resource management is a key component to run scientific applications and to attain high performance in Distributed environments such as grid and clouds. So, resource provisioning and scheduling has become an important issue in cloud resource management. To handle cloud resource management systems efficiently, it is required to design provisioning strategy in an efficient and effective way. Most operations are still largely dependent on experience based cloud user decisions. With the explosive growth of compute intensive applications, it is essential that real time information which is of use to the business, industry and scientific community can be extracted to deliver better insights by using appropriate resource provisioning strategy. In the cloud, a large number of resources are distributed over dynamic and geographically remote sites probably all over the world [3]. The main aim of Cloud computing is to provide facility of on demand resource provisioning for compute and data intensive applications. In case of an application running on a site not holding the required data, reliable and efficient resource management facilities are required to take care of the required movement across various machines.

5.2.1 Resource Provisioning for Scientific Workflows

5.2.1.1 Definition

Deployment of a cloud involves an efficient and effective management of geographically distributed, heterogeneous and dynamically available resources [3]. Cloud platforms host various independent applications on a shared pool of resources with the ability to provide computing capabilities to them as per their demand. This is possible due to the concept of virtualization which creates an emulation of hardware or software environment that appears to a user as a complete instance of that environment. Virtualization actually serves the purpose of supporting the dynamic sharing of data center infrastructure between cloud hosted applications [20]. In the context of Cloud computing, resource management can be explained as the process of allocating computing, storage and networking resources to a set of applications in a manner that intends to fulfil the performance objectives of the applications, the data center providers and the cloud resource users. It henceforth deals with all the issues and procedures involved in locating a required capability, arranging of its use, monitoring the use as well as ensuring the efficiency of use under the assigned billing model. Cloud Resource Management needs to meet QoS (Quality of Service) agreement within the constraints of SLAs (Service Level Agreements) without expecting any cooperation from the resources being managed as none of the resource is typically dedicated to a specific virtual organization. Cloud providers' main aim lies in fulfilling consumers QoS requirements while minimizing their operational costs through a well-planned and strategic resource management policy.

5.2.1.2 Requirement of Resource Provisioning Strategies

1. Efficient: Effective Resource Provisioning should work towards minimizing the overheads of scientific workflows in clouds. There should be an efficient management of cloud resources to handle scientific workflows jobs. Scheduling too is required to be very efficient and based on fair policies.
2. Efficient Resource Usage: Wastage of the valued resources should be reduced through employing wise provisioning and scheduling strategies. Scientific workflows applications waiting for some events should be made to release the resources not being used currently and be granted dynamically to the other jobs needing them.
3. Fair Allocation: There should be fair provisioning of resources and the amount of resources allocated to individual users should be independent of the number of scientific workflows jobs executed by the user.
4. Adaptability and Scalability: The scheduler should be adaptive and should totally be in tune with the resources that can be allocated to and released by the jobs thus ensuring an efficient scientific workflows execution anyhow.

5. Predictable Performance: Every application should be ensured about a certain level of guaranteed performance in terms of cost, time or several other conceivable parameters that exist within the cloud environment.

5.2.1.3 Types of Resource Provisioning

In this section, types of Resource Provisioning in cloud resource management systems have been discussed.

- Virtual Machine (VM) Provisioning: VM Provisioning is allocation of virtual machines (VM) to particular task for execution of an application. In Virtual Machine (VM) Provisioning, provider will configure Virtual Machine (VM) as per the requirement of application. Authors have proposed Virtual Machine (VM) strategies for compute intensive applications [16].
- Static Provisioning: In Static Resource Provisioning, all resources will be allocated in advance before starting the execution of application [11].
- Dynamic Provisioning: In Dynamic Resource Provisioning, provider will allocate the resources and then user releases the resources immediately after the execution of application [34].

5.3 Related Work

In this section, we have highlighted some points of the existing research papers related to resource provisioning in cloud computing. A large amount of work has been reported in concern with various provisioning strategies and their performance for IaaS Clouds. We can put these work broadly in two categories.

5.3.1 Provisioning Policies

Jiyuan Shi et al. [31] designed an elastic resource provisioning and task scheduling mechanism to perform scientific workflow jobs in cloud. The intent of their work is to handle as many high-priority workflow jobs as possible under budget and deadline constraints. Villegas et al. designed SkyMark, an evaluation framework that enables the generation, submission and monitoring of complex workloads to IaaS clouds [34]. The provisioning and allocation policies for IaaS clouds have been presented in this work. They have calculated cost and performance of provisioning policies. The concept of data storage and data transfer have not been considered in this work.

Maciej Malawski et al. have considered resource provisioning for scientific workflow ensembles on IaaS clouds [26]. They have considered both static and dynamic resource provisioning approach to handle scientific workflow. The proposed algorithms assumed that all workflow data is stored in Cloud storage systems such as Amazon S3 and the intermediate node transfer times are included in task run times. They have not considered data storage and transfer cost.

In [10], a provisioning technique that automatically adapts to workload changes according to applications has been presented for facilitating the adaptive management of system. They have used queuing network system model and workload information to supply intelligent input about system requirements to an application provisioner. The author have given emphasize on workload pattern and resource demand instead data transfer between workflow tasks and resource utilization.

Lin, Cui et al. presented SHEFT workflow scheduling algorithm to schedule a workflow elastically on a Cloud computing environment [24]. They have used a task prioritizing, resource Selection method for scheduling of applications. In cloud computing environment, the number of resources can not be automatically determined on the demand of the size of a workflow and resources assigned to a workflow usually are not released until the workflow completes an execution. It leads a long execution duration and many resources might become idle for a long time during workflow execution which leads to a waste of resources and budgets. Data transfer time and cost between workflow has not been considered.

Wu. Zhangjun et al. [35] presented a market oriented hierarchical scheduling strategy for cloud workflow systems. They have used transaction intensive workflow for scheduling. They have considered the problem of workflow systems in terms of QoS constraint based scheduling and general service oriented scheduling instead of application specific scheduling. They didn't considered the communication time between workflow.

Srinivasan et al. [32] presented one approach for VM provisioning which does help to cleanup tasks added to workflow. There is one limitation of the proposed approach is disk usage. It would not be scalable for large scale workflow applications as each VM will run hundred jobs and in result the machine will run out of disk space.

R.N Calheiros et al. [9] have enhanced IC-PCP(EIPR) algorithm for workflow application execution by meeting deadline and budget constraints. They have taken into consideration the behavior of Cloud resources during the scheduling process and also applies replication of tasks to increase the chance of meeting application deadlines. Data storage has not been considered to give optimum result to user.

Zhou, A. et al. [36] developed a scheduling system called Dyna by using A*-based instance configuration method for performance dynamics. They have considered the problem of minimizing the cost of WaaS (Workflow as a Service) providers via satisfying individual workflows. They have not designed any algorithm to run scientific application. Only one framework has been developed.

Abinaya and Harris [2] proposed an approach for provisioning of resources while considering QoS parameters. They focused to keep balance load on machine and increase resource utilization. They didn't consider workflow application with data storage and data transfer parameters.

5.3.2 The Role of Communication and Data Transfer in the Execution of Scientific Workflows

The works in [21, 23] have emphasized on the storage system that can be used to communicate data within a scientific workflow running on cloud. Juve et al. have emphasized on the storage system that can be used to communicate data within a scientific workflow running on cloud. They have considered three different applications named as Montage (Astronomy application), Broadband (Seismology Application) and Epigenome(bioinformatic application) to run different storage systems such as S3, NFS, PVFS2 and GlusterFS. In this work, only a shared storage system with multiple nodes was used to communicate data between workflow tasks. In result, S3 storage system is useful for memory intensive application; GlusterFS is useful for other I/O intensive (Montage) and CPU-intensive (Epigenome) application. The runtime of the applications tested improved when resources are added. The run-times reported in experiment do not include the time required to boot and configure. They have considered the size of input output files are constant. Resources are all provisioned at the same time and file transfer, provisioning overhead are assumed to be independent of the storage system [23]. In [21], the time required to provision and configure the VMs has not been considered. In [17], the cost is only computed according to the BTUs. However, data and communications represent other costs, such as storage and communication, that should be taken into account.

Juve et al. have studied the performance of three different workflows named as Montage, Epigenome and Broadband on Amazon EC2 [22]. In this work, they have used a single node to run a workflow. In this paper, some points have been highlighted such as the provisioning resources on EC2, the trade-off between cost and performance, the trade-off between storage cost and transfer cost. The comparison of cost of running scientific workflow applications in the cloud using Amazon's EC2 and NCSA's Abe cluster has also been done. In result, cost of data transfer is higher in comparison to acquiring resources to execute workflow tasks and storage costs. They have not considered communication between workflow tasks in a Cloud.

Bittencourt et al. [4] focussed on workflow scheduling problem and used Path Clustering Heuristic (PCH) [5] to schedule workflow application. They have not considered available bandwidth for provisioning of the resources. Although many resource provisioning strategies and scheduling algorithms for clouds have been proposed for BOTs and workflows [16, 27, 34] even analyzing the impact of VM provisioning, none have addressed the importance of the provisioning strategies considering network and storage issues for the case of workflows. For BoTs tests have shown a dependency between the two which impacts both the makespan gain and the paid cost. In [11], stochastic programming based resource provisioning algorithm has been proposed for cloud. They have considered the over provisioning and under provisioning of resources and tried to minimize the cost. They have not considered the bandwidth constraints at the time of provisioning.

Line et al. [25] have studied the effect of bandwidth at the time of task scheduling. They have designed an algorithm for scheduling of task on the VM. The major drawback is that they have studied the effect of bandwidth only for independent tasks not for workflows. We have considered the workflow application to study the effect of bandwidth at the time of resource provisioning.

5.3.3 Our Contribution

Our contribution in this paper is twofold. Firstly, we have defined a Cloud resource provisioning model for cloud resource management systems in which a Resource Provisioning (RP) manager has been incorporated. The main aim of the RP manager is to determine some strategies leading to an efficient, comprehensive and performance driven VM provisioning strategy for a Cloud environment and finding the relative benefits of increasing the network bandwidth in case of data dependency among tasks in the execution of scientific workflow and then we get the provisioned set of resources from the total available resources in the cloud. Secondly, the final decisions about mapping and execution of the job to the corresponding resource is done by using the proposed PCH resource provisioning algorithm. It should be noted that we have considered the cost (BTU) and makespan of the application's execution as the user's and resource provider's functions respectively which are components of any real-life based problems.

The main contribution of this paper is the development of a resource provisioning model that enables the execution of an application according to the users requirements. To achieve this, firstly, we generate the workload on the basis of actual experiment on schlouder. Techniques used in the traditional distributed computing systems have not been very successful in producing efficient and effective results to manage the resources. Our proposed implementation of hyper-heuristic based resource provisioning algorithm minimizes the cost and makespan simultaneously.

5.4 Resource Provisioning Model

We have proposed a resource provisioning model where resource providers give the facility of resource provisioning to the user for optimum results and better services and avoid violations of the service level guarantees. The main intent of this model is to reduce cost and time of workflow applications by considering communication time and data storage space too. The implementation of this model enables the user to analyze customer requirements and define processes that contribute to the achievement of a product or service that is acceptable to their resource consumers.

Figure 5.1 shows the resource provisioning in the model. The model executes the requests as follows:

- In resource provisioning model, first of all user tries to access the resources through cloud portal. User will submit task's requirements for execution of application.

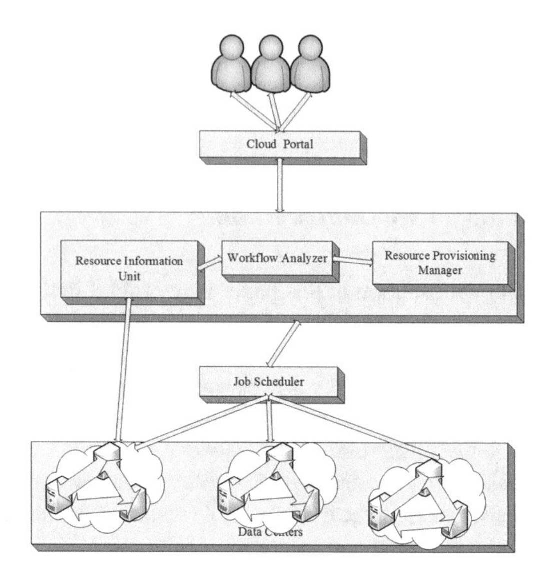

Fig. 5.1 Cloud resource provisioning model

- Once workflow application is submitted, it will be given to workflow analyzer to get to know specification of workflow. Workflow analyzer takes the information about resources from resource information unit and consult with resource provisioning manager.
- Workflow analyzer then collects the information of available resources from the Resource Information Unit (RIU). RIU contains the whole information about the resource providers resources. Resource information unit will keep track the status of resources and monitoring of resources.
- Resource Provisioning Manager (RPM) takes the information from workflow analyzer. After studying the various parameters which the user has demanded, manager checks for provisioning strategy. The selection of resources is made on the basis of workflow application's requirements.
- RPM provisions the requested resources to the user for the execution of application in the cloud computing environment only if the requested resources are capable to the complete the application's execution process on time as per user's requirements.
- If the requested resources are not available in the current time according to the resource provisioning to satisfy user's requirements then the resource provisioning unit asks user to resubmit specifications for application's execution.
- After provisioning of the resources, job is submitted through the job scheduler. After completion of application's execution; the result is then sent back to the resource provisioning manager. Finally, the user collects the result. Thus, this framework exhibits how resource provisioning can be done in the cloud environment. A resource provisioning algorithm has been identified as a part of this model.

5.5 Problem Statement

The complexity of resource provisioning model is to find ingredient resources for corresponding application. Any distributed computing environment including either Grid or Cloud needs to satisfy scalability, management and performance demands. Scientific applications require huge amount of data for processing. The cost and time of data transfer and data storage in clouds play an important role to run any scientific application. Scientific workflows run on the cloud infrastructure, which is operated by virtual machines in cloud providers' data centers and each VM has a limited resource capacity. Resource capacity constraint plays major role in the VM provisioning on the task. So, there is a need to exploit resource capacity in terms of *Bandwidth* to improve the performance of the application. In this model, provisioning problem has been considered from both client and resource provider perspectives. We have considered cost and makespan criteria to check the performance of proposed model.

To consider this problem, a scientific workflow application is represented as a directed acyclic graph (DAG) in which graph nodes represent tasks and graph edges represent data dependencies among the tasks. In a workflow, an entry task does not have any parent task and an exit task does not have any child task. In addition, a child task cannot be executed until all of its parent tasks have completed.

Thus, even though the problem can be solved by using exhaustive search methods, the complexity of generating the schedule becomes very high. The overall finish/completion time of an application is usually called the schedule length or makespan. So, the objective of provisioning techniques is to minimize the makespan and cost of scientific application by proper allocation of the resources to the tasks considering resource capacity constraint. Let us assume workflow W (T, E) consisting of a set of tasks. To consider this problem, we have taken a set of tasks $T = \{t_1, t_2, t_3, ...t_m\}$ and a set of dependencies among the tasks, $E = \{< t_1, t_2 >, < t_3, t_4 > .... < t_x, t_y >, t_m\}$ where t_x is the parent task of t_y. $V = \{v_1, v_2, v_3..........v_n\}$ is the set of active virtual machines, connected through network links. Each resource $v_i \in V$ has a set of links $L = \{l_{iv_1}.....l_{iv_k}\}$, $1 \leq k \leq n$, where $l_{ij} \in R^+$ is the available bandwidth in the link between resource v_i and v_j. The external bandwidth is usually lower than the internal bandwidth, which is common configuration in real environments. For problem formulation; few assumptions have been considered as follows:

1. First, we assume that task have unique id with known durations.
2. Users submit their application through a local resource management system, which requires users to specify a maximum runtime.
3. Second, tasks are not preemptable, i,e the migration of a running tasks is not possible and jobs can not be suspended to run another job.

The problem is the provisioning of VM to tasks so that the makespan $T_{makespan}$ is minimized. The total execution time can be calculated as:

$$minimize T_{makespan} = max_{EFT(t)} \tag{5.1}$$

$minimize(T_{makespan})$ subject to, Total Cost $\leq$ Budget. Where B is the user given constraint on cost. Estimated Finish Time $EFT(t_i, v_n)$ represent the estimated finish time of task t_i on the resource v_n.

$$EFT(t_i, v_n) = EST(t_i, v_n) + w_i n \tag{5.2}$$

$$EST(t_i, v_n) = \begin{cases} Time(v_n), & if\ i = 0 \\ \max(Time(v_n), ST_i) & otherwise \end{cases} \tag{5.3}$$

where $ST_i = \max_{\forall t\_m \in pred(t\_i)}(EST(t\_m, r\_n) + w\_mn + c\_mi)$ $EST(t_i, v_n)$ represent the earliest start time possible for the task i in the resource n at a given schedule. $Time(v_n)$ is the time when the resource n is available to execute the task i.

Let us consider t_i be the task size and p is the processing capacity of resource.

$$e_{iv} = \frac{t_i}{p} \tag{5.4}$$

$$d_{ij} = \frac{data_{ij}}{l_{r,s}} \tag{5.5}$$

e_{iv} represents the computation time of the task t_i on the resource v.
d_{ij} represents the communication time between task t_i and t_j using the link l between resource r and s. $sc(i)$ is the set of immediate successor of task t_i. $pred(i)$ is the set of immediate predecessors of task t_i. e_{iv} and d_{ij} are used to calculate the priorities to the tasks.

$$Pr_i = \begin{cases} e_{iv}, & if\ sc(i) = 0 \\ e_{iv} + \max_{\forall j \in sc(i)}(d_{ij} + Pr_j), & otherwise \end{cases} \tag{5.6}$$

5.6 PCH Based Hyper-heuristic Resource Provisioning Strategy

A hyper-heuristic employs at a higher-level of abstraction to stipulate a low-level heuristic. Low-level heuristic could be applied at any given time, depending upon the characteristics of the region of solution space currently under exploration [7]. The work presented in this paper is based on the four low-level heuristics as discussed in [6, 12]. The procedure of hyper-heuristic is split into two parts: heuristic selection and heuristic acceptance. Heuristics selection method is very simple to select the low-level heuristics. Heuristics acceptance can be further segregate into deterministic and non-deterministic. In the proposed algorithm, we have used tabu search [18] as heuristic selection method and Path Clustering Heuristic [5] as heuristic acceptance.

Path Clustering Heuristic algorithm is used to design the initial plan for VM Provisioning. For each task, a priority based on execution time and transfer times is computed. Tasks are the ordered descending by their priority rank. Tabu search is used for low-level heuristic selection with feasible solution for provisioning of resources in cloud environment. For workflow application's execution, best heuristic is found and then the procedure is repeated until all tasks have been scheduled. In this section, we have discussed solution which will be used to solve the problem.

5.6.1 The Proposed Algorithm

The proposed algorithm first initially schedule the number of jobs using path clustering heuristic. An efficient resource provisioning strategy has been designed with the task allocation using PCH as heuristic information. SCHIaaS broker works on the behalf of the users for all the details related to VM creation, destruction and number of VM allocated to that application. The main task of broker is to bind the task to particular VM. The selection of the job for rescheduling and the resource are directly influenced by the available bandwidth in the communication channels.

In this section, we present the pseudo code of PCH-based hyper-heuristic for resource provisioning in the Cloud environment. Initial schedule is a partial solution and is represented as a heuristic (e.g., select, move, swap, drop) or a sequence of heuristics. A low-level heuristic is any heuristic which is operated upon by the hyper-heuristic. Low-level heuristics can be simple or complex and are implemented as follows: (1) task selection and scheduling: the heuristics select task from the unscheduled list and schedule it in to the best available resource. (2) Try for the best combination of all tasks and resources until the best combination is found. (3) Move task t_i from its current resource/schedule. (4) Swap tasks: select the tasks randomly which can swap. (5) Remove a randomly selected task from task list already scheduled [28]. This is the only heuristic which will move the search into an unfeasible region because any task may be unscheduled. We make sure that the search can move back into its feasible region by un-scheduling job that has other valid resources so that it can move into the next iteration. The low- level heuristics are then applied so as to find an optimal solution of the problem instance. The objective of PCH is to find the best low-level heuristic that generates the best solution for resource provisioning problem. The selection process of low-level heuristic in hyper-heuristic stops after a pre-defined number of iterations. We set a fixed number of iterations to keep the computation time low. The particle rejects the new solution if it is poorer than the current solution. The pseudo code of PCH-based hyper-heuristic is given in Algorithm 1.

- Task list and a random feasible solution is initialized. Then, resource list is obtained from resource provisioning unit after provisioning of users requests.
- The task to choose best heuristic from low-level heuristics is started.

Algorithm *

KwDataA DAG G=(T,E) with task execution time and communication. A set V of V VMs with cost of executing tasks and available Budget B.
KwResultA DAG schedule S of G onto V

initialize Resource list[Number of Resources]
initialize tasklist[Number of Tasks]
Initialize a random feasible solution
tasklist= get task to schedule()
resourcelist= get available resources()

Whilethere are unscheduled tasks in the queue
Forevery resource is in resource list

get the next task from queue
Calculate the priorities of task
If$PCH_{cost} < Budget$

compute Estimated Finish Time $EFT(t_i, v_n)$ and $EST(t_i, v_n)$ of task t_i on the resource v_n
schedule the job on the resource on the basis of finishing time

select the best heuristic from non tabu list
initialize the heuristic list h[h1,h2,h3...hn]
boundary= F_current

For $m = 0 \rightarrow m = k$
i= selectheuristic(h_c)

If$F_i < F_{current}$

$Applyh_c$
$F_{current} = F_i$

addh_c$intothetabulist$
until terminating condition satisfied

Repeat the every step until terminating condition is not satisfied
PCH based Hyperheuristic Resource Provisioning Strategy

- The value of boundary is set with current feasible solution for accepting the best heuristic from non tabu list. It calls each heuristic which is not tabu.
- Path Clustering Heuristic [5] acceptance criteria has been used in this algorithm. After the selection of heuristic, its solution is compared with current solution.
- If the solution is less than the boundary value, then the heuristic is accepted for resource scheduling.

- After heuristic selection, heuristic will assign each job to resource from the queue of unscheduled jobs.
- The process of provisioning of resources to tasks is performed till there are no unscheduled task in the queue.

5.7 Existing VM Provisioning Strategies

VM provisioning deals with: allocation of VMs, reuse of existing running VMs, renting a new VM for a new task or when execution time of the task exceeds the remaining Billing Time Unit (BTU) of the available VM.

In this section, we describe VM provisioning strategies used in the experiments.

- As Soon As Possible (ASAP) minimizes the walltime, regardless of renting cost, by deploying a VM for each job that could not be serviced immediately by already provisioned VMs, or when the boot time exceeds the wait time for one VM to become available.
- As Full As Possible (AFAP) aims at utilizing the BTU as much as possible, hence minimizing the number of rented resources by delaying the execution of the submitted jobs.

Both strategies that tend to allocate VMs are better suited for tasks with larger data dependencies and file sizes. On the other hand these give larger idle times which do not provide cost efficiency.

5.8 Performance Evaluation and Discussion

SimSchlouder allows to simulate Schlouder with SCHIaaS. SCHIaaS allows to simulate IaaS with SimGrid. SimGrid implements hypervisor level functionalities (e.g. start and stop VMs) [1]. SCHIaaS implements cloud level functionalities (e.g. run and terminate instances). It also supports main VM management functions: run, terminate, suspend, resume and describe instances; description of available resources; image and instance types management; automatic VM placement on the clusters (in a round-robin way); boot and other VM life-cycle processes; cloud storage management. The SimSchlouder has been used for the evaluation due to the following reasons:

- It supports main cloud broker management functions for scientific computation.
- The type of application can be Bag-of-Tasks and workflows execution;
- Extensible provisioning and scheduling policies.
- There is no limit on the number of application jobs that can be simulated.
- Application Tasks can be heterogeneous and they can be CPU or I/O intensive.

The experimental setup for cloud resource provisioning model is considered as Amazon on demand instances with one BTU = 3, 600s. Communication cost is calculated in terms of GB. The default execution time of simulator is considered to be the one for the small instance. Transfer times are computed based on a store and forward model.

5.8.1 Performance Evaluation Criteria

The main aim of the evaluation is to check the validity of the proposed PCH based hyper-heuristic resource provisioning strategy. To do so, we have implemented resource provisioning model and the aforementioned proposed provisioning algorithms in a SimSchlouder. We have defined the performance evaluation criteria to evaluate the performance of strategy. Two matrices, namely makespan and cost is finalized for evaluating the performance of the proposed work. The former indicates the total execution time whereas the latter indicates the cost per hour resource that is consumed by the users for the execution of their applications. The makespan and cost are measured in seconds and BTU respectively.

5.8.2 Results

In this section, we present the results of our experiments for data intensive tasks. Scientific workflows are both data and computation intensive, with data up to hundreds of gigabytes exchanged between tasks. A key challenge here is to allocate tasks on specific cloud instances in order to ensure a minimum amount of data is transferred between dependent tasks. An analytical model is used for calculation of runtime by using Feitelsons results [13] Montage workflow ideally suit to this situation as approx 99% of its execution consists data transfer [33].

Montage is a workflow used in astronomical image processing. Its size varies according to the dimension of the studied sky region. In our tests we have used a version with 24 tasks. In general, a workflow application is represented as a directed acyclic graph (DAG) in which graph nodes represent tasks and graph edges represent data dependencies among the tasks on the nodes. The workflow dependencies are quite intermingled and data dependencies may exist across more than one level. This type of workflow makes it particularly interesting for locality aware data intensive scheduling. At the same time presence of a large number of parallel tasks makes it a good candidate for studying the efficiency of parallel VM provisioning policies.

MapReduce is a model first implemented in Hadoop which aims at processing large amount of data by splitting it into smaller parallel tasks. We depict a version in which there are two sequential map phases. Its original use and the large (variable) amount of parallel tasks it can hold makes it useful in the study of various parallel VM provisioning policies for data intensive tasks and in showing their benefits.

The number of tasks on each allocated VM is defined by the used provisioning and allocation policies.

CSTEM, an acronym for Coupled Structural-Thermal-Electromagnetic Analysis and Tailoring of Graded Composite Structures, is a finite element-based computer program developed for the NASA Lewis Research Center [19]. CSTEM was divided into 15 separate tasks. Its relatively sequential nature with a few parallel tasks and several final tasks makes it a good candidate for studying the limits of the parallel VM provisioning policies against the *OneVMperTask* and a particular case of *StartParExceed* in which all tasks of a workflow with a single initial task are scheduled on the same VM.

Our main aim is to make an inexpensive strategy for application's execution in Cloud computing environment. In [29], we have investigated the effect of both strategies ASAP and AFAP on increasing the size of files; we proposed a new resource provisioning model and PCH based Hyperheuristic Resource Provisioning Strategy in this paper.

Two types of experiments have been performed to check the validity of the proposed work. In the first case, we have checked the result on makespan and cost by varying file sizes of scientific workflow applications. Second case represents the result by varying number of resources for the execution of scientific workflow applications. An average of fifty runs has been used in order to guarantee statistical correctness. The simulation results have been presented using SimSchlouder discrete event simulation so as to test the performance of hyper-heuristic based resource provisioning strategy. In addition, a comparison of makespan and cost of the proposed algorithm with existing provisioning strategies such as ASAP and AFAP has been presented. In order to evaluate the performance of the proposed approach, the effects of different number of applications have been investigated.

Thus, minimization of both makespan and cost of an application may conflict with each other depending on the price of the resources. The most important characteristic applicable to real-world scenarios is that how each algorithm responds to different heterogeneity of applications and resources.

5.8.2.1 Test Case I: Effect on Makespan

In this case, the effect on makespan under three resource provisioning strategies. It can be seen from figure that as we have increased the file sizes of montage application, PCH based hyper-heuristic resource provisioning strategy gives less makespan in comparison to both strategies ASAP and AFAP. From ratio 0.25, makespan is increasing at very high rate under ASAP in comparison to AFAP. We are getting higher makespan under ASAP but there is a gradual decrease in under AFAP (cf. Fig. 5.2a). The reason of growth in the makespan is the simultaneous execution of all tasks and consumption of more power with limited bandwidth that the platform is providing. ASAP although being costly gives efficient output due to parallelism of VM employed and asynchronous data transfer contributing to parallel disk and network I/O. PCH based hyper-heuristic resource provisioning strategy gives bet-

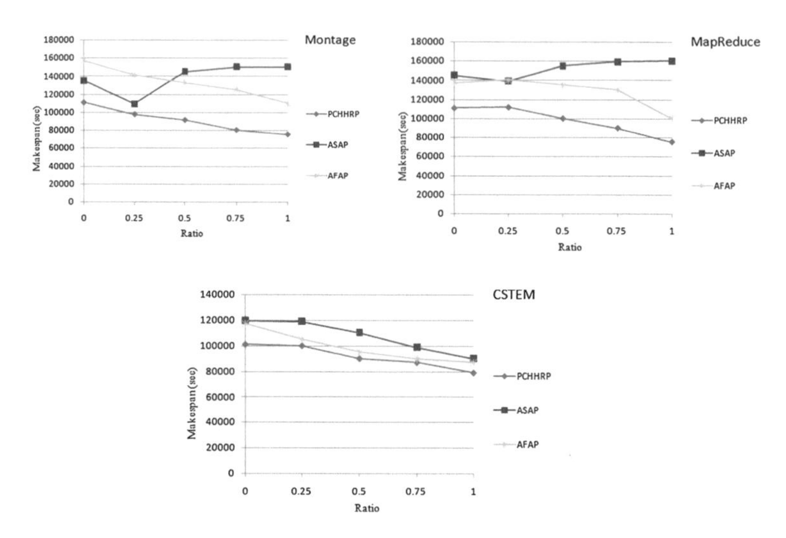

Fig. 5.2 Effect on makespan for data intensive workflows

ter result due to optimal decision as communication increases between files. The proposed strategy tries to use less number of VMs to make best utilization of the resources as per the user's requirements. This strategy tries to schedule dependent job on the same machine. At initial stage, workload analyze will analyze the requirements. The maximum communication among the dependent jobs are within single VM; we are getting less makespan.

Mapreduce is data intensive application (consisting of parallel tasks) and an increase in the file size results into the increase in makespan (cf. Fig. 5.2). Makespan for execution of mapreduce application under the proposed resource provisioning strategy is increase at initial stage when communication is increased. After ratio 0.25, makespan is decreasing. The reason is that PCH based hyper-heuristic resource provisioning startegy is meta resource provisioning strategy when communication is zero; this strategy tries to start more number of VMs to reduce waiting time but later realize communication is increasing then tries to schedule further jobs on the same already running machines. Whenever there is balance between communication and computation, the proposed strategy and AFAP starts give same results. On one hand, ASAP provisions a new VM for job, as long as it predicts that the resulting gain in performance will overcome the communication time penalty. ASAP gives result in opposite to other two strategies.

CSTEM workflow application consists of parallel and sequential tasks. For execution of CSTEM workflow, as we increase the file size, the makespan is noticeably stable under both strategies PCH based hyper-heuristic resource provisioning strategy and ASAP. Makespan is decreasing under PCHHRP and AFAP as shown in (cf. Fig. 5.2) At ratio 0.25 onwards, AFAP performs more well in comparison to PCH based hyper-heuristic resource provisioing strategy and ASAP. For CSTEM workflow, the difference between all strategies is decreasing. ASAP performs more dynamically for virtualized CSTEM workflow.

5.8.2.2 Testcase II

The cost is much lower than in comparison to AFAP and ASAP where as the performance of ASAP is more dynamic as shown in Fig. 5.3a. Figure 5.3 shows that cost per applications increases as the ration increases. The existing algorithm based application's execution resulted in a schedule which is expensive in comparison to PCH based hyper-heuristic resource provisioning strategy. For mapreduce application; the total number of BTU is stable under ASAP as we have increased the file-size. After ratio 0.3, BTU starts increasing under both strategies. A decreases of makespan under AFAP causes an increase in BTU. The main weakness of AFAP strategy is its sensitivity to runtime predictions. These predictions are used to determine the suitable time for the scheduling of the jobs to the same BTU in order to minimize the cost. It leads to more extra cost. On the other hand, AFAP postpones jobs executions in order to occupy most of BTUs created by each started VM. The cost for execution of CSTEM is stable under PCHHRP till ratio 0.25 Fig. 5.3. For CSTEM application, BTU is more as compare to AFAP. Whenever, ASAP tries to give less makespan then total number of BTU is increased under ASAP. In this case, ASAP's limits could be analyzed. The most interesting case is that of PCHHRP which produces more stable makespan and BTU throughout the different applications. It has been concluded that PCHHRPis more reliable for data intensive applications. We have identified that ASAP is more expensive strategy for data intensive application. The second observation is that makespan is higher under ASAP for every application

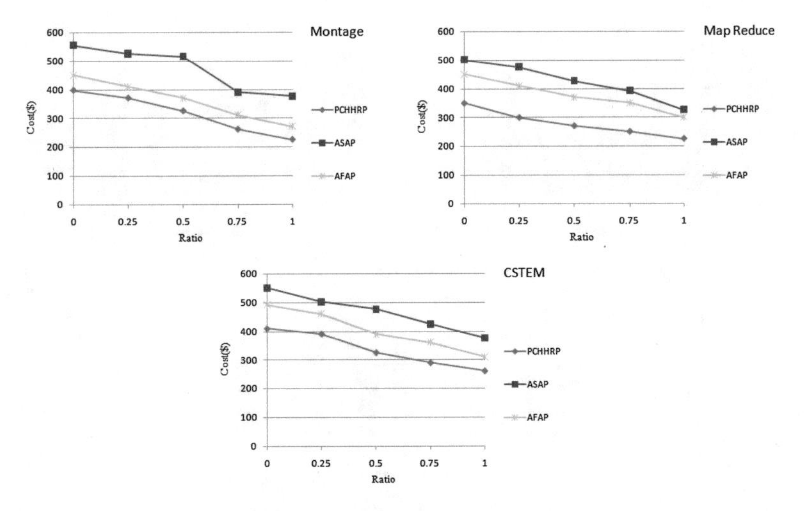

Fig. 5.3 Effect on cost for data intensive workflows

as file size increases. As we have increased the file-sizes; the makespan for both policies is increasing but makespan under AFAP is increasing at very slow rate in comparison to ASAP. ASAP's efficiency is well analyzed in the presence of parallel tasks. Result demonstrated the structure of workflow, the task execution times and file-size. Depending on what user seeks (faster executions and more savings) different combinations need to be used. The exception of this point seems to be the case of sequential application.

5.8.3 Effect of Resources

In this case, we have shown the effect of number of resources on data intensive workflows. The results depict that by increasing the number of resources, the total execution time increases thus decreasing the performance of the cloud. PCHHRP and AFAP perform better when the number of resources is less in comparison to the number of jobs. As shown in Fig. 5.4, makespan increases at high rate under ASAP in comparison to PCHHRP and AFAP. For montage scientific workflows, ASAP increases makespan at high speed but for CSTEM workflow applications, makespan is less as CSTEM workflow application is combination of parallel and sequential tasks. In this experiments, we kept the file sizes same and increased the number of resource to observe the behavior of resource provisioning strategies.

The same trend we can observe in the Fig. 5.5 the effect on cost as we kept same number of jobs. If there are more number of resources then cost increases for both data intensive workflows montage and CSTEM. For mapreduce workflow applications, cost increases at very less speed. The cost of application execution using PCHHRP is

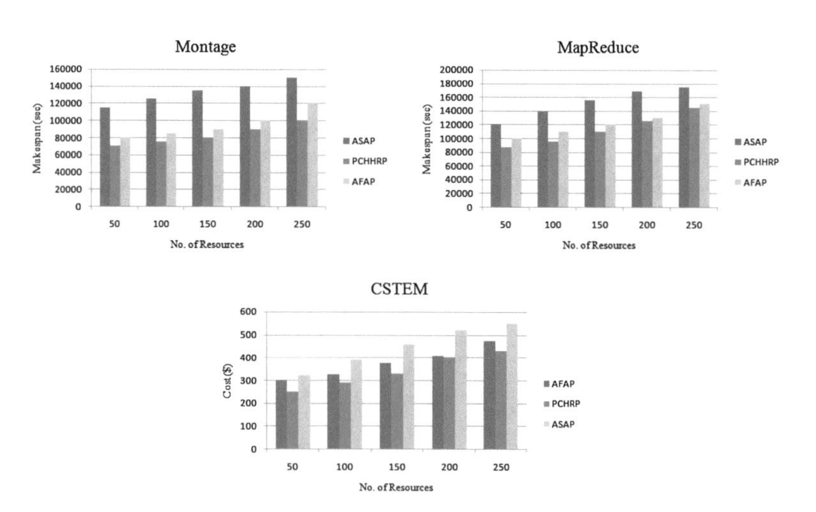

Fig. 5.4 Effect of resources on makespan for data intensive workflows

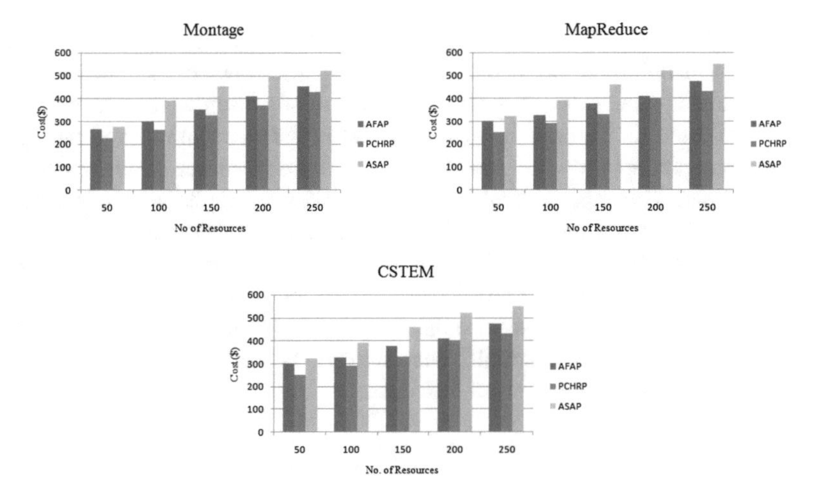

Fig. 5.5 Effect of resources on cost for data intensive workflows

much less in comparison to the execution cost using existing provisioning strategy. As the cost variations within the cloud resources are not significant so the cost benefits of only 6–9% were noticed. However, more benefits can be anticipated if the variations are higher. Thus PCHHRP outperforms all the existing provisioning strategies when the application execution cost is high, which is an important case for a large-scale cloud computing environment.

PCHHRP performs better among all three resource provisioning strategies. As full as possible is always better and more economical than the other strategy ASAP. Due to denser job packs (VM internal) coupled with less communication partners (across VMs) the communication overhead is lower. These systems are not set up with high performance interconnects and too much uncertainty is present in these allocations. This is why larger number of resources will always be worse, unless you are in a dedicated high performance network (Infiniband or similar). This observation indicates that our strategy gives an equally good performance in comparison to that given by other strategy.

Based on the observations, we concluded that a meta strategy could be expected to play a major role for execution of data intensive workflows. In cloud computing, the main goal of the providers is to minimize the makespan whereas the goal of the user is to minimize the cost for applications.

5.9 Conclusions and Future Directions

Resource provisioning model is presented for the execution of scientific workflows in cloud computing. We have considered different types of application mix of computation and data intensive application. Communication time can be consider at the selection time of resource provisioning strategy. This paper mainly focuses on the resource provisioning problems for data intensive workflows in the cloud computing environment. Result demonstrated the structure of workflow, the task execution times and filesize. Depending on what user seeks (faster executions and more savings) different combinations need to be used. QoS based resource provisioning strategy is proposed. The experimental results show that the PCH based hyper-heuristic resource provisioning strategy outperforms the existing provisioning strategies in all the cases. The proposed strategy not only minimizes cost but it also minimizes the makespan. A cloud is dynamic and diverse, which is rendered by various applications and resources. All this raises various theoretical and technical questions. In future, we'll consider data storage that is also one another important aspect for the execution of scientific application in Cloud environment.

References

1. Schiaas (2015), http://schiaas.gforge.inria.fr/simschlouder.html
2. R. Abinaya, P. Harris, A Novel Resource Provisioning Approach for Virtualized Environment (2016)
3. M. Armbrust, A. Fox, R. Griffith, A.D. Joseph, R.H. Katz, A. Konwinski, G. Lee, D.A. Patterson, A. Rabkin, I. Stoica et al., Above the clouds: a Berkeley view of cloud computing (2009)
4. L. Bittencourt, E. Madeira, HCOC: a cost optimization algorithm for workflow scheduling in hybrid clouds. J. Int. Serv. Appl. **2**, 207–227 (2011), https://dx.doi.org/10.1007/s13174-011-0032-0, https://dx.doi.org/10.1007/s13174-011-0032-0
5. L.F. Bittencourt, E.R.M. Madeira, A performance-oriented adaptive scheduler for dependent tasks on grids. Concurr. Comput. Pract. Exper. **20**(9), 1029–1049 (2008), https://dx.doi.org/10.1002/cpe.v20:9
6. E. Burke, G. Kendall, D.L. Silva, R. OBrien, E. Soubeiga, An ant algorithm hyperheuristic for the project presentation scheduling problem, in *2005 IEEE Congress on Evolutionary Computation*, vol. 3 (IEEE, 2005), pp. 2263–2270
7. E.K. Burke, M. Gendreau, G. Hyde, G. Kendall, G. Ochoa, E. O zcan, R. Qu, Hyperheuristics: a survey of the state of the art. J. Oper. Res. Soc. **64**(12), 1695–1724 (2013)
8. R. Buyya, C.S. Yeo, S. Venugopal, J. Broberg, I. Brandic, Cloud computing and emerging it platforms: vision, hype, and reality for delivering computing as the 5th utility. Future Gener. Comput. Syst. **25**(6), 599–616 (2009)
9. R.N. Calheiros, R. Buyya, Meeting deadlines of scientific workflows in public clouds with tasks replication. IEEE Trans. Parallel Distrib. Syst. **25**(7), 1787–1796 (2014)
10. R.N. Calheiros, R. Ranjan, R. Buyya, Virtual machine provisioning based on analytical performance and qos in cloud computing environments, in *2011 International Conference on Parallel Processing (ICPP)* (IEEE, 2011), pp. 295–304
11. S. Chaisiri, B.S. Lee, D. Niyato, Optimization of resource provisioning cost in cloud computing. IEEE Trans. Serv. Comput. **5**(2), 164–177 (2012)
12. P. Cowling, G. Kendall, L. Han, An investigation of a hyperheuristic genetic algorithm applied to a trainer scheduling problem, in *Proceedings of the 2002 Congress on Evolutionary Computation, 2002. CEC02*, vol. 2 (IEEE, 2002), pp. 1185–1190
13. D.G. Feitelson, *Workload Modeling for Computer Systems Performance evaluation* (Cambridge University Press, Cambridge, 2015)
14. I. Foster, Y. Zhao, I. Raicu, S. Lu, Cloud computing and grid computing 360-degree compared, in *Grid Computing Environments Workshop, 2008. GCE08* (IEEE, 2008), p. 110

15. A. Fox, R. Griffith, A. Joseph, R. Katz, A. Konwinski, G. Lee, D. Patterson, A. Rabkin, I. Stoica, Above the clouds: a Berkeley view of cloud computing. Dept. Electr. Eng. Comput. Sci. Univ. California Berkeley, Rep. UCB/EECS **28**(13), 2009 (2009)
16. M. Frincu, S. Genaud, J. Gossa, Comparing provisioning and scheduling strategies for workflows on clouds, in *Workshop Proceedings of 28th IEEE International Parallel and Distributed Processing Symposium* (IEEE, 2013), pp. 2101–2110
17. S. Genaud, J. Gossa, Cost-wait trade-offs in client-side resource provisioning with elastic clouds, in *4th IEEE International Conference on Cloud Computing (CLOUD 2011)* (IEEE, 2011)
18. F. Glover, Tabu search-part i. ORSA J. Comput. **1**(3), 190–206 (1989)
19. M.A. Iverson, F. O zguner, G.J. Follen, Parallelizing existing applications in a distributed heterogeneous environment, in *4TH Heterogeneous Computing Workshop HCW95* (Citeseer, 1995)
20. B. Jennings, R. Stadler, Resource management in clouds: survey and research challenges. J. Netw. Syst. Manag. **23**(3), 567–619 (2015)
21. G. Juve, E. Deelman, G.B. Berriman, B.P. Berman, P. Maechling, An evaluation of the cost and performance of scientific workflows on amazon ec2. J. Grid Comput. **10**(1), 521 (2012)
22. G. Juve, E. Deelman, K. Vahi, G. Mehta, B. Berriman, B.P. Berman, P. Maechling, Scientific workflow applications on amazon ec2, in *2009 5th IEEE International Conference on E-Science Workshops* (IEEE, 2009), pp. 59–66
23. G. Juve, E. Deelman, K. Vahi, G. Mehta, B. Berriman, B.P. Berman, P. Maechling, Data sharing options for scientific workflows on amazon ec2, in *Proceedings of the 2010 ACM/IEEE International Conference for High Performance Computing, Networking, Storage and Analysis* (IEEE Computer Society, 2010), p. 19
24. C. Lin, S. Lu, Scheduling scientific workflows elastically for cloud computing, in *2011 IEEE International Conference on Cloud Computing (CLOUD)* (IEEE, 2011), pp. 746–747
25. W. Lin, C. Liang, J.Z. Wang, R. Buyya, Bandwidth-aware divisible task scheduling for cloud computing. Softw. Pract. Exper. **44**(2), 163–174 (2014)
26. M. Malawski, G. Juve, E. Deelman, J. Nabrzyski, Cost-and deadline-constrained provisioning for scientific workflow ensembles in IAAS clouds, in *Proceedings of the International Conference on High Performance Computing, Networking, Storage and Analysis* (IEEE Computer Society Press, 2012), p. 22
27. E. Michon, J. Gossa, S. Genaud, Free elasticity and free CPU power for scientific workloads on IaaS Clouds, in *18th IEEE International Conference on Parallel and Distributed Systems* (IEEE, Singapour, Singapore, 2012), http://hal.inria.fr/hal-00733155
28. I. Chana, Rajni, Bacterial foraging based hyper-heuristic for resource scheduling in grid computing. Future Gener. Comput. Syst. **29**(3), 751–762 (2013)
29. A. Rajni, An empirical study of vm provisioning strategies on IAAS cloud, in *2016 IEEE 18th International Conference on High Performance Computing and Communications* (IEEE, 2016)
30. J. Sen, Security and privacy issues in cloud computing, in *Architectures and Protocols for Secure Information Technology Infrastructures* (2013), p. 145
31. J. Shi, J. Luo, F. Dong, J. Zhang, J. Zhang, Elastic resource provisioning for scientific workflow scheduling in cloud under budget and deadline constraints. Cluster Comput. **19**(1), 167182 (2016)
32. S. Srinivasan, G. Juve, R.F. Da Silva, K. Vahi, E. Deelman, A cleanup algorithm for implementing storage constraints in scientific workflow executions, in *2014 9th Workshop on Workflows in Support of Large-Scale Science (WORKS)* (IEEE, 2014), pp. 41–49
33. C. Szabo, Q.Z. Sheng, T. Kroeger, Y. Zhang, J. Yu, Science in the cloud: Allocation and execution of data-intensive scientific workflows. J. Grid Comput. 120 (2013)
34. D. Villegas, A. Antoniou, S.M. Sadjadi, A. Iosup, An analysis of provisioning and allocation policies for infrastructure-as-a-service clouds, in *2012 12th IEEE/ACM International Symposium on Cluster, Cloud and Grid Computing (CCGrid)* (IEEE, 2012), pp. 612–619

35. Z. Wu, X. Liu, Z. Ni, D. Yuan, Y. Yang, A market-oriented hierarchical scheduling strategy in cloud workflow systems. J. Supercomput. **63**(1), 256–293 (2013)
36. A. Zhou, B. He, C. Liu, Monetary cost optimizations for hosting workflow-as-a-service in IAAS clouds (2013)

Chapter 6
Consolidation in Cloud Environment Using Optimization Techniques

Ritesh Patel

Abstract The services offered by cloud computing and its usage are increasing day-by-day. Due to the elasticity characteristic of cloud computing, many organizations are now moving their services on cloud data centers. A cloud disaster recovery requires migration of a VM from one data center to another without disconnecting the user. Live VM migration is a key concept to transfer VM without disrupting services. Server consolidation and scheduled maintenance are added advantages of it. In cloud computing, moving large size of VM from one data center to other data center over a wide area network is a challenging task.

Keywords Live Virtual Machine(VM) migration · Virtualization · Network protocol stack · Cloud data center

6.1 Introduction

Process migration was hot topic in virtual computer systems since 1980s. Research community got some success to carry our process migration for the few category of applications. Milojicic [22] gave a survey and problems associated with process migration. Major problem is the residual dependencies that a migrated process has with operating system. Examples of residual dependencies include open file descriptors, shared memory segments with other processes and other local resources. Research community gradually come up with new solution: instated of process, whole operating system can be migrated. To achieve this hardware visualization is required.

Increase of bandwidth at end user has increased demand to perform operations online Indeed, it increases the requirement of efficient resource management in the data center. Many big players of Cloud Service providers such as Google, Amazon and Microsoft, are providing cloud solutions in terms of product or in terms of services

R. Patel (✉)
Computer Engineering Department, C.S. Patel Institute of Technology,
CHARUSAT, Anand 388421, Gujarat, India
e-mail: riteshpatel.ce@charusat.ac.in

B. S. P. Mishra et al. (eds.), *Cloud Computing for Optimization: Foundations, Applications, and Challenges*, Studies in Big Data 39,
https://doi.org/10.1007/978-3-319-73676-1_6

[39]. Amazon [4] and Microsoft [21] are providing services through portal. Other organizations can create their private cloud using product based solutions like VM ware, Xen and Microsoft hypervisor. Users may demand their services for Compute, Storage, Networking solutions and much more. Cloud providers are providing these services through Data Center where all the hardware elements are virtualized and made available to users based on their need. Cloud providers are able to provide these services through any one of three levels of Virtualization.

Hardware virtualization [19] is a powerful platform for making clear separation between real hardware and operating system. Applications and resources can be loaded in that operating system. It facilitates to virtualize CPU, RAM, Secondary Storage and Network. Multiple OS can be run simultaneously using hardware visualization.Migrating entire operating system instances [33] across distinct physical hosts provides ease of manageability for administrators of data centers.

6.2 Virtualization

Virtualization forms strong foundation for cloud computing. It separates strong tie up between operating system and computer hardware by purring separate thin layer of virtualization. There are three levels of virtualization exists (i) Application virtualization, (ii) OS virtualization and (iii) Hardware virtualization.

1. **Application Virtualization** where same application can be executed on different OS platform by introducing virtual execution environment.
2. **Operating System Virtualization** where functionality of operating system is abstracted into container and it can be easily ported on different machine.
3. **Hardware Virtualization** where components of computers are virtualized to run multiple operating system.

6.3 Virtual Machine Migration

Virtual machine migration is a key feature of virtualization. It enables a easier management of large variety of tasks.
Advantages of Live Migration [14] are:

1. Reduce power consumption using VM consolidation.
2. Improve reliability with business continuity and disaster recovery.
3. Efficiently manage IT infrastructure using single management console.
4. Exploitation of resource locality for performance and resource sharing.
5. Simplified system administration and mobile computing for instance as used by commuters from office to home.

6.3.1 Live VM Migration

Moving a running virtual machine between different physical machines without disconnecting the client or user application is known as live VM migration. But, it is obvious that, user may not access application for some duration of time depending on the type of migration. But that duration is negligible with respect to features provided. Live migration is a key feature [24] of virtualization. It enables a large variety of management tasks to manage very easily.

There are two approaches to live VM migration. (1) Managed migration (2) Self Migration. As shown in Fig. 6.1, in managed migration, VM controller is responsible to monitor and migrate VM on targeted host. While in self migration, migration stub [11] on host would transfer VM to the targeted host. Migration stub is responsible to find out suitable host for migration and carry out task related to migration.In either of above approach do not require to modify guest OS to support live VM migration. Major advantage of manged migration is it allows various heuristics to be applied at central controller which reduces implementation complexity.

Migrating an virtual machine involves to transfer of in-memory state and disk state. Several research problems are identified and solved to perform an efficient transfer of VM. Depending on whether VM migration suspends user processes, it is categorized under Cold and Live migration, as shown in Fig. 6.2. In cold migration, VM is shut down at source host and restart in targeted host. The user would lose the connection. In live migration, VM is continuously transferred to targeted host while the user is accessing data. Further live migration in divided into two types, Seamless migration and Non-seamless migration. In seamless migration, user would not get disturbed while migration is going on, and non-seamless migration, user processes would be suspended at source host and resumed at the targeted host [5, 29].

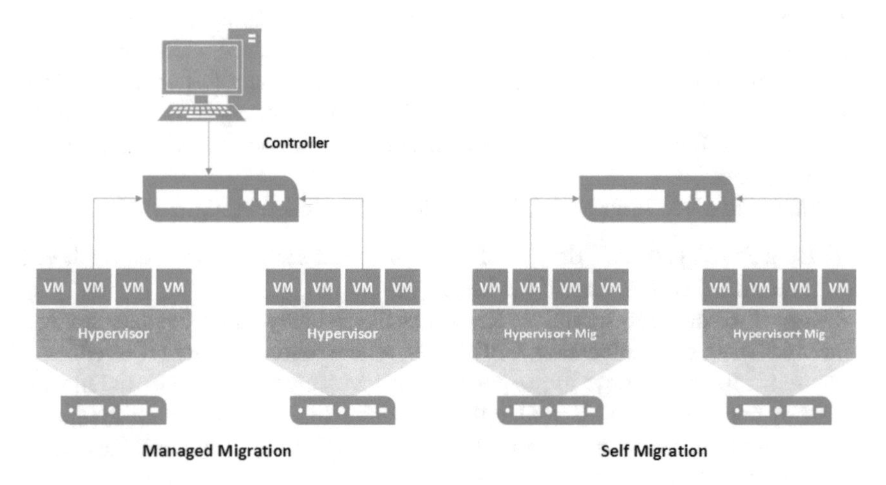

Fig. 6.1 Approaches of VM migration

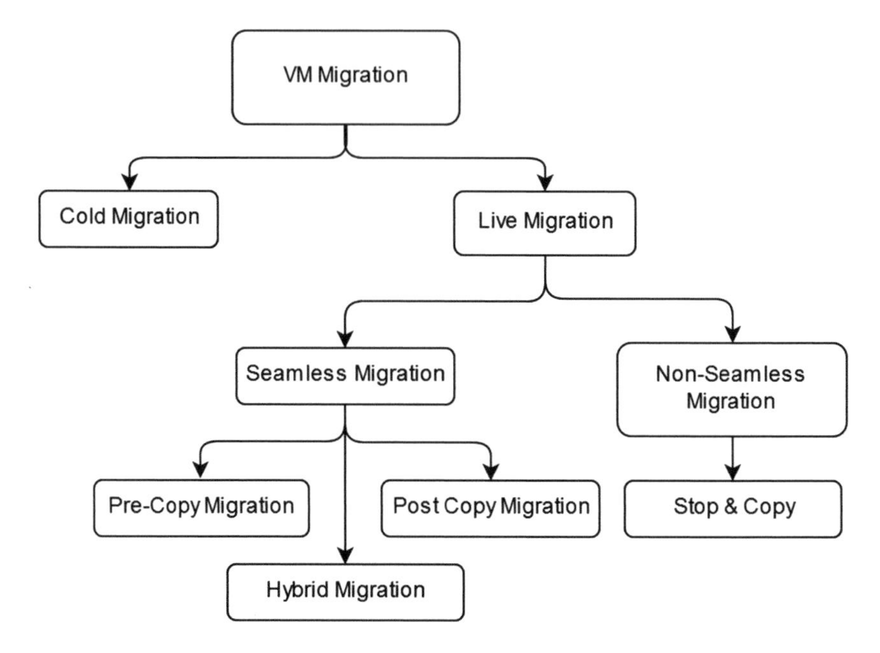

Fig. 6.2 Types of VM migration

In cold migration, VM is shut down at source host and started in targeted host. The user would lose the connection as well as current computing. In live migration, VM is continuously transferred to targeted host while the user is accessing data. This kind of scenario is suitable to applications which has characteristics like busty with small duration of time. Examples includes web traffic.

Live migration in divided into two types, Seamless Migration and Non-seamless Migration. In seamless migration, user would not get disturbed while migration is going on, and non-seamless migration, user processes would be suspended at source host and resumed at the targeted host [5, 29]. Differences among these migration are (i) downtime suffered by user (ii) whether user need to take some action to continue operation or not.

Seamless live VM Migration [31] is carried as illustrated in Fig. 6.3:

1. Phase i: Pre-Migration: Selection of VM to be transferred and Host on which VM will be shifted will be carried out here.
2. Phase ii: Resource Reservation: Resources like primary and secondary storage, network bandwidth allocation, CPU Cores, on the destination are reserved.
3. Phase iii: Iterative Copy: In-memory image VM is treated as sequential blocks. One by one block is transferred to the destination host in a number of iterations. VM is not stopped yet on the source host. Blocks which are modified in the current iteration will be transferred in the next iteration on the destination host. Number of iterations determine the efficiency of the migration process.

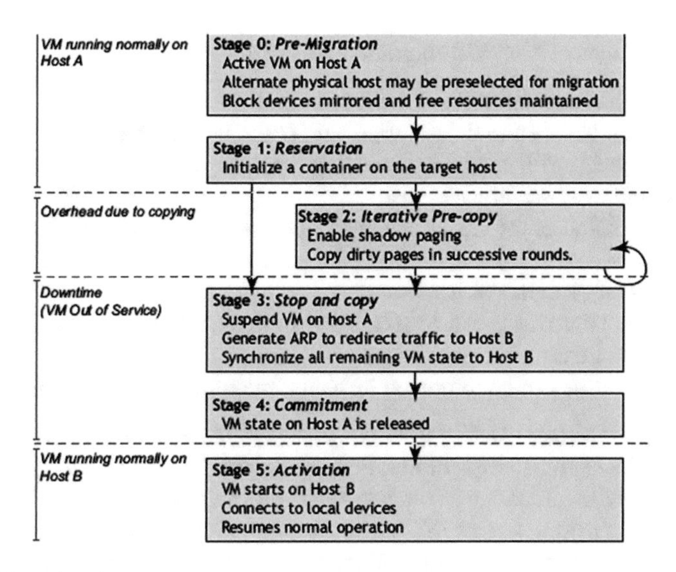

Fig. 6.3 Live migration steps

4. Phase iv: Pause-and-Copy: Once the number of modified pages reaches to the below threshold in particular iteration, VM is suspended at source, then CPU state and other remaining modified memory blocks are transferred to target host.
5. Phase v: Post-migration: Device drivers are mapped to the new VM. To inform new network attachment, it advertises MAC addresses using the ARP unsolicited message which changes the network binding of MAC address to layer 2 devices and mapping of MAC to IP address in hosts in the network.

In pre-copy migration, phase-iii will be iterated until it reaches to some threshold value. In post-copy migration, phase-v performed after phase-ii, means users will be connected to targeted host first and required pages are transferred using page fault at remote location and rest pages are transferred in single iteration.

Pause-and-Copy is an example of non-seamless migration. It is considered the simplest migration technique. But it leads to higher downtime as well as service would not be available for that duration of time. It involves stopping current VM, copying to target host and restarting VM at target host. Network also plays important role to transfer VM if network transfer is involved. As compared to Live VM migration, stop and copy takes less time of VM transfer as it does not need to support the current connections. This approach is easy to implement and control. Disadvantages is that users suffers from unavailability of VM during VM transfer. Even if migration is carried out over WAN, VM would not be available until DNS gets updated. Non-seamless migration is out of scope of thesis work so it is not discussed further.

Depending on the writable workload on the VM, current hypervisors perform live VM migration from milliseconds to few seconds. VMware [23] a proprietary solution and KVM [18] open source solution facilitates with live VM migration.

The advantages of Live VM migration are as follows:

1. Migrating VM moves residual application and all its dependencies which exist with the OS. No need to detach applications and its dependencies.
2. VM is migrated with network credentials, application level restarts or reconnect is not required at the targeted host.
3. Easy administration of virtual machine in clustered and non-clustered environment.
4. Facilitates efficient use of hardware.
5. Preventive maintenance can be scheduled without disturbing user.
6. Server consolidation becomes easy.
7. User access time can be improved by using follow-the-sun [31] scenario.

Live Migration time is mostly affected by the (1) VM memory size (2) Rate of modification of blocks in subsequent iteration (3) Application workload on VM which increases rate of modification of blocks (4) Current network bandwidth between the source and destination servers (5) Hardware configuration of the source and destination servers (6) Amount of buffer allocated to the transport layer of hypervisor.

6.3.2 Properties of the Live VM Migration

Followings are the properties of the VM migration over LAN as well as WAN [12]

- **Continuous operation**: The VM should continue to provides services while transferring its in-memory and local storage.
- **Consistency**: File system is remain consistent after VM migration on destination machine.
- **Minimum service disruption**: Migration should not significantly degrade the performance of services running in the VM.
- **Transparency**: Treatment of services running inside the virtual machine should be same as the OS is running on hardware. The VM's open network connections remain alive, even after an IP address change, and new connections are seamlessly redirected to the new IP address at the destination.

6.3.3 Performance Metrics

The following are the metric parameter [17] to evaluate any live VM migration method.

1. Preparation Time: Time between initiating migration process and transferring the VMs processor state to the target node, during which the VM continues to execute and dirty its memory.

2. Down Time: It is time that VM would not be available to the user. It is also known as non-availability time. At the end of migration, user VM is paused and all the pages are migrated to target host. This time the VM is not available to the user. It is also known as pause time.
3. Pages Transferred: VM image is divided into a number of blocks called pages [?]. VM image is transferred by picking up one page by page. So, a unit of transfer is pages. There is a trade-off between size of the page and dirtied pages. If the page size is larger than the possibility of a number of dirtied pages will be less and vice versa.
4. Resume Time: It is pause time caused by reconnection to the target host.
5. Total Migration Time: Total time taken by host machines to migrate VM from source host to target host. Total time is very important because it affects the release of resources on source node.
6. Performance Degradation: When Virtual machine is migrated from one host to another, the application would suffer a little bit pause time.

By literature survey and various experiments, it could conclude that total migration time heavily depends on the network throughput and reaction of congestion control algorithm. Following are the other parameters which also affect the performance of live VM migration.

1. Total size of VM.
2. Size and Number of pages of VM.
3. Dirtying rate while VM migration takes place.
4. Currently available network bandwidth.
5. Current Network Traffic.
6. Congestion Control Algorithm at transport layer.
7. The buffer allocated to application at Socket Layer (both at sender and receiver side).
8. Queuing mechanism at intermediate node/switch/router.
9. Network utilization during live VM migration.
10. Ingress and Outgress data traffic.

Server consolidation is one of the requirements of the data center to make energy efficient [1]. VM Migration is a key concept in server consolidation where heavily loaded and lightly loaded VMs are migrated to another host to meet the SLA violation and efficient energy consumption required in the data center. VM migration is also performed across data centers to perform disaster recovery, pre-planed data center maintenance and migrating VM as closer to users to solve the problem of response time [31].

6.3.4 Applications of Live VM Migration

According to need, VM can be migration within LAN and WAN network. Following are the applications within datacenter.

- Server Consolidation [27]: Aggregating VMs on single host is known as server consolidation. Using live VM migration multiple VM can be transferred to single host to utilize resources fully.
- Energy Conservation [27]: In host, 70% of energy is consumed by non computing elements. To conserve energy, VMs running on underutilized node are transferred to other host and under utilized hosts are shuted down. Reducing resources without affecting application performance
- Avoiding SLA Violations: In case of requirement of performance, VM can be shifted to under utilized host to avoid SLA.
- Service Load Balancing [27]: Quick instances can be created by copying in-memory copy of VM.

Following are the needs to perform VM transfer across Data Center

- Fog Computing [6, 30, 32, 38]: Cloud computing is characterized as a group of PCs and bunch of servers associated together over the Internet to form a client-server system. Today, the large number of ventures and substantial associations are starting to receive the Internet of Things, the requirement for a lot of information to be collected at the center node for the further processing. Fog Computing is a framework in which certain application procedures are managed over infrastructure and it is overseen at the edge of the system by a gadget. It is, basically, a center layer between the cloud and the equipment to empower more proficient information handling, investigation and capacity, which is accomplished by diminishing the measure of information which should be transported to the cloud.
- Follow the Sun: User applications can be brought near to user geographic area to reduce latency.
- Disaster Recovery The capability to relocate a virtual machine (VM) form one physical host to other physical host can fundamentally help an organization to enhance business reliability. It also facilitates to perform server hardware maintenance without affecting users.
- Hybrid Cloud Computing: Private data center is integrated with public cloud. VM can be migrated from public to private to cut the billing cost. Same way VM can be transferred to private to public to carry maintenance of private cloud without worrying of service availability.

6.3.5 Four Decision Pillars of Migration

1. When to Migrate: It deals with requirement of the migration. Where VM need to be migrated or not. There are many approaches and each approach have pros and cons. Simplest approach is to manual VM migration where each VM is picked up manually ans transferred to other data center. Its good for small data center wherein time consuming in larger data center.

   a. Auto
   b. Manual
   c. SemiAuto / Semi Manual
   d. Host Over loading Detection
   e. Host Under loading Detection
   f. heuristics
      i. An adaptive Utilization Threshold
      ii. Local Regression
      iii. Robust Local Regression

2. What to Migrate: It deals with proactive VM migration, where it has statistics of all VMs and applies some heuristics to decide the order of VM migration. It is also possible to transfer more than one VMs simultaneously.

   a. VM Selection
      i. The minimum Migration Time Policy
      ii. The Random Choice Policy
      iii. The Maximum Correlation Policy

3. How to Migrate: Its a reactive kind of VM migration, where decisions of migration is made during in transit VM scenario.

   a. Compression Algorithm
   b. Network aware migration

4. Where to Migrate

   a. Within same Data Center: This would transfer VM using LAN mostly using infiniband switches.
   b. Outside Data Center (Other data center): This involves transferring data from one data center to other data center using WAN. Root challenges are of high latency and higher downtime observed by user.
   c. VM Placement: It deals with in which host this VM must be placed efficient work. It also requires heuristics to decide best combination of host and VM combination.
      i. An adaptive Utilization Threshold
      ii. Local Regression
      iii. Robust Local Regression.

6.4 VM Consolidation

VM consolidation is one of the requirements of the data center to make energy efficient [1]. VM Migration is a key concept in VM consolidation where heavily loaded and lightly loaded VMs are migrated to another host to meet the SLA violation and efficient energy consumption required in the data center. VM migration is also performed across data centers to perform disaster recovery, pre-planed data center maintenance and migrating VM as closer to users to solve the problem of response time [31].

VM Consolidation is technique to reduce the number of active PMs by consolidating multiple underloaded VMs into one physical machines. VM consolidation includes

- VM Selection
- VM Placements

Following are the advantages of VM consolidation

- Reduce the amount of active physical machine,
- Reduce the data center footprints,
- Indirectly reduce power consumption,
- Energy conservation

6.4.1 VM Selection

VM selection refers to selecting VM for the migration to different host. VM selection can be initiated either by host over-load detection OR host under-load detection.

VM can be selected from one of the approaches:

- Minimum Utilization: The VMs with the lowest CPU utilization are selected for the migration.
- Maximum Utilization: The VMs with the highest CPU utilization are selected. This approach is very useful to meet SLA.
- Minimum Migration time: The VMs with the lowest calculated migration time are selected. Lowest migration time is calculated using current size of VM and current load on VM.

6.4.2 VM Placement

After VM is being selected for migration, it required to find the suitable host for the migration is known as VM placement. Once the VM is running and migrated to other host then it is known as VM placement.
Following are the approaches of VM placement:

- Power Saving based approach
- Application based approach
- Service Level agreement based Approach

Many algorithms are proposed for efficient VM Placement like first fit, single dimensional best fit, volume based best fit, dot product based fit, constraint based approach, bin packing problem, stochastic integer programming, genetic algorithm, network aware etc.

6.5 Components of Live VM Migration

Virtual machine live migration facilitates to move a virtual machine running on one physical host to another host without disturbing application availability. Live migration catches a VM's total memory state and the condition of all its processor core are collected and send it to memory space on destination host. The destination host then loads the processor cores and resumes VMs execution. All virtualization tools offer live VM migration: VMware offers vMotion, Microsoft offers Live Migration in Hyper-V, and Citrix Systems offers XenMotion.

Live Migration over WAN has some issues. First, because of the high-latency characteristic of links in the WAN, VM migration takes longer time. Second, after migration the VM will be allocated new IP address, seamless application reconnect is require.

Following are the components of live VM migration:

1. Need for the migration
2. Selection of VM to be migrated
3. Selection of host on which VM is to be migrated
4. Allocating resources at destination machine
5. Migrating CPU State
6. Migrating in-memory image
7. Migrating disk pages
8. Migrating network
    - Redirecting network to destination machine
    - Redirecting clients to destination machine
9. Modeling VM Migration
    - Cost of VM Migration.

6.5.1 Need of Live VM Migration

Live migration of VM allows transferring a VM between physical nodes without interruption and with a very short downtime. However, Live VM migration negatively impacts performance of application performance running inside VM during migration. Requirement are framed either within or outside data center migration.

As we have seen the previous section that Within Data Center Energy Conservation,Server consolidation, Mitigating SLA violations and Service Load Balancing are the major need of VM migration.

Server Consolidation and energy conservation are co-related with each other. Load on the server changes based on the customer location of geographic boundary. As servers would not be heavily loaded during night hours, those VM could be combined and put on single host.

The benefits of server grouping together include higher efficiency, reducing power and cooling costs, and the flexibility to move workloads between physical servers.

As we have seen the previous section that Across Data Center Fog Computing, Follow the Sun, Disaster Recovery and hybrid cloud computing are the major application areas.

Availability of computational resources and network resources during migration are also real time issues [10] to have control over live VM migration. Processor share of VM depends on the scheduling policy of operating system. To estimate accurate time required to transfer pages of in-memory VM, bandwidth should be constant. Fixed bandwidth of 100 Mbps is taken for consideration and probabilistic based LRU policy is used to evaluate work.

In fact, frequent VM migration would lead to increase of overhead [34]. Therefore, it is necessary to define heuristics very carefully such that it should avoid frequent VM migration. Bing wei combined heuristics of energy optimized and workload adaptive modeling for migration. Workload characteristics are analyzed to avoid the unnecessary migration. Even in case of decrease of workload, it avoids to migration VM.

6.5.2 Selection of VM to be Migrated

Selection of VM can be based on average post-migration CPU, load on the destination PM, cost of the migration, and expected load distribution in the system after the migration.

Energy guided migration [34] model selects the best VM by considering VM's current memory and utilization of CPU [13].

S. Zaman and D. Grosu [40] suggested auction based VM selection policy. Two strategies are suggested, Push and Pull. In push method, an overloaded physical machine acts as a seller. PM constitutes actioned parameters PM communicates this information to other using multicasting protocol. The underloaded PM acts as a buyers. The buyers may bid on one, several, or all auctioned VMs. The seller selects the best host from the bids.

In pull strategy, underutilized PM acts as seller and advertises free resources. The other PMs, mostly overutilized, in the system are potential resource buyers that bid with for the resources. The seller decides which bid it accepts and initiates the VM migration.

6.5.3 Selection of Host on Which VM is to be Migrated

Workload adaptive model [34] selects the best migrating target physical server with the lowest power consumption. In addition, it determines the suitable migration moment in terms of both the performance and the trend of workload.

6.5.4 Migrating In-Memory Image

In this section modeling in-memory image and predicting of migration time mathematically are discussed..

F. Checconi, T. Cucinotta, and M. Stein [10] presented technique for live migration of real-time virtualized machines. The achievement of research is very low and predictable down times, as well as the availability of guaranteed resources during the migration process.

Total time for the migration is calculated by

$$t_f = \left(\frac{P+H}{b_d}\right)\sum_{k=1}^{K} n_k + t_d \tag{6.1}$$

where n_k is number of pages in k^{th} iteration, P is page size in Bytes, H is overhead of bytes, b_d is constant bandwidth available during migration and t_d is downtime observed by user.

Downtime t_d is expressed as

$$t_d = \left(\frac{P+H}{b_d}\right) n_{k+1} \tag{6.2}$$

where $k+1$ is last iteration of the migration during which all the pages in k^{th} iteration is dirtied.

As per Eqs. 6.1 and 6.2, total migration time and downtime is inversely proposal to the bandwidth available during migration.

Assumption: Bandwidth of fixed size 50Mbps.

Future Work: Different policies can be developed and tested for reordering page to be transferred. To reduce the migration time

- Derive heuristics which reduces time by reducing dirty rate.
- Perform testing on different values of bandwidth.
- Derive heuristics which rearranges pages to be transferred in each iteration [10].

C. Clark, K. Fraser, S. Hand [11] presented that migrating the VM's from one physical host to targeted host can be can be accomplished by any one of three approaches. Approaches are assessed based on the requirements of minimizing both downtime and total migration time. The downtime is the period during which the service is unavailable to user due to pause time of VM instance. This period will be obased as service interruption by client. The migration time is the duration between when migration is initiated and client gets reconnected with VM on targeted host.

Push migration VM continues to destination server and pages are pushed by the server agent to the destination server. Pull migration has two parallel transmission. First its normal transmission and second is pulling (page on demand) transmission. Either on any types of migration, network migration remains challenging task.

6.5.5 *Migrating Disk Pages*

Following are the classification approaches of migrating disk pages

1. Replica based
2. Pull mode
3. Push mode
4. Hybrid mode

S. Al-Kiswany, D. Subhraveti, P. Sarkar, and M. Ripeanu [3] presented VMFlockMS, a migration service optimized for cross-datacenter transfer. Authors suggested two approaches to reduce migration time and minimize pause time caused by migration. First, VM deduplication: Multiple copies of VM is deployed is active-passive mode. Once the active copy is unavailable then one of the passice copy of VM will be made active. Second, Accelerated Instantiation: It crates partial set of data based the access patterns observed in last few moments. Evaluation of VMFlocks shows that it can reduce the data volumes to be transferred over the network to as low as 3% of the original VM size.

Method: Duplicate VM Image is kept on multiple data center. Data are copied frequently. Its not migration its about VM Duplication

Thought: VM Duplication is solution here. Disadvantages are resource wastage and lower down efficiency of network bandwidth.

K. Haselhorst, M. Schmidt, R. Schwarzkopf, N. Fallenbeck, and B. Freisleben [16] suggested replica based solution. Write on both the site. Multiple copies of VM are kept at multiple sites. Read operation is carried out from any one copy but write operation is performed on all the duplicated sites. For this purpose, a novel mechanism based on distributed replicated block devices (DRBD) has been presented that allows for synchronous writes on the source and the destination host.

Issue: Synchronization of VM storage for live VM over WAN. Main Challenge: writable part need to be synchronized.

B. Nicolae and F. Cappello [25] suggested to use active push prioritized prefetch strategy. It uses combination of pre-copy and post-copy to perform migration. It starts with pre-copy but end with post copy.

Challenge: Disk storage transfer under high I/O intensive workload.

K. Haselhorst, M. Schmidt, R. Schwarzkopf, N. Fallenbeck, and B. Freisleben presented VM migration over WAN, targeting to synchronize the storage on the source and destination before transferring control to the destination. They have introduces two issues:

- The same disk content may change repeatedly. This lead to frequent write on destination server. This, increases migration time.
- Rate of change of disk content is higher than the data rate of copying it to destination. Thus migration becomes impossible and it could have a negative impact on performance of application.

Problem addressed: When to transfer phase from pre-copy to post-copy.

It uses least frequently access policy during post-copy transfer. It uses sequential packet transfer in post-copy phase. Along with this it uses hybrid active push-prioritized prefetch strategy. Performance of live migration reach to 10x faster, consumed to 5x less bandwidth and resulted in high I/O throughput.

J. Zheng, T. S. E. Ng, and K. Sripanidkulchai [42] presented Workload aware scheduling of transfer of chunk. Also size of chunk depends on workload of VM. The benefits of scheduling increases as the available network bandwidth decreases, as image seize gets larger and as I/O rate gets increase. Authors validated results with five representative applications.

6.5.6 Migrating over Network

Live migration can be carried out differently in LAN and WAN.

- **Local area network redirection**: In LAN environment, migrating VM would carry its local IP address with it. There will not be a problem of transport layer connection reset. Only Layer 2 redirection is sufficient. Layer 2 redirection compromise with drop of in-flight packet which are occurring at last iteration of migration. To ensure about the network remapping of layer 2, ARP advertisement is sent in the network to assign new IP address and its corresponding MAC address.
- **Wide area network redirection**: VM migration over WAN is similar to LAN but VM migration over WAN possesses different challenges which are elaborated below.

There are some issues while dealing with WAN VM migration are as follows:

1. High latency: VM migration gets high latency while VM is migrating from source host to destination.
2. Network reconnect because of change in IP address.
3. Higher application workload (I/O intensive workload).
4. It Slow downs performance of multiple VM transfer.

Higher latency lead to lower data transfer rate. If data transfer rate is lower than the VM dirty rate, VM migration becomes endless process. It continuously involves to transfer pages which will be modified again and again. It is recommended that Higher bandwidth should be available during the transfer of VM or at least last iteration

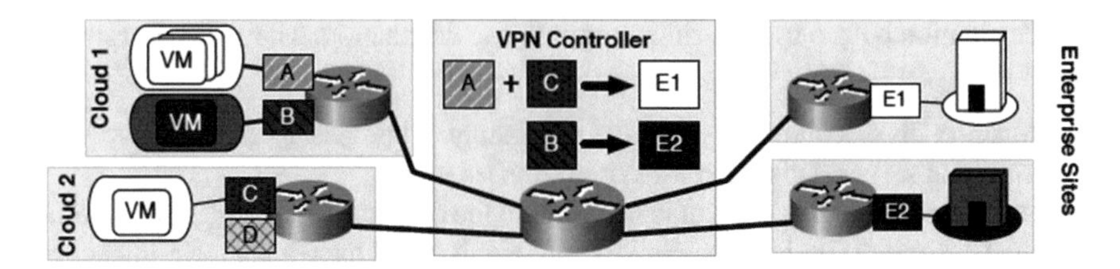

Fig. 6.4 CloudNet:VPN Proxy Based Migration

going on. Layer 3 redirection would break connection. Layer 3 redirection changes the layer 3 addressing facility. So it would reset application layer connection, network applications need to be restarted. To mitigate this issue Mobile IP [7] is a solution where a node on the home network (the home agent) forwards packets destined for the client (mobile node) in terms of care-of address on the foreign network.

VPN Proxy based method [37] To lower down downtime and migration time, [37] have suggested CloudNet, a VPN based proxy mechanism to overcome problem of WAN migration. Figure 6.4 shows a pair of VPCs, cloud 1 and cloud 2, that span multiple cloud data centers, but present a unified pool of resources to each enterprise sites. A VPN controller is responsible for the remapping of the VM to cloud site. It uses further MPLS based VPN to increase naming abstraction. Dynamic VPN reconfiguration deploy and move resources between cloud data centers. Figure 6.4 shows a pair of VPCs that span multiple cloud data centers, but present a unified pool of resources to each enterprise.

J. Cao, L. Zhang, J. Yang, and S. K. Das [9] solved the problem of connection migration at transport layer level by a secure token negotiated at connection establishment time. That token will be subsequently used during connection migration to identify the legitimate session of TCP.

DNS updates are suggested in Dynamic DNS as a means of relocating hosts after a move. In WAN VM migration, the migrated VM gets a new IP address. Temporary network redirection scheme [12] provided solution by combining IP tunneling [28] with Dynamic DNS [35].

Dynamic Rate-Limiting for Migration Traffic [11]

The downtime suffered by user can be reduced by increasing the bandwidth limit [11] with lower and upper bound. Bandwidth can be adapted dynamically during each precopy iteration as well. Fixed bandwidth could be allocated to first precopy round then subsequently, bandwidth could be increased as iteration is progressing. Bandwidth limit could be derived from the previous iteration's dirtying rate.

TCP Freeze [8]

M. V. Bicakci and T. Kunz [8] have proposed a cross-layer feasibility for virtual machine live migration between various subnets. The live migration process is straightforward to the correspondent hosts, and this is accomplished by having the

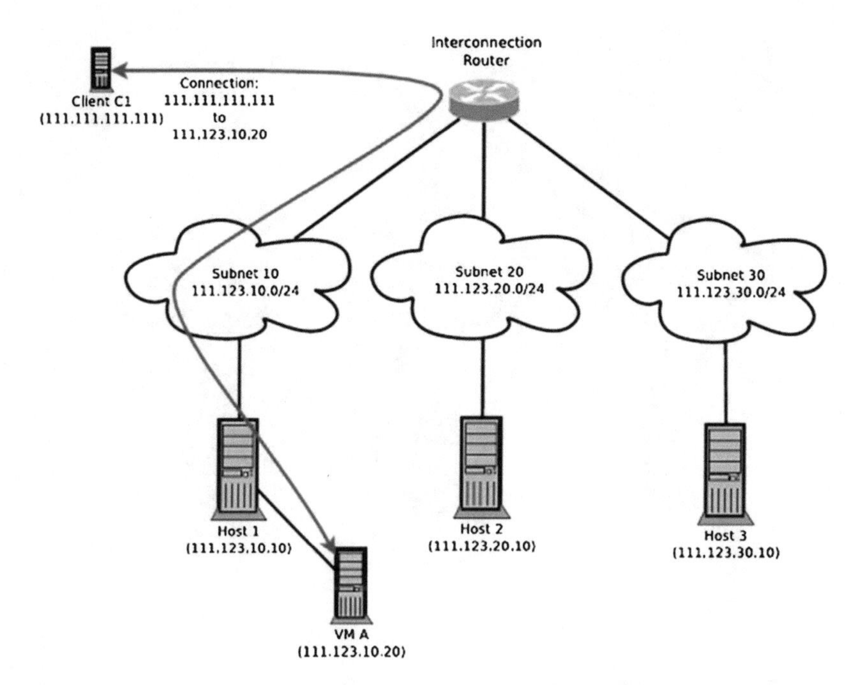

Fig. 6.5 TCP freeze: initial topology

old host of the moved virtual machine act as a proxy server at Layer 3. Further TCP freeze mechanism is used to stop the in-transit packets.

Features of TCP Freeze

1. Guest Operating system is aware of live Migration process.
2. TCP Freeze is used to stop further generation of in-transit packets.
3. Ip-in-IP tunnel is set from source host to destination host to prevent from reconnection for existing connection.

Once the virtual machine is live migrated, it will have two IP addresses: IP address on the new subnet, and IP from old subnet(by means of the IP-in-IP burrow from the source have). Figure 6.5 shows initial topology before migration take place. Figures 6.5 portray the means taken when a virtual machine is relocated more than once between various hosts. In Fig. 6.5, the virtual machine is running on Host 1, with an IP address of 111.123.10.10. A customer, C1, has a dynamic association with Host 1.

As can be found in Fig. 6.6, the virtual machine is then moved to Host 2, which is on another subnet. The new IP address of the virtual machine is 111.123.20.20. This point of time all the connection to VM a would get disrupted. Subsequently, Ip-in-Ip tunnel is created for this connection to newly created IP subnet address.111.123.20.10.

New client C2 wants to get connected to VM, but this time using the virtual machine's new IP address 111.123.20.20.

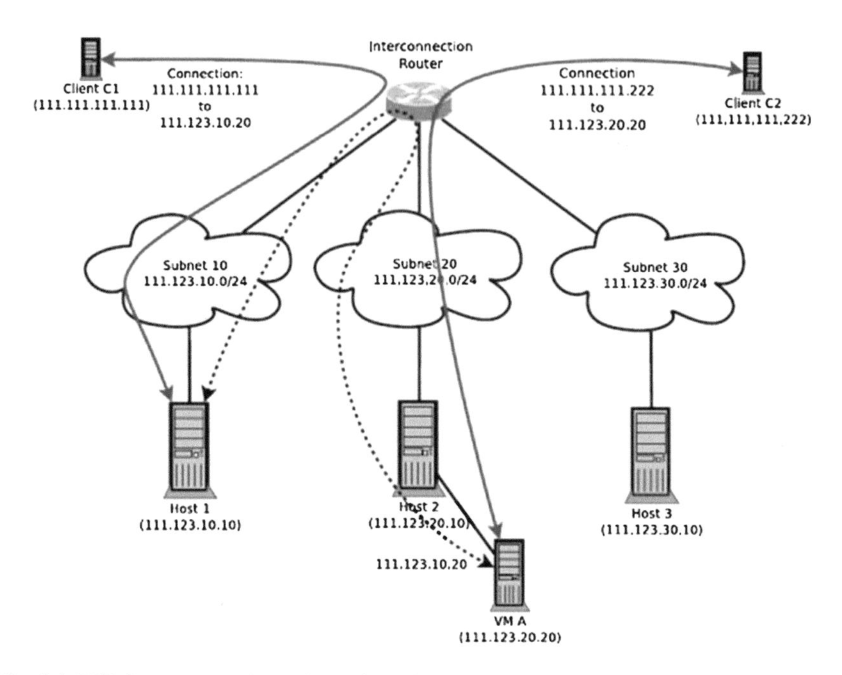

Fig. 6.6 TCP freeze: network topology after migration to network-1

Finally, the virtual machine is migrated to Host 3, as can be seen in Fig. 6.7. In this case, two tunnels will be created at VM running of host 3. Firt tunnel is created from Host 1 to host 2 to support in-flight data of client c1 and second tunnel is created from host 2 to host 3 to support in-flight data of client 2. Noe if either C1 or C2 resets the connections, tunnel is no longer available. It identifies VM with newly allocated IP address i.e. 111.123.30.20. Any further connection of client will be connected to 11.123.30.20.

VM Turntable [31]

The ability to move live VMs over a other long distanced data center addresses the issues [31] of many key serious case situations, particularly when user should not be aware about migration.

For a wide area migration of VM must have short downtime in spite of RTT. This is particularly valid for data center with hundreds or a huge number of virtual machines arranged for a hypothetical selection of VM to be transferred. Upon migration, the system endpoints be consistently adjusted to the new environment, with no earlier understanding, ans should also implement different addressing scheme than the original environment. The adjustment of the endpoints must happen transparently with respect to end users. Much of the time, the new environment must be thought to keep equal or better access than the source environment required by the virtual machine (Fig. 6.8)

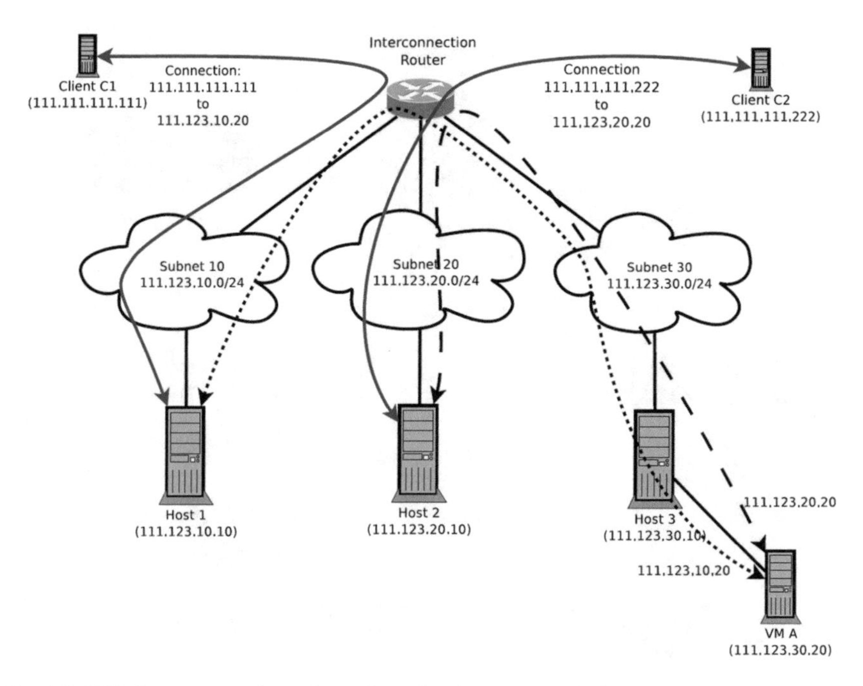

Fig. 6.7 TCP freeze: network topology after migration to network-2

VM Shadow [15]

VM Shadow considers virtual desktops and gaming application servers that required to be moved closest to the client. VM Shadow is a framework to consequently upgrade the area and execution of area delicate virtual desktops in the cloud. VM Shadow continuously performs fingerprinting of a VM's system activity as well as geo-location of other data center to conclude best VM-Datacenter pair for each clients. VM Shadow utilizes WAN-based live movement and another system association relocation convention to guarantee that the VM relocation and resulting changes to the VM's system deliver are transparent to end-clients.

Transparent VM migration is achieved in two phase. In the first phase, it identifies VM to be migrated and data center where the VM is placed. In second phase it performs transparent VM and connection migration across geo-location without disturbing current connection.

Given a multiple cloud with L locations and N VMs, the shadowing algorithm accomplish following steps. In step 1 of algorithm, it identifies good candidates to move from current data center to other data center based on location sensitive ranks of all VM. Threshold is defined to make decision whether VM need to be migrated first or not. It also counts the resources availability at other cloud sites. In step 2, it determines other data center for transferring VM based on the geo-location of each VM with respect to data center. It identifies k closet cloud sites by geographic distance and determines network distance by observing delay of user with each of k sites. All

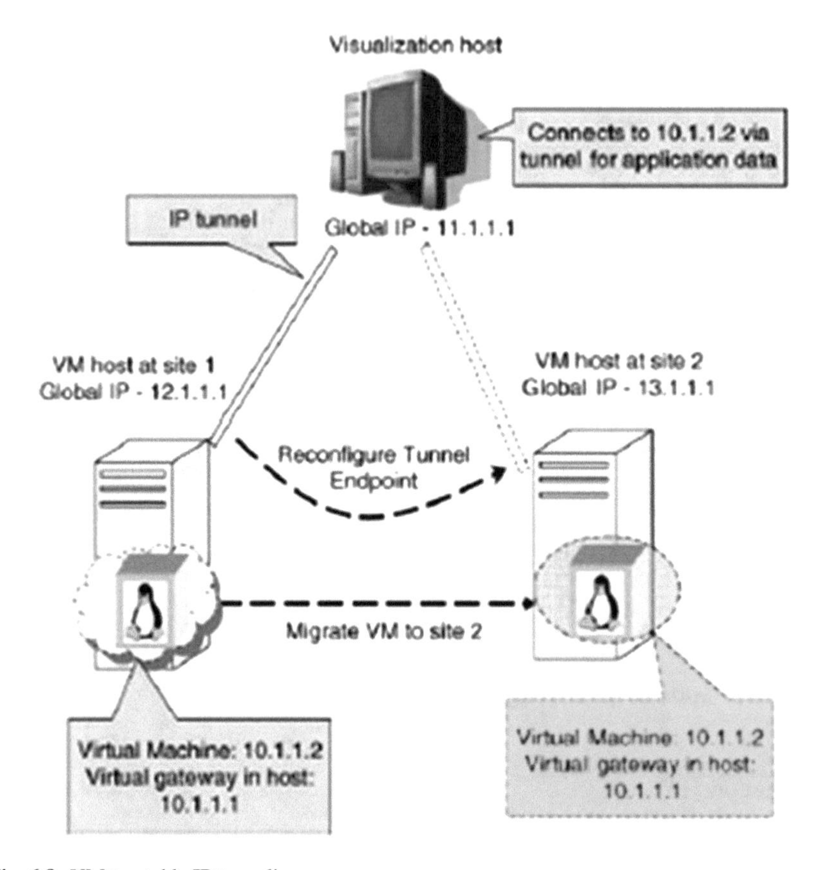

Fig. 6.8 VM turntable IP tunneling

selected k sites are ranked by their network distance. In step 3, cost-benefit analysis is carried out for the selection of VM and corresponding data center. Cost counts overhead to transfer in-memory and disk image of VM. Benefit(B) is calculated by following formula.

$$B = w_1.(l_1 - l_2) + w_2.\frac{(l_1 - l_2) * 100}{l_1}$$

where w1 and w2 are weights, l1 and l2 denotes the absolute latency increase or decrease seen by the VM due to a move and the second term is the percentage latency decrease. Percentage increase or decrease should not be considered alone to take decision [12]. Suppose l1 is of 1 unit and l2 is 2 unit. Then percentage increase of bandwidth is 100%. In other case where l1 is 50 units and l2 is 75 units then percentage increase is only 50%. Above formula counts percentage increase/decrease and also benefits of absolute value as well. Weighted are associated to increase or decrease preference over relative or absolute increase/decrease. Suggested value of w1 = 0:6 and w2 = 0:4 are 0.6 and 0.4 respectively.

Further it formulates cost-benefit trade-off by Integer Linear Program optimization. At last, it transfers VM to the selected data center. Simple rank based greedy approach is used to transfer VM one by one.

Transparent VM and Connection Migration

In the first phase, VMShadow attempts to optimize the performance of location sensitive VMs by moving them closer to their users, it is mandatory that migration should be transparent to user. In the second phase, nested hypervisor is used to relocate the VM easily in new data center without affecting application connection. Xen-blanket [36] provides nesting of hypervisor. That prevents VM from disruption of services due to change in network address. As shown in Fig. 6.9 one more layer added to the Xen hypervisor known as xen-blanket. Xen-blanket layer provides user level machine monitoring. Xen hypervisor facilitates service provider level machine monitoring.

Live WAN Migration Over WAN:

Live migration of VM in WAN has more issues than LAN migration. There is limited work done on migration over the wide area. VM shadow performs following points to carry out transparent VM migration over WAN (Fig. 6.10).

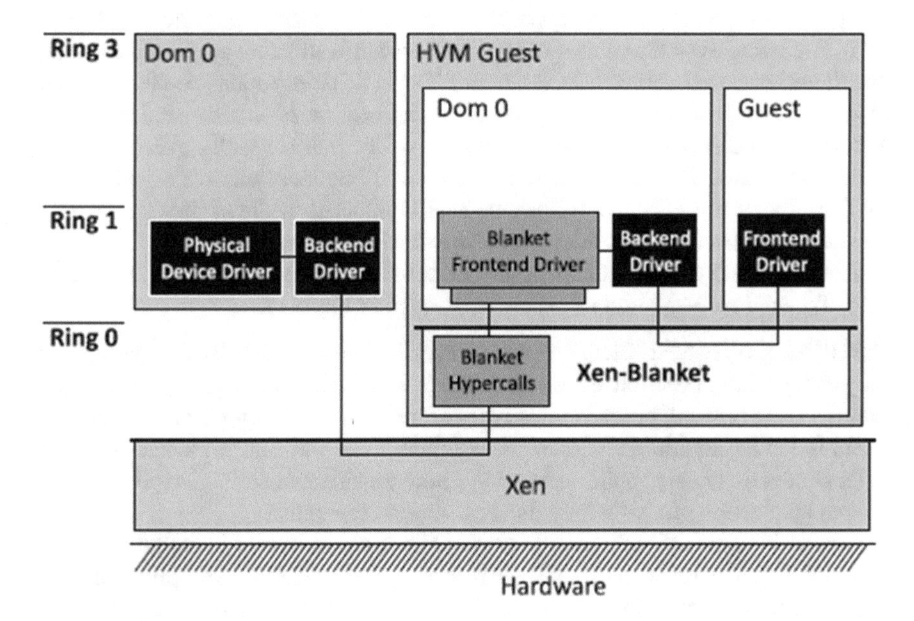

Fig. 6.9 Xen Blanket Architecture

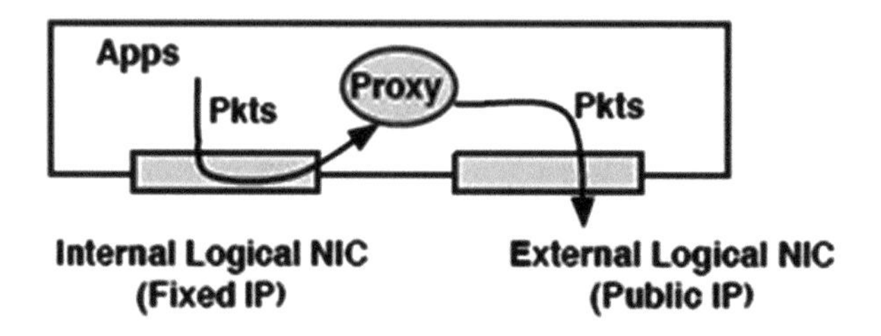

Fig. 6.10 Each VM containing two Logical Interfaces

Redirecting Network to New Location of VM:

There exists two solutions as specified below:

1. Network layer solution- IP Tunneling, VM Turntable.
2. Transport layer solution TCB(Transmission control block) change MobiSocket.
3. Application layer solution -Dynamic DNS.

Z. Pan, Y. Dong, Y. Chen, L. Zhang, and Z. Zhang [26] presented CompSC, a solution of hardware state migration that achieves live migration support on pass-through network devices.

MobiSocket-Transport Layer Solution

The problems associate of other approaches are non-transparent end-to-end operation of TCP session, as well as the requirement of new infrastructure entities (the middle agent) and triangle overhead. TCP Control Block(TCB) maintains connection state of each client. It contains information about connection sate, associated processes and state associated with state transition of connection. It usually maintained per connection basis. The MobiSocket updates new IP address with old IP address in TCB. In this sense only the TCP at the MH is affected, while at the CH the TCP session is not disturbed. Working procedure of MobiSocket:

Consider that TCP connection is going on between Mobile host and Correspondent host. Types of message exchanges are mentioned in Fig. 6.11.

MSOCKS [20]: In MSOCKS, Transport Layer Mobility is presented that permits not only change the connection to the network, but additionally to control which organize interfaces are utilized for the various types of information leaving the network. Here client is having multiple NIC card is assumed. They have implemented transport layer connectivity using split-connection proxy architecture and TCP splice is used to split connection proxy that provides end-to-end semantics.

The architecture of MSOCKS is give in Fig. 6.12. As shown in architecture, the MSOCKS architecture consists of three pieces: A user level MSOCKS proxy running in proxy server, kernel level MSOCKS proxy running in proxy server and MSOCKS library running in Mobile Node. No modification required in correspondent host.

Disadvantage: Proxy server is required and mobile node need to be modified. **TCP Redirection** TCP-R [14] is an extension of TCP which maintains application

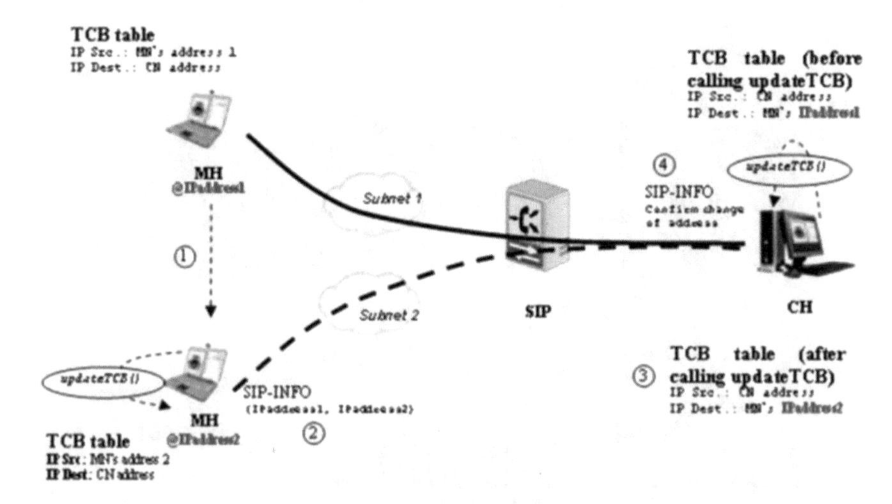

Fig. 6.11 Message exchange of Mobisocket between MH and CH

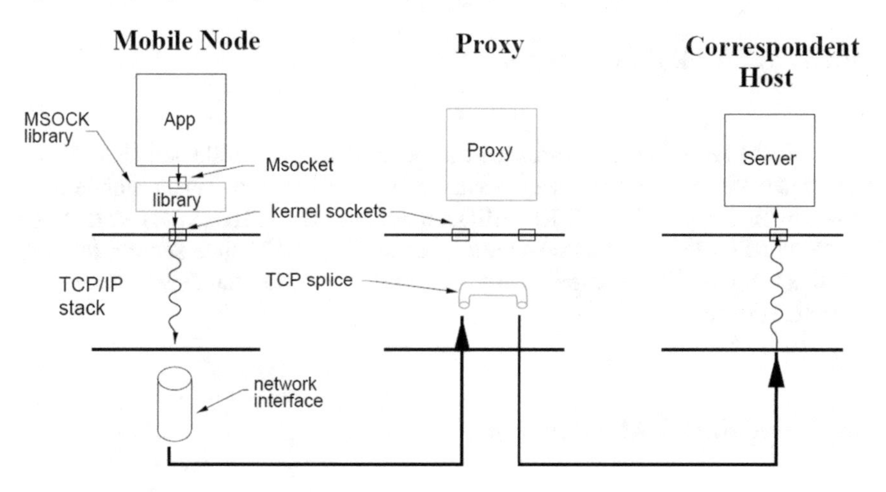

Fig. 6.12 MSOCKS architecture

level connection in case of change of IP address at network layer. It is backward compatible to TCP. It provides the efficient and secure support for mobility. Also it does not require intermediate entity to perform further action. Only sender and receiver is modified. In mobile computing environment, IP changes very frequently as users move from one region to other region. "Continuous Operation" is the center heart of TCP redirection. When mobile node moves from one location to another location it follows reconnection procedure. It devices that steps into "Continuous Operations" and "Compensative Operation".

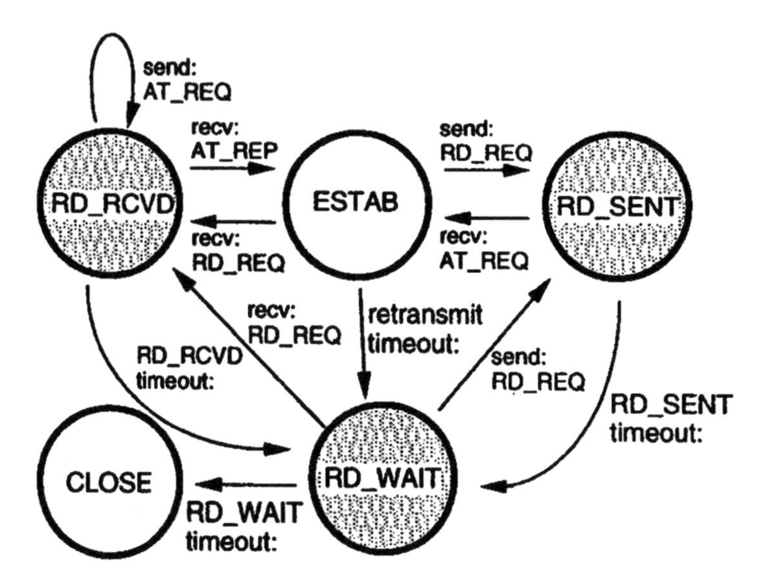

Fig. 6.13 TCP-R state transition diagram

Figure 6.13 represents TCP-R state transition diagram. Mobile node is in ESTD state whenever it is communication with correspondent host. When mobile node moves to other region it sends RD_REQ request and goes to RD_SENT state. Upon receiving RD_REQ, correspondent host goes to RD_RCVD state and it also sends AT_REQ request. Either of the side sending of REQ ensures that they have received all in-transit packets.

6.6 Modeling VM Migration

Modeling Performance Degradation

Performance degradation due to migration and downtime experienced by user depends on the application's behavior [33]. If behavior of application is frequent write, then it lead to updation of pages frequently, it forces to have a higher number of iteration for VM migration. Time required to complete the migration [?] for shared storage area network is defined as

$$T_{m_j} = \frac{M_j}{B_j} \tag{6.3}$$

where M_j is the amount of memory used by VM_j, and B_j is the available network bandwidth. keeping in the ming that B_j is constant bandwidth available during migration.

It is also found that during migration CPU cycles of source and destination hosts are utilized to the migration process. The total performance degradation by VM_j is defined as P_{d_j}

$$P_{d_j} = 0.1X \int_{t_0}^{t_0+T_{m_j}} u_j(t)dt, \tag{6.4}$$

where, t_0 is the time when the migration starts, $u_j(t)$ is the CPU utilization by VM_j.

It is also concluded that number of VM migration should be avoided to reduce the SLA violations.

6.7 Modeling Downtime, Number of Pages Transferred, Network Utilization

Analytic performance modeling [2] of live VM migration is derived in terms downtime, total number of pages transferred and network utilization for uniform dirtying rate.

Downtime observed by user can be modeled as T_{dwon}

$$T_{down} = P(n+1) * \tau = P_s * \tau * p^{n+1} \tag{6.5}$$

where τ is time required to transmit a page of VM over the network, P_s is the number of memory pages of VM, p^i is network utilization during i^th iteration of migration, n is iteration number during which pages are migrated before downtime, so $n+1$ is the final iteration for which user observes downtime and $P(i)$ is number of pages copied during iteration i.

Total migration is sum of rime required in pre-copy and time required during downtime. Pre-copy itself is iterative in nature. It is modeled as $T_{pre-copy}$

$$T_{pre-copy} = \sum_{i=0}^{n} T(i) = \sum_{i=0}^{n} P_s * \tau * P^i = P_s * \tau * \frac{1-p^{n+1}}{1-p} \tag{6.6}$$

Total migration time T_{total} is

$$T_{total} = T_{pre-copy} + T_{down} \tag{6.7}$$

$$T_{total} = P_s * \tau * \frac{1-p^{n+2}}{1-p} \tag{6.8}$$

Total number of pages migrated during migration $P_{TotalMig}$ is

$$T_{totalMig} = P_s * \left(\frac{1-p^{n+1}}{1-p} + \alpha\right) \tag{6.9}$$

Gain(G) as the ratio between the downtime without live migration and with live migration.

$$G = \frac{P_s * \tau}{T_{down}} = \frac{P_s * \tau}{P_s * \tau * P^{n+1}} = \frac{1}{P^{n+1}} \tag{6.10}$$

The utilization of network U_{net} is defined as

$$U_{net} = \frac{p - p^{n+2} + \alpha(1-p)}{1 - P^{n+2}} \tag{6.11}$$

where α is the representation of uniform dirtying rate of VM. $\alpha < 1$. Most programs under VM exhibits principle of locality reference. It means probability of accessing page is very frequent. Those pages are called Hot pages. In the first iteration of VM it copies all the number of pages of VM. Second iteration onwards it copies only the pages which are dirtied. There are two different situations. First, all the dirtied pages will be transferred in multiple iterations. Second, all dirtied pages are transferred with downtime without any more iterations. In the first situation total downtime, total pages transferred, Gain and network utilization are modeled as

$$D_{effective} = D_{nh}(1-\beta) + D_h * \beta \tag{6.12}$$

$$p_{effe} = D_{effective} * \tau \tag{6.13}$$

$$T_{down} = P_s * \tau * P_{eff}^{n+1} \tag{6.14}$$

$$P_{TotalMig} = P_s * \left(\frac{1 - P_{eff}^{n+1}}{1 - P_{eff}}\right) + \alpha \tag{6.15}$$

$$G = \frac{1}{P_{eff}^{n+1}} \tag{6.16}$$

$$U_{net} = \frac{P_{eff} - p_{eff}^{n+2} + \alpha(1 - p_{eff})}{1 - P_{eff}^{n+2}} \tag{6.17}$$

In the second situation total downtime, total pages transferred, Gain and network utilization are modeled as

$$p_{nh} = D_{nh} * \tau \tag{6.18}$$

$$T_{down} = P_s * \tau * [(1-\beta) * p_{nh}^{n+1} + \beta] \tag{6.19}$$

$$P_{TotalMig} = P_s * \left[(1-\beta)\left(\frac{1 - p_{nh}^{n+1}}{1 - p_{nh}}\right) + (\alpha + \beta)\right] \tag{6.20}$$

$$G = \frac{1}{(1-\beta) * p_{nh}^{n+1} + \beta} \tag{6.21}$$

$$U_{net} = \frac{(1-\beta)(p_{nh} - p_{nh}^{n+2}) + (\alpha+\beta)(1-p_{nh})}{(1-\beta)(p_{nh} - p_{nh}^{n+2}) + (\beta)(1-p_{nh})} \tag{6.22}$$

References

1. R.W. Ahmad, A. Gani, S.H.A. Hamid, M. Shiraz, A. Yousafzai, F. Xia, A survey on virtual machine migration and server consolidation frameworks for cloud data centers. J. Netw. Comput. Appl. **52**, 11–25 (2015)
2. A. Aldhalaan, D.A. Menasc, Analytic performance modeling and optimization of live vm migration, in *EPEW*, vol. 8168, Lecture Notes in Computer Science, ed. by M.S. Balsamo, W.J. Knottenbelt, A. Marin (Springer, Berlin, 2013), pp. 28–42
3. S. Al-Kiswany, D. Subhraveti, P. Sarkar, M. Ripeanu, Vm Flock: virtual machine co-migration for the cloud, in *HPDC* ed. by A.B. Maccabe, D. Thain (ACM, 2011), pp. 159-170
4. Amazon: Amazon elastic compute cloud (2010), Disponvel online em http://aws.amazon.com/ec2/
5. A. Bejleri, O. Shurdi, A. Xhuvani, I. Tafa, E. Kajo, The performance between XEN-HVM, XEN-PV and open-VZ during live-migration. Int. J. Adv. Comput. Sci. Appl. (IJACSA) **2**(9) (2011)
6. P. Bellavista, A. Zanni, Feasibility of fog computing deployment based on docker containerization over raspberrypi, in *ICDCN* (ACM, 2017), p. 16
7. A. Beloglazov, R. Buyya, Optimal online deterministic algorithms and adaptive heuristics for energy and performance efficient dynamic consolidation of virtual machines in cloud data centers. Concurr. Comput. Pract. Exp. **24**(13), 1397–1420 (2012)
8. M.V. Bicakci, T. Kunz, Tcp-freeze: beneficial for virtual machine live migration with ip address change?, in *IWCMC* (IEEE, 2012), pp. 136–141
9. J. Cao, L. Zhang, J. Yang, S.K. Das, A reliable mobile agent communication protocol, in *ICDCS* (IEEE Computer Society, 2004), pp. 468–475
10. F. Checconi, T. Cucinotta, M. Stein, Real-time issues in live migration of virtual machines. European conference on parallel processing (Springer, Berlin, 2009), pp. 454–466
11. C. Clark, K. Fraser, S. Hand, J.G. Hansen, E. Jul, C. Limpach, I. Pratt, A. Warfield, Live migration of virtual machines, in *NSDI* ed. by A. Vahdat, D. Wetherall (USENIX, 2005)
12. A. Feldmann, R. Bradford, E. Kotsovinos, H. Schioberg, Livewide-area migration of virtual machines including local persistent state, in *VEE'07* (ACM, San Diego, California, USA, 2007)
13. M. Forsman, A. Glad, L. Lundberg, D. Ilie, Algorithms for automated live migration of virtual machines. J. Syst. Softw. **101**, 110–126 (2015)
14. D. Funato, K. Yasuda, H. Tokuda, Tcp-r: Tcp mobility support for continuous operation, in *ICNP* (IEEE Computer Society, 1997), p. 229
15. T. Guo, V. Gopalakrishnan, K.K. Ramakrishnan, P.J. Shenoy, A. Venkataramani, S. Lee, Vmshadow: optimizing the performance of latency-sensitive virtual desktops in distributed clouds, in *MMSys* ed. by R. Zimmermann (ACM, 2014), pp. 103-114
16. K. Haselhorst, M. Schmidt, R. Schwarzkopf, N. Fallenbeck, B. Freisleben, Effcient storage synchronization for live migration in cloud infrastructures, in *PDP* ed. by Y. Cotronis, M. Danelutto, G.A. Papadopoulos (IEEE Computer Society, 2011), pp. 511-518
17. M.R. Hines, U. Deshpande, K. Gopalan, Post-copy live migration of virtual machines. Op. Syst. Rev. **43**(3), 14–26 (2009)
18. A. Kivity, Y. Kamay, D. Laor, U. Lublin, A. Liguori, in *Proceedings of the 2007 Ottawa Linux Symposium (OLS- 07*, Kvm: the linux virtual machine monitor (2007)

19. H. Liu, C.-Z. Xu, H. Jin, J. Gong, X. Liao, Performance and energy modeling for live migration of virtual machines, in *HPDC* ed. by A.B. Maccabe, D. Thain (ACM, 2011), pp. 171–182
20. D. A. Maltz, P. Bhagwat, Msocks: an architecture for transport layer mobility, in *INFOCOM* (IEEE, 1998), pp. 1037–1045
21. Microsoft: Windows azure platform (2010), Disponvel online em http://www.microsoft.com/windowsazure/
22. D.S. Milojicic, F. Douglis, Y. Paindaveine, R. Wheeler, S. Zhou, Process migration. ACM Comput. Surv. **32**(3), 241–299 (2000)
23. M. Nelson, B.-H. Lim, G. Hutchins, Fast transparent migration for virtual machines, USENIX annual technical conference, General Track, (USENIX, Berkeley, 2005) pp. 391–394
24. B. Nicolae, F. Cappello, A hybrid local storage transfer scheme for live migration of i/o intensive workloads, in *HPDC* ed. by D.H.J. Epema, T. Kielmann, M. Ripeanu (ACM, 2012), pp. 85–96
25. B. Nicolae, F. Cappello, Towards effcient live migration of i/o intensive workloads: a transparent storage transfer proposal (2011)
26. Z. Pan, Y. Dong, Y. Chen, L. Zhang, Z. Zhang, Compsc: live migration with pass-through devices, in *VEE* ed. by S. Hand, D.D. Silva (ACM, 2012), pp. 109–120,
27. P.D. Patel, M. Karamta, M.D. Bhavsar, M.B. Potdar, Live virtual machine migra- tion techniques in cloud computing: a survey. Int. J. Comput. Appl. **86**(4), 18–21 (2014)
28. C. Perkins, Ip encapsulation within ip (1996)
29. K. Rybina, A. Patni, A. Schill, Analysing the migration time of live migration of multiple virtual machines, in *CLOSER*, ed. by M. Helfert, F. Desprez, D. Ferguson, F. Leymann, V.M. Muoz (SciTePress, Portugal, 2014), pp. 590–597
30. H. Shi, N. Chen, R. Deters, Combining mobile and fog computing: using coap to link mobile device clouds with fog computing, in *DSDIS* (IEEE, 2015), pp. 564–571
31. F. Travostino, P. Daspit, L. Gommans, C. Jog, C.D. Laat, J. Mambretti, I. Monga, B.V. Oudenaarde, S. Raghunath, P. Wang, Seamless live migration of virtual machines over the MAN/WAN. Future Gener. Comput. Syst. **22**(8), 901–907 (2006)
32. B. Varghese, N. Wang, D.S. Nikolopoulos, R. Buyya, Feasibility of fog computing, in *ICDCN* (ACM, 2017), p. 16
33. W. Voorsluys, J. Broberg, S. Venugopal, R. Buyya, Cost of virtual machine live migration in clouds: A performance evaluation, in *CloudCom*, vol. 5931, Lecture Notes in Computer Science, ed. by M.G. Jaatun, G. Zhao, C. Rong (Springer, Berlin, 2009), pp. 254–265
34. B. Wei, A novel energy optimized and workload adaptive modeling for live migration. Int. J. Mach. Learn. Comput. **2**(2), 162 (2012)
35. B.Wellington, Secure dns dynamic update rfc 3007 (2000)
36. D. Williams, H. Jamjoom, H. Weatherspoon, The xen-blanket: Virtualize once, run everywhere, in *Proceedings of the 7th ACM European Conference on Computer Systems, EuroSys '12* (ACM, NewYork), pp. 113–126
37. T. Wood, K.K. Ramakrishnan, P.J. Shenoy, J.E. van der Merwe, J. Hwang, G. Liu, L. Chaufournier, Cloudnet: Dynamic pooling of cloud resources by live wan migration of virtual machines. IEEE/ACM Trans. Netw. **23**(5), 1568–1583 (2015)
38. M. Yannuzzi, R.A. Milito, R. Serral-Graci, D. Montero, M. Nemirovsky, Key ingredients in an iot recipe: Fog computing, cloud computing, and more fog computing, in *CAMAD* (IEEE, 2014), pp. 325–329
39. J.-S. Yih, Y.-H. Liu, Data center hosting services governance portal and google map-based collaborations, in *Web Information Systems Engineering - WISE 2009*, vol. 5802, Lecture Notes in Computer Science, ed. by G. Vossen, D.D.E. Long, J.X. Yu (Springer, Berlin, 2009), pp. 455–462
40. S. Zaman, D. Grosu, Combinatorial auction-based mechanisms for vm pro- visioning and allocation in clouds, in *CCGRID* (IEEE Computer Society, 2012), pp. 729–734

41. Y. Zhang, S.K. Dao, A persistent connection model for mobile and distributed systems, in *ICCCN* (IEEE, 1995), p. 300
42. J. Zheng, T.S.E. Ng, K. Sripanidkulchai, Workload-aware live storage migration for clouds, in *ACM Sigplan Notices*, vol. 46 (ACM, 2011), pp. 133–144

Chapter 7
Virtual Machine Migration in Cloud Computing Performance, Issues and Optimization Methods

Preethi P. S. Rao, R. D. Kaustubh, Mydhili K. Nair and S. Kumaraswamy

Abstract This chapter tries to broaden the reader's perspective on Virtualization and how it works at the heart of Cloud Computing. Advantageous features of virtualization such as cost effectiveness, portability, security etc. can be manipulated to effectively provide cloud services to users. Virtualization can create an image of personal servers while in reality storing, processing and manipulation of data is done on a few physical servers present at the data centres of cloud service providers. We further focus on need for virtualization in the following topics: migrate workloads as needs change, protect apps from server failure, maximising uptime, consolidation and resource optimization. That done, we want the reader to learn about the architectural design of working and storage structures of a key virtualization technology, VMWare) elaborating on their functionalities, how performance goals are met, reduction of complexity and increasing reliability, security.

Keywords Virtualization · Big data · Cloud computing · VM Migration · Optimization · vSphere

7.1 Introduction to Cloud Computing and Virtualization

This chapter tries to broaden the reader's perspective on Virtualization and how it works at the heart of Cloud Computing. Advantageous features of virtualization such as cost effectiveness, portability, security etc. can be manipulated to effectively provide cloud services to users.

P. P. S. Rao (✉) · R. D. Kaustubh · M. K. Nair · S. Kumaraswamy
M S Ramaiah Institute of Technology, Bengaluru, India
e-mail: preethi.psrao@gmail.com

M. K. Nair
e-mail: mydhili.nair@msrit.edu

P. P. S. Rao · R. D. Kaustubh · M. K. Nair · S. Kumaraswamy
Global Academy of Technology, Bengaluru, India

B. S. P. Mishra et al. (eds.), *Cloud Computing for Optimization: Foundations, Applications, and Challenges*, Studies in Big Data 39,
https://doi.org/10.1007/978-3-319-73676-1_7

7.1.1 Cloud Computing

It can be described as a technique of accessing storage and data processing servers remotely to overcome the drawbacks of local computation.

7.1.2 Virtualization

Virtualization can create an image of personal servers while in reality storing, processing and manipulation of data is done on a few physical servers present at the data centers of cloud service providers.

7.1.3 Case Study: Benefits of Virtualization Reaped by Sports Therapy, New York (a Health Care Provider)

Sports Therapy of New York, a health provider having 24 locations across the city, saw the advantage of virtualization when it was able to overcome the loss of money and time caused by repeated failure and large downtime of physical servers (Bigelov, 2011). The warranties of the physical servers bought in 2005 had expired, and were causing around 24–30 h of downtime for each crash. They selected VMware for server virtualization, and currently, 12 Virtual Machines (VMs) are hosted on 3 physical servers. Future vision is to get 10 desktop virtualization using VMware vision. This lead to growth in reliability and increased availability, thereby overcoming the requirement of high maintenance for the small-scale organization.

7.2 Introduction to VM and Its Role in Resource Management

According to Popek and Goldberg [7], a Virtual Machine is an efficient, isolated, duplicate of a real computer machine. Here, a physical computers resources are shared among different virtual environments, which are then presented to the users.

The physical resources of server include memory, CPU, disk and network. VMs help in making efficient use of physical resources of servers in the cloud. These VMs are typically used in data centres, where they can be copied, moved, reassigned among the host physical servers to optimize the utilization of the current underlying hardware. VMs provide partitioning (running multiple VMs in single server), isolation (VM is isolated from other VMs on same server), hardware independence (running VMs without modification on any server).

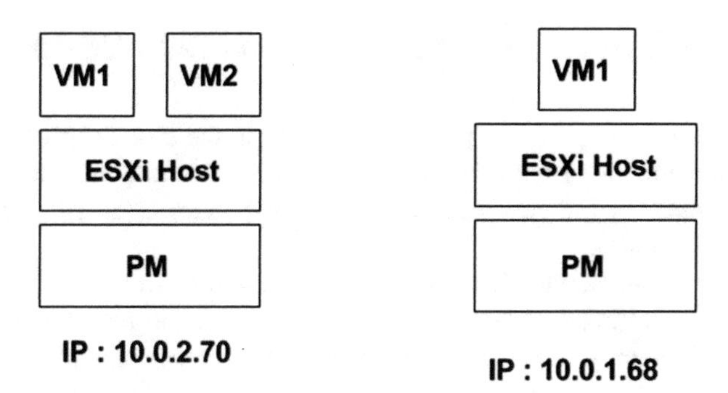

Fig. 7.1 Example of VM setup

Hypervisor or Virtual machine monitor (VMM) is required to run VMs. Resource sharing among the VMs, keeping track of the state of VMs, hardware resource emulation etc., is done by the hypervisor. Load balancing, solving resource contention among VMs, unifying multiple OSes and fault tolerance are other functions of the hypervisor. In a typical data center, just 5–15% (on an average) of the physical servers are in use, but use up all the available resources, including high electricity cost (air conditioning needed to cool down the servers). With the use of virtualization, data centers can aim at around 70–80% resource utilization, resulting in greater ROI (Return On Investment).

7.2.1 Typical VMs Setup

Figure 7.1 shows a block diagram shows set up of two ESXi hosts managed by VMWare vSphere tool discussed at the end of this chapter. ESXi host is installed in two physical machines (PM) with IP address: 10.0.2.70 and 10.0.1.68. On top of that, VM1 and VM2 are installed on the first host and a single VM is installed in the second physical host. This is an example to show usage of multiple VMs on a single host. These multiple VMs share the resources of the physical machine (PM).

7.2.2 Case Study on the Motley Fool

The Motley Fool is a company providing online financial services and is based in Alexandria, VA, USA. Traditional data centers proved to be frustrating due to low efficiency, high power consumption, less available space, and difficulty in scaling up. They decided to move to VMware vCentre and are currently about 75% virtualized, with 200 VMs running on 10 physical hosts. This lead to 33% less power consumption of servers, gain in space and improvement in flexibility.

7.3 VM Performance Overhead in Cloud Data Centers

Many companies are opting for pay-as-use model in the cloud services, where the users pay only for the specific amount of resources that they use. Investing huge amount of money in indigenous computer infrastructure is costly, especially for small-scale companies.

There exists SLA (Service Level Agreement) between the users and the cloud service providers, detailing the resources needed at peak and non-peak hours, acceptable downtime period etc. Overcommitting the cloud resources can end up in violating the SLA, whereas loss of revenue occurs in case of underutilization. In order to balance the two, the data centers opt for dynamic resource management while providing cloud based services [1].

In the data centers, multiple VMs are installed on each physical hardware, which helps in providing efficient services to the users. Virtualization based approach in cloud centres has advantages, such as - generation of less heat, reduced hardware cost, easy backup, faster recovery, reduced carbon footprint, and operational efficiency.

The performance overhead factors, reasons and solutions of VMs used in data centres are discussed below. Performance overhead might be caused due to VM migration, snapshotting of VMs, and resource isolation of VMs.

7.3.1 VM Migration

Migration means moving a VM from one host to the other. The host can be in the same or different physical server [4]. VM migration is done under the following conditions [6]:

7.3.1.1 Avoiding Server Sprawls

Server sprawls is an event when many VMs are under-utilized but are taking up much more resources than needed for their workload. By migrating the VMs to other host, the administrator can make better resource optimization. This is also called server consolidation, where the VMs are all brought together onto fewer machines, and the free hosts can be shut down to save electricity costs.

7.3.1.2 Balancing of the Load

Load on the VMs have to be evenly distributed. Sometimes, specific VMs may have severe load, whereas other VMs may be idle. In such a scenario, applications and the VM itself may be moved from one physical server to the other using VM migration.

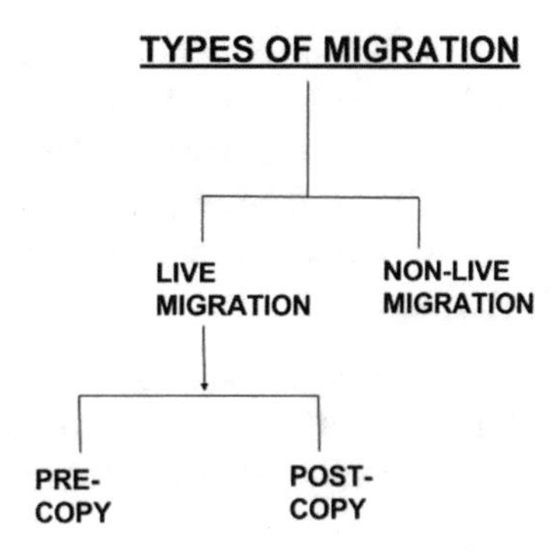

Fig. 7.2 Types of migrations

7.3.1.3 Avoiding Hotspots

Hotspot means the condition where the VM does not have sufficient resources to fulfil the SLA requirements. To remove hotspot, the resources can be provided locally by the same physical machine. Migration of VMs can also be done to remove/mitigate the hotspot.

7.3.2 Types of Migration

Figure 7.2 shows two types of migrations: live (hot) and non-live (cold). In live migration, the source VM is kept running, and migration is done simultaneously. In non-live, the user can feel the interruption while the VM is being migrated. Service will be down until the migration is complete. This non-live migration can be done for the non-essential services of the cloud, where the continuous working of application is not necessary.

7.3.2.1 Live Migration

The live migration has two components in it:

1. The control is transferred to the destination of the migration.
2. The data is then transferred to the required destination.

Algorithm for live migration

1. The hypervisor will mark all pages as dirty.
2. Algorithm will then start transferring the dirty pages iteratively across the network till either the remaining pages are under certain threshold or the maximum iterations are exhausted.
3. The pages at the source are marked as clean.

4. Since the migration is live, the source VM will be working during the migration. Hence, previously clean pages may be marked dirty again by the hypervisor, and will need to be transferred to the destination again.
5. The VM is then suspended and remaining pages are transferred completely. The VM will now resume its functionality at the destination.

Pre-copy

In this approach, pages are transferred continuously without suspension of source VM. Once sufficient pages are transferred, VM gets suspended at the source and the remaining state is transferred to the target physical machine.

Post-copy

As described above, the goal of pre-copy is to keep the service downtime as small as possible, achieving through minimizing the copying of VM state. In this pre-copy, through multiple iterations, the memory state is transferred, followed by the processor state. In contrast, post-copy method does exactly the opposite: the processor state is copied and sent to the target host, followed by transferring the memory state.

Advantage of post-copy over pre-copy is that in pre-copy, if the workload is write-extensive, more pages are dirtied and have to be retransmitted. This leads to the increase in VM downtime.

Algorithm for Post-copy

1. The processor state is copied first from source to the target.
2. VM at the target is started.
3. VM memory pages are actively pushed to the target.
4. Concurrently, the memory pages that the target faults and which are not yet pushed are demand-paged to the source through the network.

7.3.2.2 Non-live Migration (Cold Migration)

Its also known as suspend-and-copy. The VM is turned off or suspended till the migration completes.

Two Cases of Cold Migration:

1. If a suspended VM is migrated, then the compatibility issues of the CPU must be met in the new host. This is needed to enable the running of suspended VM and resume executing the instructions from the new destination when the migration is complete.
2. If a turned off VM is migrated, then CPU compatibility checks are not needed.

In cold migration, the virtual machines can be migrated to different data-store as well. Generally, the configurations files, the log files, BIOS files and suspend files are all moved to the destination storage area. The associated disk files can be moved to the destination host situated in the other data-store as well. The main advantage of cold migration is that the setup (host and the destination) need not have shared memory.

7.3.3 VM Snapshot

Snapshotting is a process, apart from VM migration, which causes overhead in VMs performance. Snapshotting of a VM means copying the virtual machine's disk file at the given point in time. (Using the snapshot, retrieved from https://www.vmware.com/support/ws4/doc/preserve_snapshot_ws.html)

In case of any errors, or system failure, the VM can be successfully restored to the earlier working state through the snapshot.

7.3.3.1 Steps in VM Snapshotting

The following steps take place when a snapshot is taken:

1. The virtual machine is paused.
2. Differencing disk for each VHD (virtual hard drive) is created and is associated to the VM. Note: Differencing disks mean a VHD (virtual hard disk) that is used for storing all the changes made to guest operating system or another VHD. The main purpose of differencing disks is to maintain record of the changes made so that It can be reversed if necessary.
3. A copy is made of the VMs configuration file.
4. The virtual machine is then resumed.
5. The VMs memory is saved to the disk once the VM resumes.
6. In this process, if the operating system attempts to make a write operation, it must be intercepted and let the original contents of memory get copied. After copy operation is done, the write operation is resumed (this write is not reflected in the snapshot)

7.3.3.2 Restrictions in Snapshotting

- Snapshot alone cannot provide backup in the cloud data centres.
- Size of the snapshot file cannot exceed the size of the original disk file.
- Snapshot requires additional disk space, thereby affecting the performance of VM.
- Reliance on snapshot for VM which has extensive data changes is not recommended, as inconsistencies might occur when the VM is restored through the snapshot.
- When a snapshot is taken, any writable data now becomes read-only. Size of snapshot file increases rapidly with write-heavy data.
- Potentially serious and unintended consequences may occur in the form of data loss if the virtual hard disk which is connected to a VM that has snapshots is edited.

VM snapshots are used mainly in testing and development stages, enabling us to quickly revert back to a previously working state of the VM. In production environment too, the snapshots can be used before undertaking a risky operation.

7.3.3.3 Deleting a Snapshot

When a snapshot is deleted, the saved state file and the configuration files are immediately deleted, and the snapshot entry is removed from the user interface.

7.3.3.4 Types of Deletion

1. If a snapshot which has no other snapshot referencing it is deleted, then all the files associated with the snapshot get deleted.
2. If a snapshot with only one descendent (snapshot which references the current snapshot) is deleted, then the snapshot differencing disks will be merged together and the configuration along with saved state files will be deleted.
3. A snapshot may have two descendants dependent on it (two or more snapshot referencing to the current snapshot). Deleting the snapshot in such a case will result in deletion of the configuration and the saved state files - but the differencing disks wont be merged till the number of descendants is one.

7.3.4 Shared Cloud Network and Contention Issues

- SAN - Storage Area Network
  This is a type of network which provides access to block level data storage and is used to enhance the performance and speed of accessibility of storage devices, like tape and disk arrays. Typically, the amount of data which can be accessed by a server is limited by the storage attached to it. Instead, a common storage utility is made accessible to multiple different types of servers. SAN provides high speed data transfer between storage and the servers which use the data.
  In SAN, the applications servers need not store all relevant data. Instead, they can be optimized for running their own applications. The shared space storage ensures the peak storage of individual servers is met easily. The physical machines use the LAN to communicate with the storage, hence potentially causing performance issues and creating bandwidth.
- Storage QoS
  Storage QoS (Quality of Service) includes bandwidth, latency and queue depth. Here, queue depth refers to the number of operations waiting to be executed to the disks. Storage QoS options differ from vendor to vendor, and third-party offer them too. In the shared network system, a spike in usage from one user can cause decrease in performance in other users of the same network. This is called "noisy neighbour effect".
  Using performance monitoring tools, the storage administrator can analyse the minimum IOPS (input output per second) required by a typical VM. These IOPS can then be reserved for each VM to avoid contention by the storage QoS. This method can also be used to put a limit to the maximum bandwidth usage per VM,

applicable for the high-demand VMs. This way, resource contention can be managed.
Also, reduction in number of VMs sharing common storage is needed to avoid performance degradation. Bottleneck needs to be identified using performance monitoring, and this can give the exact cause of the problem.

- Virtual switch
  Virtual switch will enable one VM to communicate with another. This is more than just forwarding packets of data. These switches can monitor data packets before forwarding them. This helps in moving and migrating VMs from one host to another and in expanding server virtualization. Functionality of switches also include maintenance of a table of MAC address to port. This is used to forward the frames from the ports to the correct destination.

7.4 Optimization in VM Migration

7.4.1 Optimizing VM Migration in Networking

Networking in virtualization brings together globally placed entities into a single software controlled unit. This embodies various software and hardware resources and their functionalities to form a virtual network [11, 14])

Some approaches of optimization are:

1. De-duplication
2. Throttling

De-duplication

Pre-copy is a popular method in migration where pages are migrated and copied onto the target system one after the other. Once a threshold number of pages are migrated to the destination, the state of the VM is suspended on the host and is completely migrated.

This approach obviously shows the disadvantage of having to transfer duplicate data thereby leading to an increase in:

1. Migration Time.
2. Down Time (time for which the hosts have to be suspended).

In the event of repeated updation of a page on the host machine during pre -copy, the copies of the updates have to also be stored repeatedly on the destination machine. This leads to excess memory usage in the host as well as destination.

The graph in Fig. 7.3 clearly shows the relative improvement for different chunk sizes when original disk access method was replaced. For 128 bytes it went down to just 6% and for 512, 4096 and 16384 bytes it went down to 10, 6 and 2% respectively.

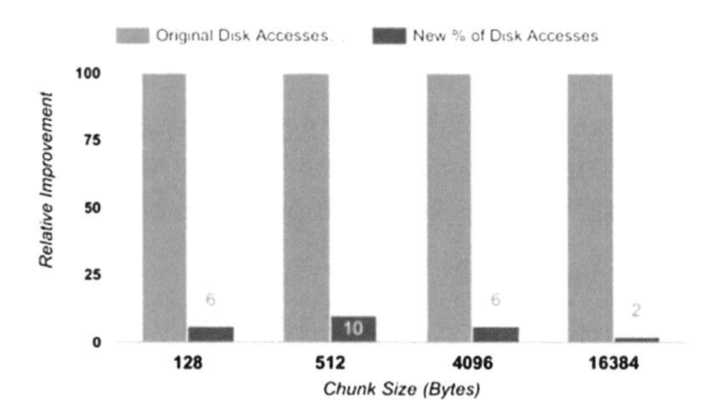

Fig. 7.3 Graph showing chunk size versus relative improvement for disk access [10]

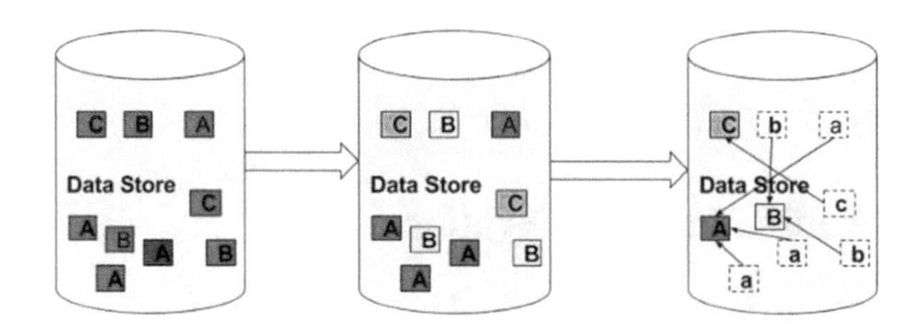

Fig. 7.4 Steps in hash based deduplication (From left to right)

De-duplication deals with this drawback by comparing data blocks/pages to identify duplicates and erase redundancy during migration. It employs **hash based de-duplication** where the data blocks are fed to a hash function which compares them against the entries in a hash table. If similarity is found the count of occurrence for that block is incremented and it is marked as redundant. If a unique block is found it is added to the hash table. Thus it now only contains non redundant data and only these are migrated. This leads to efficient storage management and reduced migration time.

In the Fig. 7.4, the memory is divided into small chunks and each is allotted a unique ID. They are compared to find duplicates and these are removed and replaced by pointers to the actual chunks. Thus only the dissimilar chunks remain and the migration process is accelerated. Data blocks/chunks in de-duplication can be:

1. **Fixed sized**: Memory is divided into fixed sized blocks. This makes it fast and simple. But smaller blocks require more processing to be done to achieve de-duplication whereas large blocks require less processing but efficiency de-duplication decreases with increase in size.
2. **Variable sized**: Takes care of disadvantages of de-duplication of fixed sized blocks but CPU overhead increases in trying to determine the size and boundary of each block.

Throttling

The network may sometimes experience an overload in migrations if all of them are triggered at the same time. In these situations, it is necessary to throttle the flow of VMs on a network. Here, throttling means controlling the number of VMs travelling towards a destination host away from a source host.

One way to achieve throttling to restrict the bandwidth available for the VMs. A maximum upper limit can be set that will signify the upper boundary of usable network bandwidth for a VM.

Bandwidth can also be allocated according to the conflicts arising between different VMs migrating at the same time. Some can get a greater bandwidth compared to others instead of having fixed bandwidth boundaries for all.

- **Transfer Throttle**
  Assume that the pages are being updated continuously at the host machine. These pages are marked dirty and should hence be updated on the destination machine. These dirty pages are transferred on the net bandwidth allotted to the VM. The rewriting rates of each of the pages on the destination machine is then measured. If this sum is more than the allotted bandwidth, the transfer of pages that remain on the source are throttled.
- **Increasing redundancy to increase network throughput**
  Bringing redundancy in network components such as the NIC can prevent stoppages due to component failure in the network. A physical NIC can be assigned to each virtual server. This prevents any interference between VMs running on the same physical machine.
  NICs can also be connected to their switches. In case of NIC or switch failure, each server can directly connect to the backbone network. [13]
  Example: In VMware, a virtual dual switch can be configured to act as an interface on a pair of NICs.

7.4.2 Optimization in Application Performance

- **Write throttling**
  Sometimes, during migration, rate at which applications create dirty pages on the host can be faster than the allotted bandwidth of migration. This has to be curbed to have a stable synchronization effect between the source and destination. In the case of "the number of dirty pages exceeding a limit", the kernel interrupts the application and stops it from creating more dirty pages. It then waits for the dirty pages to be "cleaned" at the destination. This process is write throttling.
- **Optimization through resource management**
  Application performance can be optimized by predicting their performance for a given allocation scheme of resources. This prediction can also be used for resource management in an environment where number of applications have increased rapidly and the number of resources remain the same.

Conventionally, each application would have a dedicated set of resources that it can use without having to share. But with growing number of applications on different VMs, resources must be optimally timed and shared across the network of VMs.
Traditionally, multiple VMs would work on similar applications and demand for a single type of resource by all VMs had the potential to create serious bottlenecks. An approach to optimizing application performance can be - allocating different applications to different VMs. That way all resources can be used uniformly across VMs without having to send demand for a single resource at the host by all VMs.

- **Dynamic write throttling**
  It is a technique that overcomes the drawbacks of currently deployed static sized buffers. In case of an event of congestion detection, the clients buffers are to be dynamically throttled.
  This ensures that the rate at which the client's buffer is being written at is slowed and an event of bursty traffic (very high bandwidth usage for a short time period) is avoided to optimize networking conditions and application performance on the host as well as destination increases.
- **Rate Limiting**
  Rate limiting deals with restricting to and fro traffic between applications running on VMs either present on their source machines or destination machines.
  Some Network Interface Cards on host machines supports hardware that limit rate of input and output of buffer queues. With increasing server consolidation (resources provided by the server are optimally used to have less number of servers), the need to have rate limiting hardware is important to protect servers from uncontrollable, congestion-causing, traffic.

7.4.3 Power Optimization

With virtualization it was possible for many applications to be run on a limited number of servers. This drastically reduced the number of hardware resources required. Furthermore, the possibility of being able to migrate applications from one physical machine to another leads to optimal load balancing, and while one physical machine can run more than one application, the other physical machines can be put to sleep thereby conserving power.

Virtualization and VM migration also reduces the memory power (constitutes up to 19% of the total machine power) consumed.

Clients usually have complex SLAs and cloud service providers, in order to meet them, must overprovision resources. Heavy usage of resources leads to high operational costs of servers in datacenters. The electrical energy consumed and thus the emission of carbon increases.

Optimal provisioning of resources is the solution to reducing the negative impacts and keeping up with the SLAs of the clients [11].

This is a difficult job as applications requested by the clients along with the SLAs they provide suffer from tremendous diversity.

Dynamic Voltage Frequency Scaling(DVFS) Algorithm

It is an algorithm developed to optimize energy consumption in cloud data centers. The frequency of operation of the processor is adjusted based on the current need.

Thermal design point/Thermal design power is the amount of heat dissipated by a computer system. It is measured in Watts. Now, power is a product of amount of flow of current in an electrical unit per unit time and the potential across it (voltage). Thus, a watt = amps * volts, where the current is measured in amps and the voltage is measured in volts [12].

This equation makes the power be directly proportional to the voltage under a constant amperage. Hence adjusting the voltage can directly influence the amount of power dissipated. DVFS algorithm is used to adjust the MIPS (million instructions per second) frequency and optimize the electrical energy dissipation by fluctuating the potential (voltage) as per requirement on the selected machine.

The DVFS works by selecting an appropriate Virtual Machine and adjusts the frequency. The operation of the VM is then adjusted to the required clock frequency instead of the original frequency it is required to run on.

Input: The algorithm takes the VM that is victimized and the frequency that is required to be set as the input.

Output: The output of the algorithm is consumption of energy that is optimized. It is measured in Watts.

Case Study for Power Optimization

Data centres use 40–60% of additional power for maintaining a cool environment for the servers. Some cloud service providers hence sought to bring down this energy expenditure by setting up their data centres in places with a cold climate.

1. Facebook: Lulea, in Sweden, has a Facebook data center 70 miles south of the Arctic Circle. Since it is located very close to the Arctic Circle, the climate is quite cold and the energy requirement for cooling was largely mitigated.
2. Google: The DeepMind AI (Artificial Intelligence) unit was put in charge by Google to optimize its power usage. This method was cited to reduce the electricity used in cooling by as much as 40
3. Microsoft: The idea of having underwater data centers is being explored at Microsoft.

7.4.4 Storage Optimization

The concept of virtual storage can be described as - Storage in form of uniform memory chunks are present on various physical machines either locally or remotely.

Virtualization helps in accumulating these chunks together to form a picture of contiguous memory for the benefit of the clients. Thus a client does not have to wholly rely on the constricted memory available on the local physical machine it is running on [9].

You can pick one the below two steps to optimize virtual hard disk storage:

1. The hardware used to store the virtualized hard disk must have a high throughput and good performance. A SAN (Storage Area Network) or SCSI disk (Small Computer System Interface) can be used. SCSI allows a computer to be interfaced with peripheral such as tape drives, CDS, disk drives, printers, etc. and this allows a fast and flexible data transfer rates. A SAN provides access to storage devices that are either present locally or remotely by forming a network of these devices. It overcomes the limitation of a LAN network that is usually incapable of accessing devices located outside its reach.

2. Although having fragmented chunks of memory is an advantage over large contiguous chunks, a periodic fragmentation is necessary especially in a dynamically growing VHD(Virtual Hard Disk) environment. This is because the virtual memory dynamically becomes larger and the amount of fragmentation increases with it. This makes it harder to keep track of each of the fragment and also the concept of contiguous memory becomes harder to achieve.

Storage Migration - Optimization and Best Practices

This concept deals with migrating the virtual hard disks of a VM while being available for usage by the VM.

Steps in Migrating a VHD:

1. The read-write operations begin and end at the source VHD all through the migration process.
2. The new or modified contents are then copied later to the destination VHD.
3. Any dirty or outstanding pages are committed at the destination after checking with the source VHD.
4. Complete synchronization of the source and the destination takes place and switching of VM takes over from old VHD to the new one at the destination.
5. The VHD is deleted at the source.

Best Practices

1. Knowing the form of the data in the VHD that is being migrated is important along with form it will take after migration on the destination machine.
2. Prior to moving the data, its extraction, deduplication and transformation has to be done.
3. Guidelines for migration must be followed so that the VHD is migrated in a precise and orderly manner.
4. Accuracy of the data can be checked through various validation tests.
5. Review and verification of the data should follow migration.
6. All the steps from the start of migration to its end must be documented.

7.4.5 Cost Minimization Algorithm

The cost minimization algorithm can be formulated as,

$$Min \quad C_p \sum_j [x_j P_j^0 + P_j^p \sum_i \phi_{ij}] T_e + \sum_i \sum_j z_{ij} cost_i^m + T_e \sum_i f_i^c \lambda_i \sum_j a_{ij} e^{-(c_j^p \phi_{ij} \mu_{ij} - \alpha_{ij} \lambda_i) R_i^c} \tag{7.1}$$

where, $x_j \geq \sum_i \alpha_{ij}, \forall j$

$\phi_{ij} \geq y_{ij}(\frac{(\alpha_{ij}\lambda_i - ln \frac{h_i^c}{R_i^c})}{\mu_{ij} C_j^P}), \forall i, j$

$\sum_i \phi_{ij} \leq 1, \forall j$

$\sum_j y_{ij} m_i \leq C_j^m, \forall j$

$\sum_j a_{ij} = 1, \forall i$

$y_{ij} \geq \alpha_{ij}, y_{ij} \leq 1 + \alpha_{ij} - \varepsilon, \forall i, j$

$z_{i,j} \geq y_{ij} - y_{ij}^p, \forall i, j$

and

$x_j \in 0, 1, y_{ij} \in 0, 1, z_{ij} \in 0, 1, \forall i, j$

$\phi_{ij} \geq 0, \alpha_{i,j} \geq 0, \forall i, j$

The algorithm consists of 3 steps (Table 7.1):

- **Sorting of Clients and dynamic assignment of resources**: Client's status consideration from the previous timestamp as well as resource demand estimation for next decision time period determines the sorting of clients. The result is used in creation of VMs on the available servers. But this may lead to under-utilization of servers and uncompetitive sharing of resources.
  $Min\, C_p \sum_j [(P_j^0 + P_j^p)\phi_{ij}] T_e + \sum_j z_{ij} cost_i^m + T_e f_i^c \lambda_i \sum_j a_{ij} e^{-(c_j^p \phi_{ij} \mu_{ij} - \alpha_{ij} \lambda_i) R_i^c}$
  subject to:
  $\phi_{ij} \geq y_{ij}((\alpha_{ij}\lambda_i - \ln h_i^c / R_i^c)/\mu_{ij} C_j^p), \forall j$
  $\phi_{ij} \leq 1 - \phi_j^p, \forall j$
  $y_{ij} m_i \leq (1 - \phi_j^m) C_j^m, \forall j$

  The above formula describes the minimization of the cost for the client that was chosen after the sorting process.

  Algorithm 1: Dynamic Programming Resource Assignment
  Inputs: $C_p, T_e, cost_i^m, \lambda_i, f_i^c, C_j^p, R_i^c, \mu_{ij}, P_j^0, P_j^p \phi_j^p, \phi_j^m, C_j^m, C_j^p, h_i^c$ and m_i
  Outputs: ϕ_{ij}, α_{ij} (i is constant in this algorithm)
  ga = granularity of alpha;
  for(j=1 to number of servers)
  for(α_{ij} = 1/ga to 1)
  ϕ_{ij} = optimal resource shares based on KKT conditions
  $C(j, \alpha_{ij}) = C_p T_e (P_j^0 + P_j^p)\alpha_{ij} + z_{ij} cost_i^m + T_e f_i^c \alpha_{ij} \lambda_i exp(-(C_j^p \phi_{ij} \mu_{ij} -$

Table 7.1 Variables used in the algorithm [8]

Symbol	Definition
λ_i	Predicted average request rate of the ith client
R_i^c, f_i^c	Contract target response time and penalty values for each request in the SLA contract
h_i^c	Hard constraint on the possible percentage of violation of the response time constraint in the SLA contract
μ_{ij}	Average service rate of requests of the ith client on a unit processing capacity of jth server
m_i	Required memory for the ith client
C_j^p, C_j^m	Total processing and memory capacities of the jth server
$cost_i^m$	Migration cost of the ith client
P_j^0, P_j^p	Constant and dynamic (in terms of utilization) power consumption of the jth server operation
T_e	Duration of the decision epoch in seconds
C_p	Cost of energy consumption
y_{ij}^p	Pseudo Boolean parameter to show that if the ith client is assigned to the jth server in previous epoch (1) or not (0)
x_j	A pseudo-Boolean integer variable to determine if the jth server is ON (1) or OFF (0)
y_{ij}	Pseudo Boolean parameter to show that if the ith client is assigned to the jth server (1) or not (0)
α_{ij}	Portion of the ith client's requests served by the jth server
ϕ_{ij}	Portion of processing resources of the jth server that is allocated to the ith client

$\alpha_{ij}\lambda_i)R_i^c)$
End
End
X = ga, and Y=number of servers
For (j= 1 to Y)
For (x=1 to X)
D[x,y] = infinity; // Auxiliary XxY matrix used for DP
For (z=1 to x)
D[x,y] = min(D[x,y], D[x-1,y-z]+C(j, z/ga))
D[x,y] = min(D[x,y], D[x-1, y])
End
End
Back Track to find best $\alpha'_{ij}s$ and $\phi'_{ij}s$ to minimize the cost.

- **Resource adjustment**: It is done to minimize cost of servers that are running applications for multiple clients.

$$Min\ \ C_p P_j^p \sum_{i \in I_j} \phi_{ij} + \sum_{i \in I_j} f_i^c \alpha_{ij} \lambda_i e^{-(C_j^p \phi_{ij} \mu_{ij} - \alpha_{ij} \lambda_i) R_i^c}$$

subject to:
$\phi_{ij} \geq y_{ij}((\alpha_{ij}\lambda_i - ln\ h_i^c / R_i^c)/\mu_{ij} C_j^p), \forall i \in I_j$
$\sum_{i \in I_j} \phi_{ij} \leq 1$

where, I_j is used to represent the Virtual Machines on the server numbered j.

- **Dynamic resource arrangement**: Resources are allocated as and when data is fed for computation. This way client requests are serviced and disadvantages like underutilization of resources is taken care of by turning off the servers that are underutilized. This reduces costs such as energy cost due to an idling server. The minimum cost is found through an iterative method which makes use of results acquired in the previous steps.

1. A threshold is set to compare the extent of utilization of each server. If any server falls under this or is unable to satisfy the threshold value it is deemed as underutilized and turned off.
2. Threshold can be set to a value such as 15
3. After a server is turned off, the total cost is calculated.
4. This is compared with the *least total cost that was globally calculated.* If turning off the server caused any significant changes in the cost, the state is kept as it is.
5. Else the move to turn off the server is undone.
6. This is carried on until the last server showing low utilization is identified.
7. The removed VMS are then reassigned to other running servers.

7.5 Issues, Goals and Challenges

7.5.1 *Issues and Challenges*

7.5.1.1 Big VM Migration

In recent times, IaaS (infrastructure as a service) clouds are providing VMs with huge amount of memory. Amazon EC2 (elastic cloud compute), for example, provides VM of the size of 2TB [2]. These types of large memory sizes are needed for big data analysis. Hence, VM migration becomes difficult because of the huge memory involved.

The destination host would require large amount of free space for migration. Even in cloud infrastructure, large amount of free space is hard to come by. Reserving hosts with large memory size for future migration is not cost effective.

Two issues come up while migrating large VMs:

- **Migration time**: The migration time is an important issue in migrating large sized VMs. This migration time is directly proportional to the size of the VM to be migrated. Hence, larger the size, more the migration time.

This issue can be tackled in two ways:

1. **Using high speed interconnections**: 25, 50, or even 100 Gigabyte Ethernet. This ensures the VM is migrated on faster connections, hence will save the migration time.
2. **Parallelizing VM migration**: Different portions of the data which are unrelated to each other can be migrated simultaneously to the destination though the network. This saves time and large VM migration can be achieved.

- **Destination host**: The destination host having large amount of free space must be available for migration. Having huge free memory hosts reserved all the time for future migration is costly even in the cloud infrastructure. In case large sized VMs cannot be migrated, then it will result in downtime of services and big data analysis is disrupted.

7.5.1.2 Inter VM Migration

The VMs use hypervisor to access system resources, whereas the host machine can access it directly. There exists isolation constraint between VMs, which results in a VM not knowing the existence of another VM on the same host. Communication throughput of the VMs running on the same host will be limited by this isolation constraint. Isolation among VMs of the same physical host is necessary from the security perspective. At the same time, enforcing this isolation will lead to degradation in performance, as explained below. Hence, a balance between the two is needed for effective and secure VM inter-communication.

When two processes residing in the same physical host, but different VMs, want to exchange data among themselves, it is an example of **IVMC** (Inter Virtual Machine Communication). The traditional mechanism for communication between VMs located on same host is using their VNI (virtual network interface).

Therefore, communication between VMs is done via standard network interface, as if the VMs dont share the same physical host. This brings in unnecessary overhead and degrades performance of both the VMs. Now, the data being shared has to be encapsulated, given an address, transmitted and verified through the network as well as the virtualization layer.

Problem with this is that it ends up using too much of CPU and low throughput of network because the network data-path is completely traversed.

This overhead is reduced by using shared memory channel, which is used to communicate between the VMs located on the same host. Shared memory technique has the following challenges:

1. **Providing Security**: The guest as well as the host can access shared memory. The data which is intended for a VM may be altered or maliciously intercepted by another VM. Therefore, information leakage has to be contained, so that the shared communication buffer is not exploited by malicious VM.

2. **Migration**: To perform load balancing, VMs may be migrated from one host to the other. Even when his happens, the communication between the VMs, including the one being migrated, has to continue.

Other challenges include transparency issues, migration support, data integrity, security, and denial of service.

7.5.2 Goals

7.5.2.1 Energy Management

Data centres are huge and have large operational costs. Along with this, electricity consumption is extremely high too. Thousands of servers exists in one data-centre and they need centralized AC (Air-Conditioning) to keep their temperatures down.

Each servers heats up while being switched on. Cooling systems are needed for proper working of the data centres. Hence, electricity is consumed by these cooling systems, and carbon footprint increases due to this. Management of energy is needed in data centres from an economical as well as environmental viewpoint.

Green IT

This is a type of optimization in which the goal is to reduce the carbon footprint associated with the IT (Information Technology) industry. Its aim is to reduce ill effects of IT on the environment by planning, designing, operating, manufacturing, and disposing of computers and their components safely. Decrease in the usage of hazardous materials, increasing efficiency of energy, increasing the products lifetime, and reusing the components can be effective in moving towards Green IT.

PUE (power usage effectiveness): This metric is defined as the ratio between the energy consumed during the process of computing over total energy (including cooling) being consumed for the data centre operations.

In data centres of Google and Facebook, this ratio is close to 1 (other data centres have PUE ratio close to 2).

It is possible to bring down PUE ratio to less than 1 by reusing wasted energy. Heat produced is proportional to the I/O operations (input/output) being down by the servers. The heat which is thus generated can be used to places where heat is required. According to McKinsey report, energy consumption was $11.5billion in 2010 and has doubled in 2015.

Static Power Consumption (SPC)

This is the power consumed by the components of the system. It is independent of the usage scenarios and clock rates. This depends on the type of transistors and the quality of the processors used in the servers. SPC is caused by the leaking of current by the active circuits.

This can be prevented in the following ways:

1. **SVR (Supply Voltage Reduction)**: the supplied voltage itself is reduced. This will ensure server components (like CPU, and cache etc.) have longer life.
2. **Reducing size of the circuits**: This is done by making circuits with fewer transistors. Also, the reduction or cut in power supply to the idle components of the circuit will reduce power consumption.
3. **Cooling technology**: There is less resistance to electricity in lower temperatures. And lower degradation of the chips used. Power leakages can be minimized because the circuits work faster (due to low resistance of electricity).

7.5.2.2 Fault Tolerance

Fault tolerance is among the key issues yet to be fully addressed in the cloud computing industry. It deals with all the techniques and needed to let a system tolerate errors, software faults, and prevent total system failure. Improved performance, increased efficiency, better reliability, reduced costs and failure recovery are the advantages of implementing failure recovery in cloud computing.

On different VMs, multiple instances of the application may be running. If a server goes down, then there is a need for the implementation of automatic fault tolerance techniques to reduce downtime and recover quickly.

Fault Tolerance Metrics:

Throughput - it is the number of completed tasks by the system. In every system, the throughput should be high.
Response time -This is the time taken to reply and respond to the input/query by the process. Minimum response time is required for high performance.
Scalability - The performance of the algorithm of fault tolerance must not be affected by the addition of the nodes/VMs/host machines. Adverse effect on the performance due to scalability of nodes must be prevented.
Performance - Measures the efficiency of the system. This performance metrics has to be increased while keeping the cost down.
Availability -The reliability of a resource is directly dependent on its degree of availability. It is the situation where for any epoch of time, the resource must be functioning.

Types of Fault Tolerance techniques:

- **Reactive Fault Tolerance**:
  These policies are applicable after the failure effectively occurs. They help in limiting the effects of application execution failure.
  - **Checkpoint/Restart**: Task is made to restart from the recently saved checkpoint when it fails. This avoids restarting the task from the beginning, ensuring effective fault tolerance for long running application.

- **Replication**: Multiple copies of the same application is made to run on different VMs and hosts. This ensures that even if one copy fails, other copies of the same application are running smoothly and the application is not affected.
- **Job Migration**: If a task fails, it has to be migrated to another VM, or another host.
- **Retry and Resubmission**: A task is made to retry on the same resource of the cloud when it has failed. This is one of the simplest techniques of fault tolerance. Resubmission is when a failed task at runtime is resubmitted to the same/different resource of the cloud service.
- **User-defined fault handling**: The user (client or the system administrator) gives the specific solution to the particular task failure.

• **Proactive fault tolerance techniques**:
Here, the policies are such that errors, failures and faults are avoided before they occur. Prediction and prevention is initiated on suspected components which may fail in the future. Policies like preemptive migration, software rejuvenation etc. are based on this technique and are introduced below.

- **Software Rejuvenation**: In this, the system is designed for periodic restarts. Each time the systems gets a fresh start.
- **Preemptive Migration**: Here, the system and the application is continuously checked and analyzed to give a feedback response to decide when to preemptively migrate the application from one VM to the other.

7.5.2.3 Load Balancing

Distribution of work and allocation of resources among servers in the cloud data centre is known as Load Balancing. This balancing of load helps in achieving higher performance and simultaneously reducing the operational costs. Energy considerations and constraints are fulfilled as well, and helps in overall reduction of energy and power consumption [3].

Goals of Load Balancing

1. Reduction of costs: Operational costs of servers reduces due to load balancing as the load is evenly distributed among the servers. This ensures servers work at around 80% efficiency workload whereas underutilized servers can be switched off.
2. Flexible and scalable: The data centre can be expanded and changes can be made inside the servers and clusters.

Algorithms for Load Balancing

1. Task Scheduling: This algorithm consists of two steps. First step is to map tasks to VMs and next is to map VMs to the host physical machines. Utilization of resources is high using this algorithm.

2. Round Robin: This algorithm assigns processes to the multiple processors, and each processor is allowed to work for a specified time quanta, after which the control switches over to the next processor.

Examples of Load Balancing Technologies

1. Cloud service providers like Google, Amazon Web Services (AWS), Rackspace and Microsoft Azure provide load balancing technology to their clients.
2. Elastic Load Balancer is provided by AWS for their EC2 instances. It manages the traffic and the workload in their servers.
3. Google Compute Engine is the load balancer provided by Google. It is responsible for dividing the network traffic among its VMs and servers.
4. Traffic Manager is the name of the balancer by Microsoft Azure. Across multiple data centers, the traffic is distributed by Traffic Manager.
5. Rackspace has its Cloud Load Balancer which is responsible for managing the servers for the distribution of workload.

7.5.2.4 Server Consolidation

The main goal of server consolidation is to avoid server sprawls. Server sprawls occur when many under-utilized VMs run on the cloud platform leading to high energy consumption. In this technique the number of active PMs is minimized. This is achieved by migrating the VMs in the under-utilized, lightly loaded PMs to other active PMs, so that the lightly loaded PMs can be switched off to save energy.

7.5.2.5 Hot Spot Mitigation

'Hot Spots' are created in PMs if the VMs associated with them are not having sufficient resources to run dynamic changes in traffic load of the applications hosted in them. To mitigate such hot-spots, it is required to migrate VMs from PMs with hot-spots to other PMs where the required amount of resources to run the hosted application causing the hot-spot is available.

7.6 Case Study: vSphere 6.0 Cold Migration

One of the authors belongs to Dept. of CSE, Global Academy of Technology (GAT). vSphere 6.0 lab is setup in GAT premises through VMware IT Academy of VMWare and ICT Academy, Chennai. Evaluation copy of vSphere and ESXi software is used to demonstrate the creation of VMs and cold migration of VMs.

7.6.1 *Screenshots of vSphere Client with VMs Creation and Snapshotting*

Figure 7.5 shows usage of vSphere Client, physical host on which EXSi operating system is installed is connected and the console of the VMs running in that physical host can be viewed. The vSphere Client helps the administrator gain access to the ESXi physical hosts as well as the VMs installed on top of them, and gives a GUI (graphical user interface) to the administrator.

The Fig. 7.6 shows two VMs, named vm_win7_70_1 and vm_win7_70_2, are powered on in the physical host (with IP address 10.0.2.70) and are simultaneously running. These VMs share resources of the physical host machine to carry out the applications in the VMs.

Figure 7.7 shows VMWare vSphere web client, where new VMs can be created in different connected physical hosts, snapshots of any VM of any connected physical host can be taken, and also, migration of VM from one physical machine to the other (to manage workload) can be done. In this figure, a cluster with two physical

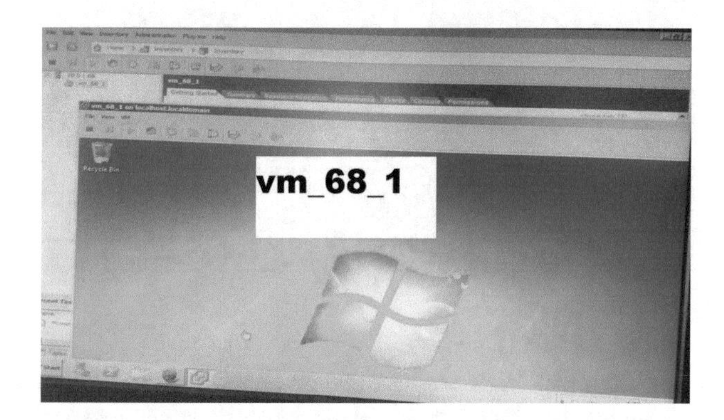

Fig. 7.5 VM installed on a physical host

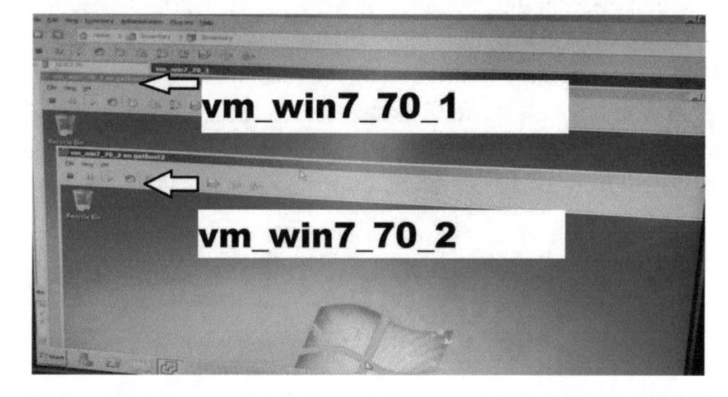

Fig. 7.6 2 VMs on a single physical host

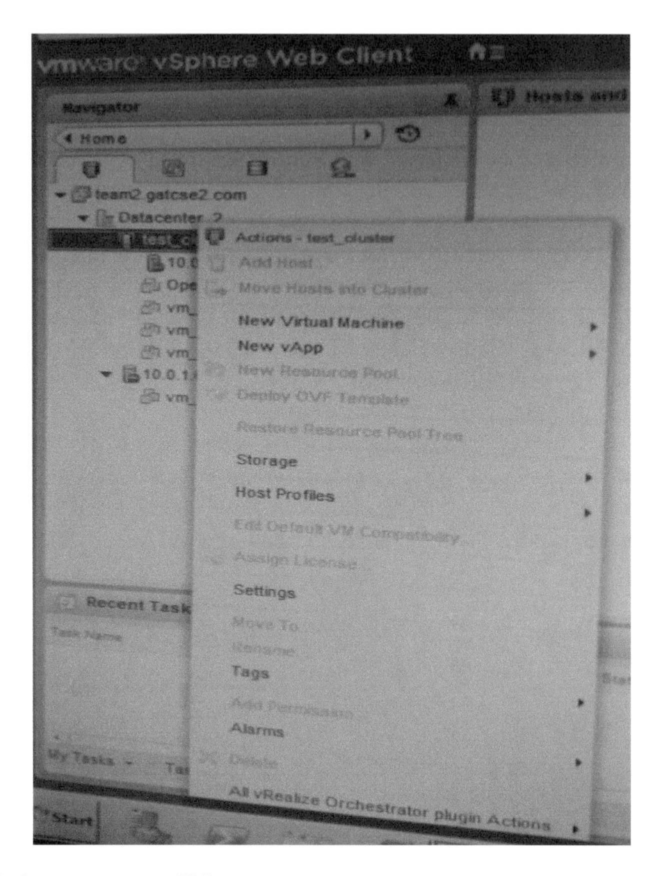

Fig. 7.7 Option to create new VM

machines is present, and 3 VMs are installed on them. Right clicking on the name of cluster gives options in which a new VM can be easily added by specifying the IP address of the physical host on which the VM needs to be placed.

In Fig. 7.8, it can be seen that by right clicking on any of the VMs, and on clicking "all vRealize Orchestrator plugin Actions" , we get various options like "migrate virtual machine with vMotion", "remove all, excess, old snapshot", "shut down and delete virtual machine". One host IP address is 10.0.2.70 and the VMs placed on it are named: vm_68_2 , vm_win7_70_1 and vm_win7_70_2. Other physical host with ESXi has IP address of 10.0.1.68 and the VM placed on it is named: vm_68_1.

7.6.2 *Screenshots of ESXi Hosts*

This Fig. 7.9 shows the display on the monitor of the physical host of IP address 10.0.2.70. A total of 16 GB of memory space is allocated to this ESXi from the

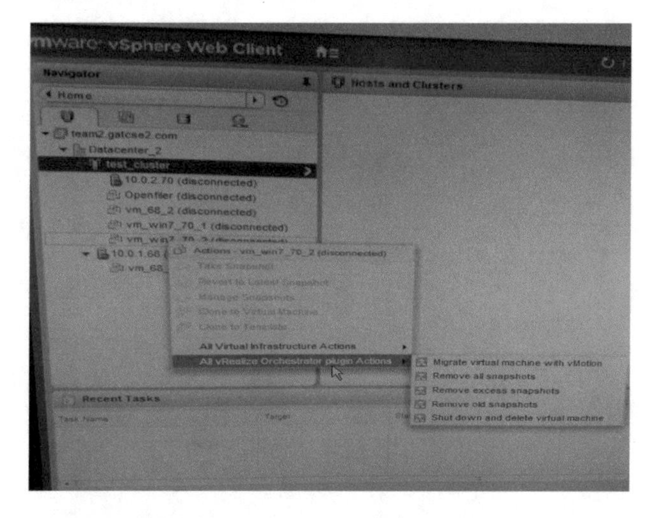

Fig. 7.8 Various options on a VM

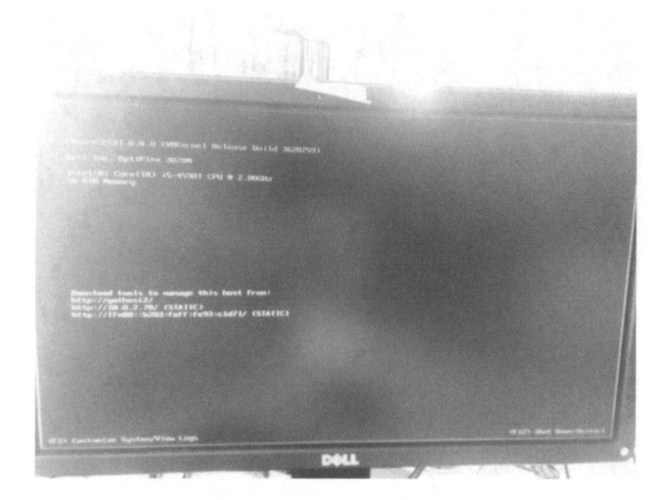

Fig. 7.9 Display on the monitor of the physical host

physical hosts memory. Using vSphere Client, VMs can be created on this ESXi host.

The Fig. 7.10 shows the host with IP address of 10.0.1.68. Memory of 16 GB is allocated to this ESXi host from the physical machine. VMs on top of this ESXi host can be created by specifying this IP address while using VMWare Web Client by the administrator.

Figure 7.11 shows the interface of the vSphere Client. On the left, there are the details of the VMs installed on the physical host with IP address 10.0.2.70. VMs vm_68-2, vm_win7_70_1 and vm_win7_70_2 are installed in the ESXi host.

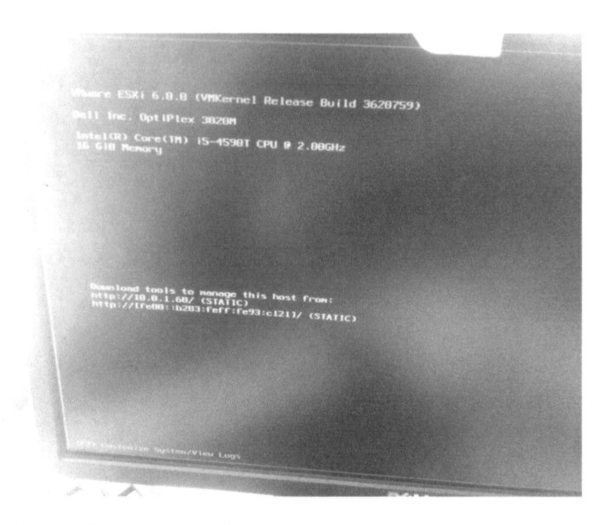

Fig. 7.10 Host with IP address of 10.0.1.68

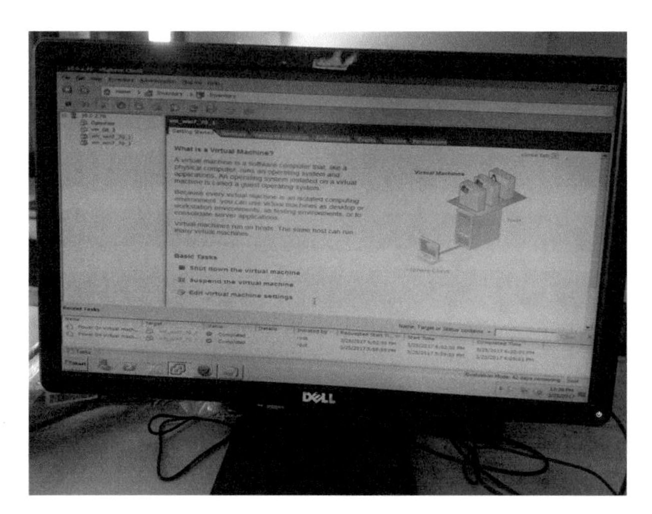

Fig. 7.11 Interface of the vSphere client

vm_68_2 is in the powered off state, while the other two VMs are in the running state (indicated by green colour). On the task bar located on the top, various options exists like "resource allocation" , "performance" , "events", "console" and "permission". Situated below, the recent tasks specify the state and actions done by the administrator. Here, name, power state, start and end time of the task etc. can be viewed.

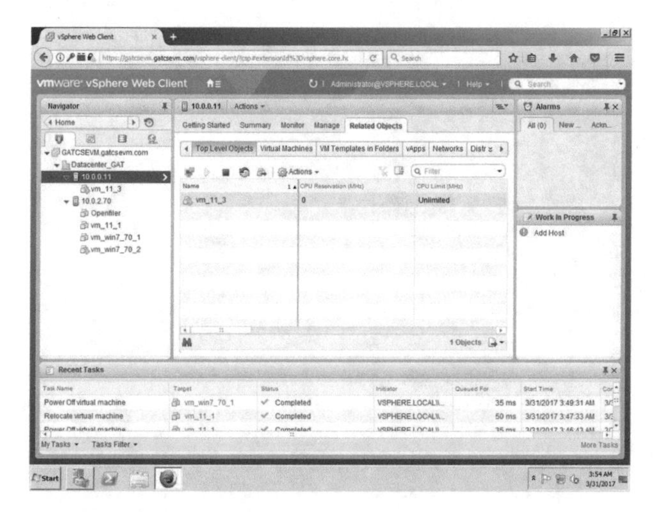

Fig. 7.12 Cluster setup before cold migration

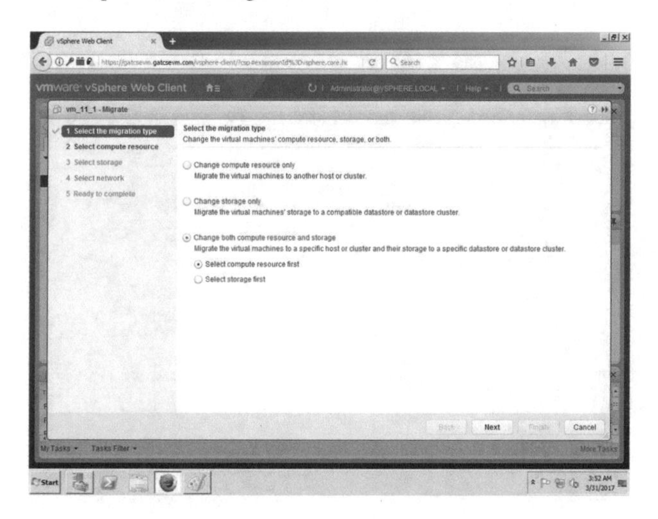

Fig. 7.13 Determining the migration type

7.6.3 *Screenshots of Cold Migration Using vSphere 6.0 Web Client*

The Fig. 7.12 shows the setup of the cluster before cold migration. ESXi server is installed on top of two physical hosts, and four VMs are installed. Host with IP address 10.0.0.11 has just one VM named vm_11_3 on it, whereas host with IP address 10.0.2.70 has three VMs, named: vm_11_1, vm_win7_70_1 and vm_win7_70_2.

The Fig. 7.13 shows the first step in cold migration which is to determine the migration type involving storage and compute resource. Three options, namely, changing

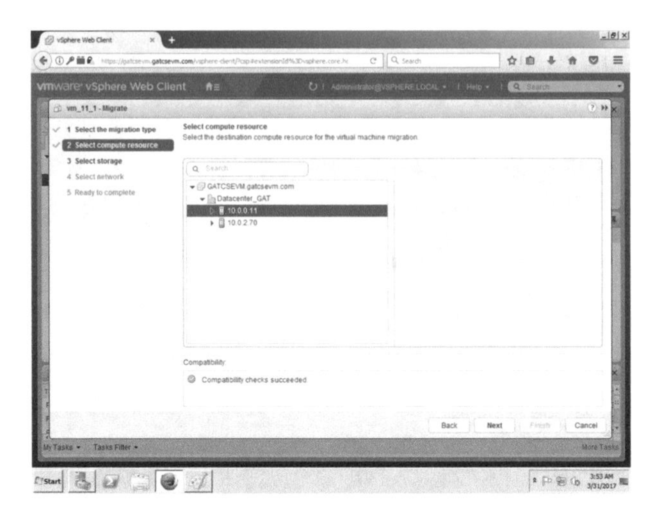

Fig. 7.14 Computing resources

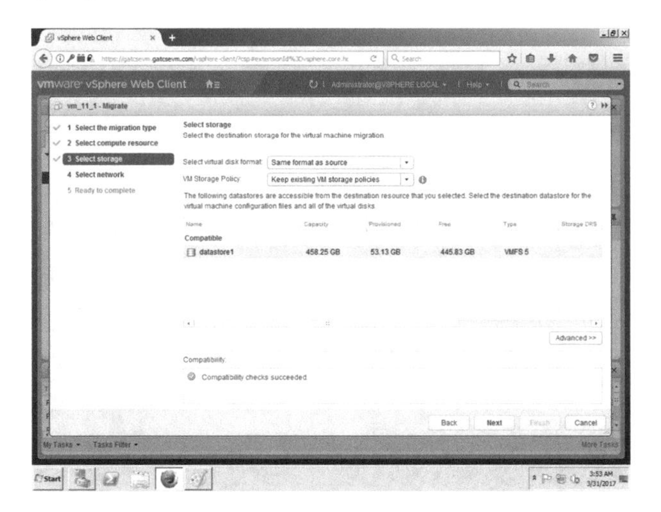

Fig. 7.15 Selecting destination storage

the compute resource alone, changing the storage alone, and changing both, are available.

This Fig. 7.14 shows the second step, which is to compute resources. Here, the destination of the VM to be migrated is selected. In this example, 10.0.0.11 is selected as the destination host.

The Fig. 7.15 shows the third step which is, selecting the destination storage for the migrating VM. Format of virtual disk can be elected, along with storage policy. In this case study, datastore1 is selected as the destination storage.

The Fig. 7.16 shows the fourth step which is to select the network destination.

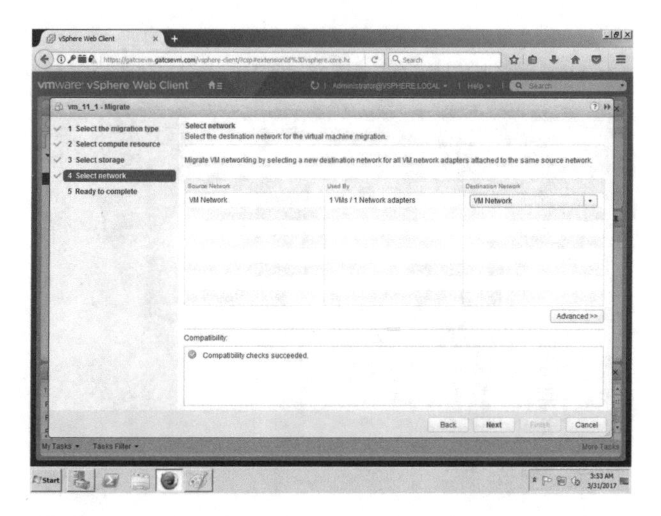

Fig. 7.16 Selecting the network destination

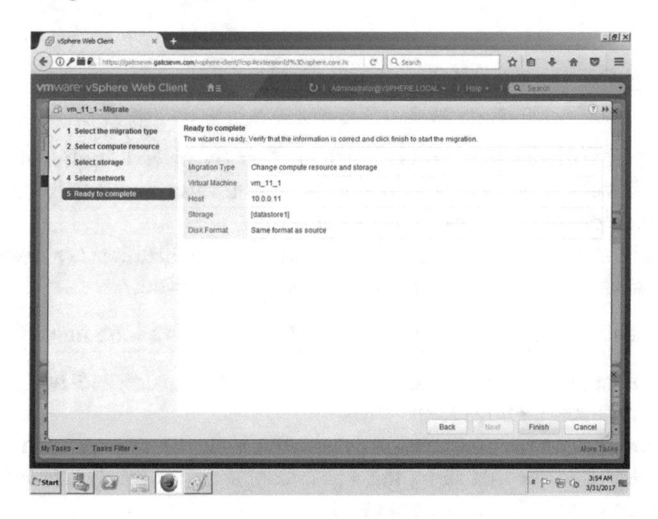

Fig. 7.17 Verifying the information about the migration type

The Fig. 7.17 shows the last step, which is to verify the information about the migration type, VM to be migrated, host, storage, and disk format. Clicking finish will start the migration process.

This Fig. 7.18 shows the setup after cold migration is performed. The VM named vm_11_1 which was on host IP address 10.0.2.70 has now been migrated to the other host with IP address 10.0.0.11.

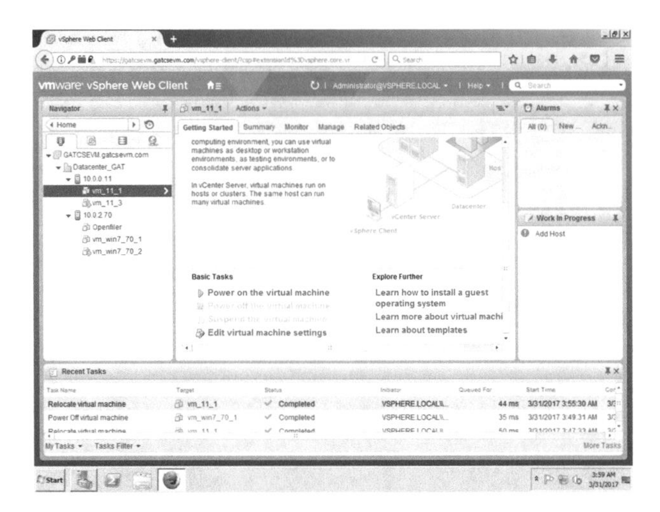

Fig. 7.18 Setup after cold migration

7.7 Optimization in Tool

Networking in virtualization brings together globally placed entities into a single software controlled unit. This embodies various software and hardware resources and their functionalities to form a virtual network.

Below a few tips for optimizing networking in virtualization are discussed and optimization strategies in some example VMs are also listed.

1. Have a Separation Between VM Network and the Management Network

Management network, here, refers to the group of servers and hosts providing resources and functionalities to the VMs.

Traditionally, a server communicates with the VMs through an interface. This is also, however, used for managing the VMs. Congestion is easily formed when the communication traffic increases even a little.

Thus a separate management network can be configured to reduce the overhead of the network and enhance its performance.

How to Achieve this for ESXi Server?

Locate the Configuration menu and choose Networking ≫ Add Networking. You'll have the option to choose between 2 network types.

1. Choose the network type, Virtual Machine, for VM generated traffic.
2. Choose the network type, VMKernel, for vSphere related traffic and for bandwidth over utilizing services, such as vMotion or iSCSI traffic (Fig. 7.19).

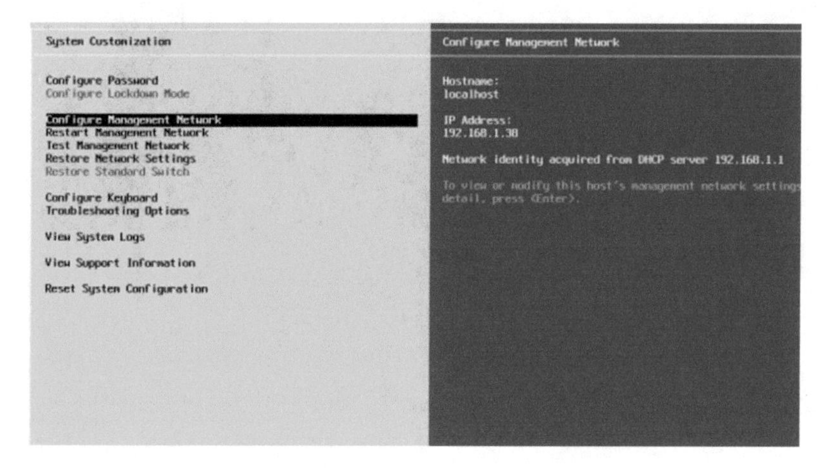

Fig. 7.19 System configuration menu

2. Have Redundancy in Management Networks

Since failure of a NIC(Network Interface Card) can cause the network to fail, having redundant network interface cards can brace us for such situations.

How to Have Redundancy?

On the Configuration Panel go to Networking ≫ Management vSwitch properties. Here select a new physical network card to be connected to vSwitch.

3. Changing Network Properties for Better Utilization of Bandwidth Between VMs

A VM is capable of using up the entire bandwidth available. This will hinder the performance of all other entities present on the network. The concept of packet shaping/traffic shaping can be used to achieve this. It involves regulating the traffic by setting an upper limit for the usable bandwidth or the density of packets travelling in unit time etc.

How to Achieve this for All the VMs?

The option of Traffic Shaping can be found under Properties of VirtualSwitch.

4. Simplifying the Network Topology

Introducing changes to the topology often leads to degradation in the performance. A simple solution is to revert the network settings to default.

How to Revert to Default?

Locate the option to restore the VMWare to default on the ESXi host panel. This resets all the characteristics of the network to initial mode of operation.

Caution: Resetting drops all the changes made and boots the server anew (Fig. 7.20).

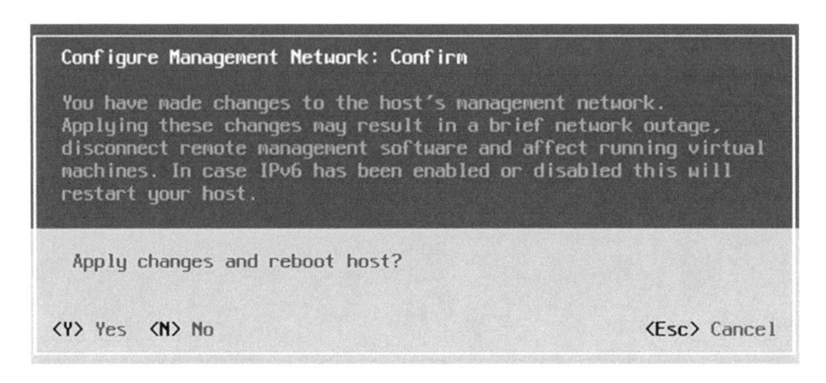

Fig. 7.20 Dialog box to confirm the changes made to host's management network

5. Using a Pass Through Card Instead of Having Dedicated Servers

VMs requiring the entire bandwidth must ideally be allotted a dedicated server that opposes the whole idea of virtualization. A solution for this is to have a pass through card. This approach allows a VM to access a physical NIC directly, bypassing the hypervisor. Thus a dedicated network board is possible for the bandwidth consuming VM.

How to Configure a Pass Through Card?

On the Configuration menu choose Configure Pass Through. Here, choose the network card to act as a pass through card and further assign this to the VM.

References

1. A. Beloglazov, J. Abawajy, R. Buyya, Energy-aware resource allocation heuristics for efficient management of data centers for cloud computing. Future Gener. Comput. Syst. **28**(5), 755–768 (2012)
2. M. Suetake, H. Kizu, K. Kourai, Split migration of large memory virtual machines, in *Proceedings of the 7th ACM SIGOPS Asia-Pacific Workshop on Systems* (ACM, New York, 2016)
3. A. Bhadani, S. Chaudhary, Performance evaluation of web servers using central load balancing policy over virtual machines on cloud, in *Proceedings of the Third Annual ACM Bangalore Conference* (ACM, New York, 2010)
4. V. Shrivastava et al., Application-aware virtual machine migration in data centers, in *INFOCOM, 2011 Proceedings IEEE* (IEEE, New York, 2011)
5. A. Beloglazov, R. Buyya, Energy efficient allocation of virtual machines in cloud data centers, in *2010 10th IEEE/ACM International Conference on Cluster, Cloud and Grid Computing (CCGrid)* (IEEE, New York, 2010)
6. M. Mishra et al., Dynamic resource management using virtual machine migrations. IEEE Commun. Mag. **50**(9) (2012)
7. J. Smith, R. Nair, The architecture of virtual machines. Comput. IEEE Comput. Soc. **38**(5), 3238 (2005). https://doi.org/10.1109/MC.2005.173

8. H. Goudarzi, M. Ghasemazar, M. Pedram, SLA-based optimization of power and migration cost in cloud computing, in *2012 12th IEEE/ACM International Symposium on Cluster, Cloud and Grid Computing (CCGrid)* (IEEE, New York, 2012)
9. Y. Du, H. Yu, G. Shi, J. Chen, W. Zheng, Micro wiper: efficient memory propagation in live migration of virtual machines, in *2010 39th International Conference on Parallel Processing* (San Diego, CA, 2010), pp. 141–149. https://doi.org/10.1109/ICPP.2010.23
10. C. Constantinescu, J. Pieper, T. Li, Block size optimization in de-duplication systems, in *Data Compression Conference. Snowbird, UT* (2009), pp. 442–442. https://doi.org/10.1109/DCC.2009.51
11. M. Anan, N. Nasser, A. Ahmed, A. Alfuqaha, Optimization of power and migration cost in virtualized data centers, in *IEEE Wireless Communications and Networking Conference. Doha 2016* (2016), pp. 1–5. https://doi.org/10.1109/WCNC.2016.7564869
12. A. Paulin Florence, V. Shanthi, C.B. Sunil Simon, *Energy Conservation Using Dynamic Voltage Frequency Scaling for Computational Cloud*, vol. 2016, Article ID 9328070 (The Scientific World Journal, 2016), p. 13. https://doi.org/10.1155/2016/9328070
13. M.K. Nair, C. Bhosle, V. Gopalakrishna, Net mobile-Cop: A hybrid intelli-agent framework to manage networks, in *Proceedings of IEEE International Conference on Intelligent and Multi-Agents (IAMA09), 21–23 July, Chennai* (2009), pp. 1–8
14. M.K. Nair, V. Gopalakrishna, Agent based web services with RuleML for network Mgt, in *Proceedings of IEEE International Conference on Networks and Communications (NetCoM 09), 27–29 Dec, Chennai, India* (2009) pp. 214–219
15. Using the VM Snapshot, Accessed from https://www.vmware.com/support/ws4/doc/reserve_snapshot_ws.html 2. Understanding VM Snapshots in ESXi, Accessed from https://kb.vmware.com/selfservice/microsites/search.do?language=en_US&cmd=displayKC&externalId=1015180
16. S. Bigelov, Case Study on Server Virtualization, Accessed from http://searchdatacenter.techtarget.com/tip/Server-virtualization-software-refresh-cycles-go-hand-in-hand (For case study 1.3 topic number)
17. Shivanshu, Resource Management in Cloud, http://www.ninjaducks.in/thesis/mainch1.html
18. D. Dias, Resource Management in VMWare, Accessed from http://techgenix.com/getting-started-resource-management-vmware-vsphere/ (2013)
19. B. Tholeti, Hypervisor, Virtualization and the Cloud, Accessed from https://www.ibm.com/developerworks/cloud/library/cl-hypervisorcompare/ (2013)
20. S. Bigelov, Case Study on Improving Efficiency Through Server Virtualization of Company Motley Fool, Accessed from http://searchdatacenter.techtarget.com/tip/Improving-IT-efficiency-with-server-virtualization-technology (2011)
21. Case Studies on Benefits of Virtualization, Accessed from https://searchdatacenter.techtarget.com/feature/Case-studies-show-the-benefits-of-virtualization (2011)
22. https://www.cl.cam.ac.uk/research/srg/netos/papers/2005-migration-nsdi-pre.pdf
23. Mishra, Das, Kulkarni, Sahoo, Dynamic Resource Management using VM Migration (2012), https://ieeexplore.ieee.org/document/6295709
24. J. Vincent, Google Data Centre Energy Bills, Accessed from https://www.theverge.com/2016/7/21/12246258/google-deepmind-ai-data-center-cooling
25. Kaur, Rani, VM Migration in Cloud Computing, Accessed from https://www.sersc.org/journals/IJGDC/vol8_no5/33.pdf (2015)
26. J. Vincent, Facebook's Data Centre, Accessed from https://www.theverge.com/2016/9/29/13103982/facebook-arctic-data-center-sweden-photos (2016)

Chapter 8
Frameworks to Develop SLA Based Security Metrics in Cloud Environment

Satya Ranjan Dash, Alo Sen, Pranab Kumar Bharimalla and Bhabani Shankar Prasad Mishra

Abstract Cloud computing, the growing technology which most of the small as well as large organizations adopt to maintain IT as it is a very cost effective organization should consider the business risk associated with cloud computing all of which are still not resolved. The risk can be categorized in several issues like privacy, security, legal risks. To solve these types of severe risks, organization might make and develop SLA for the establishment of an agreement between the customer and the cloud providers. This chapter provides a survey on the various frameworks to develop SLA based security metrics. Various security attributes and possible threats are having also been discussed in this chapter.

Keywords Cloud computing · Security metric · Service level agreement (SLA) · Cloud-hosted application SLA (A-SLA) · SLA-tree

8.1 Introduction

Cloud computing is one of the most emerging technology now a days and grabbing popularity day by day because of reduction in cost of investment as well as consumption of energy of the shared computing resources like software, servers, networking and storage. Though it enables efficient use of IT resources and increases flexibility to expand the new infrastructures quickly [1], it also has many security risks to the business that have to resolve effectively. These types of risks include handling

S. R. Dash (✉) · A. Sen · P. K. Bharimalla · B. S. P. Mishra
KIIT University, Bhubaneswar, India
e-mail: sdashfca@kiit.ac.in

A. Sen
e-mail: alosen10@gmail.com

P. K. Bharimalla
e-mail: pranab.bharimalla@gmail.com

B. S. P. Mishra
e-mail: mishra.bsp@gmail.com

B. S. P. Mishra et al. (eds.), *Cloud Computing for Optimization: Foundations, Applications, and Challenges*, Studies in Big Data 39,
https://doi.org/10.1007/978-3-319-73676-1_8

of access to privileged user, to ensure legal compliance, to ensure segregation of data, to maintain recovery of data and no assurance of guarantee of cloud service providers for long term viability [2]. For this reason, cloud customers should adopt some techniques to improve and measure their data security in the cloud. Cloud providers should make their data security metrics using a Service Level Agreement (SLA). SLA can be distinct as an concurrence among cloud providers and aloud customers to manage and control the service. Thus, SLA metrics help to assess service level between them and take a part in improving services [3]. But there are no actual standard format to prepare SLA. Different cloud service providers use their own format to prepare it, thus it leads to a huge risk to have faith on the service from the viewpoint of customers [4]. To solve this issue CSA provide STAR Repository to be included for different security issues and common rule within SLA STAR [5]. However, secure SLA should be provide to the customers so that they can have faith on Cloud Services. Many researchers proposed various security mechanisms for cloud services agreement, but a novel secure approach was introduced partially over FP7-ICT project SPECS [6] team for automatic management of secure SLA and deploying automatic secure mechanisms proposed by [7] and SPECS project [6].

This chapter has four sections which includes introduction and conclusion. 2nd Section covers some basic concepts of cloud computing, 3rd section describes service level agreement and Sect. 4 describes the security metric with the list of security attributes and threats as well as the list of different frameworks to develop SLA based security metrics in a cloud environment [8].

8.1.1 Cloud Computing Characteristics

- Self on demand service: Cloud providers enables the provisioning of cloud resources on demand whenever they are required. The resources are self provisioned without much human intervention, and the client will be charged as per the usages.
- Ubiquitous access of network: Cloud services can be accessed everywhere though standard terminals like laptops, desktops, mobile handsets, PDAs etc. with Internet connectivity.
- Easy to use: The Internet based interface offered by the cloud providers is much simpler than APIs, so user can easily use cloud environment.
- Business model: Cloud is a standard business model as it is pay for use of resource or service.
- Elastic: The illusion of infinite computing resources is provide by the cloud providers to their consumers. As a result, consumers expect to get resources in any quantity at any time from the cloud.
- Customizable: In a multi-tenant cloud environment all the resources rented from the cloud must be customizable. Customization means to allow the users for deploying special virtual appliances and to give root access to the virtual cloud servers. Other cloud service models (PaaS and SaaS) provide less flexibility thus

not suited for general-purpose computing, but provide limited level of customization [9].
- Location independent resource pooling: The cloud providers dynamically assign the virtual resources according to the consumer demand irrespective of the location.

8.1.2 Cloud Computing Deployment Models

- Public cloud: Public cloud environment service is available to the general public over Internet. The providers other consumers share the computing resources. Public cloud is a pure cloud computing model.
- Private cloud: Private cloud environment is solely operated by small or large organizations where they want to keep their services, application programs, own data in cloud by taking the responsibility for security of their data.

They implement firewall behind their cloud environment and enable pooling and sharing of resources among various departments, applications, etc. The main advantage behind this cloud is compliance, security and QoS.

- Community cloud: As the name suggests, community cloud environment is shared by multiple organizations. The third party suppliers provide the necessary computing infrastructures, or it may be provided inside.
- Hybrid cloud: Hybrid cloud atmosphere is a combination of two or more clouds. Many organizations prefer hybrid cloud by using public cloud for general computing purpose where the data of the customers are kept within other types of the cloud model. It enables portability of data application by balancing the load between clouds [10].

8.1.3 Cloud Computing Solutions

- Infrastructure as a service: This service provides virtual machines, some abstracted hardware, and operating system on the Internet.
- Software as a service: This is software on the Internet through which customers can rent full applications without purchasing and installing the software on their own computer [11].
- Platform as a service: Cloud providers provide a platform of developed environment as a service where users can run their application.
- Storage as a service: In this service, it provides database like services. Users are charged based upon their usage e.g., gigabyte per month.
- Desktop as a service: In this service, desktop environment is provisioned as a terminal or within a browser.

8.2 Service Level Agreement (SLA)

A SLA is agreed upon agreement between two parties, namely cloud environment providers and customers, which provide the guarantees of a service [12]. This guarantee includes the list of transactions need to be executed with the best procedure of execution. SLA had been defined as the what without considering the how. SLA tells the different types of service, customer should get with respect to information security though SLA do not specify by what procedure the service levels will be achieved. It also provides a common understanding between two parties on priorities, responsibilities, services, guarantees and warranties. In a cloud environment, when customers submit their vital information to the cloud service providers, then they don't bother about the storing and processing of their information. So, SLA is needed by the cloud providers to convince people to access their services and guarantees about their information security.

Following are the types of SLAs:

1. Cloud Infrastructure based SLA (I-SLA): Cloud providers offer this type of SLA to the cloud customers to give assurance for the level of quality of the cloud computing resources like performance of the server, speed of network, availability of resources, storage capacity etc.
2. Cloud-hosted Application SLA (A-SLA): This type of SLA gives assurance over the quality level for the deployed software applications on a cloud environment.

Actually, cloud customers sometimes offer such type of assurance to the software end user in classify to give promise to the excellence of different type of services like the response time of the software and data freshness. Figure 8.1 illustrates different SLA parties in the cloud environment.

8.2.1 SLA Life Cycle

A standard SLA life cycle has been proposed by Roja (Fig. 8.2) [13]:

Fig. 8.1 Different SLAs in cloud environment

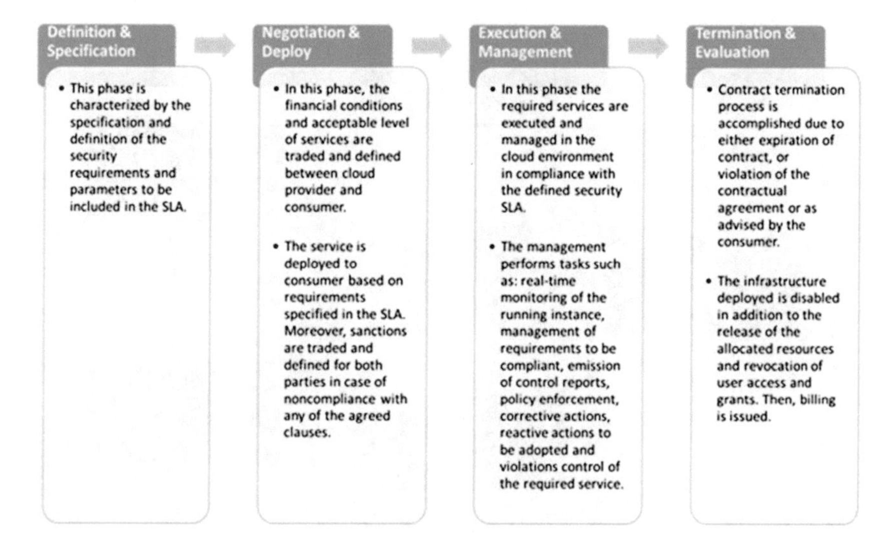

Fig. 8.2 SLA life cycle process [13]

8.2.2 Features of SLAs

SLA has the following characteristics:

- SLA format should be properly maintained so that it will be easier for a consumer to understand the details of the available services by a cloud provider.
- Explanation about the level of performance.
  - Description about monitoring of system parameter and report.
  - Penalties for the violation of rules or in the case of unable to provide all the requirements to the consumer.

There are many functional, as well as non-functional necessities which should be fulfilled to meet the expectation of the customers. All the requirements can be classified under different services provided by the cloud providers. Figure 8.3 classifies the cloud services.

As per the non-functional requirements there are many things to be considered. The most vital criteria are quality of service i.e. availability of the desired service on time. This is the assurance for the customers about the probability of the cloud services perfectly. Cloud customers can use the service or the resources and pay as per usages. Sometimes they have a need to scale up or down for some specific resources. So, there should be some assurance over the degree of scalability by which cloud customers can utilize the resource properly. There should be an accurate method for service billing. In the cloud environment, customers are willing to pay as per their use of services.

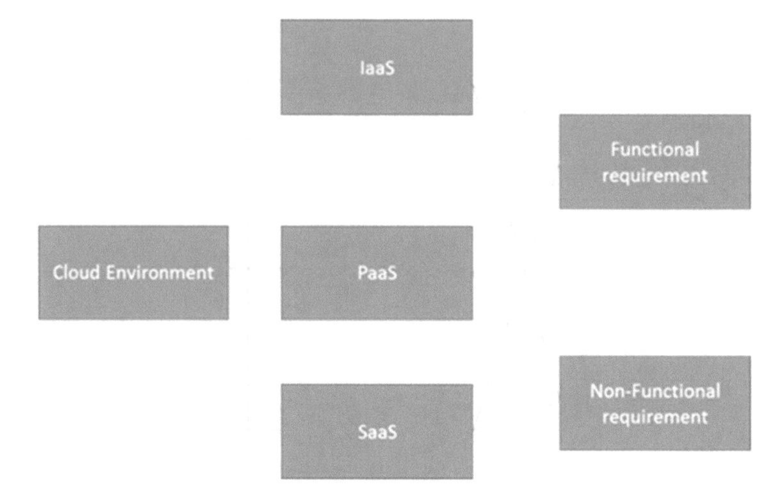

Fig. 8.3 Classification for requirements in different cloud service

So, a monthly or an annual billing system is not suitable for the service. It is also not possible to calculate the cost, according to the reservation method. So, an efficient billing method should be maintained like the billing for storage service can be made by depending on the size of data and duration of usage of the service.

Cloud environment deals with many virtual machines for each individual server. As a need, customers should get a flexible access over the VMs to get the service smoothly with a little effort of managing the configuration themselves. Secure medium should be used to accumulate the data on the server and between the duration of transferring of data between customers and cloud providers. Security features should be assured by cloud providers so that organizations will to invest on cloud environments rather than switching to spend on operating their own data centers.

8.3 Security Metrics

Security metrics are measured quantitatively in organization for providing security operations [14]. It helps the organizations to take decisions related to different types of security including the architecture of security to control structure for effective and efficient operations [15]. Security metrics are valuable to the stakeholders and managerial level of IT [16]. Sometimes, security operations demand high level of expenditures to them. NIST [8] divides security metrics according to their characteristics into three different types:

1. **Implementation metrics**: These types of metrics are for demonstrating the advancement in implementation of different programs for information security, controls of security and different procedures and policies [4].
2. **Effectiveness/efficiency metrics**: These types of metrics are for monitoring whether the program-level process and security controls of system level have been implemented properly and operating as expected and achieving the desired outcomes [4].
3. **Impact metrics**: These types of metrics are for articulating information security impact on target of organization [4].

Security metrics can also be decomposed according to their uses. The user group can be divided into quality assurance, strategic support and tactical oversight [16]. Table 8.1 demonstrate different security attributes and threats in cloud environment.

Kaiping [18], anticipated a new safe group allotment structure for public cloud. The structure ensures that, no insightful data is showing to cloud supplier and invader. It combines proxy signature, improved TGDH, alternative re-encryption to form a robust security protocol layer.

Diffie–hellman (CDH) [19] problem in Random Oracle Model. Islander et al. in 2011 proposed a mechanism for implementing authentication policies in cloud computing and they defined different levels of policy consistency constraints. Authors propose a Two-Phase rationale Committed protocol to provide a explanation which is customized version of essential Two-Phase Commit protocols [20].

8.4 Frameworks for Developing SLA Oriented Security Metrics

The SLA parameters are driven by metrics and metrics define the way to measure cloud service parameters and specify values of measurable parameters [9].

At first, we have tried to identify the various parameters necessary for SLA in different cloud environment like IaaS, PaaS, SaaS, Storage as a service and parameters for all the users for all types of cloud which are general parameters. Table 8.2 illustrates list of parameters needed in SLA for different services in the cloud environment.

We have surveyed the list of frameworks associated with SLA.

1. **Goal Question Metric (GQM)**: In GQM (Fig. 8.4), evaluation is carried out in a top-down manner. This approach first formulated Basili. GQM defines a dimension of replica on three levels [10].

Table 8.1 List of security attributes and threats in cloud environment

Serial no	Security attributes	Description	Threats
1	Accountability	This attribute keep track of the security actions and responsibilities [8]	• Ambiguous ownership for data protection • Loss of business • Audit difficulty • Regularity and legal issues • Difficult bug detection • Difficult intruder detection
2	Confidentiality	This attribute protect the sensitive information from the unauthorized users [17]	• Identify theft • Unauthorized modification • Data theft • Malware attacks • Requirement of data segregation • illegal access • Data loss • Eavesdropping • Insecure data storage • Inadequate authorization and authentication • Cloud provider espionage • Phishing attack • Insecure interfaces and APIs
3	Integrity	This attribute deals with completeness, accuracy and information validation with respect to business requirements and goals [17]	• Identify theft • Unauthorized modification • Data theft • Malware attacks • Data segregation not sufficient • Unauthorized access • Inconsistent data • Authorization and authentication is inadequate • Insecure interfaces and APIs
4	Availability	This attribute deals with the information to be fetched and accessible later when it is necessary by the business procedure as in the prospect [17]. Though, this information is not accessible to the unauthorized users	• Unauthorized modification • Malware attacks • DOS/DDOS • Lack of data segregation • Unauthorized access • Data loss • Loss of business • Inadequate authorization and authentication • Service disruption

Table 8.2 List of parameters needed in SLA for different services in the cloud environment

SLA for environment	Parameters
IaaS	• Capacity of VM CPU • Size of VM cache memory • Ready time for VM • Size of data • Maximum VM per user • Minimum VM per user • Time to add one VM • Time to deallocate one VM • Maximum VMs run per server • Service availability • Time for completion and receiving process
PaaS	• Integration with other platforms • Degree of scalability • Pay as per use of the service • Servers and browsers • Accessibility
SaaS	• Ability of operation in each case • Usage of user interface • Degree of scalability • Availability of software • Flexibility to use
Storage as a Service	• Availability of data storage location • Degree of scalability for updating storage area • Capacity of storage area • Cost of storage • Privacy and security • Information about the backup of data • Availability of recovery in case of failure • System throughput • Network bandwidth • Management of data
For all types of users	• Detail description of monitoring the system • Description about the billing • Privacy and security • Information of networking system • Clarification of support and service • The standard policy should be followed

- Conceptual Level (Goal): An objective is defined for a particular entity for several reason, with value to many excellence models, from many points of vision and with respect to a particular environment.
- Operational Level (Question): A put of questions is basically used to describe the way the appraisal or the accomplishment of a particular goal is going to be performed based on some depicting form.

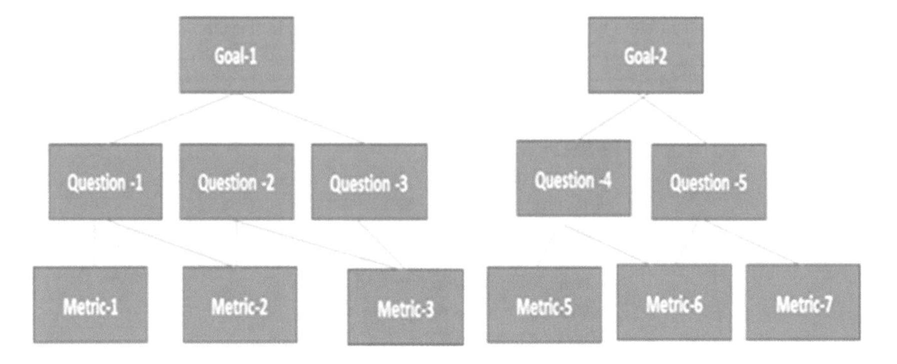

Fig. 8.4 Approach in GQM

- Quantitative Level (METRIC): A set of related and connected information with each and each query in order to respond it in a computable and quantitative way. The data can be short or long or may be objective or subjective.

2. **Electronic Bill Presentment and Payment (EBPP)**: The framework identifies security threats using STRIDE model, which is basically derived from acronym for following six threat category:- (i) Spoofing identity, (ii) Tampering with Data, (iii) Repudiation, (iv) Information Disclosure, (v) Denial of Service, and (vi) Elevation of Privilege.
   In this model, to measure the levels of security, the architectures components and the threats are combined together [21].
3. **Policy Based Metrics**: This framework is specified in NIST SP800-53 and NIST 800-53 to measure and monitor overall IT security [21, 22].
4. **Security Metric Identification Framework**: The framework provides a systematic method for the identification and development of a security metric suite for software [23].
5. **Risk Based Proactive Security Configuration Manager (ROCONA)**: This framework is to measure proactive security configuration within a network. It identifies, formulate, and validate the risk factors which severely affect the network security [24].
6. **Security Performance Framework (SPF)**: The proposed Framework (SPF) is used for the measurement and monitoring of whole Information Technology safety performance in an association using NIST SP800-80 procedures [21].
7. **Control Objectives for Information and Related Technology (COBIT)**: The proposed framework is a place of open papers for the declaration of sound IT authority in an association. COBIT covers assurance, absolute governance and manage over IT [25] & [26].
8. **Web Service Level Agreement framework (WSLA)**: The proposed framework is developed to monitor and enforce SLA in service oriented architecture (SOA) for web services [27].

9. **Framework of Service Level Agreements in the Cloud Bank Model (CBSLA)**: Cloud Bank model includes the following: (i) Resource Consumer, (ii) Cloud-Bank (CB) (iii) Cloud-Consumer-Agency (CCA), (iv) Physical layer (v) Virtual layer, (vi) Cloud-Resource-Agency (CRA) and (vii) Resource Provider. The CBSLA is proposed to capture the SLAs in Cloud Bank environment [28].
10. **Framework for Consumer-Centric SLA**: The author planned an end-to-end structure for consumer-centric SLA supervision of cloud-hosted databases. The SLA document is clearly defined by the terms of goals specific for the application requirements with some constraints [29].
11. **Collaboration-Based Cloud Computing Security Management Framework**: The framework is based on the improvement and for the support of association between cloud stakeholders to expand a cloud security measurement and implementation framework including all of their requirements. The framework has been implemented by align FISMA customary with the cloud replica and using collaboration between the stakeholders [30].
12. **SLA-Tree**: It is a framework to support efficiently SLA-based earnings tilting decision in cloud computing. SLA-tree is mostly used to reply professionally two important questions about the possible earnings modify between queries in the buffer. SLA-tree can be used for capacity planning, dispatching and scheduling, etc. [31].
13. **Framework for SLA Assurance**: This trust model framework addresses the issues of real-time QoS appraisal. The structure contains two modules that is reputation assessment module and transaction risk assessment module. Structure to deal with the real-time QoS appraisal problem. The framework/structure consists of two modules specifically, standing appraisal module and transaction risk assessment module. The authors have worked on trust-based design to decide pre interaction and matter risk-based design to determine post-interaction of QOS estimation for assessment building [32].
14. **SMICloud Framework**: This is a method to determine the excellence and prioritize cloud services. Such a structure makes considerable contact and will create competitive atmosphere between Cloud providers to convince their Service Level Agreement (SLA) and get better their quality of service(QoS). The SMI framework provide a perfect vision of QoS required by the consumers for the selection of a Cloud service provider based on: responsibility, quickness, service declaration, performance, price, privacy, safety and usability. The SMICloud structure provide various features such as choice of services based on QoS parameters and position of services on past client experience and quality of services. It is a tool for assessment, planned to evaluate Cloud services in terms of KPI parameters and client necessities [33].

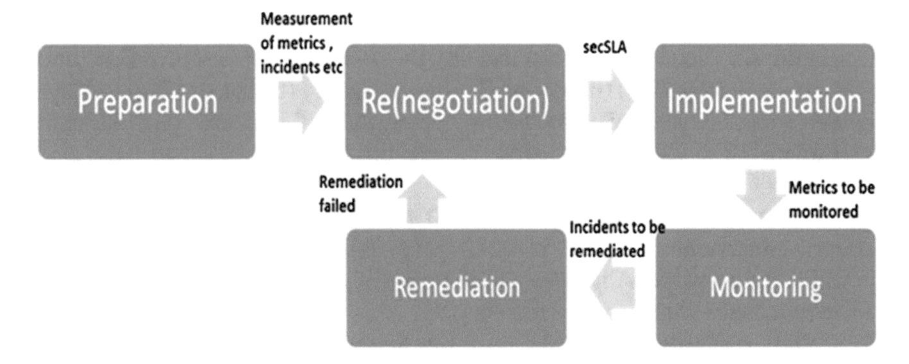

Fig. 8.5 secSLA life cycle by [34]

15. **secSLA**: The research on a secure SLA was proposed by SPECS (SecureProvisioning of cloud services based on SLA management) team for FP7 Project. This is a novel mechanism to reduce the possibility of damages with an automatic remedy by continuous monitoring of secSLAs that detect the violation occurred in any time and react to it for that agreement by [34]. Figures 8.5 and 8.6 illustrates the life cycle of secSLA with remedy.
16. **Security SLA**: MUSA, the successor team of SPECS had proposed the same design for secure SLA design but for multi-cloud environment defined the whole process of step-by-step design of secure SLA [35] in the following Figs. 8.7 and 8.8.
    In our survey paper, we had discussed about 17 different SLA based framework that should be provided by the cloud providers. Users had no choice to interact directly for defining their requirements for preventing themselves from different security issues arise in cloud service and they can have a trust on the service they are getting from cloud service providers. Now we will illustrate another framework (USRAF) that is basically user-centric.
17. **USRAF**: This is a cloud security architectural framework proposed by [19], where user will set the goals, parameters, rules and resources to increase their security requirement. User send the service request to the cloud service providers by determining four factors to get their required security.
    Then, cloud service providers identify the requirements and provide the service to the cloud customers according to customer's choice. Figure 8.9 illustrates the security procedure.
    Table 8.3 illustrates the stages of the security metric frameworks and illustrates the comparative study between the different security metric frameworks for developing SLA oriented security metrics in the cloud environment.

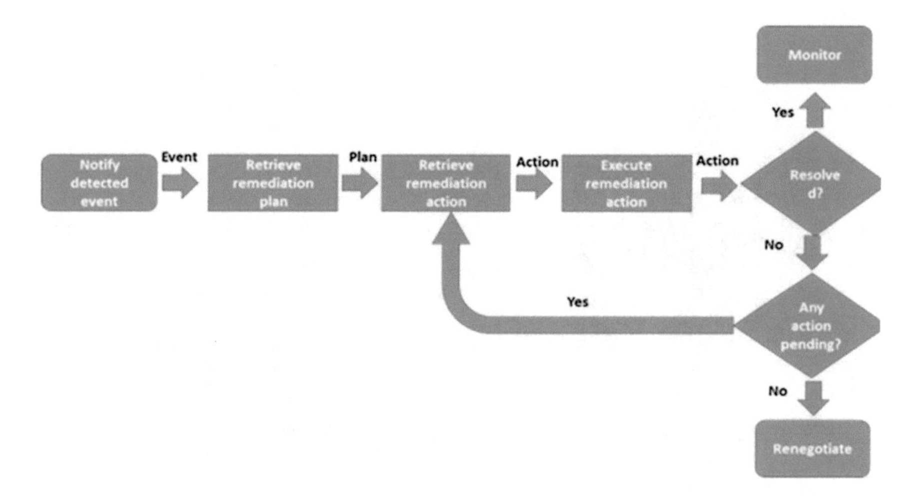

Fig. 8.6 Remediation flow by secSLA by [34]

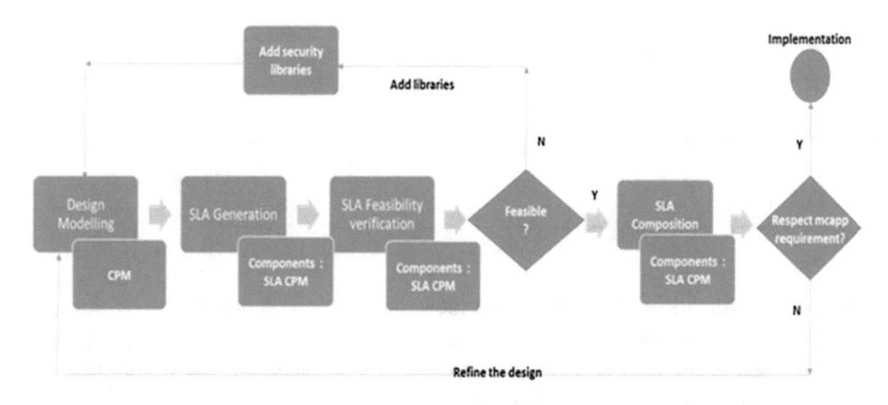

Fig. 8.7 MUSA Security-by-design cloud application development process by [35]

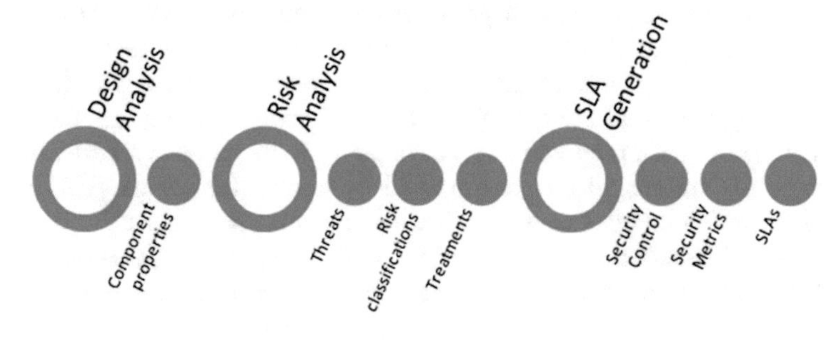

Fig. 8.8 MUSA security-by-design cloud application development process by [35]

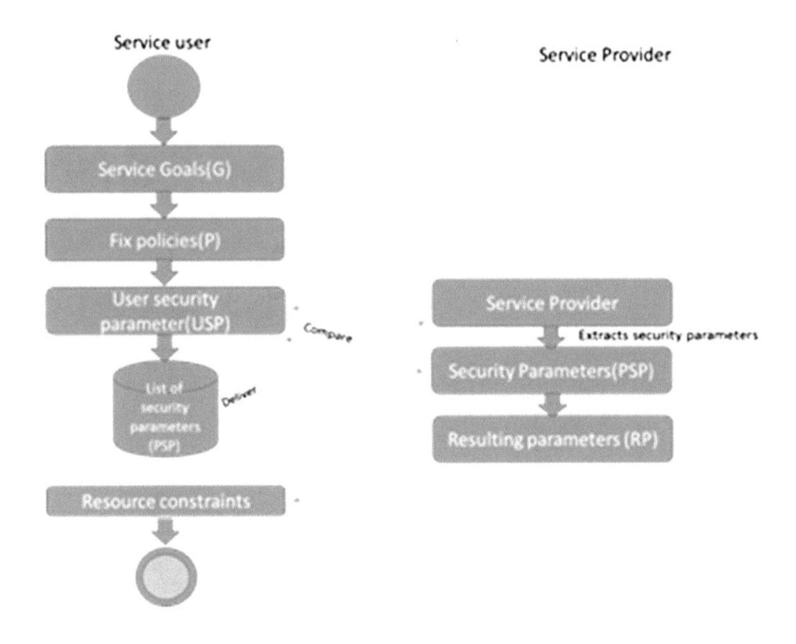

Fig. 8.9 Security procedure for USRAF by [19]

8.5 Comparative Analysis

We have discussed 16 different frameworks to recognize SLA based security metrics. Now, in this section we have done a comparative analysis of those frameworks [36] agreed that the same information security threats are also vulnerable to cloud computing as the traditional computing. But, cloud security alliance argues that new security threats are challenged to the cloud computing depends on its deployment over an organization [37]. To identify manageable number of security metrics among many proposed metrics, there is a need of proper identification of a metric framework [38]. The study shows that researches have identified a large number of security metrics among which all are not related to cloud computing security treats [39]. They should have been proposed an efficient model to derive such security metrics. Goal definition are crucial for creating security metrics framework [10], without which, it is unclear on what metrics to be used and how you interpret it.

Each one of the recognized frameworks have their individual limitations. Experiments have also revealed that bottom-up structure may not generate improved metrics as here exist also many thing exclusivity to watch. We observed that NIST SP800-[4]. SANS security metrics strategy [40] and GQM [32], all are measured to be based on top-down approach. But, they also cannot obtain all criteria of a suitable framework. Among them, GOM fulfills most of the criteria.

Table 8.3 Stages of the frameworks of SLA oriented security metrics in a cloud environment

Serial no	Name of the framework	Stages
1	GQM	• Business goals are developed. Protection • Questions are generated • Measures are specified accurately • Mechanisms are developed • Data are collected, validated and analyzed • Data are analyzed by a post-mortem way
2	NIST SP800-55 [4]	1. Process of information security measures development • Identification of interest of stakeholders • Identification of goals and objectives • Review of procedures, information security policies and guidelines • Review of implementation process • Development of measures and selection • Templates are build for measuring development •Feedback are considered in the process of measurement development 2. Process of information security measurement program implementation • Data collections are prepared • Data are collected and results are analyzed • Corrective actions are identified • Business cases are developed and resources are obtained • - Finally corrective actions are applied
3	EBPP	• For each individual components security and criticality index assessment is done • Ratings are determined for the STRIDE coercion • Map of the process current states is done for the three state transition diagram transversely mechanism • Measurement of STRIDE fear are done face by a variety of the state diagrams • Calculation are performed for vulnerabilities and normalization are completed for the risk sets for the state diagrams • Consolidation of situation vulnerabilities are performed at state diagram intensity • Origin of the concluding safety score is done finally
4	Policy based metrics	• Model the security procedures and policies • Achieve the security targets and goals • Measure the security fully • Develop the metrics and analysis it • Model the metrics and measurement detail • Analysis the report • Module for report and recommendation

(continued)

Table 8.3 (continued)

Serial no	Name of the framework	Stages
5	Security metric identification framework	• Security metrics requirements are specified • Vulnerabilities are identified • Software characteristics are identified • OO models are analyzed • Security metrics are analyzed • Security metrics are categorized • Security metric measures are specified fully • Metric development process are designed • Security metrics are designed • Finalization of metric suite
6	ROCONA	• Analysis of traffic • Analysis of service risk • Analysis of policy risk • Calculation of threat portability are performed • Analysis of risk impact • Measure the system risk • and finally, security attain and recommendations
7	SANS security metrics guideline [36]	• Defining goals and objectives of the metrics program • Decide the metrics to be generated • Define a plan for generating the metrics • Setup targets and benchmarks • Outline a plan to report the metrics • Generate an implementation map and start execution, and Set up a proper program evaluation and alteration cycle
8	SPF	This Framework follows 2 approaches: • Control-specific approach:- This approach chose a control depending upon the metric for a better design of organization chart which will represent the entire family in a better way • The cross-cutting approach:- This approach mainly based on security routine based on one or more than one being or group of control. Group of control are used mainly to analyze, develop and collection of matrices
9	COBIT	The framework is developed by the basic principle to provide the required information for the organizations to achieve objectives for business while managing investment and resources for IT department
10	WSLA	The framework is composed of a set of concepts and a XML language. Following are services of WSLA: • Measurement Service • Condition Evaluation Service • Management Service

(continued)

Table 8.3 (continued)

Serial no	Name of the framework	Stages
11	CBSLA	Cloud Bank model forms the cloud economic architecture of the market. From the local cloud, the bank stores transaction resource related to market. Then the guidance information is provided to the consumers after analyzing the resources effectively Life cycle of CBSLA: • Initial Specification • Activation/Execution • Monitor SLA Performance • Modification • Termination
12	Framework for consumer-centric SLA	• The framework inspect the database workload continuously • Keep track the execution of the application-based SLA • Assesses the state of the act policy and takes dealings when required
13	Collaboration-based cloud computing security management framework	• Aligning NIST-FISMA standard with the cloud model - Service Security Categorization(SSC) - Security Control Selection(SCS) - Security Controls Implementation(SCI) - Security Controls Assessment(SCA) - Service Authorization(SA) - Monitoring the Effectiveness of Security Controls(MESC) • Security automation
14	SLA-tree	• Converting delay to slack • Structured the slack tree S+ • Querying the slack tree S+ • Structured and query the tardiness tree S
15	Framework for SLA assurance	• For pre-interaction of qos assessment: - Reputation assessment methodology - Fuzzy logic-based approach • For pre-interaction of qos assessment: - Transactional risk assessment approach - Evaluating cost due to service level filth - Evaluating cost due to non-dependable criteria
16	SMICloud Framework	• Hierarchical organization for cloud services based on SMI KPIs • calculation of comparative weights of each QoS and service • comparative value-based weights for position Cloud services aggregation of comparative standing for each SMI attribute
17	secSLA	• Preparation • Negotiation • Implementation • Monitoring • Remediation(followed by re-negotiation)

- GQM framework measures the object for which an organization should define the goals for measurement related to business goal. It helps to find the security metrics which are relevant to cloud computing.
- NIST SP800-55 is also an goal based approach. But it is suitable to produce the metrics for selected customers. This framework is having no global acceptance because this model is specially used by USA government organizations.
- EBPP framework is developed for a specific system and has no global acceptance.
- Framework for policy based matrices is totally depended on organizational polices and it is made by US NIST guidelines.
- Security metric identification framework is a newly proposed framework is considered complicated for practical uses.
- ROCONA framework is proposed for a explicit purpose to determine the security pattern in a network. It is not validated for academic or industrial purpose.
- SANS security metrics guideline is a general framework developed for intended customers. This is also an goal based approach. There is no full documentation available for this proposed model.
- SPF framework is totally depended on organizational polices and it is made by US NIST guidelines.
- COBIT structure is extra paying attention on industry and it ensure that its function are enabling industries in planned and industry objectives while rationally running IT risks [8].
- As cloud is engineering structural design, we thus believe the COBIT structure is further suitable than other accessible protection frameworks. COBIT is a measurement-driven structure which offers numerous metrics to determine several industrial processes as well as metrics for evaluating SLA and security routine. We have also studied that the COBIT framework describes approximately 340 metrics, even though not all of the metrics described in the COBIT structure are applicable in cloud computing [34]. SecSLA uses detection and have remedy for the vulnerability occurred in agreement and have measurement level to measure the fulfillment of agreed Security Level Objectives which can be used in current Cloud service trends.

8.6 Conclusion

Cloud Computing model will gradually change the landscape of IT industry. Though, There are several benefits provided by the Cloud environment, A efficient model or framework for SLA metric for security is needed which can assure the QOS needed by the consumers. In this chapter, we have surveyed various structure suitable for SLA in cloud computing. We have explained the detail stages of those frameworks so that it will be easier for the new researchers to do research on the field of SLA and make more suitable and flexible framework for SLA security metric in the upcoming year.

References

1. R. Barga et al. Cloud computing architecture and application programming: DISC'09 tutorial, half day, in *ACM Sigact News 40.2* (22 Sept 2009), pp. 94–95
2. B.R. Kandukuri, A. Rakshit, Cloud security issues, in *IEEE International Conference on Services Computing, SCC'09* (IEEE, 2009)
3. I.S. Hayes, Metrics for IT outsourcing services level agreements (2004), www.Clarity-consulting.com/metricsarticle.Htm
4. E. Chew et al., Performance measurement guide for information security. Special Publication (NIST SP)-800-55 Rev 1 (2008)
5. J. Huang, D.M. Nicol, Trust mechanisms for cloud computing. J. Cloud Comput. Adv. Syst. Appl. **2**(1), 9 (2013)
6. V. Casola et al., SLA-based secure cloud application development. Scalable Comput. Pract. Exp. **17**(4), 271–284 (2016)
7. M. Rak et al., Security as a service using a SLA-based approach via SPECS, in *2013 IEEE 5th International Conference on Cloud Computing Technology and Science (CloudCom)* (IEEE, 2013), pp. 1–6
8. J. Feng, Yu. Chen, Pu. Liu, Bridging the missing link of cloud data storage security in AWS, in *Consumer Communications and Networking Conference (Ksons)* (2010), pp. 1–2
9. M. Alhamad, T. Dillon, E. Chang, Conceptual SLA framework for cloud computing, in *2010 4th IEEE International Conference on Digital Ecosystems and Technologies (DEST)* (IEEE, 2010), pp. 606–610
10. R. Van Solingen et al., Goal Question Metric (GQM) approach, in *Encyclopedia of Software Engineering*, ed. by J.J. Marciniak (Wiley Interscience, New York, 2002). Online version
11. G.B. Tanna et al., Information assurance metric development framework for electronic bill presentment and payment systems using transaction and workflow analysis. Decis. Support Syst. **41**(1), 242–261 (2005)
12. De Chaves, S. Aparecida, C.B. Westphall, F.R. Lamin, SLA perspective in security management for cloud computing, in *2010 Sixth International Conference on Networking and Services (ICNS)* (IEEE, 2010)
13. M.A.T. Rojas et al., Inclusion of security requirements in SLA lifecycle management for cloud computing, in *IEEE 2nd Workshop on Evolving Security and Privacy Requirements Engineering (ESPRE)* (IEEE, 2015), pp. 7–12
14. T. Klaus, Security metrics-replacing fear, uncertainty, and doubt (2008), pp. 62–63
15. W. Jansen, *Directions Research in Security Metrics* (Diane Publishing, 2010)
16. E. Kahraman, Evaluating IT security performance with quantifiable metrics, Master's thesis, DSV SU/KTH (2005)
17. J. Wei et al., Managing security in a cloud environment, in *Proceedings were Encapsulated ACM 2009 Workshop On Cloud Computing Security* (K ACM, 2009), pp. 91–96
18. Kaiping Xue, Peilin Hong, A dynamic secure group sharing framework in public cloud computing. IEEE Trans. Cloud Comput. **2**(4), 459–470 (2014)
19. V.R. Thakare, K.J. Singh, Users' Security Requirements Architectural Framework (USRAF) for emerging markets in cloud computing, in *2015 IEEE International Conference on Cloud Computing in Emerging Markets (CCEM)* (IEEE, 2015), pp. 74–80
20. A. Aieh et al., Deoxyribonucleic acid (DNA) for a shared secret key cryptosystem with Diffie–Hellman key sharing technique, in *2015 Third International Conference on Computer, Communication, Control and Information Technology (C3IT)* (IEEE, 2015), pp. 1–6
21. C. Martin, M. Refai, A policy-based metrics framework for information security performance measurement, in *2nd IEEE/IFIP International Workshop on Business-Driven IT Management, 2007. BDIM'07* (IEEE, 2007), pp. 94–101
22. C. Martin, M. Refai, Service-Oriented Approach to Visualize IT Security Performance Metrics Trust Management (2007), pp. 403–406

23. S. Chandra, R.A. Khan, Software security metric identification framework (SSM), in *Proceedings of the International Conference on Advances in Computing, Communication and Control* (ACM, 2009), pp. 725–731
24. E. Al-Shaer, L. Khan, M.S. Ahmed, A comprehensive objective network security metric framework for proactive security configuration, in *Proceedings of the 4th Annual Workshop on Cyber Security and Information Intelligence Research: Developing Strategies to Meet the Cyber Security and Information Intelligence Challenges Ahead* (ACM, 2008), p. 42
25. B. Von Solms, Information security governance: COBIT or ISO 17799 or both? Comput. Secur. **24**(2), 99–104 (2005)
26. A. Da Veiga, J.H.P. Eloff, An information security governance framework. Inf. Syst. Manag. **24**(4), 361–372 (2007)
27. H. Ludwig et al., Web service level agreement (WSLA) language specification, IBM Corporation (2003), pp. 815–824
28. F. Zhu, H. Li, J. Lu, A service level agreement framework of cloud computing based on the Cloud Bank model, in *2012 IEEE International Conference on Computer Science and Automation Engineering (CSAE)* (IEEE, 2012), pp. 255–259
29. L. Zhao, S. Sherif, L. Anna, A framework for consumer-centric SLA management of cloud-hosted databases. IEEE Trans. Serv. Comput. **8**(4), 534–549 (2015)
30. M. Almorsy, J. Grundy, A.S. Ibrahim, Collaboration-based cloud computing security management framework, in *2011 IEEE International Conference on Cloud Computing (CLOUD)* (IEEE, 2011) pp. 364–371
31. Y. Chi et al., SLA-tree: a framework for efficiently supporting SLA-based decisions in cloud computing, in *Proceedings were incompletely International L4th Conference On Extending Database Technology* (K ACM, 2011), pp. 129–140
32. A.M. Hammadi, O. Hussain, A framework for SLA assurance in cloud computing, in *2012 26th International Conference on Advanced Information Networking and Applications Workshops (WAINA)* (IEEE, 2012), pp. 393–398
33. S.K. Garg, S. Versteeg, R. Buyya, A framework for ranking of cloud computing services. Futur. Gener. Comput. Syst. **29**(4), 1012–1023 (2013)
34. R. Trapero et al., A novel approach to manage cloud security SLA incidents. Futur. Gener. Comput. Syst. **72**, 193–205 (2017)
35. V. Casola et al., Security-by-design in clouds: a security-SLA based methodology to build secure cloud applications. Procedia Comput. Sci. **97**, 53–62 (2016)
36. C.P. Pfleeger, S.L. Pfleeger, Security in Computing. Prentice Hall Professional Technical Reference (2002)
37. T. Jena, Disaster recovery services in intercloud using genetic algorithm load balancer. Int. J. Electr. Comput. Eng. **6**(4), 18–28 (2016)
38. A. Aich, A. Sen, S.R. Dash, A survey on cloud environment security risk and remedy, in *2015 International Conference on Computational Intelligence and Networks (CINE)* (IEEE, 2015), pp. 192–193
39. M. K. Iskander et al., Enforcing policy and data consistency of cloud transactions, in *2011 31st International Conference on Distributed Computing Systems Workshops (ICDCSW)* (IEEE, 2011), pp. 253–262
40. S.C. Payne, A guide to security metrics. SANS Institute Information Security Reading Room (2006)

Chapter 9
Security and Privacy at Cloud System

Nandita Sengupta

Abstract This chapter is focused to provide security mechanism for complete cloud system by implementing encryption and intrusion detection system. Hybrid encryption is applied on data at cloud client level so that data in medium will be safe as well as data will be stored in cloud server in safe mode. Data in server will be accessible only to the authorized users which have the decryption key. Computation for decryption becomes challenging and difficult in case of hybrid encryption. The second phase of security will be applied in cloud server by implementing intrusion detection system which will detect the anomaly traffic towards server and block the unauthorized and unauthenticated traffic. Dimension reduction techniques are also focused in this chapter to make the efficient intrusion detection system.

Keywords Artificial neural network · Cloud system · Dimension reduction · Hybrid encryption · Machine learning techniques · Security · Support vector machine

9.1 Introduction

Use of electronic data is increasing day by day in individual level as well as in corporate world. Various aspects of such data need to be considered for its effective use. Maintaining data is very important research work in computer science world. Maintaining data refers the processes like, cleaning, storing, providing security, accessing. Daily transaction data is generated in large amount which are not required to be stored as it is. Cleaning, i.e., preprocessing of data is very vital for efficient utilization of data. Cleaning data refers removal of noise, removal of redundant data, and preparation of data in required format. After cleaning, data is stored for future use in the form of data warehouse or in big data form. Large size of space is required to store such huge amount of data. Cloud system provides such storage to store data.

N. Sengupta (✉)
Department of IT, University College of Bahrain, P O Box 55040, Manama, Bahrain
e-mail: ngupta@ucb.edu.bh

B. S. P. Mishra et al. (eds.), *Cloud Computing for Optimization: Foundations, Applications, and Challenges*, Studies in Big Data 39,
https://doi.org/10.1007/978-3-319-73676-1_9

Cloud system has become popular very recently because of various advantages of using cloud. Users can use infrastructure, i.e. storage, RAM, etc. in IaaS model of cloud. They can use different types of software in SaaS model of cloud. They can use readymade platform in PaaS model of cloud. Without buying hardware and software, people can use those based on pay-per-use model as agreed through Service Level Agreement between cloud service provider and cloud consumer. There are many reasons to use cloud system even though many challenges are faced while using cloud. There are some challenges in using cloud, like uninterrupted high bandwidth internet connection, security, downtime, service quality etc. Out of these challenges, most critical issue is to maintain security in cloud system. Security of cloud includes security of data in client, in server as well as when data is in medium. Cloud client can access data in safe mode from cloud server if cloud security is maintained. Though many researchers from Computer Science field are working on divergent aspect of cloud security, it needs continuous effort for maintaining privacy of data of cloud consumers. Cloud security refers to the security which is maintained at all cloud actors level, like cloud consumer level, cloud service provider level, cloud broker level, cloud audit level and also when data at the medium. Both encryption and Intrusion Detection System play an important role in maintaining security in cloud. Designing computationally complex encryption technique for data at transmission medium and efficient real time Intrusion Detection System (IDS) at cloud consumer site, cloud service provider site contribute a lot in cloud security system. In this chapter, hybrid encryption and IDS have been strengthened for improvement of security of cloud system. Hybrid encryption algorithm is proposed for protection of data in medium and in cloud service provider. Classifier plays an important role in IDS. Designing of IDS is proposed where the performance of two classifiers, based on machine learning techniques, have been compared. Hybrid technique of encryption is consisting of three phases. In the first phase, simplest transposition technique, rail fence, will be applied on plain text. One key, which is known as rail key, will be considered for this transposition technique. In the second phase, Vernam Cipher substitution technique will be applied on cipher text developed from the first phase of hybrid encryption. Another key will be considered for this substitution technique. In the third phase, each digit of cipher text, achieved from the result of second phase, will be complemented. Once the encrypted message reaches cloud server side, the data will remain protected by implementing IDS at server side. Real life data is too big to handle where time is very important for effective performance of IDS. Data preprocessing is a mandatory step before applying any machine learning technique as a classifier. Data Preprocessing refers the processes involving data conversion, data reduction, removal of noise, removal of inconsistency, removal of incompleteness. Application of data preprocessing is very much necessary to develop an efficient intrusion detection system. IDS at cloud server will analyze the traffic in real time and if it finds the traffic as intruder, same will be notified immediately to the administrator, so that the attack can be blocked. Artificial Neural Network has been applied to determine the traffic for IDS. As a second method, support vector machine has been applied to determine any abnormal traffic. Learning system will be generated to identify the traffic. Intruder will be detected by analyzing the attributes of the traffic. Here, in this chapter, raw data refers to the data about network traffic. Normally, network traffic has many attributes and only some of the attributes will be required to determine the traffic. Values of some of the attributes can be ignored to determine the traffic. In data preprocessing stage, these attributes are identified and unimportant attributes are known as redundant attributes. In the analysis, redundant attributes are identified before applying to IDS, which makes the IDS efficient. Section 9.2 explains cloud system and its challenges, Sect. 9.3 describes hybrid encryption, Sect. 9.4 delineates intrusion detection system, Sect. 9.5 demonstrates experimental results, and Sect. 9.6 concludes the chapter and expresses future work.

9.2 Cloud System

Cloud provides many advantages through different models of cloud system. There are three types of model used in cloud worldwide. Infrastructure as a Service (IaaS) provides different types of infrastructure, like RAM, Hard Disk, Processing Power. Software as a Service (SaaS) provides accessibility of different types of software to its clients. Platform as a Service (PaaS) provides readymade platform to its clients. With affordable cost and resources, computation is achievable by using clouds [1–7]. Service Level Agreement (SLA) is signed between cloud service provider and cloud client. Depending on the usage, amount is determined which needs to be paid by cloud client to the service provider. Popularity of cloud is increasing day by day among the individuals and corporates. One of the challenges of using cloud system is to have uninterrupted internet connection which is another research area where continuous improvement is taking place. As there is a remarkable improvement in communication, use of cloud is increasing. Continuous research work is required to provide complete solution for security of cloud computing [8–20]. The National Institute of Standard and Technology has framed a clear roadmap for cloud security. In any of the deployment model, private cloud, public cloud, community cloud, hybrid cloud, NIST has encouraged to implement security at its highest level. Five common actors are involved in cloud system, cloud consumer, cloud provider, cloud auditor, cloud broker, cloud medium. Though in this chapter, security at cloud consumer, at cloud medium and at cloud server are focused, but cloud security also involves security at cloud auditor and at cloud broker level. The objective of maintaining security refers maintaining confidentiality, availability, integrity, accountability. To protect the cloud system, some countermeasures should be taken like, authentication, identification, intrusion monitoring detection prevention, authorization, access control, cryptography, auditing. In this chapter, some countermeasures are explained in detail with proposed model, like cryptography, intrusion monitoring, detection and prevention.

9.3 Encryption

Encryption is a technique which is used for converting plain text into cipher text. Encryption technique is developed in such a way that the encryption computation becomes very difficult. Time complexity and computation complexity should be too difficult for decryption to provide robust cloud, security of data and system [21–23]. If such difficult encryption [24–27] is applied before the data leaves cloud client, data will be secured in medium as well as in server. To make encryption difficult, encryption with three phases is proposed for client data. There are lots of substitution techniques and transposition techniques are used for encryption. In the first phase, transposition technique, rail fence is applied on plain text. In the second phase, Vernam Cipher is applied on the cipher text achieved from the first phase. In the third phase, each digit is complemented achieved from the second phase.

9.3.1 Rail Fence

Rail fence is one of the effective transposition techniques used in encryption. Some sorts of permutation is applied in this technique. Rail fence is also called zigzag cipher. Rail key is considered to make the computation difficult. Depth of zigzag varies based on the rail key. Suppose the plain text is "security is the biggest challenge for cloud computing". Consider rail key is 4, i.e., depth of zigzag is 4 and also considers the blank spaces for plain text. Cipher text derived from the given plain text is as STTGCNRDUEIY HIG HEGO U PTCR SEBETALEFCOCMIGUI SL LON. It's explained diagrammatically in Fig. 9.1.

9.3.2 Vernam Cipher

In Vernam Cipher, plain text first gets converted into bit stream for each of the character. One key is considered which also gets converted into its bit stream. Then cipher text is formed applying XOR operation with plain text and the key. In this chapter, in the second phase of encryption, Vernam Cipher will be used on the cipher text which is achieved from the first phase of encryption, i.e., after applying rail fence.

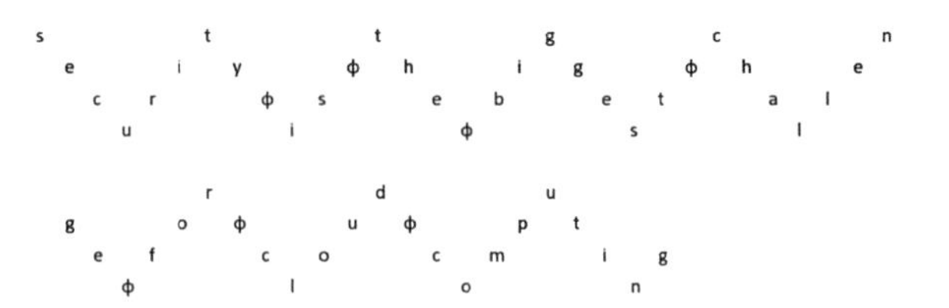

Fig. 9.1 Rail fence encryption with rail key 4

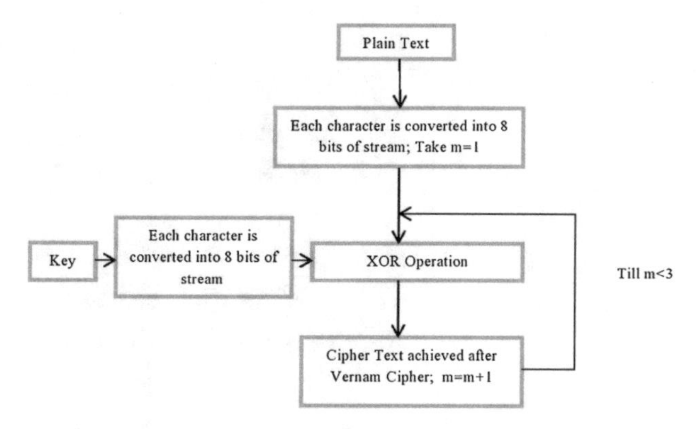

Fig. 9.2 Flowchart for applying Vernam Cipher

In normal Vernam Cipher, only one time this XOR operation is done. But here Vernam Cipher will be achieved after applying XOR operation with 3 different keys. This is applied to make the Vernam Cipher difficult from the general one. Computation time will be 3 times of the general Vernam Cipher. This proposed method is explained through flowchart in Fig. 9.2.

9.3.3 Digital Computation

This is the third phase of the encryption method proposed in this chapter. This method will be applied after getting cipher text from the second phase. Cipher text of second phase will be plain text for this phase of encryption. Each of the letters of plain text will be converted into corresponding bit stream, if they are not already in bit form. Each bit of this plain text will be complemented as cipher text. This method is explained through flowchart in Fig. 9.3.

9.3.4 Algorithm for Applying Encryption

Step 1: Take input as plain text data
Step 2: Apply Rail fence encryption algorithm
Step 2.1: Take input as rail key (n) where $n > 1$
Step 2.2: Consider blank spaces in the plain text
Step 2.3: Take the letter from plain text and write in the first row of zigzag rail
Step 2.4: Take next letter from plain text and write in the second row of zigzag rail
Step 2.5: Repeat the same n times for coming down via the rail from left corner of zigzag

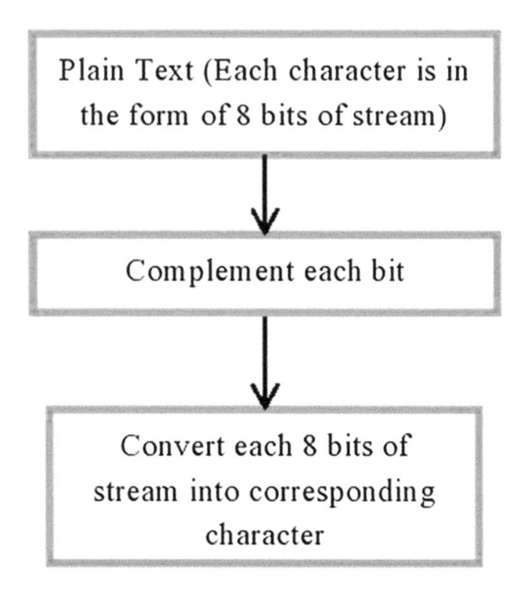

Fig. 9.3 3rd phase of hybrid encryption with simple digital computation

Step 2.6: Repeat the same (n − 1) times for going up via the rail from zero depth of zigzag to right corner of zigzag.
Step 2.7: Repeat the above steps from 2.3 to 2.6 until entire plain text gets over
Step 2.8: Write the letters in order from the first row for developing cipher text
Step 3: Consider cipher text developed from rail fence as plain text for Vernam Cipher
Step 3.1: Consider ASCII codes for each characters
Step 3.2: Convert each letter of plain text into bit streams
Step 3.3: Consider m = 1
Step 3.4: Consider a key
Step 3.5: Repeat the key and make the length of the key as same as plain text
Step 3.6: Convert the entire length of key into equivalent bit streams
Step 3.7: Apply XOR operation for each bit of plain text with each bit of key
Step 3.8: Result of XOR operation will be the cipher text developed from Vernam Cipher from m number of key
Step 3.9: $m = m + 1$
Step 3.10: Repeat above steps from Step 3.4 to Step 3.9 till $m < 4$
Step 4: Consider cipher Text developed from Vernam Cipher as plain text for digital computation
Step 4.1: Each bit of this plain text will be complemented and the result will be of set of 8 bit streams.
Step 5: Each set of 8 bit streams to be converted into the corresponding characters as assumed in Step 3.1.

9.4 Intrusion Detection System

The correct IDS [28–40] monitors the traffic, applies detection method on the traffic to identify any suspicious traffic. Mainly two types of detection techniques are used, misuse detection and anomaly detection. In misuse detection technique, patterns are generated from past records. IDS checks the pattern of new traffic with the pattern stored in database. If the pattern of new traffic matches with the pattern of anomaly traffic stored in the database, IDS generates the alarm for the administrator. In anomaly detection technique, IDS tries to identify the traffic which behaves differently than the normal traffic. IDS generates the alarms, quality and quantity of these alarms are very sensitive issue. Too many false alarms or missing alarms are not desired, in such cases IDS fails and system administrator looses the trust. Misuse detection model are not able to find the new attack. Anomaly detection model also finds difficulty in learning new attack and it generates many false alarm due to addition of new components in the network. Therefore, its very difficult to design a fixed IDS model. Dynamic design of IDS is the only choice to reduce false alarm. Different statistical models are used in IDS. Only through experience, learning can be more accurate. There are various types of IDSs, Network based IDS (NIDS) [41] which monitors entire network on which it is loaded, Protocol based IDS (PIDS) detects the traffic based on the protocol, Host based IDS (HIDS) [42] is loaded on the host which needs to be protected from attack traffic. Here, in this chapter, HIDS and PIDS are referred for cloud security. IDS has two components, IDS Sensor and IDS Manager. IDS Sensor identifies the traffic as normal or anomaly by classification. Different types of classifiers are used in IDS classification. Performance of classifiers depends on system data, noise, number of attributes, etc. Different types of classifiers, artificial neural network, support vector machine, genetic algorithm, rough set theory, etc. are used in IDS. IDS Manager informs the administrator about anomaly traffic. In this chapter, it is proposed to implement IDS at cloud server to restrict the accessing of data. Artificial Neural Network and Support Vector Machines are used as classifier to classify the traffic. Data preprocessing is required to make the classification technique efficient.

9.4.1 Data Preprocessing

Data preprocessing means preparing system into the form which is easier for classification. Real life data can have noise, missing value. It can have the combination of discrete data and continuous data. Many classification techniques do not take both types of data. In that case, one type of data needs to be converted. Most importantly, real life data size is too big for quick processing. Classification for IDS should work in real time. Here, efficiency is considered in terms of accuracy and in terms of time, i.e., classification techniques should work quickly to produce the result. If it delivers its output with higher computation time, result may not be useful for the system even though it produces the result with high accuracy. For classification, study of attributes, i.e., determination of importance of attributes is an essential part of data preprocessing. In an object, attributes are classified in two broad categories, conditional and decisional. Decisional attribute means which specify the classification and conditional attributes are those which determine the value of decision attributes. Some conditional attributes do not participate in classification. i.e., removal of such attributes does not affect the result of classification. Such attributes are known as redundant attributes. Sometimes, some objects are repetitive in the data set and in that case, repetitive objects should be removed before computation /training of classification. Therefore, preprocessing of data refers noise removal, conversion of data, removal of redundant attributes, and removal of repetitive objects. Here, in this chapter, for identifying redundant attributes, Rough Set Theory (RST) is used. Feature extraction is another method by which important attributes are identified and redundant attributes are removed for classification. Data reduction refers to dimension/attribute reduction and object/instance reduction of an information table. In this chapter, both reduction methods have been addressed to efficiently handle the huge data sets. Dimensionality reduction [43–47] compromises between selection of most predictive features for class-outcome and rejection of irrelevant features with minimal information loss. In addition, large database contains redundant objects or instances which increase system overhead and dominating the important objects results wrong prediction. Hence attribute and instance reduction is essential to reduce complexity of the system and increasing scope of accurate prediction.

In the chapter, two approaches of dimensionality reduction are presented while handling both discrete and continuous domain data sets. First, real valued NSL KDD network traffic data is discretized and then rough set theory (RST) is applied for attribute reduction using discernibility matrix and attribute dependency concepts. In the second approach, information loss due to discretization has been avoided by extending Fuzzy-RoughQuickReduct (FRQR) algorithm [48], which is applied on continuous domain data set. Genetic algorithm (GA) is integrated with FRQR to select optimum number of attributes from the continuous domain data set. The proposed algorithm (Fuzzy-Rough-GA) generates minimal attribute set, called reduct [49] consisting of information that classifies data more accurately compared to the FRQR algorithm.

Rough Set Theory (RST) has ability to deal with vague data sets and discover knowledge by processing large volume of data, has been applied in the chapter as a data mining tool for dimension reduction. Since RST cannot handle continuous data, efficient way of discretization has been focused in the work with the objective of minimum information loss while maintaining consistency in a comprehensive way. Two different discretization approaches have been proposed here to achieve the discretization with minimum loss and primarily applied on NSL-KDD network traffic data set. To retain important attributes, classification accuracy using Support Vector Machine is evaluated before and after reduction of attributes. One attribute has been reduced at a time and the effect on classification accuracy is observed in terms of percentage of error. The result of classification considering 34 continuous attributes has been shown in "confusion matrix" [50] and using "error rate". If change in error rate is more than 1% for consecutive two attributes, that attribute is considered as an important attribute and so retained.

Instance reduction has been tackled here by extending the concept of simulated annealing fuzzy clustering (SAFC) algorithm [51] with RST. First SAFC algorithm is modified to partition the objects with respect to individual attribute. Most significant cluster for each attribute is evaluated using attribute dependency concept of RST and thus less informative objects are removed to reduce instances in the information table. Finally, non-redundant and most informative instances are retained and reducts using the reduced instances are obtained. Classification accuracy of reduced data set is compared with the original NSL-KDD data set, demonstrating effectiveness of data reduction methods.

Dimension Reduction Using RST

Before illustrating the proposed algorithms, concepts of Rough set theory (RST) used in the algorithms are presented briefly in this subsection.

Preliminaries of RST

Information system is a data table consisting of objects and attributes with values. Formally, the table is represented as a set of tuple (U, A, Va) where U is the nonempty set of objects, A is a nonempty set of attributes and Va is the value of attribute a ($a \varepsilon A$) such that a: $U \rightarrow Va$. The set of attributes A is divided into two subsets, namely conditional set of attributes (C), representing features of the objects and decision set of attributes (D), representing class label of the objects.

Indiscernibility Relation

The Indiscernibility relation is denoted by IND(P) states that two objects x and y are equivalent if and only if they have the same attribute values with respect to attribute set $P \varepsilon A$, as defined by Eq. (9.1).

$$IND(P) = \{(x, y) \varepsilon U \mid \forall a \varepsilon P, a(x) = a(y)\} \tag{9.1}$$

Indiscernibility relation IND(P) is referred as P-indiscernibility relation and the class of indiscernible objects is denoted by [x]P. The partition of U, determined by IND(P) is denoted as U/IND(P) or U/P.

Lower Approximation

Lower approximation of a set X with respect to attribute set R (R´X) is the set of all objects, which are certainly or unambiguously elements of set X, as defined using Eq. (9.2).

$$R'(X) = \{x \varepsilon U, [x]_R \subseteq X\} \tag{9.2}$$

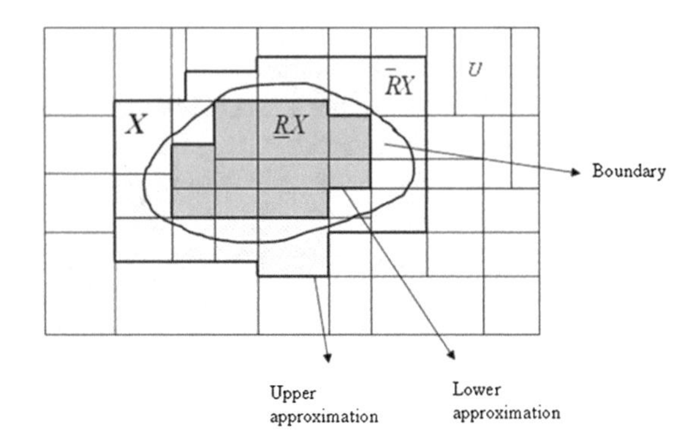

Fig. 9.4 Diagrammatical representation of rough set

Upper Approximation

Upper approximation of a set X with respect to attribute set R (R` (X)) is the union of all equivalence classes in [x]R, which have non-empty intersection with set X. It is the set of objects, possibly members of the set X, defined using Eq. (9.3).

$$R^{`}(X) = \{x \varepsilon U, [x]_R \cap X \neq \phi\} \tag{9.3}$$

Boundary Region

The boundary region of set X consists of those objects which are neither be ruled in nor ruled out of the set X. It is represented as the difference of upper and lower approximation of the set X.

Rough Set

The rough set (RS) [52–55] is represented by the tuple (R´(X),R` (X)) consisting of lower and upper approximation of the set X, as shown in Fig. 9.4.

Positive Region

The positive region of a set X contains all objects that are certainly belongs to set X, defined in Eq. (9.4).

$$POS_R(D) = \bigcup_{X \varepsilon U/D} R^{´}(X) \tag{9.4}$$

where $POS_R(D)$ is a positive region of the partition U/D with respect to R, the set of all elements of U that can be uniquely classified by means of attribute R.

Attribute Dependency

An important issue in data analysis is discovering dependencies between the attributes. A set of attributes D depends totally on a set of conditional attributes R, where the attributes of R uniquely determine the attributes of D. D depends on R to a degree k is denoted as $R \Rightarrow_k D$ and defined using Eq. (9.5).

$$k = \gamma(R, D) = \frac{\Sigma_{i=1}^{N}|R\acute{}D_i|}{|U|} = \frac{|POS_R(D)|}{|U|} \quad (9.5)$$

where POSRD is called a positive region of the partition U/D with respect to attribute set R and N is the number of labels of attribute set D. If k = 1, D depends totally on R, and if k < 1, D depends partially (to degree k) on R. k = 0 indicates the positive region of the partition U/D with respect to R is empty or D doesn't depend on R. The higher the dependency, the more significant the attribute is.

Reduct Using Discernibility Matrix

In the information system is there any attribute or set of attributes which are more important to the knowledge than other attributes. If so, is it possible to remove those less important attributes from an information system preserving its properties? There is a minimal subset of attributes which can fully characterize the knowledge of the information system as the whole set of attributes and maintains partition of data. Such an attribute set is called a reduct.

In Rough Set Theory (RST) reduct is the minimum attribute set that retains the indiscernibility relation as defined by the entire attribute set. For dimensionality reduction of information system, reduct is generated using two different approaches of RST. First, the concept of discernibility matrix [56, 57], introduced by Skrowron [58] is applied for computation of reduct.

The discernibility matrix is defined as follows: Given a decision system DS = (U, A, C, D), where U is the universe of discourse and A is the total number of attributes. The system consists of two types of attributes namely conditional attributes (C) and decision attributes (D) so that A = C ∪ D. Let the universe U = $x_1, x_2, \ldots, x_n$, then the discernibility matrix M = (m_{ij}) is a $|U| \times |U|$ matrix, in which the element m_{ij} for an object pair (x_i, x_j) is obtained by Eq. (9.6).

$$m_{ij} = \{a\varepsilon C : a(x_i) \neq a(x_j) \wedge (d\varepsilon D, d(x_i) \neq d(x_j))\}, i, j = 1, 2, \ldots, n \quad (9.6)$$

where m_{ij} is the set of attributes classifies objects x_i and x_j into different decision class labels using partition U/D. The physical meaning of the matrix element m_{ij} is that objects x_i and x_j can be distinguished by any attribute in m_{ij}. The pair (x_i, x_j) can be discerned if $m_{ij} \neq 0$. A discernibility matrix M is symmetric, i.e., $m_{ij} = m_{ji}$, and $m_{ii} = \varphi$. Therefore, it is sufficient to consider only the lower triangle or the upper triangle of the matrix.

Consider an information system of Table 9.1 consisting of 10 objects, 5 conditional attributes, namely a, b, c, d, e and one decision attribute, say f.

Discernibility matrix of the Information System (Table 9.1) is shown in Table 9.2 and corresponding discernibility function f(s) has been derived as shown.

Table 9.1 An information system

Objects	Attributes (A)					
	Conditional					Decision
	a	b	c	d	e	
O1	1	2	0	1	1	CLASS1
O2	1	2	0	1	1	CLASS1
O3	2	0	0	1	0	CLASS2
O4	0	0	1	2	1	CLASS3
O5	2	1	0	2	1	CLASS2
O6	0	0	1	2	2	CLASS1
O7	2	0	0	1	0	CLASS2
O8	0	1	2	2	1	CLASS3
O8	2	1	0	2	2	CLASS1
O10	2	0	0	1	0	CLASS2

Table 9.2 Discernibility matrix using Table 9.1

	O1	O2	O3	O4	O5	O6	O7	O8	O9	O10
O1	–	–	–	–	–	–	–	–	–	–
O2	–	–	–	–	–	–	–	–	–	–
O3	a,b,e	a,b,e	–	–	–	–	–	–	–	–
O4	a,b,c,d	a,b,c,d	a,b,d,e	–	–	–	–	–	–	–
O5	a,b,d	a,b,d	–	a,b,c	–	–	–	–	–	–
O6	–	–	a,c,d,e	e	a,b,c,e	–		–	–	–
O7	a,b,e	a,b,e	–	a,c,d,e	–	a,c,d,e	–	–	–	–
O8	a,b,c,d	a,b,c,d,e	a,b,c,d,e		a,c	b,c,e	a,b,c,d,e	–	–	–
O9	–	–	b,d,e	a,b,c,e	e	–	b,d,e	a,c,e	–	–
O10	a,b,e	a,b,e	–	a,c,d,e	–	a,c,d,e	–	a,b,c,d,e	b,d,e	–

$$f(s) = (a \vee b \vee e) \wedge (a \vee b \vee e) \wedge (a \vee b \vee c \vee d) \wedge (a \vee b \vee c \vee d) \wedge (a \vee c \vee d \vee e) \wedge (a \vee b \vee d) \wedge (a \vee b \vee d) \wedge (a \vee b \vee c) \wedge (a \vee c \vee d \vee e) \wedge (e) \wedge (a \vee b \vee c \vee e) \wedge (a \vee b \vee e) \wedge (a \vee b \vee e) \wedge (a \vee c \vee d \vee e) \wedge (a \vee c \vee d \vee e) \wedge (a \vee b \vee c \vee d) \wedge (a \vee b \vee c \vee d) \wedge (a \vee b \vee c \vee d \vee e) \wedge (a \vee c) \wedge (b \vee c \vee e) \wedge (a \vee b \vee c \vee d \vee e) \wedge (b \vee d \vee e) \wedge (a \vee b \vee c \vee e) \wedge (e) \wedge (b \vee d \vee e) \wedge (a \vee c \vee e) \wedge (a \vee b \vee e) \wedge (a \vee b \vee e) \wedge (a \vee c \vee d \vee e) \wedge (a \vee c \vee d \vee e) \wedge (a \vee b \vee c \vee d \vee e) \wedge (b \vee d \vee e)$$

Equivalent terms (elements connected by logical "OR" operation) are removed and discernibility function becomes:

$f(s) = (a \vee b \vee e) \wedge (a \vee b \vee c \vee d) \wedge (a \vee c \vee d \vee e) \wedge (a \vee b \vee d) \wedge (a \vee b \vee c) \wedge (a \vee b \vee c \vee d \vee e) \wedge (e) \wedge (a \vee b \vee c \vee e) \wedge (a \vee c) \wedge (b \vee c \vee e) \wedge (b \vee d \vee e) \wedge (a \vee c \vee e)$

Next absorption law is applied on f(s), which specifies that if one term is a pure subset of another term and connected with Boolean "AND" operation then the term with minimum number of variables is sustained. By applying the absorption law, discernibility function is derived as:

$f(s) = (e) \wedge (a \vee c) \wedge (a \vee b \vee d)$

Finally, expansion law algorithm is applied to retain attributes which are more frequently appearing in partitions compared to other attributes. AND operation is applied on attributes having highest frequency and so selected because they play important role in classification, compared to others which appear less frequently. Less frequently attributes are not rejected but apply OR operation to select any of them. Finally apply AND operation on each of the OR term so that any of them may belong to different reducts by applying distributive law.

Following steps are executed to apply expansion law.

Step 1: Find the attributes appearing most frequently (at least twice).
Step 2: Apply "AND" operation on the terms having such attributes and "OR" operation on the rest.
Step 3: Apply the connective "AND" between the "OR" terms and the term if consisting of such attribute then eliminate.
Step 4: Combine the terms obtained from Step 2 and Step 3 using "AND" operation.

For the example, most frequent attribute is a, based on which following derivations are obtained.

From Step 2: $(a \wedge e)$
From Step 3: $c \wedge (b \vee d)$
From Step 4: $(a \wedge e) \wedge (c \wedge (b \vee d))$
so, $f(s) = (a \wedge e) \wedge (c \wedge (b \vee d)) = (a \wedge e) \wedge ((c \wedge b) \vee (c \wedge d)) = (a \wedge e \wedge c \wedge d) \vee (a \wedge e \wedge c \wedge d)$ Therefore, reducts are {a, e, c, b} and {a, e, c, d}.

Reduct Using Attribute Dependency

The algorithm for attribute selection or dimensionality reduction has been proposed based on the attribute dependency concept of RST and implemented using tree data structure. Initially, the set of all conditional attributes is mapped at the root node of the tree. Gradually one by one the attributes are removed from the set and dependency of remaining attributes is evaluated. If dependency remains same to that of the root node then there is a possibility of reduct in the reduced set of attributes and the path is extended otherwise, the path is aborted.

Algorithm Reduct

Step1: Indiscernibility of the set of conditional attributes (C) and decision attributes (D) are evaluated using Eq. (9.1).
Step 2: POSC (D) is evaluated using Eq. (9.4).

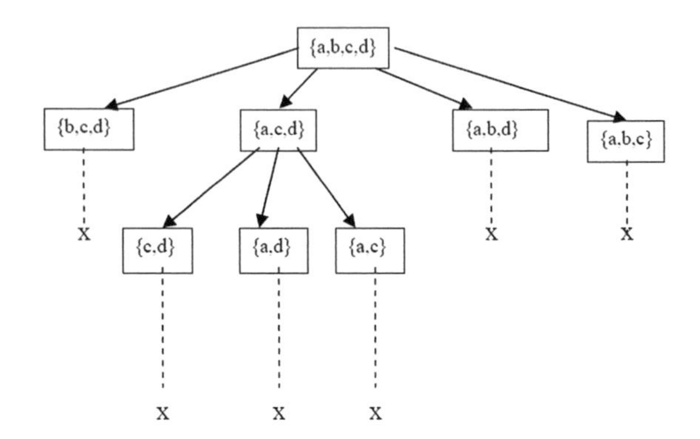

Fig. 9.5 Illustrating algorithm reduct

Table 9.3 Information system with 13 objects

Objects	Conditional attributes				Decision attributes	
U	a	b	c	d	e	f
1	2	1	1	1	1	3
2	2	1	1	0	1	3
3	2	2	1	1	1	3
4	1	1	1	0	0	3
5	1	1	1	1	0	3
6	2	1	1	2	1	2
7	2	2	1	2	1	2
8	3	2	1	2	1	2
9	3	2	2	2	1	1
10	3	3	2	2	1	1
11	3	3	2	1	1	1
12	3	2	2	1	1	1
13	3	0	2	1	1	1

Step 3: Dependency for the root node $\gamma(C, D)$ is calculated using Eq. (9.5).
Step 4: One by one attributes are removed from C and dependency for the set consisting of remaining attributes is evaluated using Eq. (9.5).
Step 5: If the dependency is same as that of the root node then go to step 6, else abort the path.
Step 6: Repeat Step 3 to Step 5 until height of the tree is equal to $|C|$.

Figure 9.5 narrates the reduct generation procedure considering information system of Table 9.3, which consists of conditional attributes {'a', 'b', 'c', 'd'} and decision attributes {'e', 'f'}.

The domain of the conditional attributes are:-

$$a \rightarrow \{1, 2, 3\}; b \rightarrow \{0, 1, 2, 3\}; c \rightarrow \{1, 2\}; d \rightarrow \{0, 1, 2\}; e \rightarrow \{0, 1\}; f \rightarrow \{1, 2, 3\}$$

Discretization is the process by which continuous variables or data get converted into discrete form. Most of the real-life data sets contain continuous variables need to be discretized to obtain qualitative data instead of quantitative. Advantage of discretization is many fold, it increases learning accuracy and processing speed while producing results in more compact and concise form. Discrete attributes are interpretable, understandable and easier to use and elaborate. But process of discretization is associated with few negative effects in data analysis. Existing methods of discretization are source of information loss resulting inconsistency in data, due to which classification accuracy gets affected. There are different discretization methods which can be applied, like cut generation method, center spread encoding method, machine learning based method.

From machine learning point of view, two different discretization processes are proposed here. Optimized Equal Width Interval (OEWI), an unsupervised method and Split and Merge Interval (SMI), a supervised method have been presented in the work. It must be observed that discretization leads to some information loss and may result in inconsistent rules. So, by discretization process strategies have been adopted to minimize number of inconsistencies that may arise after discretization. Both these discretization schemes handle inconsistency in data as illustrated below.

Optimized Equal Width Interval (OEWI)

The Optimized Equal Width Interval discretization is an unsupervised method that divides the range of conditional attributes into equal width (w) bins, as given in Eq. (9.6).

$$w = (v_{max} - v_{min})/k \tag{9.7}$$

where k is any positive integer denoting number of bins, v_{max} and v_{min} represent maximum and minimum value of the attribute (feature) in continuous domain.

Thus, the continuous data is mapped to one of the bins based on their spatial distribution where the intervals have width w and the cut points are at $v_{min} + w$, $v_{min} + 2w, \ldots, v_{min} + kw$.

However, since the size of each bin is fixed, the number of inconsistent rules generated due to discretization depends on the value of k. To minimize the number of inconsistent rules, Particle Swarm Optimization (PSO) technique has been invoked in OEWI method to optimize number of bins (k).

Particle Swarm Optimization(PSO)

Particle swarm optimization uses a population of particles where it is assumed that each particle is a potential solution. The system is initialized with a population of random solutions and searches for optima, according to some fitness function. Particles are updated over generations where particles "fly" through the N-dimensional problem search space towards the current better-performing particles. Each particle remembers its own best position X_{pbest} (the function was fittest), and searches globally best value, X_{gbest}.

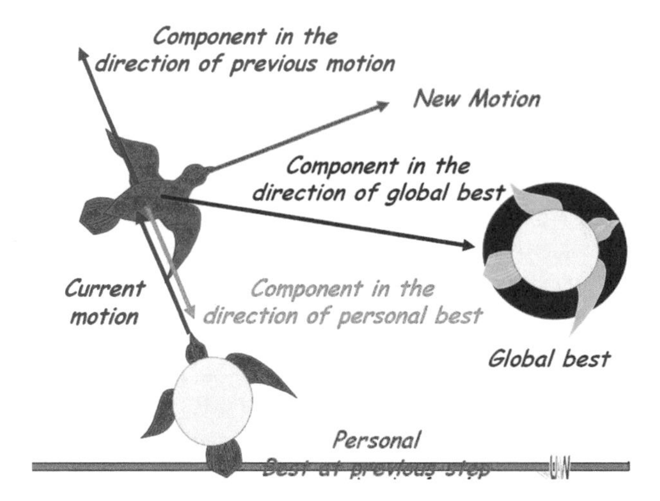

Fig. 9.6 Particle Swarm optimization

At each iteration, the particle velocity vectors V and position vectors X are updated according to Eqs. 9.7 and 9.8 respectively. Factor c1 and c2 are empirically determined and used to establish a balance between exploration and convergence.

$$V_i = w * v_i + c1 * rand(.) * (X_{pbest} - X_i) + c2 * rand(.) * (X_{gbest} - X_i) \quad (9.8)$$

$$X_i(t+1) = X_i(t) + V_i \quad (9.9)$$

where V_i, X_i, X_{pbest} *and* X_{gbest} are N-dimensional vectors and the parameter w is the inertia weight, a suitable selection of w provides a balance between global and local exploration.

After each candidate solution is generated in the intermediate course of PSO, inconsistencies present in the data set are evaluated as fitness function of PSO which is optimized. Therefore, after Equal width optimization process, optimized number of inconsistent rules are formed.

Figures 9.6 and 9.7 explains the concept of PSO and the flow of control of the OEWI Discretization process is shown by the flowchart.

Split and Merge Interval (SMI)

In this scheme, each attribute is considered individually while discretized. Say, x is the minimum difference between attribute values that differentiates two distinct classes (normal and anomaly). However, there is no specific range of the values taken by the attributes, which creates problem due to the heterogeneous nature of data distribution of different attributes. Homogeneity is achieved by dividing each attribute value with maximum of modulus values obtained while considering all

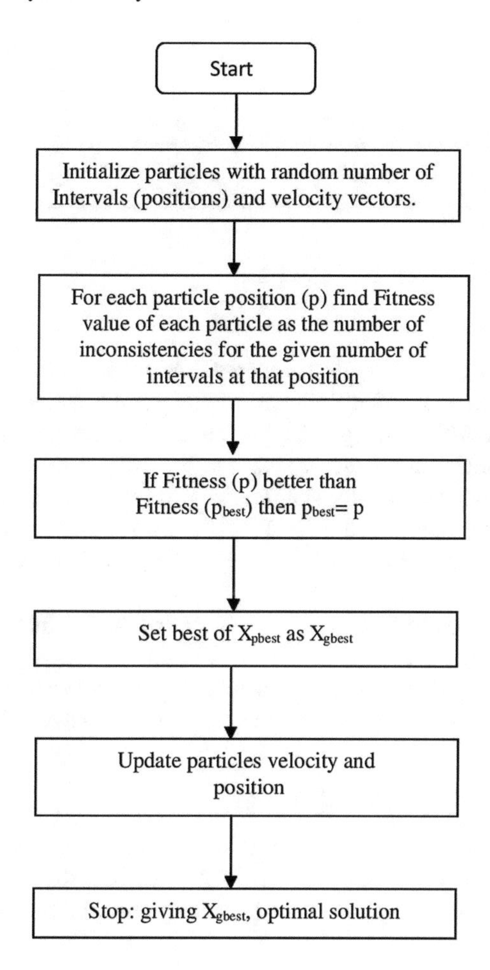

Fig. 9.7 Flowchart of OEWI discretization process

sample points. The standardization operation as described in Eq. (9.9) ensures every attribute value is maintained within −1 to +1 range.

$$Standard_{data} = (original_{data})/maximum(mod(original_{data})) \tag{9.10}$$

The whole range [−1, +1] is splitted considering ∂x as the split interval length. In this method, splitting the whole range into small intervals ensures no inconsistency after splitting. However, if there is inconsistency already present in the continuous datasets, the SMI discretization method fails to remove it.

After splitting, finitely many intervals are generated. As a next step, consecutive intervals are merged provided the objects in the two intervals have the same class label. This method reduces number of intervals and at the same time retains con-

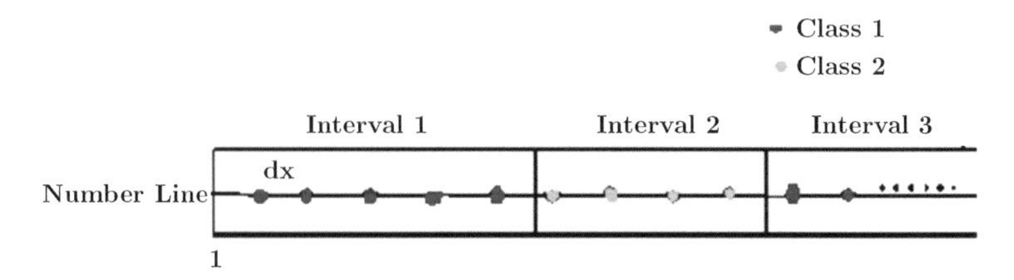

Fig. 9.8 The SMI process determine the intervals

sistency in the dataset as intervals are merged only when they belong to the same class. It must be noted that after Split and Merge processes (see Fig. 9.8), the intervals possibly are of different length. SMI discretization process is more adaptable to non-uniform data distribution and orientation. It partitions the whole data range of each attribute into distinct and disjoint intervals and thus maintaining consistency.

The intervals are represented as $[-1, K1\partial v]$, $[K1\partial v, K2\partial v], \ldots, [K(n-1)\partial v, K(n)\partial v]$ where $K1 < K2 < K3 < K(n-1) < K(n)$ *and* $K(n) = 1/\partial v$, the last interval.

The following theorem proves that the Split and Merge Interval Discretization process always preserves consistency in the data set.

Theorem I. Split and Merge Interval Discretization process preserves consistency. The prove follows in two steps, considering first the Split phase and then the Merging phase.

(i) Consistency is preserved after Splitting

Proof. Assume, after splitting an attribute with values x1 and x2 ($x1 \neq x2$) lie in the same interval but in different class labels. Also assume that the interval length is ∂v, denoting minimum difference in attribute values that differentiates two different class labels.

Therefore,$|x1 - x2| = \partial x \leq \partial v$

Now if x1 and x2 have two different class labels then minimum distance separating two class labels is ∂x as $\partial x \leq \partial v$. But we assume that v is the minimum separating distance. Therefore, assumption is incorrect and there are no distinct attribute values belonging to the same interval but with different class labels. Therefore, it is proved that consistency is preserved after splitting.

(ii) Consistency is preserved after Merging

Proof. In the merging phase consecutive intervals are merged if and only if their class label is same. Before merging (i.e. after splitting phase) consistency is preserved and the merging process itself preserves consistency by merging only the consecutive intervals with same class label. Therefore, consistency is maintained after merging. To illustrate the method, consider Table 9.4 with one attribute and two class labels.

For the given sample dataset, it has been observed that the minimum difference between attribute value that separates two classes ('Normal', 'Anomaly') is 0.10

Table 9.4 Illustration of SMI method

Attribute value	Class	Attribute value	Class
0.04	Normal	0.48	Normal
0.1	Normal	0.55	Normal
0.15	Normal	0.66	Anomaly
0.20	Normal	0.70	Anomaly
0.30	Anomaly	0.85	Normal
0.35	Anomaly	1.0	Normal

Table 9.5 Output of splitting process

Range of attribute values	Class	Range of attribute values	Class
0.0–0.1	Normal	0.5–0.6	Normal
0.1–0.2	Normal	0.6–0.7	Anomaly
0.2–0.3	Normal	0.7–0.8	Anomaly
0.3–0.4	Anomaly	0.8–0.9	Normal
0.4–0.5	Normal	0.9–1.0	Normal

Table 9.6 Output of the merge Process

Merged Range	Class
0.0–0.3	Normal
0.3–0.4	Anomaly
0.4–0.6	Normal
0.6–0.8	Anomaly
0.8–1.0	Normal

(0.30–0.20). After splitting, the result is summarized in Table 9.5 that demonstrates preservation of consistency in the dataset.

Now in the merging phase, consecutive intervals are merged provided corresponding class labels are same. The merged intervals along with class labels are given in Table 9.6, demonstrating consistency in the dataset after merging.

9.4.2 Artificial Neural Network

Artificial Neural Network (ANN) [59, 60] classifies the object based on pattern matching. For many real life complex applications, ANN works very efficiently. ANN has many characteristics. It considers some instances based on the attribute values, as its training data. It takes long time for training. In the learning data, there can be noise. Here, rules are generated based on pattern of the objects. ANN learns the system before classification. There are three types of learning, supervised learning, unsupervised learning, reinforcement learning. In supervised learning, before classifying the objects, classifier has set of inputs and the targeted outputs. Classifier algorithm works through the iterations. In each iteration, classifier classifies the objects and finds the difference between the targeted output and actual output. After every iteration, classifier decides to change its parameters/algorithms/rules to reduce the difference between the targeted output and the actual output. In unsupervised learning, targeted output is not available for the set of inputs. In such type of learning, classifier tries to learn the system as and when it receives the set of inputs and finds the output. In reinforcement learning, classifier uses some kind of feedback on every iteration. If it achieves the reward, the classifier strengthens its algorithms/rules in the same manner so that it achieves the desired accuracy within a short period of time. There can be various designs of ANN, like perceptrons, linear units, sigmoid units. Training of multilayer ANNs involves a lot of complexity like hypothesis space search, representing units and moreover the learning part. Perceptron is one type of ANN which provides the output as 1 or -1. Based on the inputs, linear combinations of the same are calculated and the sum is derived. Basically, perceptron separates the objects with the concept of a hyperplane. Some objects lie on one side of the hyperplane based on the output of the perceptron, like $+1$. Some objects lie on another side of the hyperplane for which the output of perceptron is -1. As the output of perceptron is either $+1$ or -1, perceptron can represent many Boolean functions. Perceptron rule and delta rule yield some acceptable results in learning. Perceptron rule works successfully when the objects are linearly separable. When the objects are not linearly separable, perceptron rule fails. In that case, delta rule works. ANN follows supervised learning for classification. ANN takes the inputs, assumes weight for each inputs and forms the output. In each iteration, it calculates the output and finds the difference between the targeted output and actual output. If the output is far from the desired output, ANN model changes the weights and recalculates the output. Iteration processes repeat till it achieves the desired output. ANN model is shown in Fig. 9.9.

ANN classifier assumes the values of weights and finds the output as mentioned in Eq. (9.8).

$$output = \sum_{i=1}^{n} I_i w_i \tag{9.11}$$

Algorithm for applying ANN is given as follows.

Step 1: Take inputs after dimension reduction and take desired accuracy, targeted output, Output
Step 2: Assume weights for each input
Step 3: Calculate actual Output applying Eq. (9.8)

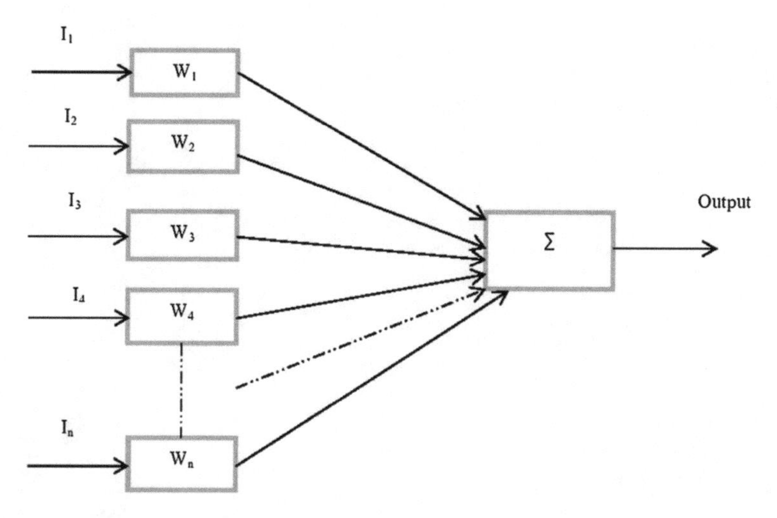

Fig. 9.9 ANN classification

Step 4: Calculate the difference between actual output and targeted output
Step 5: If desire accuracy is not achieved, change weights and repeat step 3 to step 5, else declare the accuracy.

9.4.3 Support Vector Machine

Support Vector Machine (SVM) [61–65] is an efficient classification method which uses hyperplane for classifying the objects. Though SVM performs very well in binary classification and regression problem, SVM has many challenges. Selecting parameters, pre processing of data, limiting use of SVM are a few of those. For classifying single dimensional objects, a point is required to classify. For classifying two dimensional objects, a straight line is required to classify those objects. For three dimensional objects, a plane is required to classify the objects. For n dimensional objects, $n-1$ dimensional hyperplane is required to classify those objects. Classification through SVM is shown in Fig. 9.10 for 2 dimensional objects.

The objective of SVM classification is to find out the optimum hyperplane which classifies the objects. Margin is defined as double the distance between the hyperplane and the closest point. The optimized hyperplane will be achieved when the value of margin is maximum. SVM will be efficient when number of attributes will be less. Though in Fig. 9.10, a straight line is shown which divides the objects into two classes. For two dimensional objects, separating line equation will be normal equation of a line ($y = mx + c$). But in reality, objects are with multi dimensions. In multidimensional cases, straight line equation cannot be considered as a separating line. In that case we

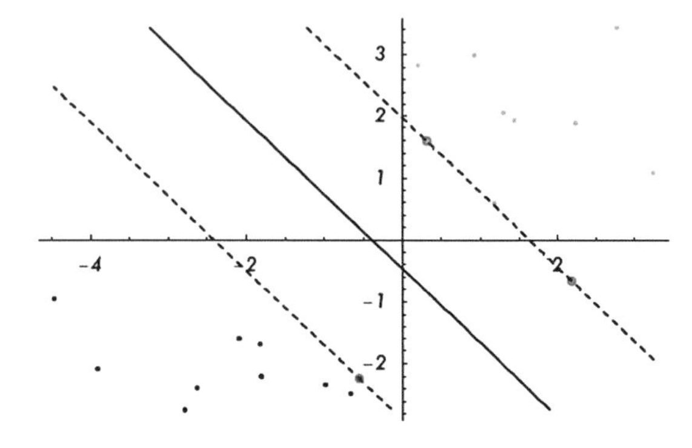

Fig. 9.10 SVM classification of two types of objects

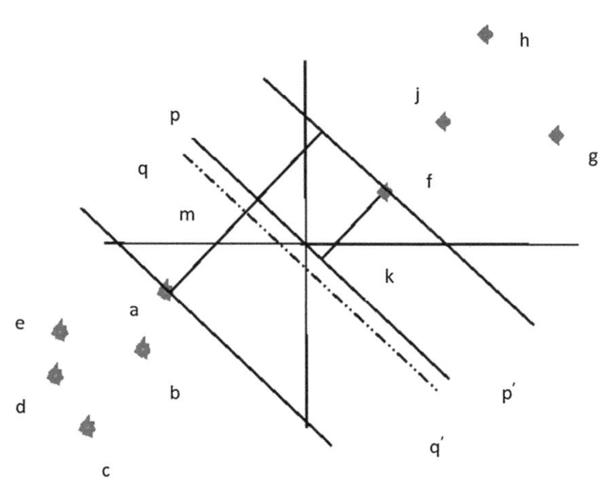

Fig. 9.11 Hyperplane showing the classifications of multidimensional objects

need to consider the equation of a hyperplane ($W^T X = 0$) for separating the objects. The method of determining hyperplane for classifying multidimensional objects is shown in Fig. 9.11.

Consider the objects a, b, c, d, e for one class and f, g, h, j are the objects of another class. These objects are considered as training objects. There can be infinite number of hyperplanes which can classify these objects into two classes. But the objective is to determine the optimal hyperplane where the margin will be the maximum. Suppose pp is the hyperplane which classifies those training objects and in this case, margin = 2Xfk. Now consider another hyperplane qq which also classifies those objects into two classes and the margin = 2Xam, which is higher than 2Xfk (clear from the drawing). So, qq is the better hyperplane than pp but optimum hyperplane needs to be determined.

Algorithm for Implementing SVM

Step 1: Get the training dataset
Step 2: Find two hyperplanes so that between these two hyperplanes no training data point should be there

Step 3: Displace these two hyperplanes in opposite direction upto the level where the distance between two hyperplanes become maximum
Step 4: Distance between these two hyperplanes will be the margin for optimum hyperplane
Step 5: Hyperplane passing through the mid point of the distance between these two hyperplanes will be the optimal hyperplane.

Different models of SVM have been designed by the researchers to improve in performance of classification. Twin Support Vector Machines are proposed by some researchers where objects are classified by using two hyperplanes. Computation cost has been reduced in such cases. Algorithms for parallel randomized support vector machine and parallel randomized support vector regression works very well for the cases of linearly non separable objects and linearly non separable regression problems. Incremental learning support vector machine uses hyper distance for learning. It uses less number of training samples as historical data is used for learning. It takes less time for training and provides better classification accuracy. One of the objective of SVM is to reduce the complexity of the decision function of SVM which can be decreased by reducing the number of training vectors. ClusterSVM provides a better model where similar types of objects are clustered and the distance from the center point of cluster to the hyperplane is measured to determine the hyperplane. As in real life IDS, time is very important, clusterSVM can yield better result. Multiclass SVM is also very useful for application of multiple classification. As in IDS, different types of attacks can be there, multiclass SVM can also be useful. Smooth SVM is another type of SVM where kernel function needs to be designed accordingly. Manifold proximal SVM is another classification which takes out maximum geometrical information from inside data. Boosting support vector machine is an effective classification method. In such method, dimension of data is reduced to help in training and in classification accuracy. This chapter tried to focus this boosted SVM for application in server IDS.

9.4.4 Classification Parameters

Confusion Matrix or table of confusion provides information about actual and predicted classification achieved by a classification system. Actual values are written column wise and predicted values are written as row wise. Confusion matrix in Table 9.7 gives the indication how actual and predicted are represented.

9.5 Experimental Results

In proposed IDS, NSL-KDD 2009 Intrusion Detection System data set [66] has been considered for experimental analysis for phase three implementation. DARPA network traffic data has been analyzed for SVM. Total 11850 objects have been

Table 9.7 Confusion matrix

	Actual value (p)	Actual value (n)	Total
Predicted value (p')	True positive(a)	False positive(b)	P'
Predicted value (n')	False negative(c)	True negative(d)	N'
Total	P	N	

Table 9.8 Confusion matrix

	Anomaly	Normal	Sum
Anomaly	9591	107	9698
Normal	491	1661	2152
Sum	10082	1768	11850

considered which have 41 conditional attributes and 1 decision attributes. Out of 41 attributes, 34 attributes are continuous and 7 are discrete. 34 continuous attributes have been applied for SVM and found out the confusion matrix as in Table 9.8.

Error rate is found as 5.05% applying SVM. For same set of data, by applying ANN, error rate is found as 6.67%. As SVM provides better classification result than ANN, SVM is proposed for implementation of IDS at cloud server.

9.6 Conclusions and Future Work

Security is one of the biggest challenges in cloud system. This chapter has been focused to provide the mechanism to overcome the above mentioned challenge of cloud system. Maintaining complete security system of cloud has been discussed in the chapter. Complex encryption has been suggested for implementation in the client cloud system so that data becomes secured before leaving the cloud client. Data will remain secure in the medium as well as in the cloud server. Practically, in the server level, two tier architecture of security has been applied, first is encryption and second is implementing IDS at cloud server level. Robust IDS with SVM classification technique is applied in server level so that it can protect the server data from any unauthorized access. In the section of SVM, various models of SVM are discussed with their merits and demerits. Cloud system provides storage place for various types of data. Single classification method or a particular model of SVM cannot yield best result of IDS for cloud server because of heterogeneous nature of data. Here, in the chapter, SVM is suggested for classification of traffic in nature. In future, for different groups of data or for different locations of cloud server can involve any SVM model which suits that environment most. Different deployment models, different business cases should be taken as an experiment for identifying classification model. In a nutshell, future work can be done for implementing powerful IDS with more efficiency.

In this chapter, only some countermeasures for cloud security is discussed, cryptography, intrusion monitoring detection and prevention. Other measures for cloud security, like, identification, authentication, authorization, access control, auditing are not explained. These areas need more research work in future. Future continuous research work can bring the solutions for security of cloud.

References

1. J. Xiong, X. Liu, Z. Yao, J. Ma, Q. Li, K. Geng, S.P. Chen, A secure data self-destructing scheme in cloud computing. IEEE Trans. Cloud Comput. **2**(4), 448–458 (2014)
2. N. Sengupta, Designing encryption and IDS for cloud security. in *The second International Conference on Internet of Things, Data and Cloud Computing (ICC 2017)*, Cambridge city, Churchill College, University of Cambridge, United Kingdom, 22–23 March 2017
3. M. Ficco, M. Rak, Stealthy Denial of service strategy in cloud computing. IEEE Trans. Cloud Comput. **3**(1), 80–94 (2015)
4. D. Talia, P. Trunfio, F. Marozzo, *Data Analysis in the Cloud: Models, Techniques and Applications* (Elsevier Science Publishers B.V. Amsterdam, The Netherlands, 2015). ISBN:0128028815 9780128028810
5. C.A. Chen, M. Won, R. Stoleru, G.G. Xie, Energy-efficient fault-tolerant data storage and processing in mobile cloud. IEEE Trans. Cloud Comput. **3**(1), 28–41 (2015)
6. T. Erl, R. Puttini, Z. Mahmood, *Cloud Computing:Concepts, Technology and Architecture* (Prentice Hall, Englewood Cliffs)
7. D. Zissis, D. Lekkas, Addressing cloud computing security issues. Futur. Gener. Comput. Syst. ELSEVIER Int. J. Grid Comput. eScience **28**, 583–592 (2012)
8. S. Thakur, G.J. Breslin, A robust reputation management mechanism in federated cloud. IEEE Trans. Cloud Comput. **PP**(99), 1 (2017)
9. Z. Zhang, M. Dong, L. Zhu, Z. Guan, R. Chen, R. Xu, K. Ota, Achieving privacy-friendly storage and secure statistics for smart meter data on outsourced clouds. IEEE Trans. Cloud Comput. **PP**(99), 1 (2017)
10. J. Baek, H.Q. Vu, K.J. Liu, X. Huang, Y. Xiang, A secure cloud computing based framework for big data information management of smart grid. IEEE Trans. Cloud Comput. **3**(2), 233–244 (2015)
11. S. Jiang, X. Zhu, L. Guo, J. liu, Publicly verifiable boolean query over outsourced encrypted data. IEEE Trans. Cloud Comput. **PP**(99), 1 (2017)
12. M.K. SIM, Agent-based approaches for intelligent intercloud resource allocation. IEEE Trans. Cloud Comput. **PP**(99), 1 (2016)
13. K. Hashizume, G.D. Rosado, F.E. Medina, B.E. Fernandez, An analysis of security issues for cloud computing. J. Internet Serv. Appl. **4**, 5 (2013). https://doi.org/10.1186/1869-0238-4-5
14. P.R. Padhy, R.M. Patra, C.S. Satapathy, Cloud computing: security issues and research challenges. IRACST Int. J. Comput. Sci. Inf. Technol. Secur. **1**(2) (2011)
15. P. Schoo, V. Fusenig, V. Souza, M. Melo, P. Murray, H. Debar, H. Medhioub, D. Zeghlache, Challenges for cloud networking security. HP Laboratories
16. H. Liu, B. He, X. Liao, H. Jin, Towards declarative and data-centric virtual machine image management in IaaS clouds. IEEE Trans. Cloud Comput. **PP**(99), 1 (2017)
17. P. Jamshidi, A. Ahmad, C. Pahl, Cloud migration research: a systematic review. IEEE Trans. Cloud Comput. **1**(2), 142–157 (2013)
18. Y. Zhang, X. Liao, H. Jin, G. Min, Resisting skew-accumulation for time-stepped applications in the cloud via exploiting parallelism. IEEE Trans. Cloud Comput. **3**(1), 54–65 (2015)
19. K. Hashizume, G.D. Rosado, F.E. Medina, B.E. Fernandez, An analysis of security issues for cloud computing. J. Internet Serv. Appl. **4**(5) (2013). https://doi.org/10.1186/1869-0238-4-5

20. N. vurukonda, T.B. Rao, A study on data storage security issues in cloud computing. in *2nd International Conference on Intelligent Computing, Communication & Convergence (ICCC-2016)*, vol. 92 (Elsevier Procedia Computer Science, 2016), pp. 128–135
21. N.P. Smart, Algorithms, key size and parameters report, European Union Agency for Network and Information Security, 2014
22. T. Olufon, E.-A.C. Campbell, S. Hole, K. Radhakrishnan, A. Sedigh, Mitigating external threats in wireless local area networks. Int. J. Commun. Netw. Inf. Secur. (IJCNIS) **6**(3), 200 (2014)
23. F. Ayoub, K. Singh, Cryptographic techniques and network security. IEE Proc. F Commun. Radar Signal Process. **131**(7) (2008)
24. U. Somani, K. Lakhani, M. Mundra, Implementing digital signature with rsa encryption algorithm to enhance the data security of cloud in cloud computing. in *1st International Conference on Parallel, Distributed and Grid Computing (PDGC - 2010)* (2010)
25. T. Cusick, P. Stanica, *Cryptographic Boolean Functions and Applications* (Elsevier). eBook ISBN: 9780128111307
26. M. Abdalla, X. Boyen, C. Chevalier, D. Pointcheval, Distributed public-key cryptography from weak secrets, in *Public Key Cryptography PKC 2009*, ed. by S. Jarecki, G. Tsudik. PKC 2009. Lecture Notes in Computer Science, vol. 5443 (Springer, Berlin, 2009)
27. U. Somani, K. Lakhani, M. Mundra, Implementing digital signature with rsa encryption algorithm to enhance the data security of cloud in cloud computing. in *1st International Conference on Parallel, Distributed and Grid Computing (PDGC - 2010)* (2010)
28. N. Sengupta, Intrusion detection system for cloud computing. in *Middle East and North Africa Conference for Public Administration Research*, Bahrain, 23–24 April 2014
29. D.E. Denning, An intrusion-detection model. IEEE Trans. Softw. Eng. **SE–13**(2), 222–232 (1987)
30. Aydin, EhsanAmiri E., Keshavarz, H., Mohamadi, E., Moradzadeh, H. : Intrusion detection systems in manet: a review. in *Elsevier, International Conference on Innovation, Management and Technology Research*, Malaysia, 22–23 Sept 2013
31. S. Roschke, C. Feng, C. Meinel, Intrusion detection in the cloud, dependable, autonomic and secure computing, in *DASC09, Eighth IEEE International Conference* (2009), pp. 729–734. E-ISBN: 978-1-4244-5421-1
32. Y. Mehmood, U. Habiba, M.A. Shibli, R. Masood, Intrusion detection system in cloud computing: challenges and opportunities. in *2nd National Conference on Information Assurance (NCIA)* (2013)
33. A. Patel, M. Taghavia, K. Bakhtiyaria, J.C. Jnior, An intrusion detection and prevention system in cloud computing: a systematic review. J. Netw. Comput. Appl. **36**(1), 2541 p (2013)
34. X. Ren, Intrusion detection method using protocol classification and rough set based support vector machine. Comput. Inf. Sci. **2**(4), 100–108 (2009)
35. Z. Yu, J.J.P. Tsai, *Intrusion Detection, A Machine Learning Approach*, vol. 3 (Imperial College Press, London, 2011). ISBN-13: 978-1848164475
36. R.P. Patil, Y. Sharma, M. Kshirasagar, Performance analysis of intrusion detection systems implemented using hybrid machine learning techniques. Int. J. Comput. Appl. (0975–8887) **133**(8) (2016)
37. A.M. Aydin, H.A. Zaim, G.K. Ceylan, A hybrid intrusion detection system design for computer network security. Comput. Electr. Eng. **35**, 517–526 (2009). Elsevier
38. D. Anderson, T. Lunt, H. Javitz, A. Tamaru, A. Valdes. Safeguard final report: Detecting unusual program behavior using the NIDES statistical component, Technical report, Computer Science Laboratory, SRI International, Menlo Park, CA, 1993
39. Lianying, L. Fengyu, A Swarm-intelligence-based intrusion detection technique. IJCSNS Int. J. Comput. Sci. Netw. Secur. **6**(7B) (2006)
40. S. Owais, V. Snasel, P. Kromer, A. Abraham, Survey: Using genetic algorithm approach in intrusion detection systems techniques. in *CISIM 2008* (IEEE, 2008), pp. 300–307
41. P. LaRoche, N.A. ZincirHeywood, 802.11 network intrusion detection using genetic programming. in *Proceeding GECCO '05 Proceedings of the 2005 workshops on Genetic and evolutionary computation* (2005), pp. 170–171

42. Y. LIN, Y. ZHANG, OU Yang-Jia, : The design and implementation of host-based intrusion detection system. Third International Symposium on Intelligent Information Technology and Security Informatics, 2010
43. R.W. Swiniarski, Rough sets methods in feature reduction and classification. Int. J. Appl. Math. Comput. Sci. **11**(3), 565–582 (2001)
44. Y. Zhao, F. Luo, S.K.M. Wong, Y.Y. Yao, A general definition of an attribute reduct. in *Proceedings of Rough Sets and Knowledge Technology, Second International Conference, RSKT 2007, LNAI 4481* (2007), pp. 101–108
45. L. Cuijuan, L. Yuanyuan, Q. Qin Yankai, Research on anomaly intrusion detection based on rough set attribute reduction. in *The 2nd International Conference on Computer Application and System Modeling* (Atlantis Press, Paris, France, 2012)
46. M. Sammany, T. Medhat, Dimensionality reduction using rough set approach for two neural networks-based applications. in *Proceedings of the international conference on Rough Sets and Intelligent Systems Paradigms* (Springer, Berlin, 2007), pp. 639–647
47. R. Jensen, Q. Shen, *Computational Intelligence and Feature Selection: Rough and Fuzzy Approaches* (Wiley-IEEE Press, Oxford, 2008)
48. J.R. Anaraki, M. Eftekhari, Improving Fuzzy-rough quick reduct for feature selection. in *IEEE 19th Iranian Conference on Electrical Engineering (ICEE)* (2011), pp. 1–6
49. P.E. Ephzibah, B. Sarojini, E.J. Sheela, A study on the analysis of genetic algorithms with various classification techniques for feature selection. Int. J. Comput. Appl. **8**(8), 33 (2010)
50. B. Zadrozny, Learning and evaluating classifiers under sample selection bias. in *International Conference on Machine Learning ICML'04* (2004)
51. Y.X. Wang, G. Whitwell, M.J. Garibaldi, The application of a simulated annealing fuzzy clustering algorithm for cancer diagnosis. in *Proceedings of IEEE 4th International Conference on Intelligent Systems Design and Application*, Budapest, Hungary, 26–28 August 2004, pp. 467–472
52. R.C. Chen, K.F. Cheng, C.F. Hsieh, Using rough set and support vector machine for network intrusion detection. Int. J. Netw. Secur. Appl. (IJNSA) **1**(1) (2009)
53. Z. Pawlak, *Rough Sets Theoretical Aspects of Reasoning about Data*, vol. 229 (Kluwer Academic Publishers, Boston, 1991)
54. Z. Pawlak, Rough set theory and its applications to data analysis. Cybern. Syst. **29**, 661–688 (1998)
55. N. Sengupta, J. Sen, J. Sil, M. Saha, Designing of on line intrusion detection system using rough set theory and Q learning algorithm. Neurocomputing (Elsevier Journal) **111**, 161–168 (2013)
56. Y. Yao, Discernibility matrix simplification for constructing attribute reducts. Inf. Sci. **179**(5), 867–882 (2009)
57. Y. Zhao, Y. Yao, F. Luo, Data analysis based on discernibility and indiscernibility. Inf. Sci. **177**, 4959–4976 (2007). Elsevier Inc
58. A. Skowron, C. Rauszer, The discernibility matrices and functions in information systems, in *Intelligent Decision Support-Handbook of Applications and Advances of the Rough Sets Theory*, ed. by R. Slowinski (Springer, Dordrecht, 1991), pp. 331–362
59. H. Li, Q. Zhang, J. Deng, B.Z. Xu, A preference-based multiobjective evolutionary approach for sparse optimization. IEEE Trans. Neural Netw. Learn. Syst. **PP**(99), 1–16 (2017)
60. J. Faigi, A.G. Hollinger, Autonomous data collection using a self-organizing map. IEEE Trans. Neural Netw. Learn. Syst. **PP**(99), 1–13 (2017)
61. A.N. Alias, M.H.N. Radzi, Fingerprint classification using support vector machine. in *2016 Fifth ICT International Student Project Conference (ICT-ISPC)*, 27–28 May 2016
62. S.K.M. Varma, K.K.N. Rao, K.K. Raju, S.P.G. Varma, Pixel-based classification using support vector machine classifier. in *2016 IEEE 6th International Conference on Advanced Computing (IACC)*, 27–28 Feb 2016
63. H.A. Sung, S. Mukkamala, Identifying important features for intrusion detection using support vector machines and neural networks. in *Proceedings of International Symposium on Applications and the Internet (SAINT 2003)* (2003), pp. 209–217

64. S.S. Keerthi, S.K. Shevade, C. Bhattacharyya, K.R.K. Murthy, A fast iterative nearest point algorithm for support vector machine classifier design. IEEE Trans. Neural Netw. **11**, 124–136 (2000)
65. C. Cortes, V.N. Vapnik, Support vector network. Mach. Learn. **20**, 273–297 (1995)
66. Nsl-kdd data set for network-based intrusion detection systems (2009), http://nsl.cs.unb.ca/NSL-KDD/

Chapter 10
Optimization of Security as an Enabler for Cloud Services and Applications

Varun M. Deshpande, Mydhili K. Nair and Ayush Bihani

Abstract The advent of cloud computing has created a paradigm shift in how people around the world communicate and do business. Its inbuilt characteristics have empowered companies to build cutting edge solutions that bring us all together than we ever were before. Cloud computing provides avenues to use storage and computing resources in metered basis to provide optimized virtual infrastructure for service providers to prosper. Service providers can concentrate on building technology rather than worrying about the infrastructure and the platform for service hosting or server maintenance. Amount of information being shared and exchanged by users is growing exponentially by the passing of each hour. People around the globe have openly embraced the era of information technology, and almost unknowingly, it has become an essential part of everyday life. In this context, securing our digital life by enabling cloud applications to perform at its fullest is of prime importance. Security engineering community is continuously optimizing security standards, tools and practices to achieve this. This chapter throws light into such methods and technologies that form the Digital Guardians of our Connected World! In any discussion related to optimization of Cloud Security, it is important to recognize the current market and research trends. This chapter adopts a case study based approach to understand the current scenario and best practices with respect to Cloud Security. We discuss the overall security objectives and challenges that developers and cloud service vendors face during life cycle of Cloud software applications. Topics related to Cloud software quality assurance including cloud penetration testing are dealt with religiously

V. M. Deshpande (✉)
Department of CSE, Jain University, Bangalore, India
e-mail: varundesh@gmail.com

M. K. Nair · A. Bihani
Department of ISE, MSRIT, Bangalore, India
e-mail: mydhili.nair@msrit.edu

A. Bihani
e-mail: bihani37@gmail.com

B. S. P. Mishra et al. (eds.), *Cloud Computing for Optimization: Foundations, Applications, and Challenges*, Studies in Big Data 39,
https://doi.org/10.1007/978-3-319-73676-1_10

in the chapter. We then propose certain tools and techniques which would help any developer or cloud security enthusiast to understand how to secure any application to make it cloud ready. This is very important; especially, with the growing complexity of web application threats. Hence, we have dedicated a section of this chapter to identify and mitigate security loopholes in the applications in a smart and focused manner.

Keywords Cloud security optimizations
Security standards and recommendations · Cloud penetration testing
Tools and techniques for assessment and mitigation of security issues
Security in big data (Hadoop) · Trusted computing and identity management
Efficient security operations management · Open forums and certifications

10.1 Introduction

This chapter is designed for a Cloud Security Enthusiast having solid foundation of Cloud Architecture and Models, having basic background about web application security, digital privacy and security management. On completion of reading, the reader would be able to appreciate and understand through case studies, the Security and Privacy aspects of Cloud Computing, Big Data and Trusted Computing and how to optimize the same in real world scenario. The reader would be equipped with knowledge of usage of Tools, Standards and Methodologies to ensure Security is used smartly to enable of Cloud applications to perform at its best potential. When we discuss the optimization of cloud technologies in providing better engineering solutions to real world problems and requirements; it is important to understand the security implications of the same. As cloud computing evolves and amount of data being handled increase exponentially, the necessity to understand the security concerns and bring out innovative solutions to these concerns is riper than ever. It is up to the able hands of the security engineering community to stay ahead and respond to the new threats to the fresh requirements of cloud architecture. This section deals with understanding the security requirements, design considerations, risks and challenges of a cloud based systems by means of strategically selected case studies - Amazon Web Services and Microsoft Azure.

10.1.1 Current Scenario of Cloud Applications and Security

Cloud technology involves orchestration of several services at difference levels of abstraction, such as infrastructure, platform and software. The service orchestration in turn provides the necessary environment for the service provider to host their applications etc. without having to worry about underlying architecture as well as maintenance of quality of service. Question comes in ones mind- as to what is the level of importance and relevance of Security in this service stack and where does it fit in? This chapter throws light on just that topic. With case study based approach, we try to postulate the title of the chapter and convey to the reader as how security can play a pivotal role in enabling cloud applications to reach its fullest potential.

Drawing parallel with real world, where a lot of funds are allocated to securing a building, locality or a state; similarly, we have been seeing a notable amount to IT budget being allocated for securing the computing elements of any organization. Security can never be an afterthought, rather it needs to be in built at every level of system. As per report produced by SANS in 2016 top business drivers for security spending in the IT industry were protection of sensitive data, regulatory compliance and reducing incidents and breaches.

10.1.2 Case Study: Amazon Web Services (AWS)

Amazon started the Amazon Web Services platform [1] in 2006 to provide services for other web sites and applications. AWS offers a host of on-demand services over the cloud. Primary components of AWS include Amazon Elastic Compute Cloud (EC2) which forms the virtualization and computation core of AWS and other is Amazon Simple Storage System (S3) which handles the scalable storage unit of AWS. Along with these, AWS now offering over 70 unique Cloud computing services such as Amazon Aurora, Amazon VPC etc. These can be availed on the fly on a metered basis or highly competitive rates. This is one of the reasons for its growing popularity in the business (Fig. 10.1).

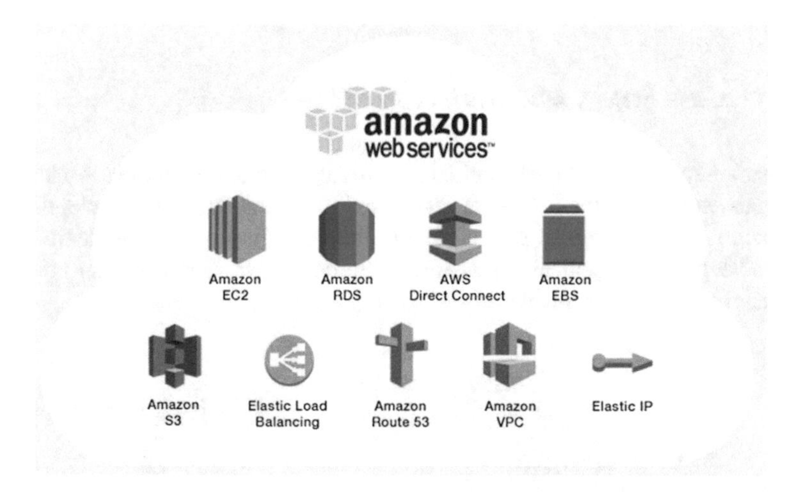

Fig. 10.1 Services provided by AWS

AWS has helped many organizations solve their challenges through web services. We will discuss two such case studies involving Avira (A Company that provides IT Security Solutions) and Alert Logic (An Amazon partner company that provides security-as-a-service).

Avira Avira offers a several IR security solutions that help its customers who are home users to large enterprises. Meaning each of the projects would require a customized deployment environment. Their local data center was simply not equipped to handle high amount of data. Also, the effort that would need to be put in to handle a huge user base would not allow the project to be launched within the required time to market. Now, Avira choose to work with AWS team to host more that 10 of their top projects on AWS. Each of these projects are isolated using Amazon Virtual Private Cloud (Amazon VPC). One of the main benefits that was highlighted during this process was that Avira engineers could spend most of their energies in building cutting edge products instead of worry about their hosting and maintenance. Lot of money was saved by Avira for not having to procure new hardware and other resources to maintain it.

Alert Logic Alert Logic has built their business providing security monitoring solutions for web and business applications. They have built a new vulnerability and configuration management solution called Cloud Insight. Cloud Insight helps its customers by identifying potential security threats and remediates them in real time. Instead of hosting this service on their own which is very costly or a small growing organization like itself, they used AWS platform features and services such as AWS CloudTrail. By doing so, Alert Logic estimate that they have saved over $1 Million in capital expenses and close to 6 months of development time!

10.1.3 Case Study: Microsoft Azure (MA)

Microsoft Azure [2] is a holistic collection of integrated cloud services stack which developers and IP professionals can use to build, deploy and manage cloud based applications via their global grid of data centers. Azure provides the flexibility to build and deploy web applications where ever the developer wants using any preferred tools, applications and frameworks of choice (Fig. 10.2).

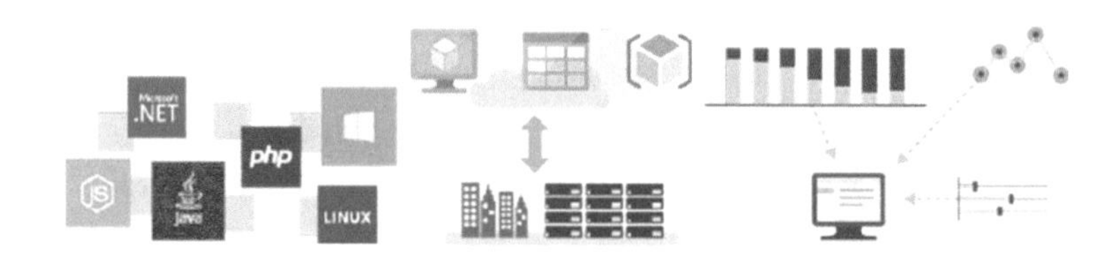

Fig. 10.2 Services provided by MA

The case studies chosen to explain the benefits of Microsoft Azure are One Horizon Group, a media and cable company from United Kingdom and Daman, a health insurance company from United Arab Emirates.

One Horizon Group a telecommunications startup company that needed a cost-effective cloud platform for hosting their VoIP service called Smart Packet. They used Microsoft Azure cloud for both software development and hosting of their solution. They could complete the product release without having to invest on highly expensive computing resources. This helped them to gain competitive advantage over others and provide aggressive rates. They also reduced operation costs and enhanced their product security. All this was possible only because of evolution and optimization of cloud computing services.

Daman is a healthcare company in the Middle East based in United Arab Emirates. The problem that they were facing is to use the data available to find creative solutions for getting business insights for managing growing healthcare costs. They needed to use decision making algorithms based on behavior monitoring, qualitative analyzing, cost control mechanisms and fraud detection. To achieve this, they essentially migrated their 3TB data warehouse from Sybase IQ to Microsoft SQL Server 2016 with Column store indexes facility provided by Microsoft Azure. With this they could 5-fold improvement in the performance and avoided $250,000 annual expenses and in turn strengthened data security and better change management.

Note: Figures taken from AWS and MA website for illustration and educational purpose only

10.2 Cloud Computing Security Objectives, Challenges

Security and Privacy has always been an essential factor in discussion related to computing. Security has always been regarded as one of non-functional requirements of a system such as reliability, maintainability etc. making it hard to measure. Providing too much security to applications may adversely affect quality of service parameters like response time, cost etc. However, providing less security may lead to security breaches of the cloud services which affect the reputation and mean time to recovery. Hence, planning smartly and providing an optimized security plays a great role in enabling the cloud applications to successfully function as designed and meet all the desired quality of service metrics. In this section, we discuss about such objective and challenges. Also Security Policy Implementation and challenges in virtual security management using VMWare ESX server case study as dealt with in the section.

Consider a scenario where a company is considering to move their products and information to cloud. Then, security of the organizations critical information in a cloud environment maintained by cloud service provider is of principal concern. Especially, when the data is being kept in geographically distributed cloud environments which are not under direct control of the company. In order to provide confidence to the companies and developers, the cloud service providers adhere to strict set of security requirement and objectives. This section [3] talks about such requirements, design constraints of secure cloud systems. We also cover topics related to challenges in policy implementation as well as go over case study of Virtualization security management using VMWare ESX server as example.

10.2.1 Requirements of Secure Cloud System

Security is a non-functional requirement. Hence, the principle requirement of a secure cloud based system is that we mitigate any known vulnerabilities in the system and make sure that system performance is not compromised when it is under external malicious attack. This requirement is not same as security functionality requirement such as identification, authentication and authorization. However, it speaks about the resilience factor of the cloud system that we are building.

The key factors that are required to be exhibited by the secure cloud system are:

- **Dependability** Software system needs to be dependable even during hostile conditions
- **Trustworthiness** Ability of not to be compromised by malicious programs and vulnerability injection
- **Resilient** Recover from any damage done gracefully and recover to full operational ability and to secure itself from further attacks. Performing this with as less damage as possible as well as in as less time as possible.
- **Availability and Fail Safe** Ensure that there is always a fail-safe state functionality which can be used when there are no other options available. This ensures that system is not compromised and up and running all the time. To achieve the above said nonfunctional requirements, organization must follow good secure development practices and life cycle. During project execution, utmost importance should be given to security during the execution of the project to avoid possible security lapses in the end. While there are several tools and techniques to design, and develop efficient cloud programs, special attention needs to be given to below aspects:
- **Sensitive Data handling and Input validation** Some data generated during computing are sensitive. Revealing of which could be harmful to the users. And leaking such information could mean facing litigations and legal actions for the service providers. For example, storing account critical information such as passwords, account recovery questions etc. in clear text/unencrypted manner in the database; storing the credit card details and other financial information in unencrypted format. Sending these parameters via URLs as parameters for easy access. These kinds of activities should be taken care of very carefully.
- **Code practices and Language Options** Reverse engineering is quite possible to trace back some loopholes present in the system and hackers can use it to exploit the low hanging fruits to gain more information on the system. This can be used to design more sophisticated exploits later. Hence using secure memory management via language options are necessary. For example, Java and C# have

option to dynamically stop buffer overflows etc. Not removing uncommented code in web visible pages as well as providing detailed version details to the user can help provide clues to hackers and they can identify loopholes if any. A way to ensure that requirements of secure cloud system are captured unambiguously is by using S.M.A.R.T.E.R. method.

- *Specific* - Aimed at specific goals which can be put on paper
- *Measurable* - We should be able to measure the extent of success both quantitatively as well as qualitatively
- *Achievable* - Do not set unrealistic goals. Achievable goals should be set.
- *Relevant* Goals should be relevant to the organizational goals and objectives
- *Time-Oriented* - The execution and results need to be time bounded activity.
- *Evaluate* - The process and result should undergo periodic evaluations and based on the evaluation, remedial measures should be taken
- *Revise* - Based on the evaluations, goals may need to revise to achieve the overall goals.

10.2.2 Risks Involved in Providing Secure Cloud System

Along with traditional risks pertaining to any IT company, cloud computing comes with its very own set of risk factors. These needs to be evaluated by developers and service providers carefully.

The 3 basic pillars of information security are Confidentiality, Integrity and Availability. These three tenets are popularly called as CIA triad. This is used as a yardstick to evaluate the security levels of a system (Fig. 10.3).

- **Confidentiality** Prevent intentional or unintentional unauthorized disclosure of contents. Need to ensure network security protocols, authentication and encryption services to maintain confidentiality of information.
- **Integrity** Guarantees that message delivered has not been altered intentionally or unintentionally from the original data received. Robust Firewall services, intrusion detection and protection systems help ensure integrity of the cloud system.
- **Availability** Assurance that the system would be stable, resilient and accessible always. Fault tolerance mechanisms, redundancy schemes using data backup etc.

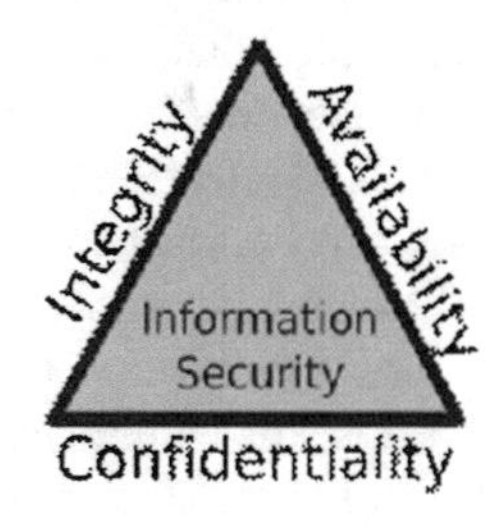

Fig. 10.3 Services provided by AWS

ensure high availability of the cloud system. Apart from CIA triad, user identification and authentication, authorization and accountability and privacy form important concepts which will result as risks if not handled properly.
- **Privacy and Legal Compliance Risks** With billions of users using cloud services on a regular basis, privacy and compliance becomes a high risk area and a pitfall for any organization using cloud computing. Identity theft resulting in privacy breach can cause lot of issues for enterprise as well as end users. We will look at privacy holistically in Sect. 10.5 of this chapter.
- **Common Threats and Vulnerabilities** Common threats to both cloud and traditional computing include eavesdropping, fraud, theft, denial of service, logon abuse, network intrusion etc.

10.2.2.1 Case Studies on Recent Security Breaches

Recent days have seen a high increase in security breaches being reported [4]. These reaches may cause downtime for the services. This costs the organizations lot of money due to loss of confidential information which eventually tarnishes their reputation. Cloud computing is one of the most valuable innovations for business, providing cheap, virtual services. The past decade has seen increasing use of cloud infrastructure to store data and provide services with major businesses and technology companies shifting to the cloud. The cloud also provides the ability to scale up or down to respond quickly to changing market conditions. Due to this, many organizations are hesitant to fully leverage the cloud, citing concerns about compliance, data loss, and unauthorized access with major security breaches having taking place in the past. Technology giants Yahoo, Dropbox, Adobe security were compromised and found millions of users credentials including private mails, passwords, and user profile data exposed.

There have been 5,754 data breaches between November 2005 and November 2015 that have exposed 856,548,312 records. Malicious attacks have gone from 200,000 a year in 2006 to 200,000 attacks a day. Major multinationals, government organizations have been at the end site of the hack. We highlight some of the recent incidents to give you an insight.

Yahoo announced in Sep 2016 that over 500 million accounts were compromised and had sensitive information stolen since 2014- over 2 years. The hacker had gained access to personally identifiable information and digital identities of the affected users. The first announced breach, reported in September 2016, had occurred sometime in late 2014, and affected over 500 million Yahoo! user accounts. A separate data breach, occurring earlier around August 2013, was reported in December 2016, and affected over 1 billion user accounts. The security lapse responsible for the hack were forged cookies to bypass security protections and access users accounts without a password.

Sony Pictures - The Sony breach began with a series of phishing attacks targeted at Sony employees. These phishing attacks worked by convincing employees to download malicious email attachments or visit websites that would introduce malware to their systems. This type of attack used social engineering, where phishing emails appeared to be from someone the employees knew, thus tricking them into trusting its source. Hackers then used Sony employee login credentials to breach Sony's network. Over 100 terabytes of data was stolen and monetary damages are estimated to be over $100 million.

MySpace - A social media behemoth before Facebook was the target of hackers in 2016 with more than 360 million emails and passwords of users being leaked online. This incident threw light at how weak passwords can be cracked easily through brute force attacks. The hacker responsible for this hack found a users password in the Myspace dump and allegedly used it to log into different accounts even though the passwords were hashed with SHA1 algorithm, which is easy to crack. Major number of hacked users had passwords such as password1, abc213, etc. which are vulnerable to brute force attack.

Dropbox - A huge cache of personal data from Dropbox that contains the usernames and passwords of nearly 70 million account holders had been discovered online. The information, believed to have been stolen in a hack that occurred several years ago, includes the passwords and email addresses of 68.7 million users of the cloud storage service. The credentials were stolen in a hack that occurred in 2012 when hackers used stolen employee login details to access a document containing the email address and passwords of users. The many reason for such hacks were largely attributed due to weak passwords set by the users which are often easy to brute force with sophisticated software. Poor security by employees of the company risked data of millions of users.

Home Depot - A security breach that attacked Home Depot's payment terminals affected 56 million credit and debit card numbers. The Ponemon institute estimated a loss of $194 per customer record compromised due to re-issuance costs and any resulting credit card fraud. Hackers first gained access to Home Depot's systems through stolen vendor login credentials. Once the credentials were compromised, they installed malware on Home Depot's payment systems that allowed them to collect consumer credit and debit card data.

10.2.3 Security Architecture Design Considerations

With all the various advantages that cloud computing brings to the table such as decreasing costs and time required to deploy new projects to market; cloud security will always remain to be a major concern. The security of the cloud system is heavily based on its security architecture. A security architecture of a cloud system is a cohesive reproducible design process which has holistic plan to address all security requirements and provides specifications on what security controls should be applied in which area.

A variety of parameters play their role in the performance and implementation of cloud security architecture. Factors related to compliance, security management and administration, controls and security awareness form the main consideration constraints for security architecture design.

Information needs to be efficiently classified to demonstrate companys commitment towards data security, identify the critical data units and to devise appropriate security measures to protect it. Information can be classified as public/sensitive/private/confidential data and provide data access to employees accordingly.

Secure execution environments is a critical design consideration for effective cloud platform utilization. In cloud atmosphere, applications need to be executed on distributed environment. When they interact with outside computing resources or even sensitive data, the computing needs to be done is a secure environment. Hence it is imperative for service provider to provide the developers with secure environments to communicate between cloud, sensitive data and client resources.

One of the advantages of cloud computing to developers or young companies is that the responsibility of providing the secure execution environment is with the service provider and not with themselves.

Cloud service providers need to be careful when developer code with potentially unsafe programming languages like C, C++ which have poor memory management capabilities. In these cases, they must perform some more compile time and runtime validations so that the coder executed does not unintentionally harm other computing resources hosted on the cloud.

CIA Triad must strictly follow in this regards so that the confidentiality, integrity and availability of the services are not compromised. Virtual private networks (VPN) are an important tool to provide secure cloud communications. It is one of the methods to ensure secure cloud environment for communications. Secure remote access to computing resources can be provided using remote access VPN. Network to network connections can be provided using network-network VPN.

VPN tunneling can be used to encapsulate data from one network to another by additional headers for security (Figs. 10.4 and 10.5).

Encryption is a key design consideration for secure communications and the information that gets exchanged between the computing resources needs to be encrypted.

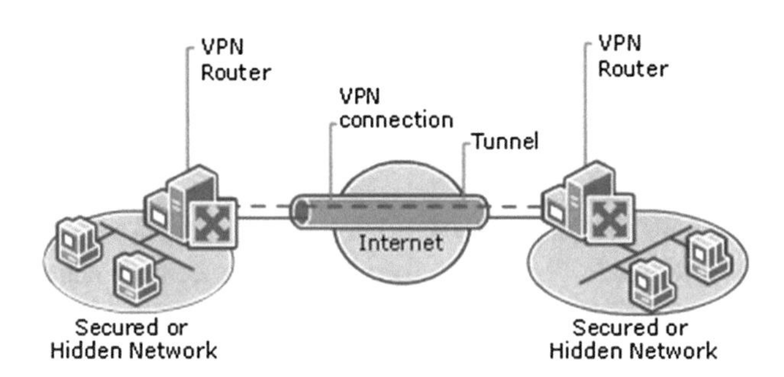

Fig. 10.4 Services provided by AWS

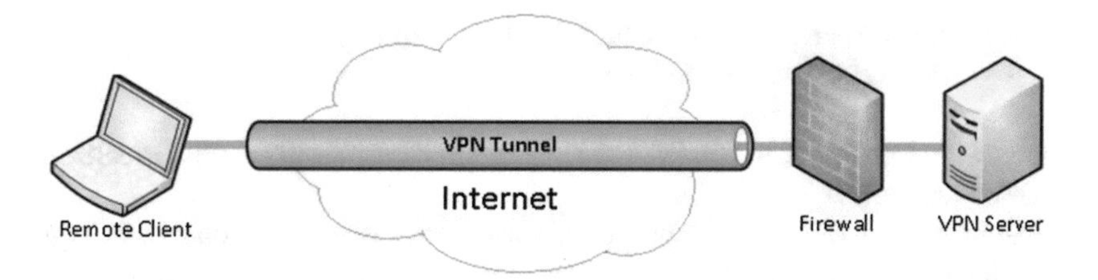

Fig. 10.5 Services provided by AWS

The server calls and communications needs to be analyzed for possible malware injections etc. In this chapter, we wont go in detail about the public and private key encryption and how it works. However, for the record we would like to highlight the usage of digital certificates, LDAP services, time stamping for creating a secure design for cloud architecture.

10.2.4 Cloud Computing Security Challenges

Some of the challenges faced by cloud based organizations can be broadly classified into: Security policy implementation, Computer intrusion, detection and response (discussed later in the chapter) and visualization management.

Security policy implementation forms the most challenging task in cloud computing for service provider. The policies include high level management policies which are mostly business oriented and lack technical specifications. The lower level policies are more specific such as regulatory policies which are related to compliance to standards, informative policies which are educating internal and external stakeholders of the company and advisory policies which are not mandatory to follow like regulation, however advisable to be followed to uphold the company defined self-standard.

Virtualization management - Virtualization describes a technology in which application, guest operating system or data storage is abstracted away from the true underlying hardware or software. The software emulates the hardware from a fully blown operating system such as Linux, windows or a specifically designed operating system such as VMWare ESX running on the virtual machine. It has allowed organizations to scale their operations with increased stability for resource fulfillment. Virtualization is now a critical part of cloud computing forming the backbone of the infrastructure as a service (IaaS) which begs the question of security risks of virtualization.

10.2.4.1 Case Study: VMWare ESX Server - Challenges in Virtualization Security Management

Some of the Security Challenges and risks involved especially with Virtualization are discussed below.

Isolation: One of the key issues in virtualization is isolation. Isolation will guarantee that one application in one VM cannot see applications running in a different VM, or that some process running in one VM cannot affect the other VMs running in the same machine. If this security assumption is broken, then an attacker can have access to other VMs in the same machine or even to the host machine. That is why it must be carefully configured and maintained.

Guest Attack: There can be some external or internal factors that can compromise isolation as mis-implementation/configuration or some bug in the virtualization software. A dangerous attack can be made if isolation between the host and the VMs is compromised. That attack is called VM escape and happens when a program can bypass the virtual machine layer from inside a VM and get access to the host machine. The host machine is the root of all the VMs, and so if a program escapes from the virtual machines privileges it will get root, allowing it to control all the VMs and the traffic between them.

Remote Management Vulnerabilities: Many VM environments have management consoles that allow administrator to manage the virtual machines through HTTP/HTTPS protocol. Older versions of VMs using HTTP stand at a greater risk of attack such as XSS and SQL injection remotely. If the attacker gains root privileges to the machine, allowing him to have access to the host, he can gain access to all the guests running.

Denial of Service: In virtual machine architecture, resources as CPU, memory, disk and network are shared between the host and the guests. The guest machine can impose a denial of service to others guest which would eventually affect the host resources. When other guests request for any resource from the host, they are denied as the resources are fully occupied by the suspicious guest. In order to prevent DOS attacks, the host allocates maximum possible resource that can be used a by guest to each of its guests.

Case Study: VMWare ESX

Hypervisors (Hosts) act as an abstraction between the underlying hardware and the guest machines running and hence is a sensitive piece of computer software. VMWare ESX [5] represents the forefront of efforts with respect to hypervisors and security which is thoroughly examined by security researchers.

VMWare ESX being one of the most common used hypervisor reported more vulnerability compared to its counters with having reported 26 vulnerabilities (Fig. 10.6).

The common vulnerability scoring system is used as a metric to determine the security lapse of a system. The figure below shows the ESX CVSS severity analysis from 2003 to 2009 (Fig. 10.7).

ESX reported 7 incidents with high vulnerability. Out of these vulnerabilities 60% were exploited over a network and 40% locally. Most High Severity attacks are Host Privilege Escalation exploitable by network. These are possible because the newer

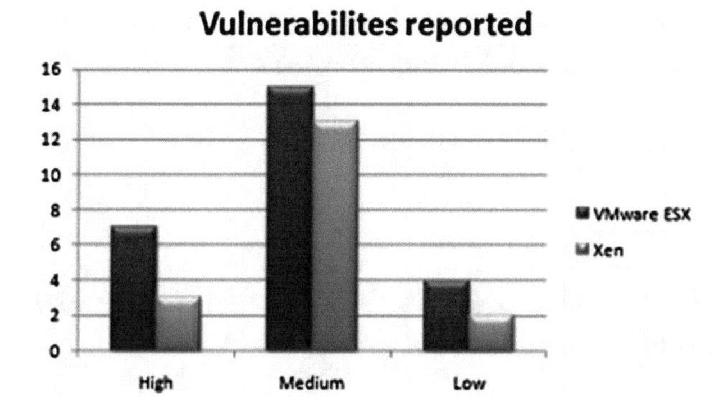

Fig. 10.6 Comparison between VMware ESX and Xen

Month/ Year	CVSS Severity	Impact Subscore	Exploitability Subscore
12/03	7,2	10	3,9
12/05	7,6	10	4,9
12/05	6,8	6,4	8,6
12/05	2,1	2,9	3,9
12/05	4,9	6,9	3,9
12/05	4,3	2,9	8,6
07/06	5	2,9	10
07/06	3,6	4,9	3,9
04/07	5	2,9	10
04/07	6,6	10	2,7
01/08	7,5	6,4	10
06/08	4,4	6,4	3,4
06/08	6,9	10	3,4
06/08	9	10	8
06/08	7,2	10	3,9
09/08	2,1	2,9	3,9
10/08	5	2,9	10
10/08	6,8	10	3,1
11/08	6,9	10	3,4
11/08	9,3	10	8,6
12/08	7,2	10	3,9
02/09	4,7	6,9	3,4
04/09	6,8	10	1
04/09	2,1	2,9	3,9
04/09	4,6	6,9	3,1
06/09	4	6,9	1,9

Fig. 10.7 Severity analysis

versions of ESX have a web server to use the client vSphere to management the server and so have a new door for attacks. For the same reason, can do XSS attacks against the ESX.

Note: Figures are taken from Internet for Illustration and educational purpose only.

10.3 Cloud Security as an Enabler for Cloud Applications: Standards and Tools for Implementation

In this section [3], we look at security elements of open standards for Cloud applications such as Organization for the Advancement of Structured Information Standards (OASIS), Open Web Application Security Project (OWASP) and Open Grid Forum (OGF) and deal with latest recommendations for implementing Web Service Security (WSS), Cross Enterprise Privacy, and Cyber-Security. The reader also is introduced to expert groups and task forces such as Computer Security and Incident Response Team (CSIRT) available in case of security incidents or activities and how to interact with them. In the same section, we dwell in detail on the Cloud Software Security Quality assurance aspects including the 3 stages of Pre-Testing. We look at the assessment methodologies and techniques for covering Cloud Penetration Testing using tools such as Visual Trace.

Cloud applications can flourish when they are empowered with necessary design and tools which help them in delivering the best in class quality of services. Cloud security is one of the main elements that enable the developers to concentrate on developing applications the meet the product requirements. To make this happen, standards need to be put in place and the process must be such that the guidelines need to be followed at each level of development cycle. In real world is not an ideal one. Hence, we need to prepare for incidents which happen and we need to respond to it. Tools that help in testing the assuring the security quality of software is available to the product delivery teams to use it and develop high quality products.

10.3.1 Standards, Guidelines and Forums

In this section, we look at some of the standard which are setup to help the developer community to use security as an enabler for their applications. Although there are several such standards and guidelines available, there is still no clear winner or widely accepted standard or 3rd party certifications that cloud security providers can affirm. We are highlighting some of the notable standards and guidelines with respect to security.

International Organizations and Standardizations (ISO) - ISO has published several series of standards pertaining to information system security and web services interoperability that are very relevant for cloud security. ISO 27001-27006

The Organization for the advancement of Structured Information Standards (OASIS) - OASIS is a standard that develops standards for security, e-commerce and web services. OASIS have created cloud related standards on SOA models, security access, data/import and export, identity management etc.

Open Grid Forum (OGF) - OGF is an international organization committed to the development and adoption of standards for applied distributed computing technologies. A sub group has been created named as Open Cloud Computing Interface (OCCI) to develop open standard APIs for Infrastructure as a service (IaaS) providing virtualization address management features. These can be used by developers to reduce possibility of vendor lock-in, develop new cloud applications, provide more cloud service options etc.

The Open Web Application Security Project (OWASP) - OWASP is an open source web community that has spent lot of time and dedicated its efforts to enhance application software community. It is one of those communities that entire security industry look up to for directions and best practices. OWASP provides standards and recommendations for cloud system security as well as application security on a regular periodic basis.

10.3.2 Incident Response Teams

For the cloud computing industry, it is critical to have incident response teams. This needs to be created using group of experts at the service provider side and well as from cloud user front. Network monitoring and management considerations have been discussed by Dr. Mydhili et al. [6]. Incident response teams are responsible for monitoring the network for any unexpected events as well as maintaining intrusion detection systems.

- **Computer Security and Incident Response Team (CSIRT)** - In a cloud based organization, creating a computer security and incident response team is required for following activities:
  Analysis of any anomaly notification
  Immediate response to an incident if analysis warrants it
  Escalating the issue based on the priority and severity
  Providing timely resolution for the issue and following up with post incident activities and remedial activities
- **NIST Special Publication 800-61** - NIST stands for National Institute of Standards and Technology. In 2004, NIST Special Publication 800-61 published a guide named as Computer Security Incident Handling Guide, Recommendations of National Institute of Standards and Technology. This guide described incident response life cycle as below:

- *Preparation* involves securing the computing arena and making sure that there are no further intrusions. Also, risk assessment needs to be done during this phase to establish incident response capabilities. NIST recommendations include user awareness training to secure networks, implementation of patch management, malicious code prevention framework etc. to be set up in the system.
- *Detection and Analysis* In the heat of the incident, detecting the actual root cause and source of the issue might not be possible or practical to search for in the available time. The next best thing would be to analyze the anomalies in input data etc. and deviation from regular system behavior. This can be achieved through analyzing the system logs for anomalous events and correlating it with events that occurred which resulted in current state of the system. Once the issue is detected, it needs to be immediately escalated to concerned teams and higher ups to get their inputs on the matter. Each of the incidents needs to be reported to chief security officer, chief information officer, legal department, system owner and public affairs team.
- *Containment, eradication and recovery* Once issue is detected, it is of highest priority to prevent the spread of the issue to any other components of the system. In some cases, the incident response teams carefully trace the hackers activity in case it will help them identify the intention of the attack and the identity of the attacker. This can be traced by looking at host IP address of the attacker, email or account details etc. Next steps involve the cleaning activity of system wherever the attackers traces are found. In many cases, even the files which are suspicious, are replaced using data backup.
- *Post incident activity* - After the incident is closed, a thorough analysis of each steps taken and time lines needs to done diligently. All parties involved need to be transparent and come up with the shortcomings, learnings and kudos for the incident response team. It is like a retrospective meeting at end of a sprint in Agile/Scrum model of application development.

10.3.3 Cloud Application Security Testing

NIST in 2001 provided certain security principles which can use used a postulate while developing any cloud applications. This was called NIST 33 Security principles. The principles address each step of software life cycle, from design to end of life of a product. Some of the highlights of these principles are:

- Principle 1 - Establish a sound security policy as the foundation for design
- Principle 2 Treat security as an integral part of overall system design
- Principle 3 Clearly delineate the physical and logical security boundaries governed by associated security policies.
- Principle 6 - Assume external systems are insecure
- Principle 16 Implement layered security; ensure there is n single point of vulnerability

- Principle 20 Isolate public access systems from mission critical resources like data etc.
- Principle 25 Minimize the system elements to be trusted
- Principle 26 Implement least privilege
- Principle 33 Use unique identities to ensure accountability

Above principles show a guiding light for cloud application security testing to ensure that the cloud applications are built the right (secured) way to achieve its goals.

Secure cloud software testing involves number of activities and techniques such as source code analysis, fault injection, dynamic code analysis, fuzz testing, vulnerability scanning, black box testing and penetration testing. Although there are no industry wide security quality assurance terminologies, ISO 9126 standard provides list of software quality standards and definitions. As per the standard, some of the attributes that need to be addressed are functionality, reliability, usability, efficiency, maintainability and portability testing.

Security testing should analyze the security properties of the software as it interacts with internal components of the system as well as external systems such as humans or other systems. SOAR Software Security Assurance State-of-the-art Report provides a guideline that each of the product security team should adhere to.

- Behavior of software should be predictable
- Should expose no vulnerabilities or weakness
- Secure exception handling and fail safe mechanisms
- Does not violate any security constraints
- Does not support reverse engineering

10.3.3.1 Cloud Penetration Testing

Cloud Penetration Testing is a methodology that is used to find the level of security of systems network security by simulating an attack from proxy malicious source. It is done at 3 levels:

- Level 1 High level assessment
- Level 2 Network valuation
- Level 3 Penetration test

Penetration testing is usually conducted without knowledge of the system specific details. So that the intruders are not influenced by the knowledge of system architecture and neglect possible issue. Penetration testing involves 3 steps - Preparation, execution and delivery.

Preparation Non-disclosure agreement is signed for legal protection of tester.
Execution Penetration testing is executed and potential vulnerabilities identified.

Delivery Results of the penetration testing along with recommendations are formally communicated to the organization.

Some of the testing tools and techniques used for penetration testing are -
Visual traceroute is a free diagnostic tool that displays the path the Internet packets take across IP networks to reach a specific destination on a network. The program requires a domain name that is the site to which the route is traced. An IP address can be used for the same.

For every trace a report identifying the approximate geophysical location of each hop and visualizing the route on a map is generated. This report helps in detecting bottlenecks, quickly track down the problem and isolate it.

Visual Trace provides a graphical interface to trace routes where the underlying commands used to trace routes are ping and tracert.

Use of visual trace in cloud:

- Improve troubleshooting
- Restore network visibility
- Connect user to apps and services

Visual Route: includes integrated traceroute, ping tests ad reverse DNS and WhoIs look up.
Smart WhoIs this tool can be used to get host of information about the target such as IP address, hostname, domain name, country etc. **SamSpade** Tool used to trace the spammers. Apart from these port scanning tools, vulnerability scanning tool, password crackers can be used for penetration testing. Some of the well-known password cracking tools are - Brutus, WebCracker, Obi wan, Burp Intruder.

10.4 Web Application Security: Security Optimization Techniques

In this section [3], we concentrate on Web Application Security Optimization Techniques and Big Data Security. We understand several men in the middle attacks (MITM) attacks such as SQL Injections, Buffer Overflows, Cross Site Scripting using Burp Suite tool. We discuss how to mitigate these risks by adopting OWASP web security recommendations and best practices taking practical examples.

10.4.1 Understanding Web Threats and Vulnerabilities Using Man in Middle Attacks

Using internet comes with the baggage of threats and vulnerabilities. Man in middle attacks are one such class of web based attack where in a malicious entity intercepts communication channel between any two entities. While doing so, he can manipulate

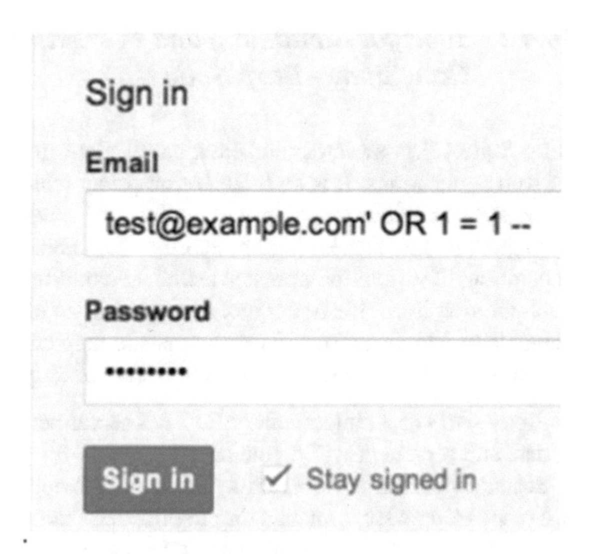

Fig. 10.8 Example for SQL injection

the messages being exchanged giving rise to vulnerabilities. **SQL Injections**: is an injection attack, where the input is not properly checked before execution. Attacker can make use of this vulnerability by embedding some SQL commands along with input. When this is executed at server side without performing proper validation of the input, the attacker can get control over the database server and cause serious damage including deleting the whole database. To ensure that SQL injections are stopped, the input needs to be parameterized and validated before execution. Also, certain access restrictions should be kept so that unintentional database query execution are avoided (Fig. 10.8).

Cross Site Scripting (XSS): When the user input such as url redirect is not properly before execution, it can result in cross site scripting. Attackers can exploit XSS by generating malicious URLs and tricking users to click on them. By clicking on the links, attackers client side scripting language like javascript and VBScript, and it gets executed in victims browser.
Example: URL http://xyz.com/users.php?register=<script>alert(Hacked!)</script>
This vulnerability can be largely avoided if the inputs are properly validated and they fall into expected set of inputs. Developers should not rely just on client side validation. They should treat all inputs as malicious and verify the correctness of input before executing them.

10.4.2 Tools for Simulating and Preventing MITM Attacks: Case Study - Burp Suite

Burp Suite [7] is an integrated Java based platform for performing security tests on web applications. It is an interception proxy which speeds up the entire testing process, from initial mapping and analysis of a web applications attack surface, through to finding and exploiting security vulnerabilities. It allows users to integrate capabilities of various tools present to find vulnerabilities such as XSS, SQL injection, configuration in cookie issues etc. faster and in an efficient manner. Burp suite is a valuable tool to detect security concerns and help organizations to rectify any bugs and optimize their code. The key components of burp are:

- Proxy -Acts as an intermediary HTTP/S server between the clients requesting for data and servers providing the data.
- Scanner - Performs vulnerability scans of web applications.
- Spider -A crawler for navigating to different links.
- Intruder - Performs powerful customized attacks to find and exploit unusual vulnerabilities.
- Repeater - A simple tool that can be used to manually test an application to modify requests to the server, resend them, and observe the results.
- Sequencer - A tool for testing the randomness of session tokens and data being cached that can be unpredictable.
- Decoder - A decoder helps to change and recognize the format of a data. It is used to transforms encoded data to raw data and from raw data to encoded and hashed forms.
- Extender - Burp suite allows the use of extensions to extend burps functionality for solving complex vulnerabilities.

Configuring Burp suite - The primary requirement of burp suite is a browser (Ex: Chrome, Firefox, Edge, Safari) which receives all the incoming traffic from the internet and establishes connection to servers on the internet. To allow burp suite to read the incoming/outgoing traffic from a web browser, a proxy server is required.

The proxy settings for the web browser is the same as the proxy setting used in burp suite. This is to make sure burp reads all the traffic through the proxy server before it reaches the browser. If the default proxy in burp suite is 127.0.0.1/8080 which is localhost, the browser must also be configured to use a proxy server with address 127.0.0.1/8080.

Working with HTTPS traffic requires additional settings since HTTPS is a secure layer HTTP protocol for secure communication which does not allow traffic to be snooped. A proxy on the other hand is a man in the middle agent which intercepts traffic, hence HTTPS traffic cannot be intercepted. To overcome this circumstance, burp provides a downloadable certificate through the link http://burp/.cert adding it as a trusted interface.

Defining target - Sitemap and scope: A sitemap is a visualization of the entire file structure of the website. It logs all the requests made by the browser. In order to

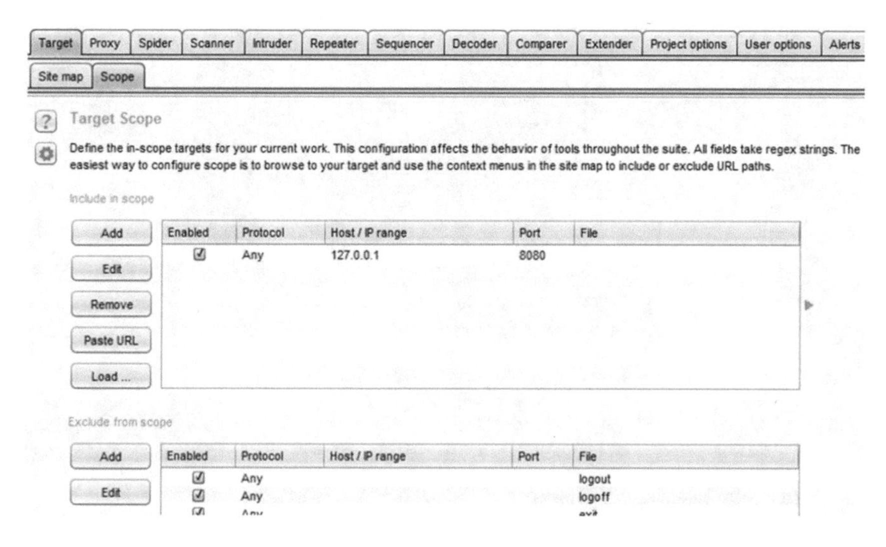

Fig. 10.9 Burp suite

concentrate only on the website we are testing, the address of the website is added to the target. This causes burp to log those requests made on the testing website (Fig. 10.9).

Using Burp Suite - Activating Burps spider will crawl the linked content on the web application, going down a depth of 5 links down by default, but these options can be configured under the Spider tab. As you interact with the web application, all the requests and responses will be logged under the Proxy tab. Each HTTP request made by your browser is displayed in the Intercept tab. You can view each message, and edit it if required. The "Forward" button is used to send the request on to the destination web server. If at any time there are intercepted messages pending, you will need to forward these for your browser to complete loading the pages it is waiting for.

Spidering the application - The spider tool is used to get a complete list of URLs and parameters for each site. The tool considers each page that was manually visited and goes through every link it finds in the testing scope. A better spider can be obtained by visiting more links manually as it gives the spider a large coverage area (Fig. 10.10).

Scanning Applications for Bugs

To find any sort of vulnerabilities in an application, all the links and URLs present in the application must be exploited. The Scanner tool gives a fine-grained control over which items get scanned, and gives immediate feedback and results for each scanned item picking out key security vulnerabilities. Burp gives a dashboard view of all the key vulnerabilities in order of their severity so that the tester can focus on the critical

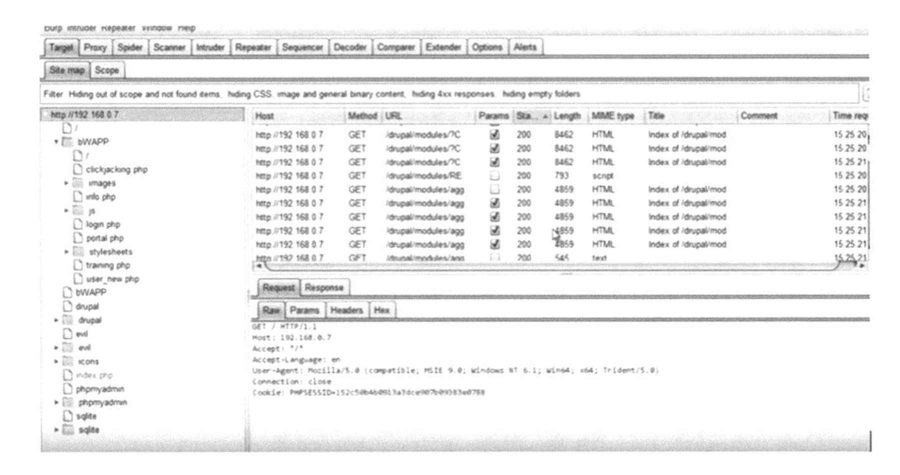

Fig. 10.10 Spidering using Burp suite

issues quickly. Three types of scans can be performed on an application, namely active scan, passive scan and user directed scans. Burp can be used to perform:

- Passive scanning of all requests and responses made through Burp Proxy, to identify shortcomings such as insecure use of SSL, and cross-domain exposure. This lets us safely find bugs without sending any additional requests to the application.
- Active scanning of all in-scope requests passing through Burp Proxy. This lets you use your browser to walk Burp Scanner through the interesting parts of the application's functionality that you want to actively scan. Burp Scanner will then send numerous additional requests to the target application, to identify vulnerabilities such as SQL injection, cross-site scripting and file path traversal.
- User-directed scanning of selected requests. Burp suite allows to select specific requests within any of the Burp Suite tools, and send these for active or passive scanning. This usage is ideal when you are manually testing individual parts of an application's functionality (Fig. 10.11).

Using the Intruder

Burp Intruder is a tool for automating customized attacks against web applications, to identify and exploit all kinds of security vulnerabilities and perform semi-targeted fuzzing. It is used against parameter of HTTP requests such as password and username. The intruder has four panels namely target, positions, payloads and options.
Target: This panel is used to specify the target host (the URL), the port to use for the connection.
Positions: The type of attack to be performed on the target. There are various types of attacks such as pitchfork attack, cluster attack, battering ram attack and sniper attack.
Payloads: This is used to configure one or more sets of payloads, which will be

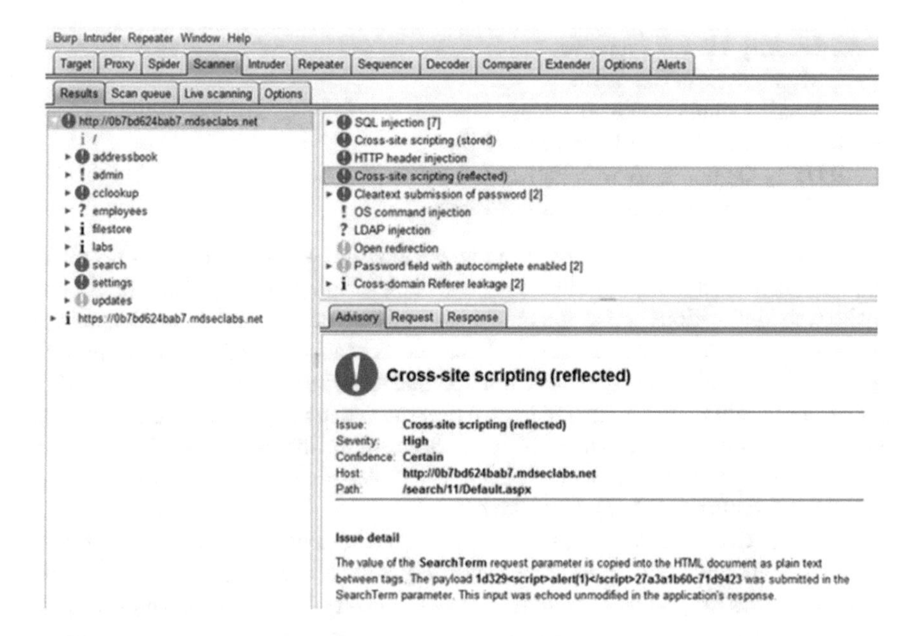

Fig. 10.11 Example of scanner output

placed into payload positions during the attack.

Options: This is used to configure numerous other options affecting the attack.

While examining the requests, Burp uses the symbol to highlight potential injection points. The potential area is given a payload, a list of data which is substituted individually in the field each made through a separate request. This can be used to crack passwords with a list of commonly used passwords or perform a SQL injection to exploit further vulnerabilities.

10.4.3 OWASP Recommendations for Web Application Security Action Plan

Several projects taken up by OWASP [8] have yielded high results in terms of industry reach and acceptance as a de-facto recommendations. Some of the major contributions of OWASP include OWASP Top Ten Project. This is an informed list of cloud and web application vulnerabilities that developer community needs to be highly vigilant on. The last published list was in 2013.

- A1 Injection.
- A2 Broken Authentication and Session Management.
- A3 Cross-Site Scripting (XSS).
- A4 Insecure Direct Object References.

- A5 Security Misconfiguration.
- A6 Sensitive Data Exposure.
- A7 Missing Function Level Access Control.
- A8 Cross-Site Request Forgery (CSRF).
- A9 Using Components with known Vulnerabilities.
- A10 Un validated redirects and Forwards.

OWASP Development Guide is another guideline which can be used for cloud application security. These guidelines provide description to developers on how to make web applications self-defensive in case of intrusion by malicious elements. It is categorized into 3 sections namely Best Practices that should be part of the web application. Secure patterns which form optional security features that can be used as guides and lastly Anti-patterns which are the patterns in the program which increase vulnerabilities (which needs to be avoided). Some of the topics addressed by OWASP development guide include secure coding principles, threat risk modelling, session management, phishing, buffer overflows etc.

OWASP Code Review Guide Security Code review is a process of auditing the code methodically to assess the security quality of the program. This OWASP guide identifies the phases of code review as Discovery, Transactional analysis, post transactional analysis, procedure peer review, reporting and presentation, laying the groundwork. In a secure code review, developers need to ensure to review for input validation, secure code design, information leakage and improper error handling, direct object reference, resource usage, API usage, best practice violation, weak session management, using HTTP Get query string.

OWASP Testing Guide- provides detailed testing techniques for security testing. Some of the aspects included in OWASP testing guide are manual inspections and reviews, threat modelling, code review, penetration testing. It also mandates some of the common security test that needs to be conducted to evaluate security controls of the system. They are as follows:

- Authentication and access control
- Input validation and encoding
- Encryption
- User and session management
- Error and exception handling
- Audition and logging.

All above guidelines and recommendations provide the developer and security community a sense of action plan to execute upon during the product development life cycle.

10.4.4 Big Data and Security

Cloud computing has been modelled by its characteristics like rapid elasticity, resource pooling capabilities, broad network access. While, Big data characterizes itself by its handling of huge volume of data, high velocity of computation and variety of data itself i.e., broad network access, elasticity and scalability. These common characteristics make Big data and cloud computing perfect partners in helping each other achieves their goals. Cloud computing plays a role of enabler with respect to Big data.

With the adoption of newer technologies, the baggage of operating challenges including security comes to the forefront of IT managers. Organizations are in a rush to adopt Big Data into their Business and data analytics without having understood complete implications of doing so. There is no question that Big data driven analytics helps gather a lot of user data. This makes it easier to make data driven decisions using new generation analytics. However, it brings along with it security implications related to the data lake that gets created soon after adoption of data analytics capabilities. Therefore, it is essential to consider data security and ensure that it is built in in the design; before scaling up the Big data initiatives. This section deals with understanding how HortonWorks is addressing these issues and helping organizations in order to reap full value of advanced analytics without exposing business to security risks.

10.4.4.1 Case Study: How HortonWorks Data Platform is Solving Hadoop Security Concerns

About HortonWorks (https://hortonworks.com): *HortonWorks is a Big Data software company that develops and supports Apache Hadoop, for distributed procession of large data sets across computer clusters.*

Hadoop is the de-facto standard framework used in the world of Big Data. Hadoop powered data lake can help provide a solid base for launching new generation of data analytics. However, it is very important to understand and consider security concerns before scaling up the Hadoop initiate. It is a widely accepted among leading companies in all industries that data is the new driver of competitive advantage.

To provide effective protection against security and data governance issues, HortonWorks (HW) has adopted 5 pillars based holistic approach [9]:

- Administration - Central management and consistent security to set policy across entire cluster.
- Authentication and perimeter security Authenticate users and system
- Authorization Provisioning access to data and provide access to what user can do
- Audit Maintain a record of data access of what a user did/accessed at what time.

Data protection Protect data stored and which is being transported across clusters using encryption over the wire.

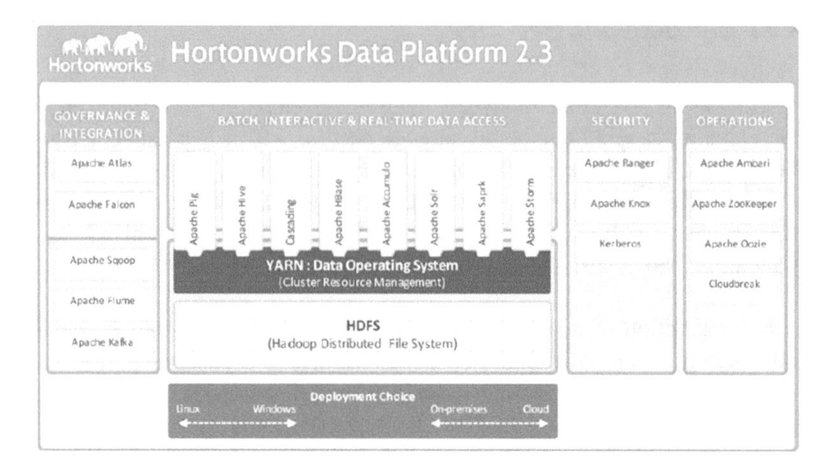

Fig. 10.12 HortonWorks data platform 2.3

Given figure describes HortonWorks Data Platform 2.3 architecture which includes main components Hadoop Distributed System (HDFS), YARN (Data Operating System), Security, Operations and Governance and Integration. As part of Hadoop Security, Apache Ranger, Apache Knox and Kerberos form the core components for enhancing security (Fig. 10.12).

Apache Ranger is a tool to enhance the productivity of the security administrations that reducing anomalous errors by helping them to define clear security policy and apply them to all the applicable components in the Hadoop stack from a centralized location. Currently, other commercially applicable solutions provide only partial support to security administrator. Hence using Apache Ranger helps the security administrators by providing centralized platform for security policy administration with respect to authentication, authorization, and auditing and data protection via encryption.

Apache Knox is a gateway that ensures perimeter wide security for HortonWorks customers. Using this solution, enterprises customers can extend the Hadoop REST APIS to new users easily with confidence and meeting the compliance standards.

Apache Knox is a pluggable framework and a new REST API service can be added and configure easily into the service stack. It provides single, simple point of access for a cluster. Its central controls ensure consistency across one or more clusters. It can be integrated with existing systems such as LDAP, AD, SSO to simplify identify maintenance.

Current vendors fail to provide a comprehensive solution in this area. Hence Knox is a market differentiator and can be adopted with high confidence.

With all these and more security features, HortonWorks provides a holistic solution with respect to administration, authentication, perimeter security, authorization, audit and data protection.

Another product developed by HortonWorks **Apache Metron** [10] can be used to identify anomalies and other issues in real time cyber security telemetry data in Big data context. Metron helps security operations professionals to identify issues in cybersecurity analytics. It can take in inputs from multiple sources and types of data, analyze and provide results. It has capabilities of log aggregation, advanced behavioral analytics and data enrichment.

Note: Figures are taken from Internet for Illustration and educational purpose only

10.5 Digital Privacy and Trusted Computing

Digital privacy and data security are an inseparable part of a successful security architecture. Current section [11] deals with Digital Privacy and Trusted Computing. Users online activity along with their accounts created on various service provider websites form their digital identity. We discuss business models used by cloud service providers which adopt data monetization for profit using case study of Facebook. There are even chances of ones digital identity being stolen by use of personally identifiable information (PII). Identity theft resulting from data/privacy breach is a serious cause of concern. Hence Digital Identity Management has become more relevant than ever before in this era driven by information exchange.

Data privacy breaches have been taking place time to time alarming the organizations, consumers and security researchers. We have already highlighted such privacy and security breaches earlier in the chapter with examples. This brings to center stage, the point that no one is immune to data security concerns and it needs to be addressed with all seriousness. We discuss this in detail in the context of trusted computing and cloud solutions for social networking.

10.5.1 Business Models of Cloud Service Providers: Case Study: Google, Facebook

When we are dealing with cloud service providers especially with SaaS model, a complete end-to-end user experience is provided by the service provider and the end user is given a simple client interface such as a desktop/mobile application or a browser to access the services. Some examples that come under this service model are social networking websites such as Facebook, Google Plus. Social networking has reunited countless friends and families who were separated by time and space. They include e-mail/messenger service providers such as Google, Yahoo. Email services help us to send and receive mails to and from anyone in the world via our email id.

Both Google and Facebook rely heavily on advertisements and in being advertisement platforms for generating revenue. The user generated data becomes their input for providing specific tailor made advertisements to the end users. Now, that Google, via google chrome has a browser product, they can piece in information of users online activity and build a digital identity out of it. They can provide contextual advertisements based on users search history, websites that he or she visited, products that he or she bought etc. Similarly, Facebook tries to analyze user preferences with respect to the content which he viewed, his friends interest, are he is residing, profession, age, gender, relationship status etc. to provide customized yet automated advertisements that are too hard to resist for the end users. And hence, these two giants have could get hold of market share and become big players in the business area. The user feels that the he is getting the service free of charge, but the service providers are making money via advertisements based on user data. So, essentially, users are paying the service providers back with their self-generated content and online activity.

10.5.2 Advent of Social Networking, Data Sharing Policies and Concerns

In many instances, SaaS service providers provide service to end users at free of cost to attract more number of users of the service. Social networking companies and email service providers rely hugely on personalized advertisements to monetize. This is how they can sustain and make profit of a free service that they provide to millions of customers online who use their services extensively. E-commerce websites make their earnings from commission it gets when an online purchase is made using their platform. In any case, service providers who provide free SaaS services have a lot riding upon the number of users using their platform. Also, the amount of information that is being shared, content created or amount of business transactions that happen online via their services have impact on their revenue.

As the service providers state the end user license agreement policies and the data sharing policies, the users are left with no choice but to agree to it if they want to use the service. By the mere urge and want to use the service, users agree to the terms based on the product reputation and peer recommendations. Concern with respect to data sharing policy is that the service providers have huge level of control over how they are going to use the user data. They can share it with 3rd party, advertisers data analytics companies etc. When some of these data are aggregated, and used using Big data analytics, they can be correlated and can reveal great deal of information about a person and his preferences. This can, in best can scenario lead to targeted advertisement, and in worst case: identity theft and fraud. There are laws governing the use of these data and regulations on what data can be stored and used. Still, they are vague and in many cases, technically incorrect and having legal loopholes. Many a times, the privacy policy breach is found after the fact when the damage is already done.

10.5.3 Towards Trusted Computing

End users are generally exhibit different kind of personas (traits). Few people are early adopters of new service, technologies. A bulk of users waits for the review for technical people and then decide whether to use the product or not. Some segment are very technical critics who use the product to understand the pros and cons of the product and then decide whether to use of exit from the product. The main binding force which ensure that the people who started using the product to stay engaged with the product is the factor called as Trust. Trust is very important to ensure that customer is retained and engaged. Hence, the service provider needs to have a vested interest in ensuring that the customers trust the product and service provider. This is possible by providing a holistic user engagement starting from first point to contact/sale to after closure of the contract. The data security policy and the EULA details need to be as transparent as possible. A clear and open channel of communication and a smooth user experience goes a long way in gaining end users trust. Several ongoing works towards providing trustable software solutions for secure cloud based applications are underway currently giving the consumer the center stage of trusted computing. Allowing customer to define certain terms of engagement with service provider [12], selecting the best service for the user based on quality of service requirements such as level of data security and privacy [13]. Proposal of framework for privacy preserving advertisement framework for social networking [14] are some of ongoing research works in this area. All of which aimed at moving towards trusted computing.

10.6 Efficient Security Operations and Management for Securing Cloud Data Centers

Discussion of cloud security is incomplete without security operations and management. This section is dedicated for this topic in context of optimal practices for securing the cloud data centers. A process oriented approach for managing day-to-day activities of a security of an organization is critical for its sustainability. Establishing effective communications channels for managing security incidents, and implementation of security and access controls form the basic framework of security operations management of an organization.

One of the important responsibilities of security operations team is to identify and define what components need to be protected and what it needs to be protected from. It also involves predictive analysis of high risk areas which needs more attention. Another important aspect with respect to security operations is the Business contingency plan. They are designed to ensure that we are prepared for unexpected circumstances. As any organization grows, the need for having disaster recovery plans and fail safe mechanisms using redundancy schemes increases.

In this section [15], we talk about how security can be efficiently managed in cloud data centers adhering to legal requirements. Security management is very important because it form the first line of defense against the attackers, without which the underlying security loses context.

Traditionally, physical security has always been kept away from IT paradigm. However, the industry trend is hinting at convergence of IT security and physical security aspects. It is important for an IT security manager to understand all the aspects of physical security and employ efficient methodologies to secure the IT resources including the cloud data centers. This chapter introduces the reader to physical security related topics and provides insights on how to efficiently manage the security operations to secure cloud data center and thereby enable cloud applications.

10.6.1 Security Operations Management

Security operations and management deals with management of security day to day basis of security functions. It is the management of ground level tasks including setting up and implementing security controls, managing security incidents. Top down approach of management is enforced where each of the details can be monitored by the higher ups to ensure swift decisions are made.

Security Administration - Communications and Reporting are two of the most important tasks that arise in a security operations department. This is because, the responsibility of providing the right measurements and metrics indicating the performance of the services provided by company rests with security operations team. Administration in an organization with respect to security operations is often driven by key performance metrics. These metrics such as measuring the set S.M.A.R.T.E.R goals (Refer Sect. 10.1) provide insights on how the overall performance is faring.

Few examples of the key performance indicators as percentage and number of resolved intrusion incidents measured in a periodic scale (month/quarter/year), percentage and number of blocked attacks from outside through firewall capabilities of the system, percentage and number of total machines in the organization of cloud cluster where OS patches are successfully installed and antivirus and other security features working as designed etc. It is a part of responsibility of security administrators to ensure that security telemetry data are captured and it is part of the design of the cloud system.

Change management is yet another important task which falls in the purview of security administration. It essentially involves carefully assessing the changes done to the system and ensuring the all the risk factors are manages and mitigated before the release to make sure that there is no negative impact due to change. This is required to maintain the availability of servers and to ensure that there are no security backdoors or loopholes created due to unintended misconfigurations during deployment of changes such as patches, updates or new feature implementations. Service Oriented framework is a preferred design paradigm in information technology as per Dr. Mydhili et al. [16]. Information Technology Infrastructure Library (ITIL) is a well-known service oriented framework for service delivery in IT sector.

It provides recommendations of best practices for change management. The change management steps involve

- **Describe** Describe the change being done in the system and which are all the touchpoints for the same. This involves steps Record (the change), Accept (accept or reject if the changes are required and valid) and classify (if it is a patch, software update, new features etc.)
- **Prepare** Prepare for deployment of change. This step involves Plan (plan for verifying) and Approve (Approve the validity and correctness of changes)
- **Execute** Ensure to have a pre-production environment which replicates production environment to execute the changes done. Build (generate the deployable files and configurations) and Test in the pre-production environments. Maintaining the health of these environments to mimic production settings is a separate challenge all together and something which most of the companies struggle with. Having an efficient response system which proactively updates all the pre-production environments post production deployment is key for ensuring smooth functioning of the system.
- **Check** Checking the changes with a set of validation test cases which are regularly maintained and updated is a crucial step in change management and administration. Automation needs to be built to evaluate the system for overall functionality of the system.

We mentioned earlier that security administration is usually enforced top down from the organization. While security administration is important, we also need to ensure that the administrators and managers and utilizing their privileges to see over security operations in an acceptable manner. Companies need to formulate acceptable use constraints and enforce it on to all the employees in order to comply with various security operations administrative management standards. Some examples related to this:

- dont store, share confidential information to unauthorized personnel
- dont access or misuse information resources where there is no specific job requirement.
- Security principles that help in avoiding abuse of administrative power include limiting authority and separation of duty.

Management Practices and Accountability - Management practices and enforcing accountability controls can help in efficient security operations management. Some of the management practices:

- creating user groups for providing controls to computing resources.
- Restricting access and privileges for remote access to consoles
- Use automation for regularly patching the employee OS and have onto virus and firewalls installed in them.
- centrally control the software installations done on each machine via endpoint management.

10.6.2 Business Contingency Planning

Uncertainty of life extends to the binary world of cloud as well. In this regard, it is highly recommended as well as mandated to have a solid Business continuity plan (BCP) as well as disaster recovery plan (DRP). These are by extremely important to formulate and implement and mandated as per compliance standards. BCP and DRP involves formulation of process that ensure that business critical computational resources are safeguarded from major system and network failures.

Cloud providers need to address few concerns such as data backup operation during service interruptions, capabilities for remote execution of business-critical processes at alternate site, ensuring reliability of the alternate systems, automating all decision-making processes to remove human dependency during emergency.

An effective BCP and DRP addresses all the cloud resources being utilized including LAN, WAN servers, telecommunication links, workstations and work spaces, applications and data. BCP needs to be tested on a frequent basis and upgraded whenever needed. Top level management have a big role to play in this regard. They need to be face of the organization to the employees and all other stakeholders. They need to provide all necessary resources to team to recover to a manageable state as soon as possible. When any incident occurs, the BCP team takes the following steps: Scope and plan initiation, business impact assessment, business continuity plan development, plan approval and implementation. During the business impact assessment, the issues are prioritized based on cost and downtime. Based on this, resources are allocated to mitigate it.

10.6.3 Physical Security

Drawing parallel to the real-world scenario, to live peacefully with calm mind, one needs a place that can be called home. It needs to be secured by doors, windows and locks. The perimeter of the house needs to have some amount of protection. No one should be allowed more access than they require to interact with us. These characteristics of physical security are applicable to the binary world of cloud computers as well.

The cloud data centers including the computer equipment, communication and technical equipment, storage media, company assets and records needs to be well protected. An initial physical vulnerability assessment must be done of the area. The buildings, computing resources and peripherals, documents and records need to be safe guarded. Physical location of the site also needs to be decided based on environmental and neighboring construction and landscape factors.

Entry controls in form or building access control systems, man traps and providing employee badges as entry pass, use of biometric features, having security guards in all the entry, exit and business critical locations form the 2nd line of defense.

Even if the physical security is breached, there should be efficient ways to detect the intrusion and take appropriate measures. Some of the examples of such tools and techniques are use of closed circuit television (CCTV), burglar alarm systems, temperature monitoring, motion sensors etc. These needs to be followed to meet compliance standards. COBIT and ISO 27002 are the industry accepted standards in this area.

10.6.4 Forensics Analysis and Compliance

We have highlighted in multiple sections in this chapter that a data breach or evidence of noncompliance is found usually after the fact. Hence, to gather the evidence, forensic analysis is very important. Forensics basically involves collection of evidence, examining it, and analyzing it to provide report on the incident.

Forensic evidence is usually used during legal proceedings to find level of conformance with said standards. Evidence acquisition is very important step in forensic analysis. It can be classified as host based evidence that can be found on the host machine and network based evidence that can be collected from network based system such as IDS or firewall.

While working with a live system to collect data, capturing system contents becomes very crucial in evidence collection. CPU activity and running processes can be traced from task manager in windows and **ps** command in Unix system. Network connections can be documented using **netstat** and **arp** command. Open files can be found using **lsof** command in Unix. In Windows, a host of forensic activities can be done using System internal tools which can be downloaded from www.sysinternals.com. Likewise, there are several other tools that can be used to examine file system for hidden data, alternate data streams, deleted data, encrypted and compressed data.

10.7 Parting Words and Useful Information

This section is reserved for our concluding message as well as providing an insight into current and future research trends in cloud security. We discuss the most important questions (related to security) to consider for any organizations before switch to cloud. Through this chapter, we want to encourage cloud researchers and professionals to think about security from a holistic perspective, understand the various issues and incorporate smart ways to tackle them. We want all stakeholders to use optimized security models, tools and techniques to enable cloud services and applications. We share some useful details on how researchers and academicians can benefit by joining hands in the cloud security community by joining open forums such as Cloud Security Alliance (OCA), Open Cloud Consortium (OCC). For a professional aspiring to create a career in Cloud Security, we provide key information on acquiring industry recognized certifications such as Certified Cloud Security Professional (CCSP) and Certificate of Cloud Security Knowledge (CCSK) and how he can apply the knowledge in create a positive impact in the organization.

10.7.1 Questions to Consider Before Shifting to Cloud

The in-built characteristics of Cloud computing such as rapid scalability, metered usage and fail safe make it very tempting to switch to cloud without a second thought. However, a reasonable thought needs to be given to ascertain if migrating to cloud is the best idea for the company or specific project. We earlier provided brief case studies on companies that adopted cloud technology and became successful.

To introspect what services should be moved to cloud, the various factors need to be considered. Such as cost to shift to cloud. Cloud expenses include storage charges, computational changes, bandwidth consumption etc. We also need to assess if the cloud provider can meet the quality of service that is required for hosting the product. Another important factor that needs to be considered is if the clouds alternate environment in case of outage, is reliable or not.

In cases where companies to be use cloud optimally to make best use of available resources, they may choose to use cloud for data backup and computational resources only during the peak seasons. Interoperable cloud providers are always preferred over others. This means that cloud applications can be built in such a way that the developer can choose from which cloud provider, it can be executed.

10.7.2 Getting Involved in Open Forums and Useful Certification Programs

As researchers, academicians, security professionals or cloud security enthusiasts we need to continuously improve our knowledge base and if possible contribute to the research community. One of the open forums *Cloud Security Alliance (OCA)* [17] works to promote the best practices for providing security assurance within cloud computing, and provide education on the uses of cloud computing to help secure all other forms of cloud computing. They also offer a certification program - Certificate of *Cloud Security Knowledge (CCSK)* which helps security professionals demonstrate their level of knowledge and get recognized in global platform. *Open Commons Consortium (OCC)* [18] formerly known as Open Cloud Consortium is a non-government organization which helps in maintaining and operating cloud infrastructure to support scientific, medical health care and environmental research. They are supported by universities around the world. They also take care of preparing reference implementations and standards for cloud. *Cloud Security Professional (CCSP)* [19] and related certifications are leading certification body for cloud security in the world. They offer several other certifications such as CISSP for Information security, SSCP Systems security, HCISSP Healthcare security etc. We recommend that the reader evaluates these forums and certifications as it would help them to move to leadership positions in security career.

10.7.3 Concluding Remarks and Research Trends

Topic of the chapter reads as Optimization of Security as an Enabler for Cloud Services and Applications. The aim of the chapter was to show how cloud services and applications can to be enabled to perform and fair better by optimizing the cloud security. This chapter is designed to help users to understand security in a holistic perspective and how managing security will help them develop great products.

To start with, we discussed via case studies about benefits of adopting cloud computing. Later, we understood the security requirements, challenges and objectives of a secure cloud system. We discussed in detail about available standards and guidelines for ensuring cloud security. We learnt how Cloud application security testing, incident response team provide use with tools and processes us to mitigate any security related issues.

Web applications are the most vulnerable of the all the cloud based services. We took examples of tools and case studies and showed how to prevent from such vulnerabilities. Big data security was also covered with example of HortonWorks data platform. Digital privacy issues and concerns were discussed and we recommended the community to move towards open standards and certification of service providers for conformance of data security policies.

We then moved to security operations management and how it plays a role in creating environment for developers to perform at their best and make sure cloud applications are not affected even in case of emergencies and unscheduled outages.

In the end, we gave call of duty to all the researchers and professionals to take up certifications based on their interests and join hand with open forums to contribute towards cloud security community.

Security has always been a silent unsung hero that ensures the high availability of the servers. Providing too much security to applications may hinder the performance parameters and response time of the service. However, reducing the minimal amount of security could mean a detrimental blow for cloud service and application. Hence, it is only wise for us to continuously optimize the methods and techniques of providing cloud security to enable cloud applications to perform to their fullest of potential.

References

1. Amazon Web Services, https://aws.amazon.com/ec2/
2. Microsoft Azure, https://azure.microsoft.com
3. R.L. Krutz, R.D. Vines, Cloud Security- A Comprehensive Guide to Secure Cloud Computing (2013) (Primary Reference Book for section 2, 3, 4)

4. Privacy Breaches, https://www.opswat.com/blog/8-largest-data-breaches-all-time
5. VMWare ESX, http://www.vmware.com/products/esxi-and-esx.html
6. M.K. Nair, V. Gopalakrishna, Cloud-cop: putting network admin on cloud nine, in *Proceedings of IEEE International Conference on Internet Multimedia Services Architecture and Applications, Bangalore (IMSAA'09), 09–11 Dec, India* (2009), pp. 1–6
7. Burp Suite, https://portswigger.net/burp/
8. OWASP Recommendations, https://www.owasp.org/index.php/Main_Page
9. A HortonWorks White Paper, Solving Hadoop Security. A Holistic Approach to Secure Data Lake (2015)
10. Apache Metron Project, https://hortonworks.com/apache/metron/
11. S. Pearson, G. Yee, Privacy and Security for Cloud Computing. Computer Communications and Networks (2013). ISBN: 978-1-4471-4188-4
12. V.M. Deshpande, Dr. M.K. Nair, B. Sowndararajan, Customer driven SLA in cloud based systems, in *Proceedings of International Conference of Emerging Computations and Information Technologies, SIT, Tumkur, Karnataka (India), 22–23 November 2013* (Elsevier, 2013), pp. 508–518, http://www.elsevierst.com/ConferenceBookdetail.php?pid=35. ISBN: 978-93-5107-176-1
13. V.M. Deshpande, Dr. M.K. Nair, Anveshana search for the right service, in *Proceedings of International Conference of Convergence of Technology, Pune, Maharastra (India)* (IEEE, 2014). ISBN 978-1-4799-3759-2
14. V.M. Deshpande, Dr. M.K. Nair, A novel framework for privacy preserving ad-free social networking, in *Proceedings of 2017 2nd International Conference for Convergence in Technology (I2CT)* (2017). ISBN: 978-1-5090-4307-1
15. M. Rhodes-Ousley, The Complete Reference: Information Security (2013) (Primary Reference Book for section 6)
16. M.K. Nair, S.M. Kakaraddi, K.M. Ramnarayan, V. Gopalakrishna, Agent with rule engine: the Glue for web service oriented computing applied to network management system, in *Proceedings of IEEE International Conference on Services Computing (SCC 09), 22–25 Sep 2009, Bangalore, India* (2009), pp. 528–531
17. Cloud Security Alliance (OCA), https://cloudsecurityalliance.org/
18. Open Commons Consortium (OCC), http://occ-data.org/
19. Cloud Security Professional (CCSP), https://www.isc2.org/ccsp/default.aspx

Chapter 11
Internet of Cloud: Security and Privacy Issues

Allan Cook, Michael Robinson, Mohamed Amine Ferrag, Leandros A. Maglaras, Ying He, Kevin Jones and Helge Janicke

Abstract The synergy between Cloud and IoT has emerged largely due to the Cloud having attributes which directly benefit IoT and enable its continued growth. IoT adopting Cloud services has brought new security challenges. In this book chapter, we pursue two main goals: (1) to analyse the different components of Cloud computing and IoT and (2) to present security and privacy problems that these systems face. We thoroughly investigate current security and privacy preservation solutions that exist in this area, with an eye on the Industrial Internet of Things, discuss open issues and propose future directions.

Keywords Internet of cloud · Cloud computing · Security · Authentication Intrusion detection · Privacy

11.1 An Introduction to Cloud Technologies

According to forecasts from Cisco Systems, by 2020 the Internet will consist of over 50 billion connected devices, including, sensors, actuators, GPS- and mobile-enabled devices, and further innovations in smart technologies, although this forecast is disputed [53]. New projections talk about 20–30 billion connected devices which is again a huge number [20]. These revolutionary devices are predicted to integrate to form hybrid networks based upon concepts such as the Internet of Things (IoT), Smart Grids, sensor networks etc., to deliver new ways of living and working. Underpinning such operating models will be 'cloud computing' technology that enables

---

A. Cook · L. A. Maglaras (✉) · Y. He · H. Janicke
De Montfort University, School of Computer Science
and Informatics, Leicester, UK
e-mail: leandrosmag@gmail.com

M. Robinson · K. Jones
Airbus Group Innovations, Newport, Leicester, UK

M. A. Ferrag
Department of Computer Science, Guelma University, Guelma, Algeria

B. S. P. Mishra et al. (eds.), *Cloud Computing for Optimization: Foundations, Applications, and Challenges*, Studies in Big Data 39,
https://doi.org/10.1007/978-3-319-73676-1_11

convenient, on-demand, and scalable network access to a pool of configurable computing resources. This remote access to high levels of processing power and storage provides a complementary platform on which to augment the low-power, low-storage characteristics of IoT devices, providing an integrated environment to provide ubiquitous capabilities to end users.

Cloud computing further offers a range of attractive benefits to organisations wishing to optimise their IT resources, such as increases in efficiency and organisational ability, reduced time to market (TTM), and a better balance between capital expenditure (capex) versus operational expenditure (opex) [30]. However, to achieve such returns on investment, organisations require a clear understanding of cloud technologies to drive their strategy, and in particular, the issues surrounding security and privacy. This chapter introduces the reader to cloud computing technologies in general, then proceeds to explain the emerging Internet of Cloud (IoC) before discussing the security and authentication issues of IoT and finally exploring the issues related to the preservation of privacy in the IoC.

11.1.1 A Definition of Cloud Computing

Cloud computing is a technological and operational model for ubiquitous, on-demand network access to a shared pool of configurable infrastructure, processing, storage and application services that can be provisioned and released for use with minimal system management effort or service provider interaction [48]. Many of the technologies that underpin cloud computing environments are not new, as they comprise existing virtualisation, processing and storage capabilities that have existed for many years. It is the operating model that surrounds the use of these technologies that delivers the revolutionary services, where ownership of physical resources rests with one party, and the service users are billed for their use [12].

As such, it is necessary to consider the essential characteristics, service models and deployment models of cloud computing.

11.1.2 Characteristics of Cloud Computing

Cloud computing environments comprise five essential characteristics; On-demand self-service, broad network access, resource pooling, rapid elasticity, and measured service [48]. We shall now review each of these in turn.

- **On-demand Self-service**: In cloud environments, a consumer can request and deploy processing and storage capabilities, such as server capacity and storage space, through the use of automated provisioning services that require no negotiation with the cloud provider [48]. This allows connected devices to remotely exploit such resources and extend their processing capabilities as necessary.

- **Broad Network Access**: The services of a cloud are made available over the network using thick or thin clients, allowing devices using different operating systems and platforms to access common capabilities [48].
- **Resource Pooling**: The computing resources of a cloud service are pooled into a model to serve multiple consumers, with different physical and virtual resources dynamically assigned and reassigned according to demand requirements, irrespective of their geography. The customer is typically unaware of the exact location of the provided resources, although they may be able to define high-level constraints such as country or data centre [48].
- **Rapid Elasticity**: Elasticity is the ability of a cloud provider to scale up or down dependent upon consumer demand, allocating or freeing resources as necessary. To the consumer, the capabilities provided often appear to be unlimited.
- **Measured Service**: In cloud models, consumers pay for the services they use, so it is necessary to monitor, control and report upon the consumption of the infrastructure. This allows usage to be optimised, and provides a transparent understanding to both the provider and consumer [48].

11.1.3 Cloud Service Models

There are various levels of service model available to consumers when they adopt cloud services, each with their own operating paradigm, offering software, platforms, or infrastructure as a service.

- **Software as a Service (SaaS)**: In this model, the consumer uses the provider's applications that run on the cloud infrastructure. The consumer accesses these applications without any knowledge of the underlying infrastructure, and does not request or provision any associated services. They provision and consume application resources, typically against an agreed service level agreement (SLA) that determine performance, and the cloud provider scales the infrastructure to meet its obligations [48].
- **Platform as a Service (PaaS)**: In a PaaS environment, consumers deploy their own (or their acquired) applications, services, libraries or tools, which they control. The cloud provider's role is to provision and maintain sufficient computing, network and storage resources to meet the agreed SLAs for the consumer-deployed elements [48].
- **Infrastructure as a Service(IaaS)**: This service model allows consumers to provision processing, network and storage resources as necessary, onto which they can deploy whichever applications and services the require. The consumer does not control the underlying hardware infrastructure, but can determine the technical detail of what is deployed, such as operating systems etc. [48].
- **Cloud Deployment Models**: The provision of SaaS, PaaS or IaaS is dependent upon the cloud provider's business model. Similarly, the scope of the cloud itself, whether private, community, public, or hybrid mix of these three, allows consumers

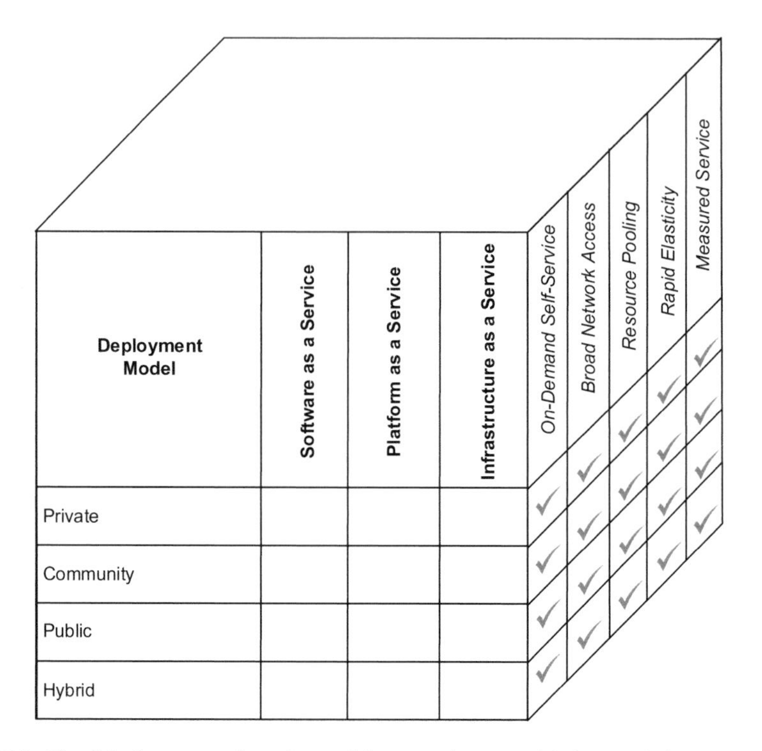

Fig. 11.1 Cloud deployment and service models mapped to essential characteristics

to constrain the exposure of their information. Irrespective of the combination of these choices however, the provider should offer the five essential characteristics of cloud services, as illustrated in Fig. 11.1.

- **Private Cloud**: The service infrastructure is provided for exclusive use by a single organisation. It may be owned, managed, and operated by the organisation, a third party, or some combination thereof, and may exist on or off the organisation's premises [48].
- **Community Cloud**: The cloud is made available for use by a specific community of consumers with shared requirements. The service may be owned, managed, and operated by one or more of the community organisations, a third party, or some combination, and be located either on or off the premises of the community [48].
- **Public Cloud**: The cloud infrastructure is provisioned for use by the general public. The infrastructure is deployed on the premises of the cloud provider [48].
- **Hybrid Cloud**: The cloud infrastructure is a mix of two or more cloud deployment models (private, community, or public) [48].

11.1.4 Enabling Technologies

As previously discussed, cloud services are based upon a common set of underpinning enabling technologies that were developed before cloud computing emerged as a business model. We shall now consider a key subset of these technologies in the context of their operation within a cloud.

- **Virtualisation**: Virtualisation is the ability to deploy multiple host operating environments on one physical device. Typically, operating systems are encapsulated within a 'virtual machine' (VM), a number of which are deployed onto a single physical server (a 'real machine'). A 'hypervisor' that abstracts the VMs from the real machine, accessing hardware components of the server as required by each VM. Hypervisors also allow VMs to be redeployed to other real machines, permitting them to be reallocated to servers with greater or lesser processing capacity as required by the consumers demands [16].
- **Storage**: Storage within cloud environment can be characterised as either file- or block-based services, or data management comprising record-, column- or object-based services. These typically reside on a storage area network (SAN) that provides a persistence platform that underpins a data centre. For file- or block-based services, the cloud ensures that sufficient capacity is provided to support the elasticity of the service, expanding or contracting as required. Record-, column- or object-based services, however, focus on database persistence and the performance of the data used by applications. As data expands within a large database it becomes necessary to optimise the storage based on frequency of access and location of consumers. Data within the cloud can be easily replicated to provide temporary copies in caches etc. that improve performance, as well as reducing the impact of backup services on production data. Similarly, where multiple data centres are used, these local caches can be optimised to focus on the datasets most frequently accessed in each location. The underlying file system elastically supports these replicas of data, expanding and contracting as necessary to support performance SLAs [27].
- **Monitoring and Provisioning**: The ability of a cloud provider to automatically provision services is an key element of its offering. Automated provisioning is typically based on a catalogue, from which consumers can select the extension or contraction of a service over which they have decided to maintain control. The nature of the service they maintain control of is dependent upon the service model they operate within (SaaS, PaaS, IaaS). Similarly, for the cloud provider, they require the ability to modify the execution environment in line with agreed SLAs, with or without human intervention. The provisioning is typically managed by a service orchestration layer that interacts with the monitoring service to determine the levels of performance of cloud elements in line with SLAs, and coordinates the manipulation of the infrastructure, deploying and redeploying resources as necessary to maintain a balanced and cost-efficient use of the available architecture [39].
- **Billing**: Given the differing service and deployment models that cloud providers can offer, the billing service must be integrated with the monitoring and

provisioning to ensure accurate accounting of consumption. The billing services, in some cases, support both prepay and postpay models, requiring the billing service to decrement or accrue respectively. The service must also only account for consumption as it occurs, and be cognisant of the elasticity of deployment and release. As the nature of the cloud service provided to consumers may differ, the billing service must support multiple, and in many cases, complex pricing models to ensure accurate accounting [19].

Cloud computing is based on a mix of technologies brought together in differing service and deployment models to provide cost-effective utilisation of IT resources. The ubiquity of access to these resources allows low-power, low-storage capacity IoT devices to extend their capabilities by leveraging these services on-demand. This integration of IoT and cloud into the Internet of Cloud (IoC) provides opportunities to provide a revolution in the use and exploitation of smart devices.

11.2 Internet of Things (IoT) and Internet of Cloud (IoC)

The Internet of Things is a term that has rapidly risen to the forefront of the IT world, promising exciting opportunities and the ability to leverage the power of the internet to enhance the world we live in. The concept itself is not a new one however, and it is arguable that Nikola Tesla predicted the rise of IoT back in 1926 when he stated:

> When wireless is perfectly applied the whole earth will be converted into a huge brain...and the instruments through which we shall be able to do this will be amazingly simple compared with our present telephone [36].

What Tesla had predicted was the Internet of Things (IoT), which today has been defined as the pervasive presence in the environment of a variety of things, which through wireless and wired connections and unique addressing schemes are able to interact with each other and cooperate with other things to create new applications/services and reach common goals [76]. Put more simply, things are anything and everything around us that are internet connected and are able to send, receive or communicate information either to humans or to other things. There are three primary methods in which things communicate. Firstly, a sensor can communicate machine to machine (M2M). Examples here include a sensor feeding data to an actuator which opens a door when movement is detected. Secondly, communication can be Human to Machine (H2M), such as a sensor which can detect human voice commands. Finally, machine to human (M2H) communication provides the delivery of useful information in an understandable form such as a display or audio announcement. When considering the number of things in our world, and the number of potential combinations of connecting them, the only limit for thinking up valuable use cases is our own imagination. Some well established use cases for the IoT are as follows:

- **Healthcare**: The use of sensors and the internet to monitor the medical variables of human beings and perform analyses on them. A real world example is NHS England's Diabetes Digital Coach Test Bed [51], which trialled the use of mobile health self-management tools (wearable sensors and supporting software). This trial leveraged the IoT to realise a number of benefits. Firstly it enabled people with diabetes to self-manage their condition through the provision of real time data and alerts based upon the data from their sensors. Secondly, the sensors were able to notify healthcare professionals if there was a dangerous condition that was not being corrected by the patient. Thirdly, the data from the sensors could be aggregated to provide a population-wide view of the health status of people with diabetes.
- **Smart Cities**: The use of connected sensors to improve city life and create an urban information system [34]. For example, detecting the amount of waste in containers to schedule a pick-up or the use of sensors to detect city-wide availability of parking and direct drivers to an available space [59]. This has the potential to not only save citizens frustration, but also to reduce congestion, pollution and fuel consumption.
- **Smart Buildings**: Both places of business and people's homes can benefit from the rise of the IoT. Buildings consume 33% of world energy [76], and there is real potential for the IoT to bring this usage down. Sensors can turn off lights when they are not needed, and appliances remotely switched off. Heating can be optimised based upon sensors detecting occupancy and weather conditions. Aggregations of this data can be used by energy providers to plan and optimise their operations.
- **Smart Transport**: The connecting of multiple transport related things. For example, sensors in roadways and vehicles to provide a full view of traffic flow and dynamically alter traffic light sequences, speed limits, information signs or satellite navigation systems to suggest quicker routes.
- **Smart Industry**: Intelligent tracking of goods and components, smart factories and innovative retail concepts such as Amazon Go [2], which offer a checkout-less experience for customers.

A summary of IoT projects around the world ranked by application domain is provided in Fig. 11.2. As the graphic shows, the most active domains for IoT (as of Q3 2016) are connected industry and smart cities. However, all of the domains are showing an upward trend that is likely to continue into the future as new and innovative use cases are developed and the value of IoT becomes increasingly apparent to actors in each domain [6].

11.2.1 IoT Technologies

Behind the things which make up the IoT are a number of essential core technologies. The literature identifies five overall technologies as follows: [41]:

- **Radio Frequency Identification (RFID)**: Passive RFID chips have no battery, and provide information to a reader when it is placed in proximity of the chip. Active

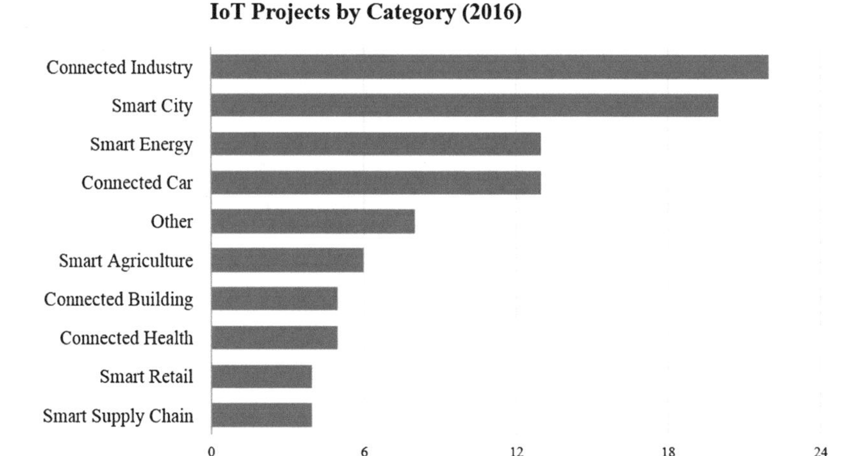

Fig. 11.2 IoT projects by category

RFID chips can initiate communication, and output information in response to a change in its environment (e.g. changes in temperature or pressure).

- **Wireless Sensor Networks (WSN)**: Wireless sensor networks are defined as collections of stand-alone devices which, typically, have one or more sensors (e.g. temperature, light level), some limited processing capability and a wireless interface allowing communication with a base station [29].
- **Middleware**: Middleware sits between the things and the raw data they generate to provide interoperability between things and developers who code the applications for interesting use cases. It provides a level of abstraction, allowing developers to work with sensors without having to know the specifics of their implementation.
- **Cloud Computing**: The cloud provides the seemingly limitless storage and processing power necessary for IoT use cases to become a reality.
- **IoT Applications**: The software that provides end users with a useful product - e.g. A smartphone app through which a driver can find and reserve a free parking space.

The focus of this section is on the cloud aspect of IoT technology, in particular how cloud computing and the IoT have found what appears to be a mutually beneficial relationship and led to the term Internet of Cloud (IoC).

11.2.2 Internet of Cloud (IoC)

The synergy between the cloud and the IoT has emerged largely due to the cloud having attributes which directly benefit the IoT and enable its continued growth. In the IoT's infancy, things either had some local computing resources (storage, processing) to produce a useful result, or they sent their data to a mainframe which had the necessary computing resources to process the data and generate an output. In effect, the "brain" as Tesla envisioned in 1926 was either highly distributed amongst the things, or it was centrally located with the things simply acting as sensors. Both of these approaches have disadvantages. The mainframe's weaknesses are that it is expensive to maintain and presents a central point of failure. The highly distributed approach whereby things communicate and perform some local computation provides better resilience to failure, but increases the cost of each thing in the network. This additional cost is both financial (the cost of equipping each thing with suitable resources and replacing failed things) and logistical (including such resources required the thing to be physically larger and consume more power). As use cases become more advanced, and the goals more complex, the demand for more complex computation has only increased.

The IoT is not only expanding in its need for more demanding computation resources. Gartner has predicted that the number of internet connected devices will reach 20.8 billion by 2020 [26], suggesting that the IoT is not only expanding in computational complexity, but also in the sheer amount of data that needs to be processed. In effect, the IoT generates big data [58], which places the demand for smaller and cheaper things directly into competition with the demand for more computing resources. Traditional approaches to the IoT cannot satisfy both demands - either the things become more expensive and complex, or limits on their computation resource needs are imposed. However, the cloud presents a solution with the potential to satisfy both demands.

11.2.3 Cloud as a Solution

The rise of cloud computing has provided an alternative solution, presenting the IoT with a virtually limitless source of computing power, easily accessible via the internet, with better resilience and at a lower cost than utilising a mainframe or including computing resources at the thing level. The cloud allows IoT developers to be freed from the constraints of limited resources, and enables the use case to be realised at reduced cost. In effect, things only require the bare minimum of hardware to perform their function (e.g. sense something, actuate something) and to communicate with the cloud. The cloud performs all computation and communicates back the result to the things. This pairing of cloud computing and the IoT has led to the term Internet of Cloud (IoC), and numerous literature reviews of this new paradigm are available [9, 17].

11.2.4 Sensor-Clouds

Cloud infrastructure is not only valuable for taking on the burden of heavy computation and storage, it has also been identified as valuable in forming what are known as Sensor-Clouds [1]. In traditional sensor networks, the deployed sensors provide data for one purpose - to fulfil the purchaser's use case. Unfortunately, this leads to an element of wastage, since the data being collected could be useful for other purposes but is not readily accessible by other organisations or third party developers. For example, if a local council deployed sensors to measure traffic flow in the city centre, a third party may wish to access the sensor data to improve their satellite navigation system and direct travellers away from congested roads. Sensor-Clouds address this scenario by making the sensor data available to multiple parties in the cloud. In effect, they offer what could be termed sensors as a service. This scenario brings a number of benefits. Firstly it allows developers of IoT applications to avoid the burden of manually deploying sensors and focus upon developing interesting use cases through the use of existing sensor networks. Secondly, sensor owners such the local council can recoup some of the cost of deployment and maintenance by charging these third parties to access the data.

A Sensor-Cloud can be visualised in three layers [1]. At the lowest layer, physical sensors around the world upload their data in a standardised format to the Sensor-Cloud. This standardisation of data allows users of the service to use the data without concern over differences in areas such as protocols and formatting. At the second layer, the Sensor-Cloud allows users to create virtualised groups of sensors for use in their applications. These virtual sensors are based upon service templates, which are defined by the sensor owners. At the top layer, application developers can plug these virtual sensors into their applications. This three layer architecture for Sensor-Clouds is shown in Fig. 11.3.

11.2.5 Ongoing Challenges

It has been noted that the cloud brings some valuable attributes to the IoT, but it is not a perfect solution which can solve all of the IoT's problems. In fact, the use of the cloud can present some new and interesting challenges as follows [17]:

- **Security and Privacy**: Cloud security is a well documented challenge, but the pairing between cloud and the IoT presents additional concerns. For example, when considering the sensitive nature of some use cases such as smart health, additional care must be taken to ensure that confidentially, integrity and availability of data is not violated. Confidentiality breaches could result in personal health data being stolen, integrity breaches could be fatal if data is tampered with and a lack of availability could fail to alert to a life threatening condition.
- **Addressing**: If Gartner's predictions on the rapid growth of the IoT are correct, IPv4 will quickly become inadequate to address all of the things. IPv6 has the

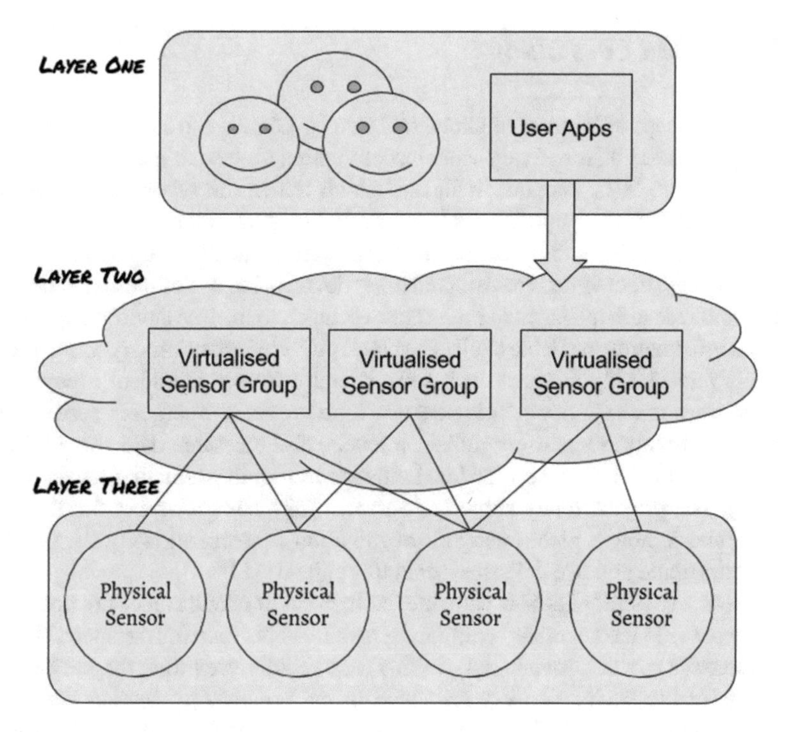

Fig. 11.3 Sensor-cloud layers

potential to address this concern, but it is not yet widely adopted. It has been proposed that an efficient naming and identity management system is required to dynamically assign and manage unique identity for an ever increasing number of things [37].

- **Latency and Bandwidth**: The cloud may provide limitless computing resources, but it cannot necessarily ensure low latency or unlimited bandwidth since this relies upon the public internet which is outside of the cloud provider's control. This challenge has led to the rise of what is termed "fog computing", where computing resources are placed as close to the things as possible in order to act as an intermediary. This intermediary can quickly service time critical processing that is latency sensitive whilst forwarding on non-time critical data for cloud processing [8].
- **Interoperability**: Due to the high number of things from multiple vendors, cloud computing alone cannot solve the issue of interoperability. The IoT would benefit from standards which describe a common format for both handling and communicating data.

11.2.6 Japan Case Study

While the concept of Internet of Cloud and Sensor-Clouds can seem abstract, it has been implemented in some very valuable use cases. One such use case was in the aftermath of the 2011 tsunami in Japan, which led to the second-largest nuclear emergency since Chernobyl. With a lack of reliable government information on the radiation threat, private individuals and organisations donated hundreds of Geiger counters to the affected region. As standalone devices, the use of these counters was limited, and researchers began to investigate methods to link the devices together and make the information available to all. The cloud provided the necessary infrastructure and agility to quickly connect each sensor, and a Sensor-Cloud of around 1000 radiation sensors was formed. This Sensor-Cloud provided emergency services with essential information regarding radiation levels, and the same data was leveraged to produce a smart phone app for local citizens to monitor radiation levels in their area [74]. The project, today known as Safecast [62], has grown beyond the initial use case and is now a global network of radiation sensors, allowing the public to access, contribute and use the sensor data through an API.

This use case highlights how valuable the Internet of Cloud can be not only for its ability to provide the necessary computing resources but also for the speed at which such resources can be provisioned - in this case rapidly providing the back end for a potentially life saving service. As stated in the previous subsection the pairing between cloud and the IoT presents additional concerns in terms of security and privacy. However, for these services to be adopted they must be trusted. Therefore, we must now consider the security and privacy implications of such integration.

11.3 Security and Authentication Issues

IoT ecosystem creates a world of interconnected thing, covering a variety of application and systems, such as smart city systems, smart home systems, vehicular networks, industrial control systems as well as the interactions among them [28]. Cloud computing is a technology that is configured to enable access to a shared pool of resource including servers, data storage, services and application [3]. It has become an important component of the IoT as it provides various capabilities to manage systems, servers and application and performs necessary data analytics and aggregation.

It is undeniable that IoT provides benefits and convenience to our everyday life, however, many of the IoT components (e.g. low cost digital devices and industrial systems) are developed with little to no consideration of security [35, 38, 63]. A successful exploit can propagate within the IoT ecosystem that could render the loss of sensitive information, interruption of the business functionalities, and the damage to critical infrastructure.

We have already seen the security concerns of the cloud services. These include but are not limited to malware injection (malicious code injected into cloud and run as SaaS), vulnerable application programming interfaces (API), abuse of data and cloud service, insider threats and the newly emerging Man In Cloud Attack [32]. Cloud involves both service providers and consumers; therefore, cloud security is a shared responsibility between the two.

Security issues are still yet to be solved for IoT and Cloud respectively. IoT adopting Cloud services could complicate this situation and raise more security concerns. This section focuses on the security issues on IoT adopting Cloud services and makes recommendations to address those issues.

11.3.1 Data Sharing/Management Issues of IoT Cloud

Within a cloud context, no matter public, private or hybrid, data security management involves secure data storage, secure data transmission and secure access to the data. During transmission, the Transport Layer Security (TLS) cryptography is widely used to prevent against threats. During processing, the cloud service provided applies isolation between different consumers. The isolation [80] is applied at different levels such as operations system, virtual machine or hardware. The secure access to data sometimes depends on the isolation level. It may sometimes separate completely from other resources or may have shared infrastructures and softwares that rely on access control policies. Within an IoT context, one of the benefits is open data sharing among different things. If the data is isolated as is currently offered in the cloud services, open wide data aggregation and analytics would become impossible. A balance needs to be found between data protection and sharing.

There is existing work such as Information Flow Control (IFC) [4, 54] defining and managing the requirements for data isolation and sharing. People can specify to what extent they want to share or protect their data. Other work is related to data encryption, encrypting the things before uploading to the cloud. This would limit the users access to the data, which again affects the IoT's data sharing and analytical capability. There are some solutions to analyse encrypted data. However, this approach is not mature to be applied in practice at this stage [31, 50].

11.3.2 Access Control and Identity Management

Within a cloud context, access control rules are strictly enforced. The service providers use authentication and authorisation services to grant privileges to the consumers to access storage or files. Within an IoT context, there are interactions between different devices that are owned by different people. Access control is usually leveraged through device identity management and configuration [46]. Existing identity management includes identity encoding and encryption [77].

When IoT uses Cloud services, access control involves the interactions among the applications and cloud resources. Policies and mechanism need to be flexibly defined to accommodate the needs of both and resolve the conflicts of different parties. There is existing work on grouping the IoT devices to enable common policies [64]. However, cares need to be taken to ensure the flexibly defined policies do not introduce vulnerabilities to the system.

11.3.3 Complexity (Scale of the IoT)

One of the benefits of Cloud service adoption is the reduction of cost through elastically resource scaling mechanisms. The increase of IoT devices, data volume, and variety has become a burden for the Cloud. The failure to coordinate and scale the "things" will impact the availability of the data. Security mechanism will bring extra burden that can impact the performance of IoT Cloud [60, 80].

Logging is an important aspect of security as it provides a view of the current state of the system. Logging within the Cloud is centralised and it is an aggregation of the logs from different components such as software applications and operation [68]. IoT logging tends to decentralise it among different components. There are some existing work on logging centralisation (e.g. design analytics tools to collect and correlate decentralised logs from "things" [57]) and decentralisation (e.g. enable logging to capture information in a data-centric manner [56]). A balance needs to be found between logging centralisation and decentralisation.

11.3.4 Protection of Different Parties

IoT Cloud raise security concerns to both service providers and consumers. Cloud service providers used to apply access control to protect data and resources. Now, the "things" can directly interact with the providers. Attacks can be easily launched by compromised "things". We have already seen some real world exploits of smart home applications, that are designed with poor security considerations [61].

From the consumers' perspective, "things" needs to be validated before it can be connected. If the "things" are determined to be compromised or malicious, alerts will be sent either in a human readable or machine-readable format. The determination can be based on reputation, trustworthy network node evaluation [52, 85] and so on.

11.3.5 Compliance and Legal Issues

Cloud demonstrated compliance using contract through service-level agreement (SLA). A typical method to assess compliance is through auditing. Within the area of Cloud, Massonet has proposed a framework that can generate auditing logs demonstrating that they are compliant with the related policies/regulations [47]. There are also frameworks designed in the area of IoT to demonstrate compliance using auditing logs.

IoT tends to be decentralised in isolated physical locations. The centralisation of cloud allows the data to flow across geographic boundaries, which has raised legal and law concerns of the data across national borders. There are some existing work on constrain data flow geographically by applying legal and management principles to the data [66]. Again this will have an negative impact on data sharing capability of IoT and Cloud.

11.3.6 Cloud Decentralisation

An emerging trend is the Cloud decentralisation in order to accommodate the IoT and big data analytics. Typical method is decentralised computing such as Fog computing [8] and grid computing [25]. Cloud decentralisation helps reduce the typical Cloud attacks such as denial of service (DoS) attack; it also raises new security concerns. Instead of targeting on the Cloud services, the attacks are directed to individual service providers and consumers. There is on-going research in securing the decentralised Cloud through coordinating the communication of things to things, things to clouds and clouds to clouds [67, 68]. Finally cloud decentralization can provide more flexible management.

Following the trend of decentralized cloud deployment, security mechanisms that must be developed for the IoC may also be decentralized. This deployment can have multiple advantages and in this context Intrusion Detection Systems for the IoC are analyzed on Sect. 11.3.7.

11.3.7 Intrusion Detection Systems

Intrusion detection systems (IDS) can be classified into centralized intrusion detection systems (CIDS) and distributed intrusion detection systems (DIDS) by the way in which their components are distributed. In a CIDS the analysis of the data is performed in some fixed locations without considering the number of hosts being monitored [40], while a DIDS is composed of several IDS over large networks whose data analysis is performed in a number of locations proportional to the number of hosts. There are numerous advantages of a DIDS compared to a CIDS. A DIDS is highly scalable and can provide gradual degradation of service, easy extensibility and scalability [14].

Novel intrusion detection Systems (IDS) must be implemented that need to be efficient both in terms of accuracy, complexity and communication overhead, false alarm rate and time among others. IDSs that have been developed for other systems, e.g. Industrial Control Systems [15, 45], wireless sensor networks [11], or cloud environments [49] can be used as a basis for developing new detection systems for the IoC area. Adaptivity of the IDS on changes in the network topology, which will be a frequent situation for an IoC, is an aspect that needs to be addressed when new IDS are going to be designed [70].

11.4 Privacy Preserving Schemes for IoC

In this subsection, we review the privacy preserving schemes for IoC. Based on the classification of authentication and privacy preserving schemes in our three recent surveys [22–24], the privacy preserving schemes for IoC can be classified according to networks models and privacy models. The summary of privacy preserving schemes for IoC are published in 2014, 2015, and 2016, as presented in Tables 11.1, 11.2, and 11.3, respectively. In addition, Fig. 11.4 shows the classification of privacy preservation models for IoC.

Cao et al. [13] defined and solved the problem of multi-keyword ranked search over encrypted cloud data while preserving strict systemwise privacy in the cloud-computing paradigm. Specifically, the authors proposed a preserving scheme, called MRSE, using the secure inner product computation. The MRSE scheme is efficient in terms of the time cost of building index and the time cost of query. Worku et al. [81] proposed a privacy-preserving public auditing protocol in order to provide the integrity assurance efficiently with strong evidence that unfaithful server cannot pass the verification process unless it indeed keeps the correct data intact. Wang et al. [78] proposed a brand new idea for achieving multi-keyword (conjunctive keywords) fuzzy search. Different from existing multi-keyword search schemes, the scheme [78] eliminates the requirement of a predefined keyword dictionary. Based on locality-sensitive hashing and Bloom filters, the scheme [78] is efficient in term of the Bloom filter generation time for a single file. Wang et al. [10] a privacy-preserving public auditing mechanism, called Oruta, to verify the integrity of shared data without retrieving the entire data. Oruta uses ring signatures [7] to construct homomorphic authenticators. In addition, Oruta can achieving following properties: (1) Public Auditing, (2) Correctness, (3) Unforgeability, and (4) Identity Privacy. Yuan and Yu [33] proposed the first secure and practical multi-party the back-propagation neural network learning scheme over arbitrarily partitioned data. Based on two phases, namely, (1) privacy preserving multi-party neural network learning, and (2) secure scalar product and addition with Cloud, the scheme [33] can support the multi-party scenario and efficient in terms of collaborative learning and communication cost compared to both schemes in [73] and [5]. Sun et al. [72] proposed an idea to build the search index based on the vector space model and adopt the cosine similarity measure in the Cloud supporting similarity-based ranking. Based on the vector space model, the scheme [72] is efficient in term of time cost for generating encrypted query. Dong et al. [18] considered four parties in a network, namely, the data owner, the data consumer, the cloud server, and the private key generator. Then,

Table 11.1 Summary of privacy preserving schemes for IoC (Published in 2014)

Scheme	System model	Privacy model	Goals	Main phases	Performances (+) and limitations (-)
Cao et al. [13]	A cloud data hosting service involving three different entities, namely, the data owner, the data user, and the cloud server	Data privacy Index privacy Keyword privacy	Achieving the multi-keyword ranked search with privacy-preserving	- Setup - BuildIndex - Trapdoor - Query	+ Efficient in term of the time cost of building index + Efficient in term of the time cost of query + Resistance to the known ciphertext model and known background model - No consideration for checking the integrity of the rank order
Worku et al. [81]	Three different entities, including, cloud server, user, and third party auditor	- User privacy	Achieving public verifiability, storage correctness, batch auditing, blockless verification, and privacy preserving	KeyGen SigGen ProofGen VerifyProof	+ Efficient in term auditor and server computation overhead compared to the scheme in [79] + Secure in the random oracle model - No threat model presented
Wang et al. [78]	Three different entities, including, data owner, cloud server, and user	- File content privacy - Index privacy - User query privacy	Support multi-keyword fuzzy search	- KeyGen - Index Enc - Query Enc - BuildIndex - Trapdoor - Search	+ Efficient in term of the Bloom filter generation time for a single file + Multi-keyword fuzzy search + Resistance to the known ciphertext model and known background model - No comparison with related schemes

(continued)

Table 11.1 (continued)

Scheme	System model	Privacy model	Goals	Main phases	Performances (+) and limitations (-)
Wang et al. [10]	Three parties: the cloud server, a group of users and a public verifier	- Identity privacy	Achieving following properties: (1) Public Auditing, (2) Correctness, (3) Unforgeability, and (4) Identity Privacy.	- KeyGen - RingSign - RingVerify	+ Efficient in terms of signature generation and communication cost + Efficient in term of auditing time + Efficient in terms of privacy and batch auditing with incorrect proofs - Traceability is not considered
Yuan and Yu [33]	Three major parties: a trusted authority (TA), the participating parties (data owner) and the cloud servers (or cloud)	- Multi-party privacy-preserving	Protecting each participant's private dataset and intermediate results generated during the back-propagation neural network learning process	- Privacy Preserving Multi-Party Neural Network Learning - Secure Scalar Product and Addition with Cloud	+ Efficient in terms of collaborative learning and communication cost compared to both schemes in [73] and [5] + Support the multi-party scenario + Scalable, efficient and secure - Does not allow multiparty collaborative learning without the help of TA

(continued)

Table 11.1 (continued)

Scheme	System model	Privacy model	Goals	Main phases	Performances (+) and limitations (-)
Sun et al. [72]	Three entities: the data owner, the data user, and the cloud server	- Search privacy - Index confidentiality - Query confidentiality - Query unlinkability - Keyword privacy	Achieving high efficiency and functionality (such as expressive/usable queries)	- Setup - GenIndex - GenQuery - SimEvaluation	+ Efficient in term of time cost for generating encrypted query + Help users ensure the authenticity of the returned search results in the multi-keyword ranked encrypted text search scenario - Traceability is not considered
Dong et al. [18]	Four parties in a network: the data owner, the data consumer, the cloud server, and the private key generator	- Backward secrecy - User privacy	Achieving fine-grained access control	- System initialization - Encryption - Key generation and distribution - Decryption	+ Efficient in terms of computation complexity, communication cost, and cost of revocation operation + Fully collusion secure + User access privilege confidentiality - Adversary's model is limited

Table 11.2 Summary of privacy preserving schemes for IoC (Published in 2015)

Scheme	System model	Privacy model	Goals	Main phases	Performances (+) and limitations (-)
Zhou et al. [87]	Cloud-assisted wireless body area networks	- Identity privacy - Location privacy	Detecting two attacks, namely, time-based mobile attack and location-based mobile attack	- Pairwise key establishment - Group key agreement	+ Efficient in terms of storage, computation, and communication overhead compared to the scheme in [75] - Traceability is not considered - No consideration for the patients' selfishness
Zhou et al. [88]	Three components: body area networks (BANs), wireless transmission networks and the healthcare providers equipped with their own cloud servers	- Identity privacy - Data confidentiality	Achieving data confidentiality and identity privacy with high efficiency	- Setup - Key Extract - Sign - Verify	+ Efficient in terms of computational overhead, communication overhead, and storage overhead compared to the scheme in [42] - Location privacy is not considered
Liu et al. [43]	Three main network entities: users, a cloud server, and a trusted third party	- Data anonymity - User privacy - Forward security	Achieving authentication and authorization without compromising a user's private information	- Ideal data accessing functionality - Ideal authority sharing functionality	+ Considers the data anonymity - Need an evolution in terms of computational overhead, communication overhead, and storage overhead - Traceability is not considered - No comparison with related schemes

Table 11.3 Summary of privacy preserving schemes for IoC (Published in 2016)

Scheme	System model	Privacy model	Goals	Main phases	Performances (+) and limitations (-)
Xia et al. [83]	Four different types of entities: the image owner, image user, cloud server and watermark certification authority	- Data privacy	Protecting the privacy of image data in content-based image retrieval outsourcing applications against a curious cloud server and the dishonest query users	- KeyGen - IndexGen - ImgEnc	+ Privacy of image content + Privacy of image features + Privacy of trapdoors + Leakage of similarity information + Efficient in term of time consumption of the trapdoor generation - The proposed watermarking method cannot be regarded as a very robust one
Xia et al. [84]	Three different types of entities: the image owner, image user and cloud server	- Image privacy	The plaintext data needs to be kept unknown to the cloud server	- The generation of unencrypted index - The index encryption	+ The privacy of the image + The privacy of the image features + The privacy of the trapdoors + The leakage of the similarity information + Efficient in four terms, namely, (1) time consumption of the index construction, (2) time consumption of the trapdoor generation, (3) time consumption of the search operation, and (4) storage consumption of the index - The feature extraction in encrypted image

(continued)

Table 11.3 (continued)

Scheme	System model	Privacy model	Goals	Main phases	Performances (+) and limitations (-)
Pasupuleti et al. [55]	Consisting of three main entities, including, data owner, cloud service provider, and authorized users	- Index Privacy - Data privacy	Proposing an efficient and secure privacy-preserving approach with following goals: privacy preserving, index privacy, and data integrity	- Key generation - Index creation - Privacy-preserving - Trapdoor generation - Ranked keyword search - Data decryption	+ Detect the modifications or deletions of data and maintain the consistency of data + Efficient in terms of computation cost and communication cost - Dynamic data updates
Xia et al. [82]	Consisting of three main entities, including, data owner, data user and cloud server	- Index confidentiality - Query confidentiality - Trapdoor unlinkability - Keyword privacy	Supporting multi-keyword ranked search and dynamic operation on the document collection	- Index Construction - Search Process	+ The search precision on different privacy level + The efficiency of index construction, trapdoor generation, search, and update - The users keep the same secure key for trapdoor generation - Location privacy and identity privacy are not considered

(continued)

Table 11.3 (continued)

Scheme	System model	Privacy model	Goals	Main phases	Performances (+) and limitations (-)
Song et al. [69]	Consisting of three main entities, including, data owner, authorized data user, and cloud server	- Data privacy - Query privacy	Providing the full-text retrieval services with privacy-preserving	- Document processing - Index structure and maintenance mechanism - Full-text retrieval algorithm over encrypted data	+ The index space cost + Time cost for inserting a new document + Query efficiency with different number of documents + Query precision with different number of documents - Traceability is not considered
Zhu et al. [89]	Four parts: trusted authority, location based services (LBS) provider, LBS user, and cloud server	- Location privacy	Providing privacy-preserving LBS data and user's location information with accurate LBS for users.	- System initialization - Cloud server data creation - Privacy-preserving location based services	+ The user query location is privacy-preserving in the proposed EPQ scheme + The proposed EPQ scheme can achieve confidential LBS data + The authentication of the LBS query request and response are achieved in the proposed EPQ scheme + Efficient in term of computation complexity compared with the FINE scheme [65] - Traceability is not considered

(continued)

Table 11.3 (continued)

Scheme	System model	Privacy model	Goals	Main phases	Performances (+) and limitations (-)
Lyu et al. [44]	Four parts: social network, data owner, members, and cloud storage server	- Data privacy	Providing the fine-grained access control	- System initiation - Privacy-preserving data construction - Interested data acquisition - Dynamic attribute management	+ Data confidentiality + Fine-grained access control + Collusion attacks resistant + Efficient in terms of communication overhead and computation cost compared to the scheme in [71] - Location privacy and identity privacy are not considered
Yu et al. [86]	Four parts, key distribution center, cloud user, cloud server and third party auditor	- Data privacy	Formalize the security model of zero knowledge privacy against the third party auditor	- Setup - Extract - TagGen	+ Perfect data privacy preserving + Efficient in terms of costs regarding computation, communication and storage - No comparison with related schemes

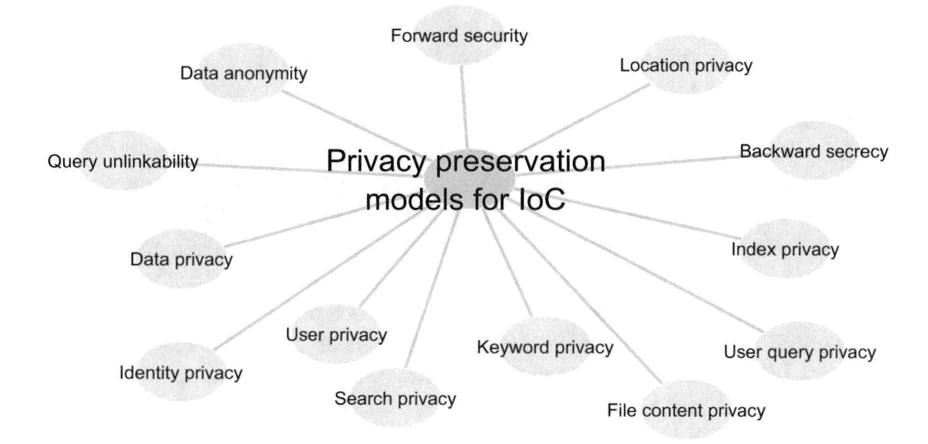

Fig. 11.4 Classification of privacy preservation models for IoC

the authors [18] proposed an idea that the cloud can learn nothing about a user's privacy or access structure, as such the scheme is fully collusion resistant.

To resilient to both time-based and location-based mobile attacks, Zhou et al. [87] proposed a secure and privacy-preserving key management scheme for cloud-assisted wireless body area networks. Based on the body's symmetric structure with the underlying Blom's symmetric key establishment mechanism, the scheme [87] is efficient in terms of storage, computation, and communication overhead compared to the scheme in [75], but the traceability is not considered. Zhou et al. [88] proposed a patient self-controllable and multilevel privacy-preserving cooperative authentication scheme, called PSMPA, to realize three levels of security and privacy requirement in distributed m-healthcare cloud computing system which mainly consists of the following five algorithms: Setup, Key Extraction, Sign, Verify and Transcript Simulation Generation. Based on an attribute based designated verifier signature scheme, PSMPA is efficient in terms of computational overhead, communication overhead, and storage overhead compared to the scheme in [42], but the location privacy is not considered. Therefore, Liu et al. [43] proposed a shared authority based privacy-preserving authentication protocol, named, SAPA, for the cloud data storage, which realizes authentication and authorization without compromising a user's private information. SAPA protocol [43] applied ciphertext-policy attribute based access control to realize that a user can reliably access its own data fields. In addition, SAPA protocol [43] adopt the proxy re-encryption to provide temp authorized data sharing among multiple users.

Xia et al. [83] proposed an idea to protect the privacy of image data in content-based Image retrieval outsourcing applications against a curious cloud server and the dishonest query users. Specifically, the idea is the first work that proposes a searchable encryption scheme, considering the dishonest query users who may distribute the retrieved images to those who are unauthorized. Similarly to the scheme [83],

Xia et al. [84] proposed an idea which the secure k-nearest neighbor algorithm is employed to protect the feature vectors in order to enable the cloud server to rank the search results very efficiently without the additional communication burdens. Based on the pre-filter tables, the scheme [84] is efficient in four terms, namely, (1) time consumption of the index construction, (2) time consumption of the trapdoor generation, (3) time consumption of the search operation, and (4) storage consumption of the index. Pasupuleti et al. [55] proposed an efficient and secure privacy-preserving approach using the probabilistic public key encryption technique to reduce computational overhead on owners while encryption and decryption process without leaking any information about the plaintext. Based on the idea of integrity verification, the scheme [55] can detect the modifications or deletions of data and maintain the consistency of data, also is efficient in terms of computation cost and communication cost. Therefore, Xia et al. [82] proposed two secure search schemes, namely, (1) the basic dynamic multi-keyword ranked search scheme in the known ciphertext model, and (2) the enhanced dynamic multi-keyword ranked search scheme in the known background model. Based on the searchable encryption scheme, the idea in [82] can supports both the accurate multi-keyword ranked search and flexible dynamic operation on document collection. Song et al. [69] defined and solved the problem of full-text retrieval over encrypted cloud data in the cloud computing paradigm. Based on a hierarchical Bloom filter tree index, the scheme [69] can protect user query privacy. Zhu et al. [89] proposed a privacy-preserving location based services query scheme in outsourced cloud, called EPQ, for smart phone. EPQ scheme can achieve confidential location-based services (LBS) data and is efficient in term of computation complexity compared with the FINE scheme [65].

Lyu et al. [44] design an efficient and secure data sharing scheme, named DASS. Based on multi-attribute granularity for social applications, DASS can support searchable encryption over data, and is efficient in terms of communication overhead and computation cost compared to the scheme in [71], but location privacy and identity privacy are not considered. Yu et al. [86] investigated a new primitive called identity-based remote data integrity checking for secure cloud storage. In addition, the scheme [86] showed that it achieves soundness and perfect data privacy. In a recent work, Ferrag and Ahmim in [21] proposed an efficient secure routing scheme based on searchable encryption with vehicle proxy re-encryption, called ESSPR, for achieving privacy preservation of message in vehicular peer-to-peer social network.

11.5 Summary

The synergy between the cloud and the IoT has emerged largely due to the cloud having attributes which directly benefit the IoT and enable its continued growth. IoT adopting Cloud services has brought new security challenges. We have identified key security issues in data management, access control and identity management, complexity and scale, the protections of different parties, compliance and legal issues, as well as the emerging Cloud decentralisation trend. There is existing work addressing these issues, however future work should primarily focus on the balance between centralisation and decentralisation, data security and sharing as well as associated policy issues.

Regarding privacy preservation, it is not a problem that can be treated in isolation for a system, but interdependencies among different users and platforms must be also analyzed. Also the combination of privacy metrics can help improve the level of privacy by combining the positive aspects of different methods while keeping the total cost, in terms of storage, computation and delay, relatively low.

References

1. A. Alamri, W.S. Ansari, M.M. Hassan, M.S. Hossain, A. Alelaiwi, M.A. Hossain, A survey on sensor-cloud: architecture, applications, and approaches. Int. J. Distrib. Sens. Netw. **9**(2), 917923 (2013). https://doi.org/10.1155/2013/917923
2. Amazon: Amazon go. Online (2017), https://www.amazon.com/b?node=16008589011
3. M. Armbrust, A. Fox, R. Griffith, A.D. Joseph, R. Katz, A. Konwinski, G. Lee, D. Patterson, A. Rabkin, I. Stoica et al., A view of cloud computing. Commun. ACM **53**(4), 50–58 (2010)
4. J. Bacon, D. Eyers, T.F.M. Pasquier, J. Singh, I. Papagiannis, P. Pietzuch, Information flow control for secure cloud computing. IEEE Trans. Netw. Serv. Manag. **11**(1), 76–89 (2014)
5. A. Bansal, T. Chen, S. Zhong, Privacy preserving Back-propagation neural network learning over arbitrarily partitioned data. Neural Comput. Appl. **20**(1), 143–150 (2011)
6. J. Bartje, The top 10 iot application areas - based on real iot projects (2016), https://iot-analytics.com/top-10-iot-project-application-areas-q3-2016/
7. D. Boneh, C. Gentry, B. Lynn, H. Shacham, Aggregate and Verifiably Encrypted Signatures from Bilinear Maps, *Intenational Conference on the Theory Applications of Cryptographic Technology* (Springer, Berlin, Heidelberg, 2003), pp. 416–432
8. F. Bonomi, R. Milito, J. Zhu, S. Addepalli, Fog computing and its role in the internet of things, in *Proceedings of the First Edition of the MCC Workshop on Mobile Cloud Computing* (ACM, 2012), pp. 13–16
9. A. Botta, W. de Donato, V. Persico, A. Pescap, Integration of cloud computing and internet of things: a survey. Future Gener. Comput. Syst. **56**, 684–700 (2016), http://www.sciencedirect.com/science/article/pii/S0167739X15003015
10. W. Boyang, L. Baochun, L. Hui, Oruta: privacy-preserving public auditing for shared data in the cloud. IEEE Trans. Cloud Comput. **2**(1), 43–56 (2014)
11. I. Butun, S.D. Morgera, R. Sankar, A survey of intrusion detection systems in wireless sensor networks. IEEE Commun. Surv. Tutor. **16**(1), 266–282 (2014)
12. B.V., I.N.: CompTIA Cloud Essentials Certification Study Guide (Exam CLO-001) (McGraw-Hill, New York, 2014)
13. N. Cao, C. Wang, M. Li, K. Ren, W. Lou, Privacy-preserving multi-keyword ranked search over encrypted cloud data. IEEE Trans. Parallel Distrib. Syst. **25**(1), 222–233 (2014)
14. M. Crosbie, E.H. Spafford, Active defense of a computer system using autonomous agents (1995)
15. T. Cruz, L. Rosa, J. Proença, L. Maglaras, M. Aubigny, L. Lev, J. Jiang, P. Simões, A cybersecurity detection framework for supervisory control and data acquisition systems. IEEE Trans. Ind. Inf. **12**(6), 2236–2246 (2016)
16. J. Daniels, Server virtualization architecture and implementation. Crossroads **16**(1), 8–12 (2009)
17. M. Díaz, C. Martín, B. Rubio, State-of-the-art, challenges, and open issues in the integration of internet of things and cloud computing. J. Netw. Comput. Appl. **67**(C), 99–117 (2016). https://doi.org/10.1016/j.jnca.2016.01.010

18. X. Dong, J. Yu, Y. Luo, Y. Chen, G. Xue, M. Li, Achieving an effective, scalable and privacy-preserving data sharing service in cloud computing. Comput. Secur. **42**, 151–164 (2014)
19. E. Elmroth, F.G. Marquez, D. Henriksson, D.P. Ferrera, Accounting and billing for federated cloud infrastructures, in *Eighth International Conference on Grid and Cooperative Computing, 2009 GCC'09* (IEEE, 2009), pp. 268–275
20. A. Ericsson, Ericsson mobility report: On the pulse of the networked society. Ericsson, Sweden, Technical Report EAB-14 61078 (2015)
21. M.A. Ferrag, A. Ahmim, Esspr: an efficient secure routing scheme based on searchable encryption with vehicle proxy re-encryption for vehicular peer-to-peer social network. Telecommun. Syst. 1–23 (2017). https://doi.org/10.1007/s11235-017-0299-y
22. M.A. Ferrag, L. Maglaras, A. Ahmim, Privacy-preserving schemes for ad hoc social networks: a survey. IEEE Commun. Surv. Tutor. **19**(4), 3015–3045 (2017)
23. M.A. Ferrag, L.A. Maglaras, H. Janicke, J. Jiang, A Survey on Privacy-preserving Schemes for Smart Grid Communications (2016), arXiv:1611.07722
24. M.A. Ferrag, L.A. Maglaras, H. Janicke, J. Jiang, Authentication Protocols for Internet of Things: A Comprehensive Survey (2016), arXiv:1612.07206
25. I. Foster, Y. Zhao, I. Raicu, S. Lu, Cloud computing and grid computing 360-degree compared, in *Grid Computing Environments Workshop, 2008. GCE'08* (IEEE, 2008), pp. 1–10
26. Gartner, Inc: Gartner says 6.4 billion connected "things" will be in use in 2016, up 30 percent from 2015 (2015), http://www.gartner.com/newsroom/id/3165317
27. R.L. Grossman, Y. Gu, M. Sabala, W. Zhang, Compute and storage clouds using wide area high performance networks. Future Gener. Comput. Syst. **25**(2), 179–183 (2009)
28. J. Gubbi, R. Buyya, S. Marusic, M. Palaniswami, Internet of things (iot): a vision, architectural elements, and future directions. Future Gener. Comput. Syst. **29**(7), 1645–1660 (2013)
29. C. Guy, Wireless Sensor Networks, in *Sixth International Symposium on Instrumentation and Control Technology: Signal Analysis, Measurement Theory, Photo-Electronic Technology, and Artificial Intelligence*, ed. by J. Fang, Z. Wang, (eds.) SPIE-International society for optical engineering, vol. 6357 (2006), pp. 63571I–63571I–4. https://doi.org/10.1117/12.716964
30. K.K. Hausman, S.L. Cook, T. Sampaio, Cloud Essentials: CompTIA Authorized Courseware for Exam CLO-001 (Wiley, New York, 2013)
31. D. Hrestak, S. Picek, Homomorphic encryption in the cloud, in *2014 37th International Convention on Information and Communication Technology, Electronics and Microelectronics (MIPRO)* (IEEE, 2014), pp. 1400–1404
32. M. Jensen, J. Schwenk, N. Gruschka, L.L. Iacono, On technical security issues in cloud computing. in *IEEE International Conference on Cloud Computing, 2009. CLOUD'09* (IEEE, 2009), pp. 109–116
33. J. Yuan, S. Yu, Privacy preserving back-propagation neural network learning made practical with cloud computing. IEEE Trans. Parallel Distrib. Syst. **25**(1), 212–221 (2014)
34. J. Jin, J. Gubbi, S. Marusic, M. Palaniswami, An information framework for creating a smart city through internet of things. IEEE Internet Things J. **1**(2), 112–121 (2014)
35. Q. Jing, A.V. Vasilakos, J. Wan, J. Lu, D. Qiu, Security of the internet of things: perspectives and challenges. Wirel. Netw. **20**(8), 2481–2501 (2014)
36. J.B. Kennedy, When woman is boss: an interview with nikola tesla, in *Colliers* (1926)
37. R. Khan, S.U. Khan, R. Zaheer, S. Khan, Future internet: the internet of things architecture, possible applications and key challenges, in *2012 10th International Conference on Frontiers of Information Technology (FIT)* (IEEE, 2012), pp. 257–260
38. M.T. Khorshed, A.S. Ali, S.A. Wasimi, A survey on gaps, threat remediation challenges and some thoughts for proactive attack detection in cloud computing. Future Gener. Comput. Syst. **28**(6), 833–851 (2012)
39. J. Kirschnick, J.M.A. Calero, L. Wilcock, N. Edwards, Toward an architecture for the automated provisioning of cloud services. IEEE Commun. Mag. **48**(12), 124–131 (2010)
40. S. Kumar, Classification and detection of computer intrusions. Ph.D. thesis, Purdue University (1995)

41. I. Lee, K. Lee, The internet of things (iot): applications, investments, and challenges for enterprises. Bus. Horiz. **58**(4), 431–440 (2015), http://www.sciencedirect.com/science/article/pii/S0007681315000373
42. Li, M., Yu, S., Ren, K., Lou, W.: Securing personal health records in cloud computing: patient-centric and fine-grained data access control in multi-owner settings, in *International Conference on Security and. Privacy in Communication Systems* (Springer, Berlin, Heidelberg, 2010), pp. 89–106, http://link.springer.com/10.1007/978-3-642-16161-2_6
43. H. Liu, H. Ning, Q. Xiong, L.T. Yang, Shared authority based privacy-preserving authentication protocol in cloud computing. IEEE Trans. Parallel Distrib. Syst. **26**(1), 241–251 (2015)
44. C. Lyu, S.F. Sun, Y. Zhang, A. Pande, H. Lu, D. Gu, Privacy-preserving data sharing scheme over cloud for social applications. J. Netw. Comput. Appl. **74**, 44–55 (2016)
45. L.A. Maglaras, J. Jiang, T.J. Cruz, Combining ensemble methods and social network metrics for improving accuracy of ocsvm on intrusion detection in scada systems. J. Inf. Secur. Appl. **30**, 15–26 (2016)
46. P. Mahalle, S. Babar, N.R. Prasad, R. Prasad, Identity management framework towards internet of things (iot): Roadmap and key challenges, in *International Conference on Network Security and Applications* (Springer, 2010), pp. 430–439
47. P. Massonet, S. Naqvi, C. Ponsard, J. Latanicki, B. Rochwerger, M. Villari, A monitoring and audit logging architecture for data location compliance in federated cloud infrastructures, in *2011 IEEE International Symposium on Parallel and Distributed Processing Workshops and Phd Forum (IPDPSW)* (IEEE, 2011), pp. 1510–1517
48. P. Mell, T. Grance et al., The NIST Definition of Cloud Computing (2011)
49. C. Modi, D. Patel, B. Borisaniya, H. Patel, A. Patel, M. Rajarajan, A survey of intrusion detection techniques in cloud. J. Netw. Comput. Appl. **36**(1), 42–57 (2013)
50. M. Naehrig, K. Lauter, V. Vaikuntanathan, Can homomorphic encryption be practical? in *Proceedings of the 3rd ACM Workshop on Cloud Computing Security Workshop* (ACM, 2011), pp. 113–124
51. NHS England: Digital diabetes coach (2015), https://www.england.nhs.uk/ourwork/innovation/test-beds/diabetes-digital-coach/
52. M. Nitti, R. Girau, L. Atzori, Trustworthiness management in the social internet of things. IEEE Trans. Knowl. Data Eng. **26**(5), 1253–1266 (2014)
53. A. Nordrum, Popular internet of things forecast of 50 billion devices by 2020 is outdated. IEEE Spectrum, http://spectrum.ieee.org/tech-talk/telecom/internet/popular-internet-of-things-forecast-of-50-billiondevices-by-2020-is-outdated. Accessed 18 2016
54. T.F.M. Pasquier, J. Singh, J. Bacon, Clouds of things need information flow control with hardware roots of trust, in *2015 IEEE 7th International Conference on Cloud Computing Technology and Science (CloudCom)* (IEEE, 2015), pp. 467–470
55. S.K. Pasupuleti, S. Ramalingam, R. Buyya, An efficient and secure privacy-preserving approach for outsourced data of resource constrained mobile devices in cloud computing. J. Netw. Comput. Appl. **64**, 12–22 (2016)
56. A. Rabkin, M. Arye, S. Sen, V.S. Pai, M.J. Freedman, Making every bit count in wide-area analytics, in *HotOS* (2013), p. 6
57. A. Rabkin, R.H. Katz, Chukwa: a system for reliable large-scale log collection. LISA **10**, 1–15 (2010)
58. B.B.P. Rao, P. Saluia, N. Sharma, A. Mittal, S.V. Sharma, Cloud computing for internet of things & sensing based applications, in *2012 Sixth International Conference on Sensing Technology (ICST)* (2012), pp. 374–380
59. Rico, J., Sancho, J., Cendon, B., Camus, M.: Parking easier by using context information of a smart city: enabling fast search and management of parking resources, in *2013 27th International Conference on Advanced Information Networking and Applications Workshops* (2013), pp. 1380–1385
60. J.W. Rittinghouse, J.F. Ransome, Cloud computing: Implementation, Management, and Security (CRC press, Boca Raton, 2016)

61. R.J. Robles, T.h. Kim, D. Cook, S. Das, A review on security in smart home development. Int. J. Adv. Sci. Technol. **15** (2010)
62. SafeCast Project: Safecast project website (2017), http://safecast.jp/en/
63. P. Samarati, S.D.C. di Vimercati, S. Murugesan, I. Bojanova, Cloud Security: Issues and Concerns (Wiley, New York, 2016)
64. R.M. Savola, H. Abie, Metrics-driven security objective decomposition for an e-health application with adaptive security management, in *Proceedings of the International Workshop on Adaptive Security* (ACM, 2013), p. 6
65. J. Shao, R. Lu, X. Lin, FINE: A fine-grained privacy-preserving location-based service framework for mobile devices, in IEEE INFOCOM 2014 - IEEE Conference on Computer Communications (IEEE, 2016), pp. 244–252
66. J. Singh, J. Bacon, J. Crowcroft, A. Madhavapeddy, T. Pasquier, W.K. Hon, C. Millard, Regional clouds: technical considerations, University of Cambridge, Computer Laboratory, Technical Report (2014)
67. J. Singh, J. Bacon, D. Eyers, Policy enforcement within emerging distributed, event-based systems, in *Proceedings of the 8th ACM International Conference on Distributed Event-Based Systems* (ACM, 2014), pp. 246–255
68. J. Singh, T. Pasquier, J. Bacon, H. Ko, D. Eyers, Twenty security considerations for cloud-supported internet of things. IEEE Internet Things J. **3**(3), 269–284 (2016)
69. W. Song, B. Wang, Q. Wang, Z. Peng, W. Lou, Y. Cui, A privacy-preserved full-text retrieval algorithm over encrypted data for cloud storage applications. J. Parallel Distrib. Comput. **99**, 14–27 (2017)
70. B. Stewart, L. Rosa, L.A. Maglaras, T.J. Cruz, M.A. Ferrag, P. Simoes, H. Janicke, A novel intrusion detection mechanism for scada systems that automatically adapts to changes in network topology (2017)
71. J. Sun, X. Zhu, Y. Fang, A privacy-preserving scheme for online social networks with efficient revocation, in *2010 Proceedings IEEE INFOCOM* (IEEE, 2010), pp. 1–9
72. W. Sun, B. Wang, N. Cao, M. Li, W. Lou, Y.T. Hou, H. Li, Verifiable privacy-preserving multi-keyword text search in the cloud supporting similarity-based ranking. IEEE Trans. Parallel Distrib. Syst. **25**(11), 3025–3035 (2014)
73. T. Chen, S. Zhong: Privacy-preserving backpropagation neural network learning. IEEE Trans. Neural Netw. **20**(10), 1554–1564 (2009)
74. University of Southampton, Southampton researchers develop new tool to provide radiation monitoring in Japan (2013), http://www.southampton.ac.uk/news/2013/05/radiation-monitoring-in-japan.page
75. K. Venkatasubramanian, A. Banerjee, S. Gupta, PSKA: usable and secure key agreement scheme for body area networks. IEEE Trans. Inf. Technol. Biomed. **14**(1), 60–68 (2010)
76. O. Vermesan, *IERC Cluster Book 2016*, Innovation and Deployment. European Research Cluster on the Internet of Things, IoT Digital Value Chain Connecting Research (2016)
77. O. Vermesan, P. Friess, P. Guillemin, S. Gusmeroli, H. Sundmaeker, A. Bassi, I.S. Jubert, M. Mazura, M. Harrison, M. Eisenhauer et al., Internet of things strategic research roadmap. Internet Things-Global Technol. Soc. Trends **1**, 9–52 (2011)
78. B. Wang, S. Yu, W. Lou, Y.T. Hou, Privacy-preserving multi-keyword fuzzy search over encrypted data in the cloud, in *IEEE INFOCOM 2014 - IEEE Conference on Computer Communications* (IEEE, 2014), pp. 2112–2120
79. C. Wang, S.S. Chow, Q. Wang, K. Ren, W. Lou, Privacy-preserving public auditing for secure cloud storage. IEEE Trans. Comput. **62**(2), 362–375 (2013)
80. L. Wei, H. Zhu, Z. Cao, X. Dong, W. Jia, Y. Chen, A.V. Vasilakos, Security and privacy for storage and computation in cloud computing. Inf. Sci. **258**, 371–386 (2014)
81. S.G. Worku, C. Xu, J. Zhao, X. He, Secure and efficient privacy-preserving public auditing scheme for cloud storage. Comput. Electr. Eng. **40**(5), 1703–1713 (2014)
82. Z. Xia, X. Wang, X. Sun, Q. Wang, A secure and dynamic multi-keyword ranked search scheme over encrypted cloud data. IEEE Trans. Parallel Distrib. Syst. **27**(2), 340–352 (2016)

83. Z. Xia, X. Wang, L. Zhang, Z. Qin, X. Sun, K. Ren, A privacy-preserving and copy-deterrence content-based image retrieval scheme in cloud computing. IEEE Trans. Inf. Forensics Secur. **11**(11), 2594–2608 (2016)
84. Z. Xia, N.N. Xiong, A.V. Vasilakos, X. Sun, EPCBIR: An efficient and privacy-preserving content-based image retrieval scheme in cloud computing. Inf. Sci. (Ny). **387**, 195–204 (2017)
85. Z. Yan, P. Zhang, A.V. Vasilakos, A survey on trust management for internet of things. J. Netw. Comput. Appl. **42**, 120–134 (2014)
86. Y. Yu, M.H. Au, G. Ateniese, X. Huang, W. Susilo, Y. Dai, G. Min, Identity-based remote data integrity checking with perfect data privacy preserving for cloud storage. IEEE Trans. Inf. Forensics Secur. **12**(4), 767–778 (2017)
87. J. Zhou, Z. Cao, X. Dong, N. Xiong, A.V. Vasilakos, 4S: a secure and privacy-preserving key management scheme for cloud-assisted wireless body area network in m-healthcare social networks. Inf. Sci. (Ny). **314**, 255–276 (2015)
88. J. Zhou, X. Lin, X. Dong, Z. Cao, PSMPA: patient self-controllable and multi-level privacy-preserving cooperative authentication in distributedm-healthcare cloud computing system. IEEE Trans. Parallel Distrib. Syst. **26**(6), 1693–1703 (2015)
89. H. Zhu, R. Lu, C. Huang, L. Chen, H. Li, An efficient privacy-preserving location-based services query scheme in outsourced cloud. IEEE Trans. Veh. Technol. **65**(9), 7729–7739 (2016)

Chapter 12
A Novel Extended-Cloud Based Approach for Internet of Things

Amitabha Chakrabarty, Tasnia Ashrafi Heya, Md. Arshad Hossain, Sayed Erfan Arefin and Kowshik Dipta Das Joy

Abstract The Internet of Things(IoT) is the future Internet evolution towards a network of interconnected smart objects such as computers, smart phones, smart watches, smart televisions, smart cars and many more. It is a matter of concern that our current infrastructure may not be able to handle large amount of data efficiently involving the growing number of smart IoT devices. As a solution, in this paper we came up with proposing a hybrid model of IoT infrastructure, as compared to the existing infrastructure to overcome its challenges. Our proposed infrastructure will be a collaboration of Fog computing combined with intelligent use of Service Oriented Architecture(SOA) which will be serving as a machine to machine communication protocol. This model will be able to transfer data reliably and systematically with low latency, less bandwidth, heterogeneity and maintaining the Quality of Service(QoS) befittingly.

12.1 Introduction

Timeline of Internet is not that much old, it has been around to people for some decades. Internet has been serving to community of people, making life much easier.

First four authors contributed equally to this chapter

A. Chakrabarty (✉) · T. A. Heya · Md. A. Hossain · S. E. Arefin · K. D. D. Joy
Department of Computer Science and Engineering, BRAC University,
66 Mohakhali C/A, Dhaka 1212, Bangladesh
e-mail: amitabha@bracu.ac.bd

T. A. Heya
e-mail: tasnia.heya@gmail.com

M. A. Hossain
e-mail: arshad.antu@gmail.com

S. E. Arefin
e-mail: erfanjordison@gmail.com

K. D. D. Joy
e-mail: kou-shikjay66@gmail.com

B. S. P. Mishra et al. (eds.), *Cloud Computing for Optimization: Foundations, Applications, and Challenges*, Studies in Big Data 39,
https://doi.org/10.1007/978-3-319-73676-1_12

Before the birth of Internet, travelling or sending letters across the ocean or next to home were the paths to transfer information. Internet has taken a truly revolutionary step in this regard. With emerging time, Internet has become a much more complexed system and is gradually extending to Things. Things with Internet creates a digital fabric and with the birth of the concept of Internet of Things (IoT), communication technology has involved things where things can share feelings with other things because they are uniquely addressable. Moreover, they are able to make successful media of information across of the globe producing prosperous communication link among humans, objects and other systems.

Data processing with lots of information remotely, seems to become problematic as information growth is exponential relative to time. According to recent statistics in 2016, number of devices outstretched 22.9 billion which will range over trillion in number within 20201 and as an outcome, more than quintillions of data will be transmitted through the network. Our present infrastructure is inadequate to have a hold on this much amount of data transaction in a disciplined manner and may end up with a clutter communication system.

Fog computing is a geo-distributed computing system which let applications operate as close as possible to enormous data from machine to machine communication, coming out of individuals, processes and things. Fog computing is actually an extension of cloud computing that generates automated replies with desired values at the edge of the network but its adaptability and geographically closest accessibility differs it from typical cloud computing. Because of these features, fog computing ensures mobility support, location awareness and low latency enhancing the Quality of Service(QoS) which are the requirements of IoT along with real-time interpretation of big data and distributed compact accumulation of data and processing unlike cloud computing [1]. Moving all data from IoT to the cloud for computation would demand gigantic amount of bandwidth. Volume, assortment, and velocity of huge data that the IoT generates are not aimed by the present cloud infrastructures [2]. Exchanging information via network through Internet without any human collaborations or machine associations, is the supremacy of IoT which incorporates elements, for example, versatility support, extensive variety of geo-distribution, availability of wireless support and expansive number of nodes takes Fog computing to a superior stage for a particular number of IoT services [3].

Considering the ad-hoc network, the communication between different nodes in a particular network and resolving their next destination network are confined within a specific local group where all nodes are defined in two categories, (i) smaller size with low popularity and (ii) with many local contacts and higher popularity and to divide bandwidth in equal parts the traffic through each node can be routed assuming three diverse circumstances, (i) Nodes that are in transmission mode, (ii) Nodes in relay mode and (iii) Nodes that are in receive mode, where each nodes transmit one flow each time and carries traffic within maximum supportable traffic [4]. Though, location awareness with less bandwidth, low latency and geo-distribution is one of the center necessity of IoT which is not by any stretch of the imagination conceivable to handle through conventional data communication with cloud by taking after this structure as it is extremely tedious for vast data, causes issue of bandwidth at remote spots, depending servers are placed which causes lagged response time and versatility [5].

On grounds of having a significant ramification, cloud computing is an anarchical technology as it has a few issues in regards to service-level agreements (SLA) with security and energy efficiency. Uncertainty of availability of utilization in need, inducing unauthorized application parts and getting victimized by various security issues are the major downsides of Software as a Service (SaaS), Platform as a Service (PaaS) and Infrastructure as a Service (IaaS), which are three relevant conveyance models of Cloud Computing [6–8]. Trust administration principle in the agent based SOA level (third level) of Fog computing model of our proposed infrastructure [9] is a convincing solution for these obstacles.

An infrastructure which is composed of Fog computing converging with active Machine-to-Machine(M2M) communication corresponding with the combination of the service composition model and Agent based composition which are based upon Service Oriented Architecture(SOA) is proposed for a better approach towards data computation in case of IoT. This model will have the capacity to exchange data dependably and efficiently with low latency, less bandwidth, heterogeneity in least measure of time maintaining the Quality of Service(QoS).

12.2 Proposed Infrastructure

Fog computing uses several layers of nodes based on its usage. In the context of our proposed infrastructure as shown in the Fig. 12.1, we have put better performance utilization of the concept of fog computing by dividing it into four layers.

All the devices reside in the lowest layer of our Infrastructure. These are all consumer devices which can be cellphone, desktop, home thermostat and so on. From now on these devices will addressed as End Device. These end devices may or may not have any built-in intelligence, it can be a simple low powered sensor. End devices communicate with the next two following layers which are ad-dressed as the Middle Ware. Devices that are involved with it in the same geographical location, can be designated as the regions. Our lower one of the Middle Layer utilizes the Service composition model. It is based on Service Oriented Architecture using a very light weight model that suites devices with lower resources. By sending service request, end point devices communicate with Service Oriented Architecture model (SOA).

For large scale or complex service composition we have the upper middleware. This layer utilizes the SOA using agents, so that services can be composed intelligently if a service composition was not possible by the lower middleware the re-quest would be passed to the upper middleware.

If the agents do not have required information and simple services to compose a service it will request to the Cloud service. Cloud services will provide simple services and the information or metadata of a complex service that indicates the required simple service which are to be used to compose a complex service.

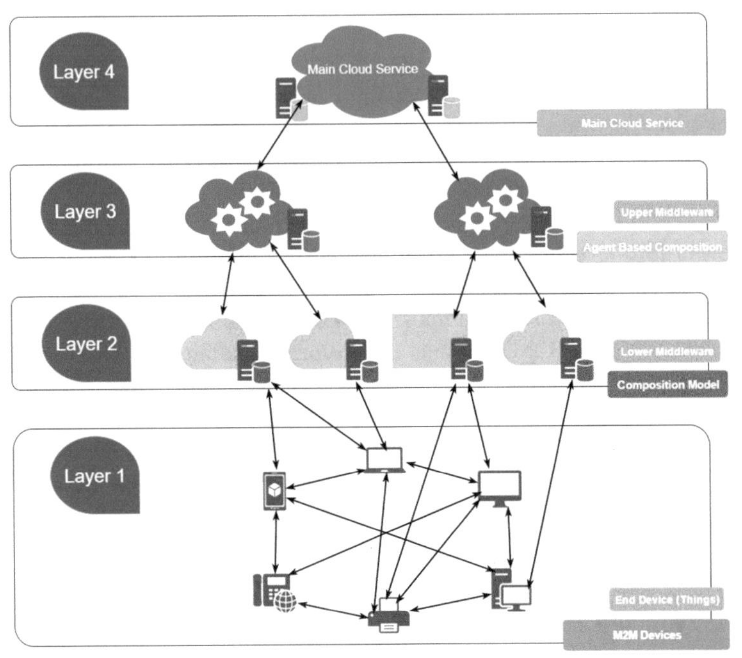

Fig. 12.1 IoT infrastructure model with FOG implementation

12.3 Literature Review

Before we dive into explaining what the proposed infrastructure does to validate the benefits that comes with it, we need to know about the concepts that it is built upon with and the significance that these concepts has on the proposed infrastructure. A brief explanation of each of the concepts that made this infrastructure possible is given below.

12.3.1 Fog Computing

CISCOs recent idea of Fog computing is to empower applications of billions of IoT devices to run from the edge of the network [10]. Fog computing extends cloud and cloud services to the edge of the network. Fog provides data, computations, storage and service applications to the end users similarly as Cloud computing but faster and in real-time as everything is done on the edges close to the end devices [11]. Fog computing is a developing technology that is fundamentally made to provide support for IoT. Fog is a distributed computing model, transporting centralized data storage,

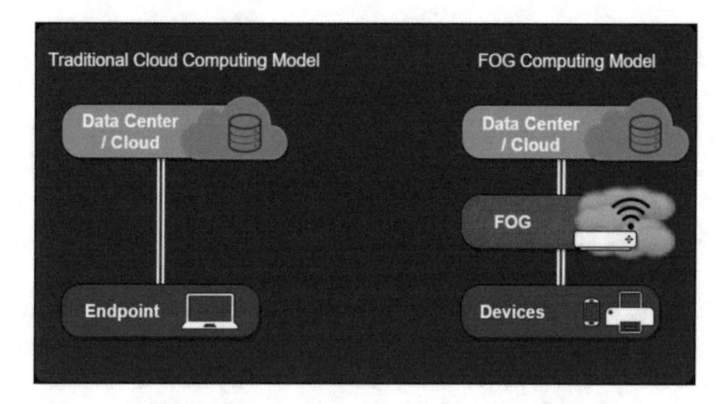

Fig. 12.2 Traditional cloud computing model versus Fog computing model

processing and applications to the edge of the network, locally hosted near the end devices [12]. This reduces the traffic as a result of low usage of band-width, reduces latency and increase the Quality of Service (QoS) (Fig. 12.2).

12.3.2 Real Life Implementation of Fog in IoT

Now industrial protocols are being used instead of IP for machine to connect with controller which will bring about billions of new IoT devices. Rapid data analyzation is the main reason for Fog computation making a substantial extension of cloud computing producing services and applications for enhancing data management and analytics [13]. The fundamentals of cloud such as, data computation, data storage, establishing network with the end-devices and data centers are interrelated by Fog computing. Fog operation is less time consuming as it processes data at particular access points rather than sending data over the main cloud and as a result demand for bandwidth falls creating faster response time and scalability for servers [14].

Internet of Things can interrelate customary devices with physical addresses. Control Packet Switches helps to assemble PCs and information delivery with proficiency by its built frameworks. As an example, the usages of this advancement are empowered by remote sensors placed to measure temperature, humidity, or different gases in the building atmosphere [15, 16]. During this situation, data can be traded amid all sensors in a location and their analysis can be consolidated to generate unfailing estimations for a smart building. Low latency and context awareness of Fog provides localization by its nodes where cloud provides centralization. Both of these are needed for big data and analytics. Fog collects data from grid sensors and devices from edges which needs real-time processing related with protection [17].

12.3.3 End Device Communication

At the very lowest part of each infrastructure lies the end devices with users. Here devices become both constructor and client of data and within these devices, each one of these will learn and acquire knowledge to solve everyday life complications directly with the data acquired, which forms the foundation2 of IoT. Now, connecting all the devices that are connected would create much more data. This vast volume of data processing becomes much more critical for existing technologies. Multiple data streams [18] can generate anyplace around the world, which is reachable via the Internet if it is made public, which in fact can possess a security threat along with an imminent chance of clogging the network [19]. Therefore, data streams have to be administered efficiently providing real-time monitoring. This is where the Machine-to-Machine (M2M) protocols come into play. M2M denotes direct communication between devices using specific simple and light communication protocols for both wireless and wired devices. Thus, M2M has a major play in IoT using smart communication orientation objects, as they are easily reachable and manageable [20]. From the beginning of working with IoT and M2M, the devices that are used for this purpose are usually low powered, short ranged and interconnected and with predefined functions and purpose. Therefore, general Internet protocols could not be used for these devices. This introduced some new protocols MQTT, CoAP, OMA LWM2M and many more, keeping in mind of the devices low energy consumption reducing the payload to keep these running for a long time [21]. These protocols generally depend on message passing for the use of MQTT (Message Queue Telemetry Port) various REST APIs used in CoAP (Constrained Application Protocol) and OMA LWM2M (Open Mobile Alliance Light Weight M2M). For each device to be identified in the network, each has different sets of configurations in their metadata. it saves name, model, hardware type, version and timestamp locally to the sensor values creating metadata for devices [22]. A preliminary configuration of the device and its endpoints is set up by an API with JSON file containing the static description [23]. This API gets the configuration file via GET request and the configuration of the device for the API must have certain attributes to be recognized such as,

- Name - Name of the device.
- Id - Unique identification of the device.
- Value - Gives the reading or value of the hardware.
- Location - Location can be designated by GPS co-ordinates.
- Protocol - Type of request.
- Gateway-in - URI to a device connected with the sensor.
- Gateway-out - URI to a device connected with the actuator.

Then the configuration of the endpoint for the API to be recognized has the attributes as below:

- Name - Name of the endpoint.
- Password - Unique password of the endpoint.
- Token - non-cryptographic token for unique identification.

Through this API, the devices manage a constant connection with the higher layer of Fog nodes and requests necessary services for real time problem solving. This huge amount of data generated from these devices may need to be transferred to other devices or even data processing may take place in the second layer of fog computation model. With this, the second layer of our fog computing model works as a middleware in real time for the devices by which a M2M protocol is always working for the end devices to provide proper services.

12.3.4 Service Oriented Architecture (SOA)

For a proficient distributed architecture, the primary concern becomes the communication process of the end devices between themselves and to the upper layer. The way devices are going to communicate is a challenging matter. This is where a middleware is required by creating nodes which will convey the services to the up-per layers of the Fog architecture. A middleware is what connects different components or software to enable multiple processes running on one or more devices to interact across a network.

Now as we know, in Internet of Things these Things can be anything such as human, car, watch, household things, vehicles etc. Here the entity which is basically the end device, the focus of interactive experience by clients and it involves a medium that can monitor the activities of these entities. For this monitoring purpose, the software or medium that is used, is called resource that can also control and give information to the entity. A service provides a standardized interface which offers all required functionalities to interact with the entities as the resource is highly dependent on the entity. The services represent the functionality of a device by accessing its resources. As a result, each device may have a different type of service. Therefore, the link between a service and a resource needs to be in such a way that will provide interoperable and automated machine-readable representations [24]. As described in M2M communication portion each entity as in each device can have some properties such as, domain value, location and temporal values which also can have several sub values for each of these properties [25]. Location can be global or Local. For these location values a kind of URI (Uniform Resource Identifier) named URN (Uniform Resource Name) is used and for distinguishing each IoT service nodes individually name or id is applied [26]. On the other hand, resources represent these entities in digital forms. The resources are specified by access interface by an interface type which is a set of instances used in distributed technologies, for example, REST, SOAP, RPC.

This way a fully service model can provide functionality to a IoT structure. Thus, as it is the service based approach so far for the IoT context, Service Oriented Architecture (SOA) is used as a designing approach to create schemes built from independent services and in which application components offer services to other components by a communications protocol over a network. The principles are independent of any business, product or technology [27]. The aptitude of SOA is enabling agile service via open standards.

For the nodes that are not at the end but is in a lower level of the fog architectures middleware the composite model is implemented based on the Service Oriented Architecture. Not all distributed software architecture supports a network of heterogeneous devices so the solution is to a middle layer application that can handle heterogeneous devices running different services on different platforms, which provides a dynamic distributed system assuring flexibility and interoperability along with improving robustness, reliability, availability and scalability if existing SOA lack proper settings of non-functional requirements. This distributed service composition model is used for improvised data acquisition which will convert basic existing heterogeneous devices into better software units along with complex functionality. This functionality is added with corresponding Quality of Service (QoS) features following the soft-real time restrictions by the most appropriate sampling time of specific services.

Services of IoT can be represented by five-tuple. We can show them with this definition:
$IoTService_i =< Id_i, Ps_i, Ip_i, Ir_i, At_i >$

Here, Id_i= Identification, Ps_i= Purpose, Ip_i=Provided Interface, Ir_i= Required Interface, At_i= Attributes. This equation characterizes IoT services from rest of the services on the network [28]. Here each IoT service needs to be identified (Id_i) uniquely with an Id or name, has a purpose (Ps_i) such as, giving a value of current temperature, provided (Ip_i) and required (Ir_i) interface of same API such as REST API using JSON and attributes (At_i) as values.

The operation of an IoT service can be either simple or composite. Simple operation defines a service which does not depend on other services for having full resources. On the contrary, composite services are depended on other services. Moreover, IoT services can act as both provider and consumer using different interfaces creating a systematic synchronized mechanism and services may use parameters based on its configuration when it acts as a provider.

12.3.5 Agent Based SOA

For the nodes in the upper middleware we use the agent based compositions to make complex compositions. The agent based model does not need to be light-weight, as a result these nodes have high resources and computation capabilities.

An agent based model signifies that, the interactions by this model is done by autonomous agents which can be a single entity or a collection of agent working together to assess the system. The goal of agents is to serve a particular purpose with the collective knowledge of each group of agents where each group is following a predefined rule. Due to recent researches with ontology to make machines represent human like knowledge, agent based model got a significant amount of interest in various sectors of science and business. Thus, in this model services are managed by agents to determine which services should be selected when its requested. This choice is determined autonomously by learning the priority of the services requested by the end devices [29].

An agent based composition needs to have three actors, service provider, business process manager and users to make autonomous agents cooperate to attain its goals. These agents also must have the capability to do certain activities such as, building workflows, make web services and monitor performance. This architecture is built upon with Society of Agents and made up with two components [30]:

1. Component Manager: This is in charge of interacting and communicate with web services via API in case of this model, JSON messages within the agent society and vice versa providing flexible services based on some rules given by the uses as required through the interface.
2. Workflow Manager: The goal is to support clients to build the workflows, compose web services and monitor their execution. As this is a complex activity its done by two processes [31]:
    - Predefined workflow: From previous communications many other templates are generated. Therefore, when a user makes a new communication process, they can get help from these templates by using them directly. Here the workflow manager works by matching services, which is possible be-cause of common background knowledge of the agents based on shared ontology.
    - Dynamic workflow: Another alternative process of making a new communicating process is by using dynamic workflow where new workflow is created based on user requirements and compose available atomic services, with the help of a planner [32]. After its creation, it replaces the failed or deprecated or unavailable web services.

Local names and the credentials are the main factors for the Trust Management principles. These structures avoid central based system and works as a distributed system which unlocks the way to build peer-peer networks with each node responsible for security as it provides proper credentials to access other nodes resources.

Local names and the tokens for each request are the main aspects for the Trust Management Principle. The Fog model work as a distributed system building peer to peer network with each node responsible for security as it checks and provides proper credentials to access another nodes resources. The authorization is very critical and important as it helps building up the trusted network without a central control following least privilege [33].

12.3.6 Cloud Service

The cloud services hold a number of resources that a cloud-service provider distributes to clients over the network. this context now has become known as the Cloud. Features of cloud services consist of autonomy and elasticity. Meaning, the clients can use services on an on-demand basis and terminate them if desired, scale it on demand and maintain data redundancy and backups [34].

Usage of cloud services has become limited to some common cloud services3, such as, SaaS, PaaS and IaaS. SaaS is a software delivery model where applications are given by a service provider and made available to clients in the method of web services [35], for example, G Suite, Microsoft Office 365, Send Grid etc. PaaS offers the platform as in, the OS and related services [36]. Which helps users to deploy and use operating system level services without investing in the infrastructure involved. Examples include Microsoft Azure, Amazon Web Services, Elastic Beanstalk etc. IaaS involves having full access to the components used to support operations, which includes storage, servers, networking modules etc. which are made accessible to clients over the Internet [37] such as, Amazon Web Services, IBM Bluemix and Microsoft Azure.

Apart from this, there are some other form of services too. For example, BaaS known as the Backend as a service and FaaS is defined Function as a service etc. Therefore, we have the main Cloud in the top or the fourth layer of the infrastructure where each cloud services are kept. Here these services are ready to receiving data from Fog nodes, summaries are collected, analyzed on IoT data and data from other sources to generate business insights and depending on these insights new application rules can be conveyed to the fog nodes [38].

12.4 Workflow of the Infrastructure

In the Fig. 12.3, the workflow of the infrastructure is described with some sample service requests and responds. Here we have end devices D1, D2, D3, D4, D5 and D6 where D1, D2, D3 and D4 is in region 1 an D5 and D6 is in region 2. Moreover, here A and B represents two cloud services. At first D5 requested for the service A to the lower middleware or SOA. SOA is regionally dependent and does not consume a lot of power like network servers. SOA is used to listen constantly for service requests from end devices. When a request is found, it is parsed and service results and values are searched locally. End devices can request two types of services, simple service and complex service. If the service requested is a simple service, SOA fetches all the values and starts preparing JSON reply. When a complex service is requested, SOA composes the simple services under this complex service and for each of simple service found, it fetches the details. Now as it got a request of A, it is searched locally but could not find it.

Things start changing when a service is not found in SOA. It is not particularly possible to take care of all the services by a low powered device like SOA. End point devices may request an infinite number of services. So, when a requested service is not found, it makes a new request to its parent which is Service based Agent (3rd layer of our model) or simply known as Agent. After getting reply from agent if the service is present in Agent database, SOA updates its storage for serving that request in future. Agent society consists of Reply Agent, Fellow Agent and Update Agent.

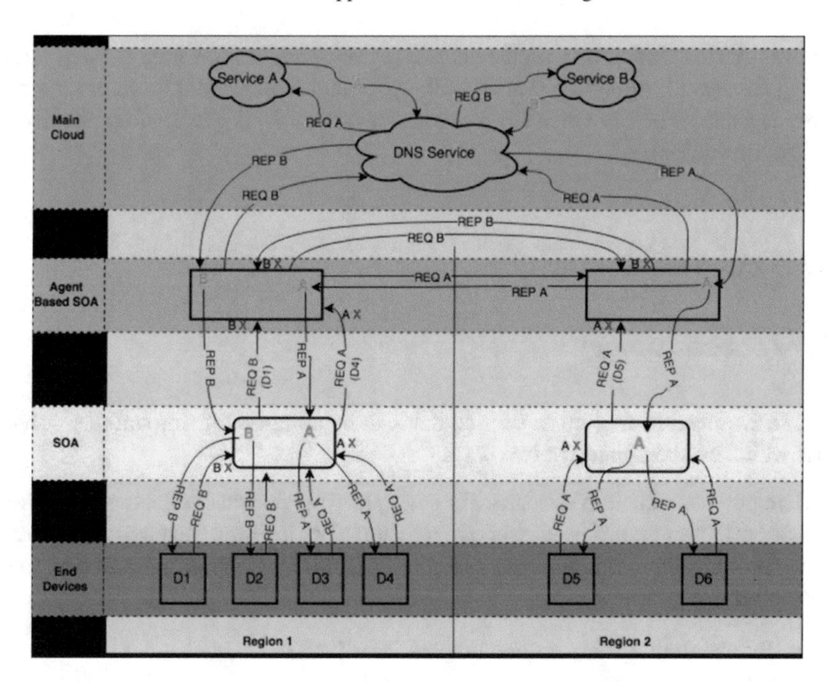

Fig. 12.3 Sample workflow of the infrastructure

For implementation purpose, we have put Reply Agent and Fellow Agent together in Checker Agent. When a SOA request is found, checker agent parses it and forwards it to the reply agent. Reply agent works just like SOA, searches for that particular request locally. If it finds with in it, then it gives back all the requested value to the SOA. One major distinguishes between SOA and Reply agent server is, Reply agent does not compose any service when requested but SOA does.

Therefore, now Reply agent searches for A and same as SOA it failed to retrieve it. As some service may not be available to Reply Agent, it calls fellow agent which resides in the same server as Reply Agent does. Fellow agent has got an IP table of other Agent Servers and the main server. Thus, it requests its fellow agent in region 1 but with another failed attempt it requests the main cloud. Here DNS Service redirects to the Cloud service that is requested and provides A to the upper middleware. Now A is stored locally by the Reply Agent and again relied to lower middleware and also stored there. Finally, service A is sent to the End device. This is what we call a worst-case scenario where the middleware does not have the required service and the Cloud is requested. But now as D6 now requests for the same service A lower middleware or SOA directly replies with the service thus giving us the best-case scenario.

In this exact same way, in region 1 when D1 requests for B we have our worst case again and for D2 the best case is seen. Furthermore, when D4 requests for A, the request goes to the lower middleware then to the upper but this time Fellow Agent retrieves A from the upper middleware in region 2 and replies step by step to D4 storing the service along the way to both upper and lower middleware in region 1. Thus, as same as before when D3 requests for service A, is gets replied directly by the lower middleware.

12.5 Implementation

12.5.1 Methodology

At first, we need to find out a service that can be composed using another service parts which can be created in two ways -

1. Using our own custom services. For example, a service that collects environment data and gives that to users who are registered for it. Here, the environment data can be called environment data as a service (EdaaS) which is composed of some core services such as:
    - Temperature. (Temperature as a service - TaaS),
    - Humidity (Humidity as a service - HaaS),
    - Wind Flow (Wind flow as a service - WFaaS) and
    - Air quality (Air quality as a service - AQaaS)
2. Using the existing services to compose a service which is required by the user. Then we have to create modules using JAVA and Cloud technologies in order to mimic our infrastructure. Here we chose JAVA for our implementation purpose. We could also use other languages for like PHP and C++ if necessary.

Finally, after that we have to set up a test environment to run a prototype of the model and get results for a number of devices from different test cases. The preferred test cases are,

- Different types of devices asking for same service request from the same region.
- Different types of devices asking for same service request from different regions.
- Different types of devices asking for different service requests from the same region.
- Different types of devices asking for different service requests from different regions.

Here, we have to have support for homogenous devices. That is why for the devices, we can implement each with VMs of Windows or Linux mimicking any environments. Moreover, we can use two actual single board devices like Arduino and Raspberry Pi or three android based devices. Therefore, the variation will not contain all same type of devices.

Some of the possible outcomes of the test case that we can expect here are -

1. When a device request for a service for the first time in a region, the service is provided by climbing up to the cloud services and then getting the results and respond with the composed service. Which can be considered the worst case in our infrastructure.
2. Once, a service has been composed, the upper and lower part of the middleware knows how to compose that. Now devices requesting the same service or other services which has common simple services can be composed without climbing to the cloud services.
3. For the best case, it can be considered that, a second device requesting for the same service which can be composed directly at the lower part of the middle layer and respond to the device directly.

A note to remember here is that, for VMs to work with the context Different Regions, we will be provisioning VMs in different data centers situated in different regions. So, even if it is a virtual machine, it is provisioned in different geo-positions or regions of the world. This makes sure that, even if the VMs are using different Storage Area Networks (SAN) of some other region, the actual VMs location stays the same as it states in the VMs provisioning configuration.

Finally, to summarize, to make the infrastructure work properly we need to implement it in a large-scale environment. However, to prove that this infrastructure can reduce the usage of traffic and improves QoS, we can just simulate the infrastructure in the cloud by provisioning virtual machines in different geographically available data centers by assigning the VMs to represent different layers of the infrastructure and then simulate the requests and responds with cloud services. Here we will be using the public network so that we can recreate the real situation of the network usability instead using the clouds internal private network or any Virtual networks. After that, we are going to monitor and record the data usage of the virtual network interface card assigned to each of the VM to get the exact volume of the data transfer and then finally compare it with the traditional approach of the cloud services to validate our proposal.

12.5.2 Creating the Prototype Model

We can either use existing cloud services from any cloud service provider like Microsoft Azure or Amazon AWS or we can create our own cloud services in order to implement our infrastructure and then test it out. But we cannot use the current cloud service only to implement our infrastructure because there are a lot of layers that needs to be implemented physically to test out our infrastructure. If we create our own cloud service, we can leverage the Clouds potential and add an extra layer in this case a virtual machine, which will enable us to mimic the hardware that do not need to be physically present but exist in the cloud.

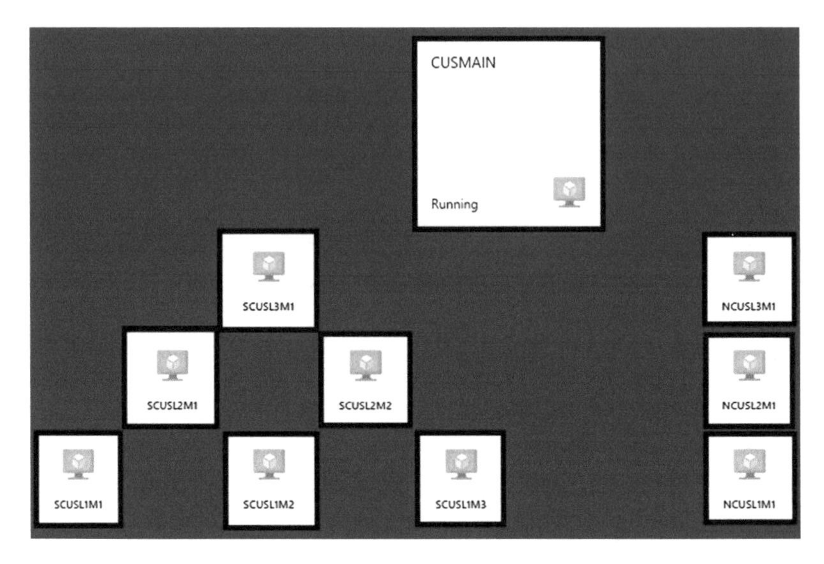

Fig. 12.4 Deployed virtual machines for test run of our infrastructure

In our Experimental setup, end devices, SOA Architecture and Agent based Architecture have been represented using Virtual Machines (VM). For this purpose, we have chosen Microsoft Azure as an implementation structure. Azure datacenters were situated in different geographical positions, this is really efficient and convenient to perform some test runs. Initially we planned to use two different geographical positions: North Central US, South Central US and Central US. The VMs represented the SOA, Agent based SOA and machines which were in the same geographically available data centers. The main cloud service could be deployed in any region.

Given Fig. 12.4 shows the deployed infrastructure in Azure using VMs in different layer. In the Fig. 12.4, SCUSL1M1, SCUSL1M2 and SCUSL1M3 are VMs which represent the layer 1 (End devices) in the South-Central US region and NCUSL1M1 belongs to layer 1 in the North Central US region. Next, in the second layer (SOA), SCUSL2M1 and SCUSL2M2 are in the South-Central US region and NCUSL2M1 belongs to the North Central US region. In the third layer (Agent based SOA), SCUSL3M1 is in the South-Central US region and NCUSL3M1 is in the North Central US region. Finally, CUSMAIN is the main cloud server (Fig. 12.4).

12.5.3 Technical Properties of the Infrastructure

As mentioned before in this paper, our infrastructure has four layers from end devices to main cloud server. In case of implementation, communication between these layers was established with different algorithm and pseudocodes as the structure and mechanism of each layer follows different approaches. But the request of data requested by any of the VM and response of that requested data replied by any server are represented through a JSON format. Request format is uniform for any request from any layer and response format is also uniform for every reply in any layer.

Request JSON format:


```
{
“REQ”:
“authentication”:{
“USERID”:“user.name”,
“password”:“password123”
},
“token”:“2b2c5f9e6655ce42740584f4c25c85b6”,
“ser-vice”:{
“name”:“environment”,
“components”:“temperature,humidity”
}
}
}
```


Response JSON format:


```
{
“’RES”:{
“Token”:“2b2c5f9e6655ce42740584f4c25c85b6”,
“C.Service”:{
“Service-Name”:“environment”,
“provider”:“metro”
},
“B.Service”:{
“ServiceName”:[
“temperature”,
“humidity”
],
“value”:[
“32”,
“55”
],
“optionalParameters”:
“protocol”:[
“http”,
“http”
],
“url”:[
“http://www.example.com/temperature”,
“http://www.example.com/humidity”
```

],
"ttl":[
"500",
"500"
],
"timestamp":[
"2016-10-08 08:26:27",
"2016-10-08 08:26:27"
]
}
}
}
}

In our implementation process, we have defined services in two phases: Simple and Complex. Complex services are combined with several basic or simple services. For example, from our JSON format we can observe that "environment" is a complex service which consists simple services such as "temperature" and "humidity". There can be many independent simple services as well which are not generalized in a particular complex service. The main advantage of this generalization of simple services as a specific complex service is, there can be plenty of simple services with same name but different category or mechanism for instance "temperature" can be of many categories such as environment, food, room, water and many more. So, if a machine wants to request for environment temperature, it is easier to fetch the data value from the server as a complex service called "environment" which consists temperature.

12.6 Pseudocodes

To Implement our proposed infrastructure, at first, we built some algorithms for layer to layer communications for each layer and applied them in datadog in order to generate graphs to compare the results. Datadog provides monitoring as a service and to use that we need to integrate datadog agents in azure VMs which sends metric of the azure VMs to the datadog dashboard. But datadog agents can have delay up to 2 min to send the data to datadog dashboard which may cause a bit delay in the generated graphs.

12.6.1 Communication from End Devices to SOA

In our communication from End devices to SOA algorithm, SOA is always listening for incoming requests from end devices. When a request for service is received its saved as a String serviceName. Then it searches for the values of the complex services. Next, for each basic services of the complex service, it tries to retrieve all the components or parameters of the basic services of that service. Otherwise, if the query result was null, that means the serviceName does not belongs to any complex service but it can be an independent simple service and therefore it searches for that service in the simple services and if the result is null, that means SOA layer does not consist this data, so, it will send an http request to an Agent in next layer which is the third layer (Agent based SOA) of our infrastructure or else if found it will send the value to the device. The algorithm is given as follows:

Pseudocode 1. SOA

/* SOA data fetch, wait for request from M2M, Received request */
valueresult ← getComplexServiceValues(serviceName)
if valueresult not null then
Reply valueresult to device

getComplexServiceValues(serviceName):

result ← query for basic services of serviceName
Call getSimpleServiceValues(serviceName)

getSimpleServiceValues(serviceName):

result ← query for values of serviceName
if result is null then
value ← serviceValue from agent based on that region
else
value ← serviceValue
Return value

12.6.2 Communication from SOA to Agent Based SOA

We have divided our Agent based SOA in three parts, Reply Agent, Update Agent and Fellow Agent. Prioritizing the services based on how frequently they are requested, success rate and up time, the agents decides as an Artificial Intelligent, which services should be served at the first place. A Reply Agent always keeps listening requests sent from layer two(SOA) and an Update Agent also keeps updating the simple services through HTTP request based on priorities.

In our communication from layer two (SOA) to layer three (agent based SOA) algorithm, a reply Agent of layer three is always listening for incoming requests from layer two as mentioned before it works as same as the layer two SOA. An update Agent also continuously updates simple service values via HTTP request based on priority where priority is fixed depending on request or update counts which increased every time a service is requested.

The fellow agent is called when an agent based SOA receives null after querying in database. This agent goes through the lookup table for IP of nearby agents and request for the service value from those agents. The used algorithm is described as follows:

Pseudocode 2. ReplyAgent
/* Agent based SOA data fetch, wait for request from SOA, Received request */
Call getComplexServiceValues(serviceName)
Call increaseUpdateCount(serviceName);
Call increasePriority(serviceName);
valueresult ← getComplexServiceValues(serviceName)
if valueresult not null then
Reply valueresult to device

increaseUpdateCount(String serviceName):
Increase and update update_count of serviceName by 1 in Database

getComplexServiceValues(serviceName) and **getSimple-ServiceValues(serviceName)** is same as explained in SOA

Pseudocode 3. UpdateAgent
/* Update Agent always runs updateTable() in the background */
updateTable():
result ← query for all complex service by priority
for the size of result
Call UpdateComponents(serviceName)

UpdateComponents(serviceName):
Make threads to send http request and get new value from Cloud
Call updateValues(serviceName, newValue)

updateValues(serviceName, serviceValue):
Update simple service value to serviceValue of the serviceName

increasePriority(String serviceName):
Increase priority count of serviceName

Pseudocode 4. fellowAgent
/* Agent based SOA data fetch, wait for request from SOA, Received request */
ip = ip of other agents from lookup table
Get serviceName value from fellow agent

12.6.3 Communication from Third Layer to Main Server

When second and third layer is unable to fetch the requested service or data, then agent sends an http request to the main server. After getting the request from third layer, the main server call the getComplexServiceValues(String serviceName) method which follows exactly similar algorithm described in the communication from End device to SOA along with the getSimpleServiceValues(String service-Name) using the main cloud server(main_server database) except the method named ttlCount(String serviceName, String timestamp) and the main server wont need to send any http request.

12.6.4 Traditional Cloud Computing

Figure 12.5 shows the deployed present infrastructure in Azure using VM. Among the VMs, in the first layer end devices TSCUSMACHINE1, TSCUSMACHINE2 and TSCUSMACHINE3 are situated in the South-Central US region and TNCUSMACHINE1 is in the North Central US region. Finally, TCUSMAIN is the main cloud server.

In case of implementing present computing infrastructure, it is a server client model where each client request hits the server and the server sends a reply accordingly. Therefore, here the main server directly gets the request from end devices in same JSON format described before. Afterwards, the main server fetch service from the traditional_main_server database and send reply to the end devices through JSON response format.

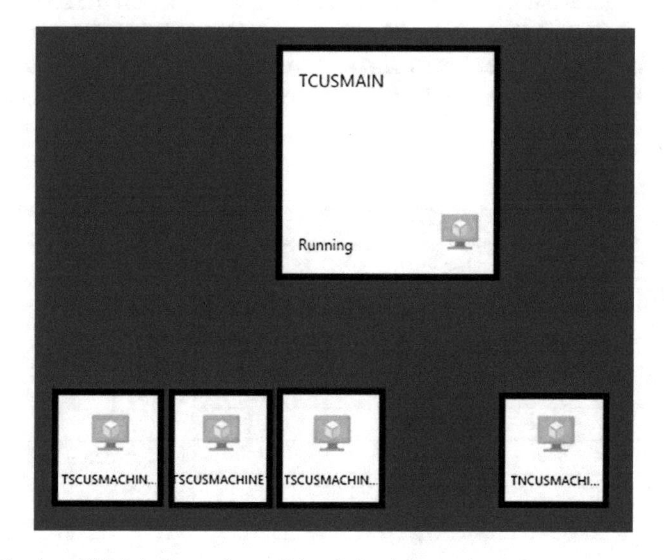

Fig. 12.5 Deployed VMs in Azure for traditional cloud computing infrastructure

12.7 Result

12.7.1 Result Graphs for Each VMs

We made a small-scale experiment to monitor the network usage of each of the VMs for our infrastructure by requesting and responding with JSON generating up to a few hundreds of kilobytes. We have to considered that for a large-scale deployment the network traffic will exceed by millions and network needs to be adjusted to cope up to deliver terabytes of data.

12.7.2 Result Graph of South Central US

The graphs below show the network usage of VMs of South Central US which were involved with the test environment while the experiment was conducted (Table 12.1).

For a trial within the first layer, request for the same service was sent from every device, SCUSL1M1 at 2:31:40am (Fig. 12.6), SCUSL1M2 at 2:36:00am, SCUSL1M3 at 2:36:00am. In the Fig. 12.6 we can see a spike in the network curve of 249kb at 2:31:40am. This curve indicates the request that was sent. For the other small curves, those are the network usage of the VM that were being conducted for update and other purposes which in this case, we are ignoring by acknowledging those as noise. The previous network spike curves are also being excluded as we started recording the network activity from 2:30:00am and this is applicable for all the graphs for our proposed model. Initially, VM of the second layer SCUSL2M1 received the request but it did not have the service. So, request of the service was sent to SCUSL3M1 at 2:31:00am (Fig. 12.7) in the third layer. When the service was not even found in the third layer it was sent to the main server CUSMAIN at 2:30:40am (Fig. 12.8). Later, from the main server the result was saved and sent back to the

Table 12.1 VM information

VM name	Layer	Location
SCUSL1M1	Layer 1	South Central US
SCUSL1M2		
SCUSL1M3		
SCUSL2M1	Layer 2	
SCUSL2M2		
SCUSL3M1	Layer 3	
NCUSL1M1	Layer 1	North Central US
NCUSL2M1	Layer 2	
NCUSL3M1	Layer 3	

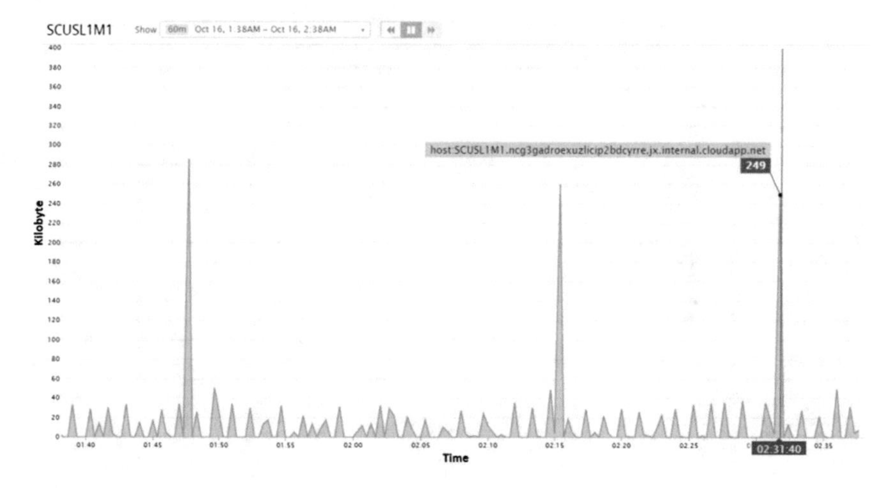

Fig. 12.6 SCUSL1M1

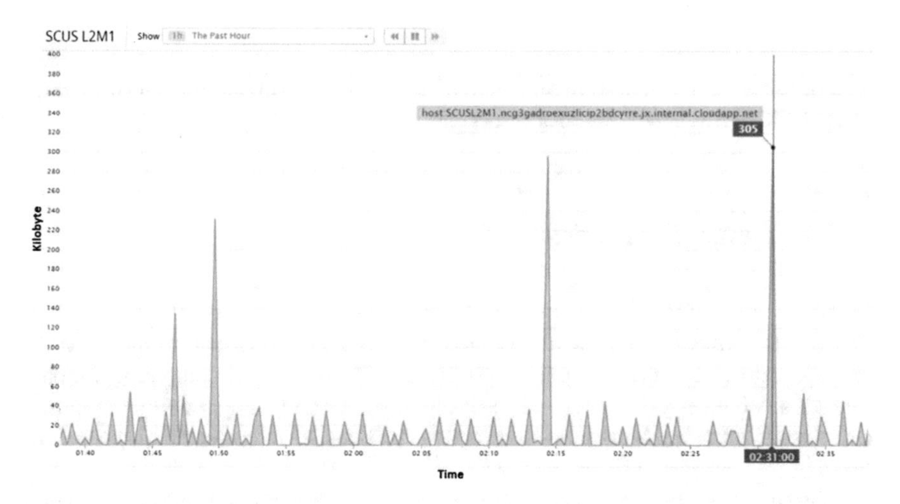

Fig. 12.7 SCUSL2M1

third layer and after that in the second layer. A note to remember here, we can see some difference in seconds in the network graph, this occurred due to the difference in regional placement of the VMs in Microsoft Azure.

We can observe from SCUSL2M1 (Fig. 12.7) and SCUSL3M1 (Fig. 12.8) where only request from SCUSL1M1 at 2:31:40am was sent as we see a network spike but not from the other two devices as the result was already saved in the second layer while processing for SCUSL2M1. In the main server, request received and requested

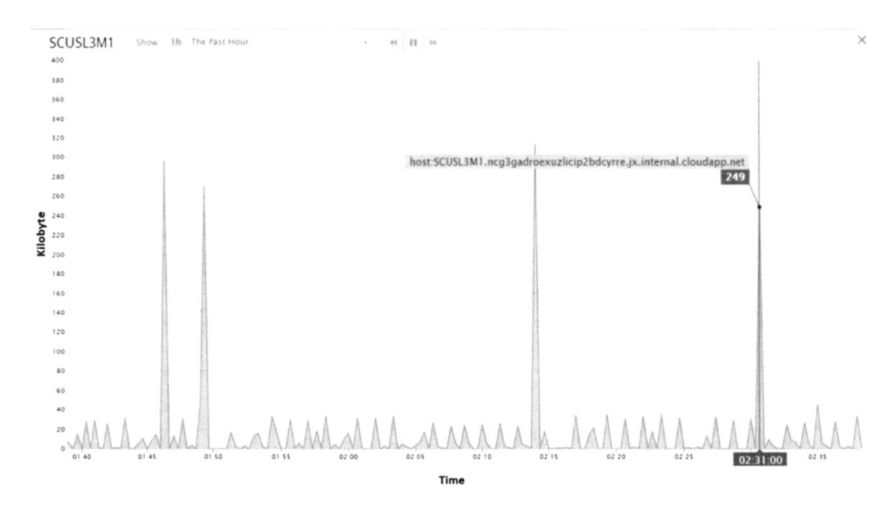

Fig. 12.8 SCUSL3M1

at 2:36:00am was not sent from SCUSL1M2 and SCUSL1M3 but from North Central where NCUSL1M1 requested for a service.

12.7.3 Result Comparison

As mentioned before, two experiments were conducted in two different test environments. Among them, one represented our proposed infrastructure and the other one represented the conventional infrastructure. For the sake of computing the data transactions between the VMs and main cloud server and comparing our proposed and present infrastructure, same service was re-quested from four end-devices of different regions as mentioned before.

After finishing the data processing, our result was projected through different graphs. Figure 12.9 shows the results of the conventional infrastructure where four devices have requested for services where first three requests are for the same service and fourth request is for different service in between 3:33:00am to 3:38:00am. If we notice on the graph, we can observe that the total amount of data both received (darker blue) and sent (light blue), shows a constant data consumption.

On the contrary, Figs. 12.10 and 12.11 symbolizes our infrastructure where the same scenario was imposed, in between 2:31:00am to 2:36:00am where we can see in the received data graph (Fig. 12.10) two request was received and sent (Fig. 12.11) to the second layer during that time. As indicated before, from the kilobyte/time graph we can see that there is a big drop of data consumption in the middle both while receiving and sending data. This data was recorded in at most 15 s interval,

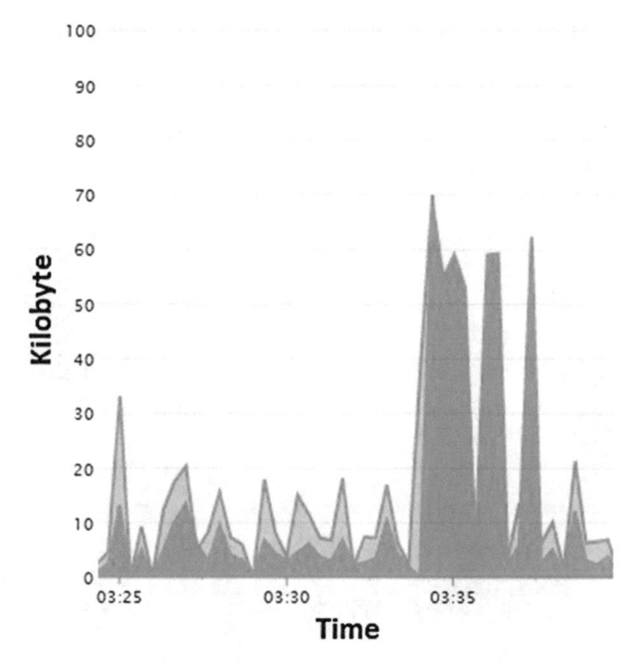

Fig. 12.9 Conventional infrastructure result

which gives this inconsistent growth of the graph. Now, comparing the graphs we can distinctly comprehend that our proposed infrastructure has a very low amount of network usage as it has a highest usage of 14–11 kb where as in the conventional infrastructure it reaches 70–60 kb within that time limit.

From these results of the described algorithms along with the comparison with present infrastructure, it can be ensured that our proposed infrastructure surpasses the traditional one in giving better performance, less traffic along with less band-width, reliability through trust management by providing token authentication and heterogeneity maintaining the Quality of Service(QoS).

12.7.4 Result Comparison Between Traditional and Proposed Infrastructure

1. Local Backup: If the server gets down for any reason, local databases of second and third layer can be used as local backup for the end devices. Thus, end devices can still get required results through reply from the local backups. On the contrary, in the same scenario, the whole communication between end devices and main

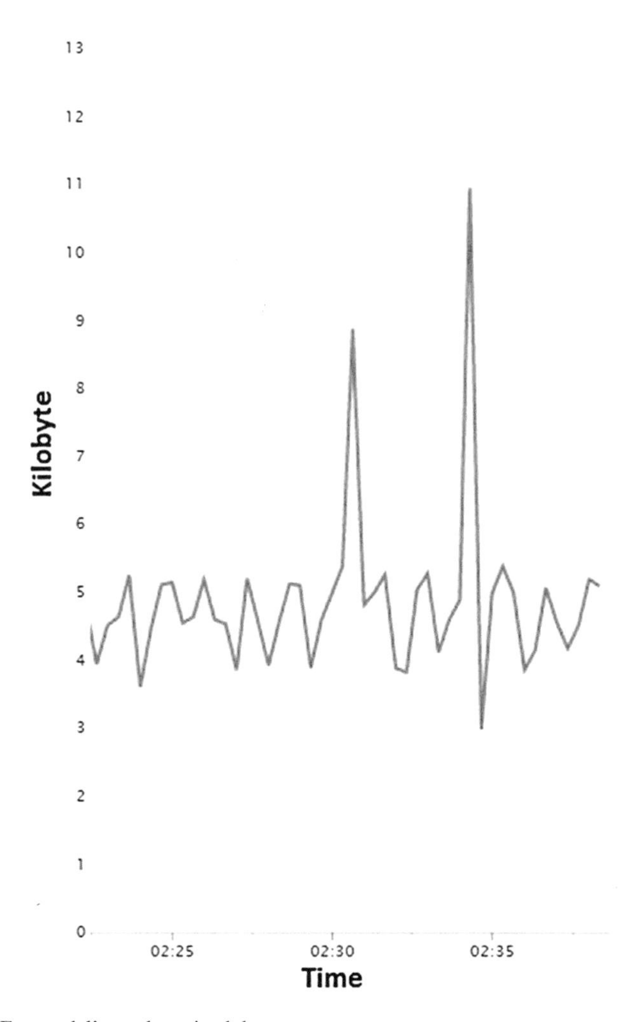

Fig. 12.10 Fog model's total received data

server gets interjected in the traditional infrastructure because of the lacking of local backup.

2. Less Bandwidth and Traffic: Less bandwidth and less traffic can be assured in our proposed infrastructure as all the request are restricted to the distributed locally updated local backups (second and third layer) instead of requesting the main server again and again for the same service.
3. Data Redundancy and latency: Form the result of TCUSMAIN, we can see that for the same service x request the main server had to fetch and reply thrice to TCUSMACHINE1 and TCUSMACHINE2 (Fig. 12.9). This increases the latency of the infrastructure. In other hand, the proposed infrastructure gives less latency

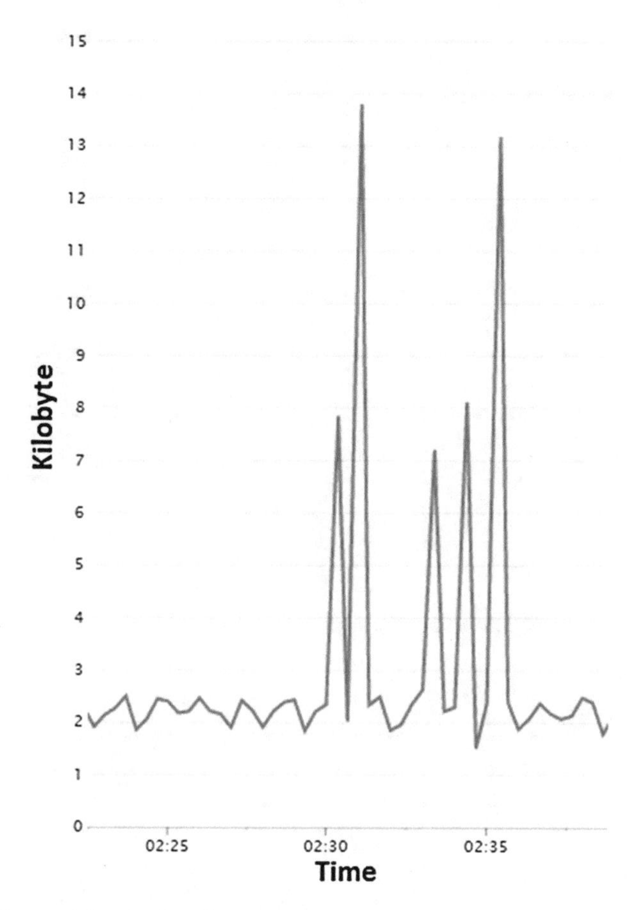

Fig. 12.11 Fog model's total sent data

since in case of the second request for service x by SCUSL1M2, the response directly came from layer two instead of the main server (Fig. 12.10) and therefore, we can say that all of the requests are not handled by the main server, instead, was handled by layer 1 and layer 2.

12.7.5 Efficiency of Our Infrastructure

In the graph (Fig. 12.12) given below, we plotted the amount of data consumed by each layer of the infrastructure for each request. In our implementation, we re-quested for two types of services (x and y) as 6 requests in total. We can observe the increase of request leads to lower data consumption eventually. For experimental purpose, we

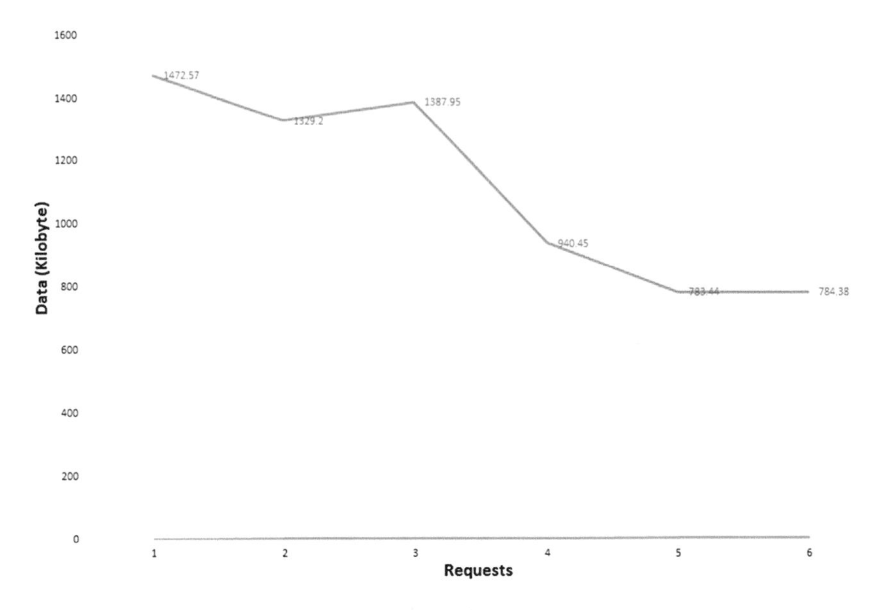

Fig. 12.12 Data consumption versus request graph

Table 12.2 Data consumption

Requests		x1	x2	y3	y4	x5	y6
SCUSL1M1	In/out	192.28/ 153.31		148.3/ 203.53			
SCUSL1M2	In/out		452.01/ 370.7		251.35/ 277.32	247.65/ 274.5	232.05/ 274.45
SCUSL2M1	In/out	256.6/ 208.29	249.22/ 95.9	356.48/ 98.43	144.56/ 91	144.84/ 90.79	144.41/90.94
SCUSL3M1	In/out	122.94/ 115.41	97.34/ 54.03	202.83/ 256.38	96.61/ 56.63	10.1/0.56	12.33/ 11.2
CUSMAIN	In/out	202.34/ 221.4	0/ 10	82.66/ 39.24	12/ 11	0/ 15	19/ 0
Total		1472.57	1329.2	1387.95	940.45	783.44	784.38

generated only 6 requests but in future when 20 billion devices will be involved with the IoT infrastructure, the number of data consumption will get lower because the Layer 3 Agents will get more intelligent with the passage of time (Table 12.2).

12.8 Conclusion

Traditional approach deals with direct communication to the cloud in order to work with IoT devices which inefficient as we have already discussed. Our infrastructure relies on Fog computing model where nodes neighboring to the IoT devices are used for calculating and providing data to the local nodes. Our infrastructure supports a light weighted SOA which will be running in the fog nodes, that will provide services to the IoT devices locally as the second layer of our distributed model. For complex service compositions, the third layer will be providing agent based services as artificial intelligence where the more devices will be involved with our infrastructure, the more mature the agents of the third layer will become over the time. In addition, monitoring the traffic of the cloud channels, data is periodically synced with the cloud where the period is determined by the agent. Moreover, the third layer can implement miniature versions of services like stream analytics, bus services, machine learning and many more as services. That makes our infrastructure more flexible, proficient and versatile from the ongoing researches regarding fog computing and IoT.

Some other benefits of using this model will be, firstly, working as local backup of second layer can maintain the data transactions even if the main server gets down which results in no data loss or jeopardized data communication. Secondly, businesses can grow based on the infrastructure. For example, big companies which needs to serve IoT services to their clients can deploy second, third layer devices and they can provide other companies to use their nodes for their services based on some business model. This way users can get their services easily and this will also open up new opportunities for companies to grow. It can be deployed on the current growing infrastructure without changing the way devices work at present with addition of trust management, low latency, less bandwidth preserving the Quality of Service(QoS).

References

1. Stojmenovic, S. Wen, The fog computing paradigm: scenarios and security issues, in *2014 Federated Conference on Computer Science and Information Systems*, Warsaw, 2014, pp. 1–8. https://doi.org/10.15439/2014F503
2. Cisco, Fog Computing and the Internet of Things: Extend the Cloud to Where the Things Are (2015)
3. N.N. Khan, Fog computing: a better solution for IoT. Int. J. Eng. Tech. Res. **3**(2), 298–300 (2015)
4. B. Azimdoost, H.R. Sadjadpour, J.J. Garcia-Luna-Aceves, Capacity of wire-less networks with social behavior. IEEE Trans. Wirel. Commun. **12**(1), 60–69 (2013). https://doi.org/10.1109/TWC.2012.121112.111581
5. D. Christin, A. Reinhardt, P.S. Mogre, R. Steinmetz, Wireless sensor networks and the internet of things: selected challenges, in *Proceedings of the 8th GI/ITG KuVS Fach-gesprch Drahtlose sensornetze*, 2009, pp. 31–34

6. P.R. Jaiswal, A.W. Rohankar, Infrastructure as a service: security issues in cloud computing. Int. J. Comput. Sci. Mob. Comput. **3**(3), 707–711 (2014)
7. N.S. Patel, B.S. Rekha, Software as a Service (SaaS): security issues and solutions. Int. J. Comput. Eng. Res. **4**(6), 68–71 (2014)
8. S. Subashini, V. Kavitha, A survey on security issues in service delivery models of cloud computing. J. Netw. Comput. Appl. **34**(1), 1–11 (2011). https://doi.org/10.1016/j.jnca.2010.07.006
9. T. Devi, R. Ganesan, Platform-as-a-Service (PaaS): model and security issues. TELKOMNIKA Indones. J. Electr. Eng. **15**(1), 151–161 (2015)
10. F. Bonomi, Connected vehicles, the internet of things, and fog computing, in *The Eighth ACM International Workshop on Vehicular Inter-Networking (VANET)*, Las Vegas, USA, 2011
11. S. Sanjeev, S. Thusu, A survey of fog computing and its applications. Int. J. Adv. Res. Ideas Innov. Technol. **3**(2), (2017)
12. N. Patel, FOG computing and its real-time applications. Int. J. Emerg. Technol. Adv. Eng. **5**(6), (2015)
13. T.H. Luan, L. Gao, Z. Li, Y. Xiang, G. Wei, L. Sun, Fog computing: focusing on mobile users at the edge (2015), arXiv:1502.01815
14. F. Al-Doghman, Z. Chaczko, A.R. Ajayan, R. Klempous, A review on Fog Computing technology, in *2016 IEEE International Conference on Systems, Man, and Cybernetics (SMC)*, Budapest, 2016, pp. 001525–001530. https://doi.org/10.1109/SMC.2016.7844455
15. I. Stojmenovic, Fog computing: a cloud to the ground support for smart things and machine-to-machine networks, in *2014 Australasian Telecommunication Networks and Applications Conference (ATNAC)*, Southbank, VIC, pp. 117–122 (2014). https://doi.org/10.1109/ATNAC.2014.7020884
16. K.P. Birman, R. van Renesse, W. Vogels, Spinglass: secure and scalable communication tools for mission-critical computing, in *DARPA Information Survivability Conference and Exposition II, 2001. DISCEX '01. Proceedings*, vol. 2, Anaheim, CA, 2001, pp. 85–99. https://doi.org/10.1109/DISCEX.2001.932161
17. F. Bonomi, R. Milito, J. Zhu, S. Addepalli, Fog computing and its role in the internet of things, in *Proceedings of the first edition of the MCC workshop on Mobile cloud computing*, ACM, Aug 2012, pp. 13-16
18. Y. Qin, N. Falkner, S. Dustdar, H. Wang, When things matter: a data-centric view of the internet of things, in *CoRR*, 2014, arXiv:1407.2704
19. G. Fortino, W. Russo, Towards a cloud-assisted and agent-oriented architecture for the internet of things, in *WOA@ AI* IA* 25 Nov 2013, pp. 60-65
20. T. Xue, L. Qiu, X. Li, Resource allocation, for massive M2M communications in SCMA network, in *IEEE 84th Vehicular Technology Conference (VTC-Fall)*, Montreal, QC, Canada, 2016, pp. 1–5. https://doi.org/10.1109/VTCFall.2016.7881198
21. M. Shanmugasundaram, M2M/ Iot protocols: an introduction. Happiest minds, 3 Feb 2016, http://www.happiestminds.com/blogs/m2m-iot-protocols-an-introduction/
22. F.H. Bijarbooneh, W. Du, E.C.H. Ngai, X. Fu, J. Liu, Cloud-assisted data fusion and sensor selection for internet of things. IEEE Internet Things J. **3**(3), 257–268 (2016). https://doi.org/10.1109/JIOT.2015.2502182
23. S.K. Datta, C. Bonnet, N. Nikaein, An IoT gateway centric architecture to provide novel M2M services, in *2014 IEEE World Forum on Internet of Things (WF-IoT)*, Seoul, 2014, pp. 514–519. https://doi.org/10.1109/WF-IoT.2014.6803221
24. Microsoft, 2014, Chap. 1: Service Oriented Architecture (SOA), http://Msdn.microsoft.com. Retrieved on May 30, 2014
25. S.K. Datta, C. Bonnet, Smart M2M gateway based architecture for M2M device and Endpoint management, in *2014 IEEE International Conference on, Internet of Things (iThings), and IEEE Green Computing and Communications (GreenCom), and IEEE Cyber, Physical and Social Computing (CPSCom)*, 2014, pp. 61–68
26. R. Garcia-Castro, C. Hill, O. Corcho, SemserGrid4Env Deliverable D4.3 v2 Sensor network ontology suite (2011)

27. E. Newcomer, G. Lomow, *Understanding SOA with Web Services (Independent Technology Guides)* (Addison-Wesley Professional, Reading, 2004)
28. S. Rodrguez-Valenzuela, J.A. Holgado-Terriza, J.M. Gutirrez-Guerrero, J.L. Muros-Cobos, Distributed service-based approach for sensor data fusion in IoT environments. Sensors **14**(10), 19200–19228 (2014). https://doi.org/10.3390/s141019200
29. W. Shen, Q. Hao, H.J. Yoon, D.H. Norrie, Applications of agent-based systems in intelligent manufacturing: an updated review. Adv. Eng. Inform. **20**(4), 415–431 (2006). https://doi.org/10.1016/j.aei.2006.05.004. ISSN 1474-0346
30. A. Poggi, M. Tomaiuolo, P. Turci, An agent-based service oriented architecture, in *WOA*, Sept 2007, pp. 157–165
31. M. Koskimies, Composing services in SOA: workflow design, usage and patterns, in *Research seminar on Service-Oriented Software Engineering*, Oct 2006, http://www.cs.helsinki.fi/u/thruokol/opetus/2006/sose/pa-pers/koskimies_workflows.pdf
32. C. Lin, A reference architecture for scientific workflow management systems and the VIEW SOA solution. IEEE Trans. Serv. Comput. **2**(1), 79–92 (2009). https://doi.org/10.1109/TSC.2009.4
33. A.M. Khan, A. Mishra, R. Agarwal, Security framework based on QoS and net-working for service oriented architecture, in *2016 3rd International Conference on Computing for Sustainable Global Development (INDIACom)*, New Delhi, 2016, pp. 1106–1109
34. N.K. Pandey, S. Chaudhary, N.K. Joshi, Resource allocation strategies used in cloud computing: a critical analysis, in *2016 2nd International Conference on Communication Control and Intelligent Systems (CCIS)*, Mathura, India, 2016, pp. 213–216. https://doi.org/10.1109/CCIntelS.2016.7878233
35. L. Wu, S.K. Garg, R. Buyya, Service level agreement(SLA) based SaaS cloud management system, in *2015 IEEE 21st International Conference on Parallel and Distributed Systems (ICPADS)*, Melbourne, VIC, 2015, pp. 440–447. https://doi.org/10.1109/ICPADS.2015.62
36. D. Gesvindr, B. Buhnova, Architectural tactics for the design of efficient PaaS cloud applications, in *2016 13th Working IEEE/IFIP Conference on Software Architecture (WICSA)*, Venice, 2016, pp. 158–167. https://doi.org/10.1109/WICSA.2016.42
37. Y. Govindaraju, H. Duran-Limon, A QoS and energy aware load balancing and resource allocation framework for IaaS cloud providers, in *2016 IEEE/ACM 9th International Conference on Utility and Cloud Computing (UCC)*, Shanghai, China, 2016, pp. 410–415
38. C.F. Lai, D.Y. Song, R.H. Hwang, Y.X. Lai, A QoS-aware, streaming service over fog computing infrastructures, in *Digital Media Industry Academic Forum (DMIAF)*. Santorini, pp. 94–98 (2016). https://doi.org/10.1109/DMIAF.2016.7574909

Chapter 13
Data Sources and Datasets for Cloud Intrusion Detection Modeling and Evaluation

Abdulaziz Aldribi, Issa Traore and Belaid Moa

Abstract Over the past few years cloud computing has skyrocketed in popularity within the IT industry. Shifting towards cloud computing is attracting not only industry but also government and academia. However, given their stringent privacy and security policies, this shift is still hindered by many security concerns related to the cloud computing features, namely shared resources, virtualization and multi-tenancy. These security concerns vary from privacy threats and lack of transparency to intrusions from within and outside the cloud infrastructure. Therefore, to overcome these concerns and establish a strong trust in cloud computing, there is a need to develop adequate security mechanisms for effectively handling the threats faced in the cloud. Intrusion Detection Systems (IDSs) represent an important part of such mechanisms. Developing cloud based IDS that can capture suspicious activity or threats, and prevent attacks and data leakage from both inside and outside the cloud environment is paramount. One of the most significant hurdles for developing such cloud IDS is the lack of publicly available datasets collected from a real cloud computing environment. In this chapter, we discuss specific requirements and characteristics of cloud IDS in the light of traditional IDS. We then introduce the first public dataset of its kind for cloud intrusion detection. The dataset consists of several terabytes of data, involving normal activities and multiple attack scenarios, collected over multiple periods of time in a real cloud environment. This is an important step for the industry and academia towards developing and evaluating realistic intrusion models for cloud computing.

A. Aldribi (✉) · I. Traore
The Department of Electrical and Computer Engineering, University of Victoria, Victoria, BC, Canada
e-mail: aaldribi@uvic.ca

B. Moa
Compute Canada-WestGrid, University Systems, University of Victoria, Victoria, BC, Canada

B. S. P. Mishra et al. (eds.), *Cloud Computing for Optimization: Foundations, Applications, and Challenges*, Studies in Big Data 39,
https://doi.org/10.1007/978-3-319-73676-1_13

13.1 Introduction

Since the 1990s to recent days the internet has changed the computing world in a dramatic way. It has evolved from the concept of parallel computing to distributed computing, to grid computing and recently to cloud computing. Although the idea of cloud computing has been around for quite some time, it is an emerging field of computer science. Cloud computing has become one of the popular terms in academia and IT industry. The two main characteristics of cloud computing are: on demand service, and flexible cost model that allows users to pay according to their needs. Cloud is just like water, electric, and gas utilities that can be charged according to the amount used. It integrates and offers several shared computational resources to users as a service on demand. Furthermore, the cloud model offers enticing features such as greater scalability, cost reduction, and flexibility. It can provide various levels of services and enable the completion of different tasks over the Internet. These features and services enabled the cloud to gain tremendous popularity in IT.

This growing popularity, however, comes with an increasing level of concern over the security of the cloud computing infrastructure. Shared resources, virtualization and multi-tenancy, which are the essential components of cloud computing, provide a number of advantages for increasing resource utilization, and on demand elasticity and scalability. However, they raise many security and privacy concerns that affect and delay the wide adoption of the cloud. Compared to the traditional IT systems, cloud computing introduces new and different security challenges that vary from privacy threats and lack of transparency to increasing likelihood of remote and inside attacks. Furthermore, these concerns are exacerbated by the fact that there is often some uncertainty or lack of transparency about the type of security measures implemented by the cloud service provider to protect the information, as well as the effectiveness of such measures. Moreover, since cloud computing relies on the Internet to operate, the security problems that exist in the Internet are not only found but are also amplified in the cloud. Likewise, all existing threats and vulnerabilities faced by traditional computer networks apply to cloud-based infrastructures.

To overcome these concerns and establish a strong trust in cloud computing, there is a dire need for effective security monitoring and defense systems adequate for the cloud. While traditional IDS can address some of the cloud security threats, they are severely limited not only by the specificity of some of the threats involved but also by the regulatory, legal, and architectural considerations unique to cloud computing. Therefore, several research proposals have been recently published on developing cloud-specific intrusion detection systems, also referred to as *cloud IDS*. Developing such systems, however, is very challenging as cloud IDS researchers are faced with one of the greatest hurdles: the lack of publicly available datasets collected from a real cloud computing environment. Most existing projects and works on cloud IDS use proprietary datasets, which they do not share with others. Sometimes the closeness of such datasets is justified by privacy concerns. However, this makes it difficult for other researchers to leverage or replicate the published results, hindering advances in cloud IDS development. Furthermore, some of the datasets used in cloud IDS development are synthetic datasets, which obviously miss important characteristics of real-world cloud environments.

In this chapter, we describe the ISOT Cloud Intrusion Dataset (ISOT-CID), which to our knowledge is the first publicly available dataset of its kind. The dataset was collected in a production cloud environment, and involves multiple attack scenarios, covering both remote and inside attacks, and attacks originating from within the cloud. It has a mixture of both normal and intrusive activities, and is labeled. It contains diverse logs and data sources that will enable the development of diverse intrusion data models, feature sets and analysis models. The dataset was collected over many days and involves several hours of attack data, culminating into a few terabytes. It is large and diverse enough to accommodate various effectiveness and scalability studies.

The remainder of this chapter is organized as follows. Section 13.2 presents the current cloud computing models and architectures. Section 13.3 discusses cloud security threats and monitoring for the cloud. Section 13.4 outlines the evaluation approaches used in the literature and the underlying datasets. A particular emphasis is placed on the publicly available datasets. Section 13.5 presents our dataset by describing the data collection environment and procedures. The involved attack scenarios are discussed in Sect. 13.6, and the main characteristics and components of the ISOT-CID will be presented in Sect. 13.7. Finally, the conclusion and future work are presented in Sect. 13.8.

13.2 Cloud Computing Models and Architectures

13.2.1 Cloud Computing Characteristics

The idea of cloud computing can be traced back to 1969. In fact, Leonard Kleinrock [8], one of the chief scientists of the original Advanced Research Projects Agency Network (ARPANET) project that seeded the Internet, envisioned the spread of 'computer utilities', similar to that of the electric and telephone utilities. Nowadays, cloud computing has become a trend in IT, and is able to move computing and data away from desktop, portable PCs and dedicated remote servers into a large number of distributed and shared computers. By doing so, the customers do not need to own the hardware or the infrastructure required to offer their services. They free themselves from dealing with the infrastructure, its installation and maintenance, and focus on developing their services on a shared infrastructure offered by another party in a pay-as-you-use manner. This paradigm also solves the limitations in storage, memory, computing power and software on desktops and portable PCs.

In cloud computing, data, applications or any other services are accessed by users through a browser or a terminal regardless of the device used and its location. A key characteristic of the cloud is its large scale. From the customer perspective, the cloud is of unlimited resources and should be ready to satisfy their computational needs no matter how big they are. Therefore, cloud providers have a very large number of servers to accommodate user requests. The other key characteristic of cloud computing is virtualization, which allows the cloud to achieve its agility and flexibility by "segmenting" the shared resources and running any operating system on them. This makes it easy for the cloud customers to offer their heterogenous services on the virtualized infrastructure without worrying about the specificity of the underlying hardware.

13.2.2 Cloud Computing Architecture

Looking at the literature, there is no standardized or universal cloud architecture reference model available. The cloud computing industry and academia are working in parallel to define and standardize reference models, frameworks and architectures for cloud computing [16]. Therefore, cloud reference models are in the development stage, and are expected to stay that way as long as the cloud computing models are evolving. According to [44], Forrester Research was able to count more than 75 different parties working on cloud standards in 2010. Today, the number of industry and academia involved has clearly increased significantly. From the literature, however, cloud computing architectures are better classified by service and deployment models. In the next section most common cloud computing service and deployment models are discussed.

13.2.3 Cloud Computing Service Models

Cloud service models try to classify "anything" providers offer as a service (XaaS), where X means any service (e.g., infrastructure, software, storage). Cloud service models are typically classified as: *Infrastructure as a Service* (IaaS), *Platform as a Service* (PaaS) and *Software as a Service* (SaaS). These three models (depicted in Fig. 13.1) are the basis of all services provided by cloud computing. In IaaS, the customers get access to the infrastructure to run and manage their own operating systems and applications. However, they do not control the underlying infrastructure, and can typically only launch virtual machines with pre-specified flavors and operating systems. PaaS provides a computing environment on the cloud infrastructure to the customers to create and control their own applications by using programming languages and runtime tools offered and supported by the PaaS provider. The customers do not, however, manage or control the operating systems or the underlying cloud infrastructure. In SaaS, vendor offers an instance of a software application as a service accessible via an Internet browser The service provider hosts the software on

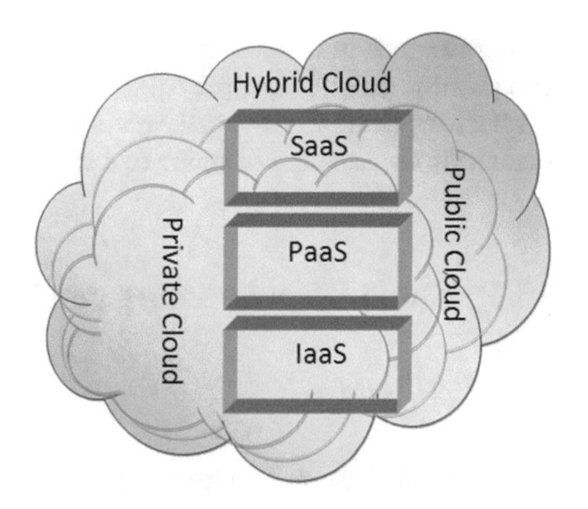

Fig. 13.1 Cloud Computing Service and Deployment models

the SaaS environment so that the customers do not have to install, maintain and run the software on their own resources. By doing so, the customers are only concerned with using it and, therefore, benefit from reduced software purchases by on demand pricing.

It is worth emphasizing that there are other proposed cloud service models including Storage as a Service, Hardware as a Service, Data as a Service, and Security as a Service.

13.2.4 Cloud Computing Deployment Models

Based on the NIST definition framework [29], there are four main cloud computing deployment models. These models are classified as public, private, community, and hybrid deployments as shown in Fig. 13.1.

In a *public cloud*, the pool of resources owned by the cloud provider are available to and shared among all the customers. They can rent parts of the resources and can typically scale their resource consumption according to their demands [34, 40]. Since multiple customers are sharing the resources, the major challenges facing the public cloud are security, regulatory compliance and Quality of Service (QoS) [5]. A *private cloud*, also known as *internal cloud*, is a dedicated cloud serving only the users within a single organization. It relies on internal data centers of an organization not made available to the general public, whereby all resources are reserved to the organization's private use. Although the private clouds are more expensive, they are more secure than the public clouds [3, 49, 52]. Moreover, the compliance and QoS are under the control of the organization [21, 53]. A *hybrid cloud* is a combination of public and private clouds, and tries to address their limitations while maintaining their features [37, 46]. The main purpose of a hybrid cloud is usually to provide extra resources in cases of high demands, for instance, to enable migrating some computation tasks from a private to a public cloud [51, 53]. A *community cloud* is similar to a private cloud with the particular characteristic of being shared by multiple organizations, usually belonging to a specific community [2, 34, 46].

13.3 Cloud Security Issues and Monitoring

13.3.1 Context

Security concerns are inherent in cloud computing and are delaying its adoption. In fact, when customers move their information to a cloud infrastructure owned by another party, there is a high risk of losing full control over their data as well as a potential leakage of their private information as the resources are not only owned by someone else but also shared with many unknown users. Moreover, when running a data processing service on the cloud (via IaaS or PaaS), the provider can obtain full control on the processes belonging to the service and, therefore, compromise its integrity [20].

Since the cloud computing system relies on the Internet to operate, the security problems that affect the Internet can be not only found but are also amplified in the cloud computing environments. Moreover, given that the cloud computing leverages the existing IT infrastructure, operating systems and applications, it is susceptible to the traditional IT attacks as well. In fact, almost all existing attacks could target cloud-based infrastructure. Moreover, cloud computing systems may face new security issues as they combine many new technologies that lead to new exploits such as, cross-virtual machines exploits, and inter-processes and cross-application vulnerability attacks [17].

There are many academic institutions, research organizations, cloud vendors and product development enterprises working on different cloud computing security issues and their solutions. From the literature, these issues are classified into various categories based on different perspectives. For example, D. Högberg [18] pointed out that the cloud computing security issues can be divided into two broad categories: security issues faced by cloud providers including Saas, Paas and Iaas providers, and security issues faced by their customers. ENISA, which developed a comprehensive detailed research in cloud computing security, classified these issues into three categories: Organizational, Technical and Legal. The Cloud Security Alliance (CSA) identified 15 areas of concerns and grouped them into three general areas: Compliance, Legal, Security and Privacy Issues [1]. V. Amandeep and K. Sakshi [43] classified the cloud computing security issues based on the delivery and deployment model of the cloud. Other researchers categorized security issues based on the different layers involved in a cloud infrastructure, which are application level, network level, virtualization level, data storage level, authentication and access control level, trust level, compliance level, and audit and regulations level [31]. Some of the security issues from the above classification will be highlighted next.

13.3.2 Cloud Security Issues Based on Different Layers

13.3.2.1 Virtualization Level Security Issues

Virtualization is a core technology in cloud computing and an essential element for a cloud to achieve its characteristics, especially scalability, location independence, and resource pooling. In a virtual environment, multiple Operating Systems (OSs) run at the same time on a single host computer using a hypervisor (also called Virtual Machine Monitor (VMM)) [38], and physical servers contain multiple virtualized servers on which several virtual machine instances may be running [26, 31].

A major function of the virtualization layer is to ensure that different Virtual Machine (VM) instances running on a same physical machine are logically but fully isolated from each other. However, the isolation technologies that the current VMMs offer are not perfect, especially with the full access and control the administrators have on host and guest operating systems. This leads to many security issues related to virtualization [50].

In cloud computing, there are mainly three types of virtualization: OS level virtualization, application-based virtualization, and hypervisor-based virtualization. In the OS level virtualization, a hosting OS can run multiple guest OSs and has visibility of and control over each guest OS. In such a configuration, if an attacker compromises the host OS, the entire guest OSs could be controlled by that attacker. Bluepill, SubVirt, and DKSM are some well-known attacks for such a configuration and it is still an open problem to prevent such threats [31]. In the application based virtualization, virtualization is only enabled at the top layer of the host OS, and each VM has its specific guest OS and related applications. This kind of virtualization also suffers from the same vulnerabilities as in the traditional OS case.

Hypervisor or virtual machine monitor (VMM) is viewed as any other code embedded on the host OS. Furthermore, this code is available at boot time of the host OS to control multiple guest OSs. As with any other code, it can harbour vulnerabilities that attackers can exploit.

Also, cross VM side-channel attacks and denial of service attack (DoS) attacks can be performed by the attacker.

However, as mentioned earlier, vulnerabilities in a hypervisor allow a malicious user to gain access and control of the legitimate user's virtual machine. Furthermore, a shared resource environment introduces unexpected side channel and covert channel attacks. An attacker if successful in neighboring a target can then use various methods for intercepting data being sent and received from the other resource [31].

The cloud can also be overloaded when trying to serve a huge number of requests, which can ultimately lead to DoS or distributed denial of service (DDoS). This type of attacks usually floods the cloud with many large number of requests via zombies. The

DoS attack against BitBucket.org is an excellent example. This attack suspended this web site over 19 hours during a DoS attack on the Amazon Cloud infrastructure [30].

According to [25], if a hacking tool is installed on a virtual machine, the other virtual machines and hypervisors can also be attacked. Hsin-Yi et al. [41] infer that VM hopping is a reasonable threat in cloud computing since it has been shown that an attacker can obtain or determine the IP address using standard customer capabilities. Moreover, because several VMs can run at the same time on the same host, all of them could become victim VMs. VM hopping is thus a crucial vulnerability for PaaS and IaaS infrastructures.

13.3.2.2 Data Storage Level Security Issues

The security of data on the cloud computing is another important research topic [45, 52]. One of the cloud computing principles is to move and store data in the cloud. Once the data is stored in the cloud, the full control of the data is transferred from the data owner to the hands of cloud computing providers. In addition, the data is moving from a single, private and dedicated tenant to a shared multi-tenant environment where untrusted users can harbor their VMs. Gaining access to private data in such a shared environment can be very catastrophic if the data storage is not done securely and properly. Therefore, according to [34], data stored in the cloud needs policies for physical, logical and personnel access control.

When the data resides on the cloud and is served from there only, there is a potential security risk that it might not be available, especially when needed the most.

The data security risks on the cloud are rising as the number of parties, devices and applications involved in the cloud infrastructure increases. In the literature, data stored in cloud, data leaks, data remanence, data availability, data location, data privacy and data breaches are examples for data security risks that are still open challenges.

For instance, confidentiality is one of the most difficult aspects of data to guarantee in cloud computing. Compared to the traditional computing, an enterprise has no control over storage and transmission of data once it is in the cloud. Therefore, to ensure that the data will not be able to accessed by unauthorized party, proper encryption techniques should be applied [21]. However, increased number of encryption and decryption operations reduces the application performance and increases the consumption of cloud computing resources [9].

In IaaS, variety of users can access computing resources from a single physical infrastructure. To achieve a proper confidentiality level, it is required to isolate resources among the multiple users and ensure that no user can view the state of the other users' resources [41]. Moreover, since PaaS relies on IaaS virtualization, protecting the state of the resources used by a user from the rest of users is also a security challenge in PaaS.

There are a lot of challenges in a public cloud computing environment to ensure data confidentiality. There are several reasons for that, for example, the needs for elasticity, performance, and fault tolerance lead to massive data duplication and require aggressive data caching, which in turn increase the number of targets that could be targeted by data theft [43].

Given all of the security issues and risks we outlined above, it is clear that cloud security monitoring is of paramount importance to both cloud service provider and cloud service customer. The next section is dedicated to this topic.

13.3.3 Cloud Security Monitoring

Since security is the most significant obstacles to the widespread adoption of cloud computing, there is a dire need for a proper monitoring system to gauge the security on the cloud. This need has attracted many researchers and cloud providers who are trying to develop security monitoring systems that can capture suspicious activity or threats, and prevent attacks and data leakage from both inside and outside of the cloud infrastructure.

Due to an increase in the amount of organized crime and insider threats, proactive security monitoring is crucial nowadays [13]. Moreover, to design an effective security monitoring system, a variety of aspects should be taken into account such as: previous knowledge of threats and their specifications in the new environment, handling a large number of incidents, cooperation among interested parties and their privacy concerns, product limitations, just to name a few.

In the literature, there is much research conducted to apply security monitoring systems to cloud computing. These systems are deployed in public and private cloud platforms and at different layers, and have different properties but each mainly focuses on a subset of security protection types and security risk mitigations.

Although researchers have approached the cloud security monitoring from different perspectives, the most common challenge faced in implementing real monitoring systems is the availability of dataset collected from real cloud computing environments. The importance of the availability of such dataset clearly appears when developing new cloud security monitoring systems or evaluating the existing ones. Our first attempt to address this challenge is to introduce the ISOT Cloud Intrusion Dataset (ISOT-CID), which to our knowledge is the first publicly available dataset of its kind.

13.4 State of the Art in Cloud IDS Evaluation

Cloud computing introduces a new era of integration of heterogeneous data in varying formats and from diverse sources and operating systems. This requires an improved understanding of the concept of a cloud dataset and its related concepts, such as the formats and the sources of data. By reviewing cloud intrusion datasets in the literature, there is a floating confusion about what can really be counted as a cloud dataset. This raises the need to define what a real cloud dataset is, since, not only there is no or limited publicly available cloud dataset available for researchers, but most of the datasets that are mentioned in the literature for cloud computing are not suitable and do not reflect the real cloud environment. These deficiencies are the main reasons why there is no effective real cloud intrusion detection datasets yet.

We define the cloud intrusion detection datasets as *data collected from real cloud computing environments and gathered from either one or different data sources. The dataset may comprise one or more data formats, and can contain strings of characters, numbers, binary or any combination of them.*

This section will briefly present the available works on cloud IDS datasets and discus the insufficiency and limitation arising when adopting them to develop, deploy or evaluate cloud IDSs.

13.4.1 Cloud Intrusion Detection Dataset (CIDD)

Kholidy and Baiardi [22] introduced a Cloud Intrusion Detection Dataset (CIDD), which is based on the Defense Advanced Research Project Agency (DARPA) dataset. The DARPA dataset was collected from a simulated network for an air force base connected to the Internet. This dataset contains simulated user activities and injects attacks at well defined points, and then gathers tcpdumps, process and filesystem information. CIDD was built by assigning users to different VMs according to their login session times and the applications that they ran. Using this method, CIDD was created by extracting and correlating user audits from DARPA dataset, and was presented to be a dataset for cloud computing.

13.4.2 Cloud Storage Datasets

Drago et al. [12] collected network traffic related to personal cloud storage service namely Dropbox. They gathered TCP flows with the commands executed by the Dropbox clients from two university campuses and two Points of Presence (POP) in two different European countries during 42 successive days. The analyzed dataset provides a better understanding for the use of cloud storage in different environments, which have different user habits and access to technology.

13.4.3 Cloud Intrusion Detection Datasets (CIDD)

Moorthy and Rajeswari [32] used the same name CIDD for their dataset as [22] but the datasets are different in content. They simulated the cloud environment by setting up a cloudstack environment using two machines: one running the management services and the other running a kernel-based virtual machine KVM hypervisor. The CIDD was collected based on the traffic, which flow to/from specific ports that were opened for communications in the simulated cloud. The malicious traffic was generated based on attack signatures that are prepared manually and scored using a common vulnerability scoring system. A systematic approach was used to generate CIDD based on profiling intrusions, protocols, or lower level network entities.

13.4.4 Denial of Service Attack Dataset

Another work on the generation of cloud intrusion dataset is the work done by Kumar et al. [24] as a part of developing their DoS attack IDS. They stimulated IaaS cloud environment using the open source cloud platform Eucalyptus. They launched 15 normal VMs, 1 target VM and 3 malicious VMs. The normal VMs are used to generate normal traffic while the malicious VMs are used to perform 7 different DoS attacks on the target VM. The attacks were run using 4 different scenarios, including 1 scenario depicting normal activities, and 3 different attack scenarios. The traffic between each VM was captured but was limited to 4 IP-header fields (i.e. source, destination, transferred bytes and protocol type) without any information about the payload. Although this work collected different internal traffics between VMs, the published data does not contain the raw data but, instead, it contains the data obtained after transformation and feature extraction [23, 42].

13.4.5 Limitation and Deficiencies of Current Datasets

The cloud intrusion datasets described above have limitations and deficiencies that hinder their adoption for a broad range of cloud intrusion detection investigations. The most significant limitation is that they were not collected from a real cloud environment, which is heterogeneous in nature. Therefore, there is a lack of certain data types that need to be collected from specific sources in the cloud environment. For example, some datasets are limited to network traffic traces that are only collected from the host side, and do not include any data collected at the hypervisor level, instance's OS level or internal network traffic. This affects the richness of the datasets and their size, hindering their use in training and testing a wide range of cloud IDS models. Moreover, current datasets do not reflect cloud IDS characteristics because they were either collected organically to evaluate the performance of a cloud service or were transformed and adjusted to be suitable for cloud IDSs. Furthermore, they do not include some attack types that are new and specific to the cloud computing environment. But most importantly, these datasets are not publicly available for researchers to access and use to build, evaluate and compare cloud IDSs.

To overcome these limitations and deficiencies, and help the cloud security community in its effort to build practical IDSs, we opted to collect and share a real cloud dataset called *ISOT-CID*. This dataset, which was collected from an OpenStack cloud production environment, is publicly accessible and available for cloud security researchers. ISOT-CID is described next.

13.5 ISOT-CID Collection Environment and Procedure

13.5.1 Overview

The analysis and assessment of cloud system based on data collected during production provide valuable information on actual cloud behaviour, and can be used to identify normal behaviours from intrusions, quantify security measures, and perform quantitative evaluations and verifications for the IDS analytical models. Therefore, it is not surprising that the availability of cloud dataset has gained increasing momentum within cloud IDS community, and can help in providing the empirical basis for researching and developing effective models for cloud IDS.

To this date, there is a lack of a cloud IDS dataset that is collected from a real production cloud environment, publicly available and rich enough to satisfy the needs of the cloud security community. The reason for this lack is that collecting a real cloud intrusion dataset is a burdensome task and there are many challenges and barriers associated with it. Privacy is one of the foremost of these challenges as cloud environments are built upon sharing resources among many users and any attempt to collect data on the cloud could reveal private information about the users, their instances and their data. For a cloud provider, protecting the users's privacy is important and revealing some of the data collected on the cloud to researchers is not negotiable. Moreover, most of the cloud systems are commercial by nature, and opening their closed infrastructures for external parties could sabotage their business by breaching privacy agreements they have with their customers. The damage can even be worse if the data collected revealed a silent breach in the provider's cloud environment.

In addition to privacy, the cloud has many different layers such as hypervisors, networks and guest hosts with different OS environments as well as many components, and collecting real data from these heterogeneous sources is difficult and can increase the complexity of the data correlation for the cloud IDS. The final challenge worth emphasizing is the data size. When collecting data on the cloud you have to be ready for enormous amount of data that will be gathered from different computing resources (more than 8 TB for ISOT-CID dataset).

We were able to tackle these challenges and difficulties, and collect a real cloud computing dataset to be the first of its kind and paving the way for cloud security communities for more research and finding.

The ISOT cloud intrusion detection dataset is an aggregation of different data collected from a production cloud computing environment and gathered from a variety of cloud layers, including guest hosts, hypervisors and networks, and comprising data with different data formats and from multiple data sources like memory, CPU, system and network traffic. It contains different attack traces like masquerade attacks, denial of service, stealth attacks, anomalous user behaviour and data attacks from outside and inside the cloud.

The purpose of ISOT-CID is the embodiment of a real cloud dataset for researchers so that they can build, evaluate and compare their works. It is designed to serve different and broad IDS models developments and evaluations. Moreover, ISOT-CID is essentially "raw" and has not been transformed, manipulated or altered. It is structured and prepared for the cloud security community, and is available for downloading through ISOT website along with the documentation and metadata files, which explain its production and use in order to ensure an unambiguous meaning of the data.

13.5.2 Target Network Architecture

13.5.2.1 OpenStack-Based Production Cloud Environment

The dataset was collected through Compute Canada, a non-profit organizations that serves the computational needs of the Canadian researchers (see https://www.computecanada.ca for more details). To address the needs of the non High Performance Computing researchers, Compute Canada recently bought two cloud environments: one in the east (east cloud) located at the University of Sherbrooke, and the other in the west (west cloud) hosted at the University of Victoria. With more than 16000 virtual cores and 1.5 PB of storage, the west cloud is the largest academic cloud in Canada dedicated to all Canadian researchers and their collaborators.

The dataset was collected on the west cloud, which is based on OpenStack. With more than 600 supporting companies, OpenStack is becoming the de facto platform for public and private cloud deployments. It allows the cloud providers to control large pools of resources, including compute, storage and network. OpenStack has a modular architecture composed of many components, each one is usually a full project on its own with a large development team behind it. Currently, OpenStack has the following main components:

- Nova: It is a service that runs on the compute nodes (i.e. hypervisors also called Nova-compute) and manages the instances on them, and on the controller nodes to provide the scheduling and the API functionalities (Nova-api, Nova-scheduler and Nova-conductor).
- Keystone: This is the identity management service for OpenStack, which controls the authentication and authorization of the OpenStack users.
- Neutron: This service manages the network services in Openstack. It allows the providers to virtualize all the network elements, including routers, subnets, ports, etc.
- Glance: A dedicated service for storing and managing images and snapshots.

- Horizon: This is the dashboard of OpenStack. It provides a web-based interface to OpenStack.
- Ceilometer: This service is the main monitoring service for OpenStack. It offers many statistics that allow both users and administrators to monitor their resources.
- Heat: This service allows the user to write recipes for deploying and orchestrating their virtual environments.
- Cinder: The service that manages the block storage on OpenStack. Cinder is designed to support different kinds of backend storage including Ceph. Using it, users can create and attach storage volumes to their instances.
- Swift: The cloud object service that allows users to store and retrieve their objects similar to Dropbox.

In addition to the dashboard web interface, OpenStack offers two other interfaces to manage the resources: a command line interface and a programming API. The communication between different OpenStack services is asynchronously orchestrated using RabbitMq.

The west cloud architecture is show in Fig. 13.2. The services Horizon, Keystone, Glance and RabbitMq are running on 3 controller nodes (Nefos1, Nefos2, and Nefos3) designed in a HA (high-availability) configuration. Two high-bandwidth nodes (Neut1 an Neut2) are dedicated to the Neutron service. All the rest of compute nodes are hypervisors that are running Nova Compute service. The back end storage used by Cinder is Ceph.

13.5.2.2 ISOT-CID Collection Environment

The ISOT-CID project aims at setting up a real production cloud environment to collect the necessary data from all three cloud layers (i.e.,VMs, Hypervisors and Network traffic). Our lab environment for performing this collection processes consists of two separated environments, namely ISOT-lab environment and ISOT-cloud environment.

The ISOT-lab environment contains the server that receives the forwarded dataset from different sources and stores it in a secure way. The ISOT-cloud environment runs on the west cloud. The ISOT-cloud environment has all the three sources involved in the dataset generation (i.e.,VMs, Hypervisors and Network traffic), and is composed of two hypervisor nodes and 10 instances launched in three different cloud zones named as A, B and C (Fig. 13.3 depict the ISOT-cloud environment).

The instances located in these zones are organized as follows: 5 instances launched in zone A, 4 instances launched in zone B, and 1 instance launched in zone C (instances specifications and hypervisors specifications are shown in Table 13.1 and Table 13.2).

All these instances are accessible from outside world through one public VLAN (namely VLAN3337), except one instance, *isotvm_in*, that is hidden under a private internal network and only accessible from specific instances via the internal router. We designed the cloud environment that way on the belief that generating a real

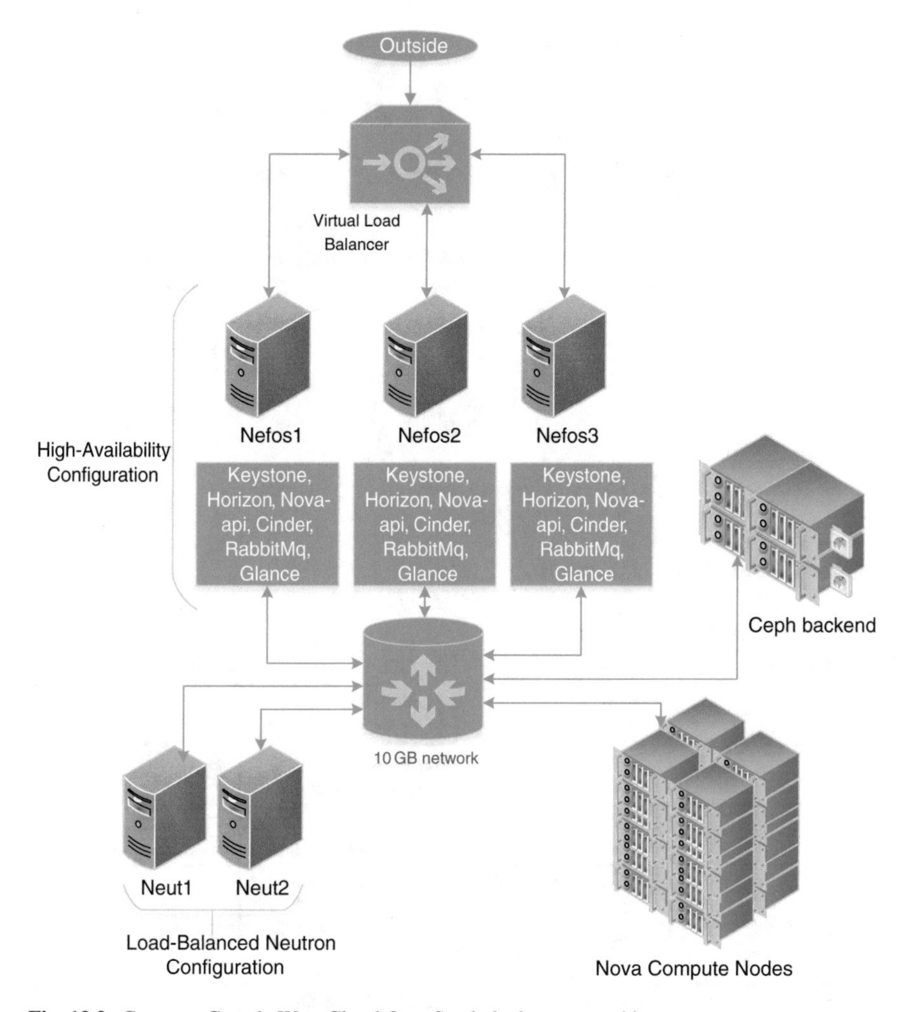

Fig. 13.2 Compute Canada West Cloud OpenStack deployment architecture

cloud computing dataset needs to consist of heterogeneous data sources (i.e. VMs, Hypervisors and Network traffic) to represent the real cloud environment. The ISOT-cloud network topology is depicted in Fig. 13.3.

13.5.3 Data Collection Procedure

The ISOT-CID project started as a response to our need to build and develop practical cloud IDSs. This need was then amplified by not finding any source for real cloud datasets, and ignited a formal data collection procedure structured as follows:

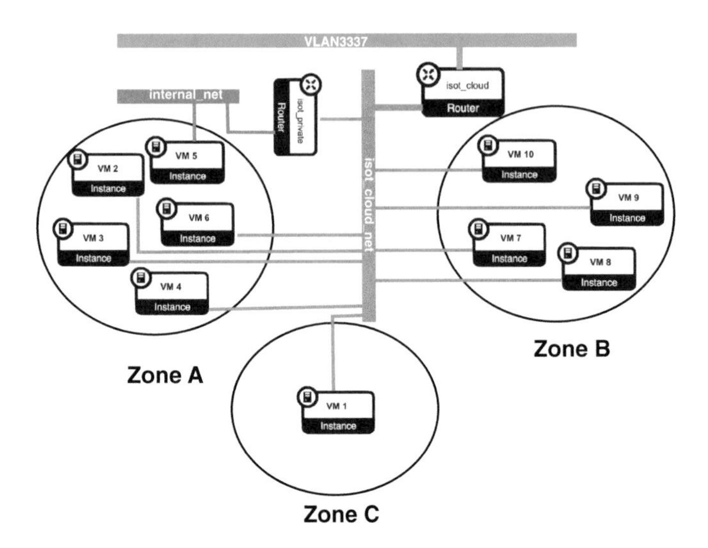

Fig. 13.3 ISOT-CID project environment

Table 13.1 Instances specifications

VM Number	VM's OS	Specification	Zone
VM 1	Centos	RAM: 6 GB Disk: 0 GB VCPUs: 4	C
VM 2	Centos	RAM: 7.5 GB Disk: 20 GB VCPUs: 2	A
VM 3	Debian	RAM: 7.5 GB Disk: 20 GB VCPUs: 2	A
VM 4	Windows Server12	RAM: 7.5 GB Disk: 20 GB VCPUs: 2	A
VM 5	Ubuntu	RAM: 7.5 GB Disk: 20 GB VCPUs: 2	A
VM 6	Centos	RAM: 7.5 GB Disk: 20 GB VCPUs: 2	A
VM 7	Ubuntu	RAM: 7.5 GB Disk: 20 GB VCPUs: 2	B
VM 8	Ubuntu	RAM: 7.5 GB Disk: 20 GB VCPUs: 2	B
VM 9	Windows Server12	RAM: 7.5 GB Disk: 20 GB VCPUs: 2	B
VM 10	Centos	RAM: 7.5 GB Disk: 20 GB VCPUs: 2	B

Table 13.2 Hypervisors specifications

ID	OS	Specification
A	Centos	RAM: 125 GB CPU: 32 cores Clock Speed: 2.00GHZ
B	Centos	RAM: 125 GB CPU: 32 cores Clock Speed: 2.00GHZ
C	Centos	RAM: 264 GB CPU: 55 cores Clock Speed: 2.40GHZ

1. Elicitation of the requirements related to the data collection targeted to cloud IDSs.
2. Official agreement between ISOT lab and Compute Canada to approve the ISOT-cloud environment on the Compute Canada West Cloud and obtain the real datasets thereafter. The agreement took few months to be finalized and approved.
3. Designing and setting up the appropriate ISOT-cloud environment for the data collection.
4. Setting up and configuring the ISOT-lab environment.
5. Designing and implementing the data collection agents.
6. Collecting the data for normal and attack scenarios.
7. Finalizing the data collection.
8. Publishing the data via a secure webserver on the ISOT-lab environment.

In addition to using syslog as our main logging mechanism, we designed and wrote agents for collecting specialized data syslog does or can not include. Since we intended to collect heavy data (e.g. memory and tcp dumps), we deployed three types of agents:

- Light weight agents that run very often (every ~5 s) and use minimal resources on the targeted machines.
- Medium weight agents that run less often (every ~60 s) but collect large amount of data.
- Heavy weight agents that are intensive, run even less often (every 120 s), and collect an even larger amount of data.

The specialized collector agents were implemented as shell and powershell scripts, and ran in the background at specific times during normal and attack periods. The time interval for the data collection in different agents can be adjusted to avoid missing the attack period. The data gathered at the agent level is either forwarded to the syslog daemon, or lz4 compressed and transferred directly to the logging server using *scp*. The type of data collected by the agents is described in Sect. 13.7. Figure 13.4 shows the general architecture of the logging platform.

After verifying and validating the collector agent scripts, we embedded them into and ran them simultaneously on the ISOT-cloud machines. The body of each script is

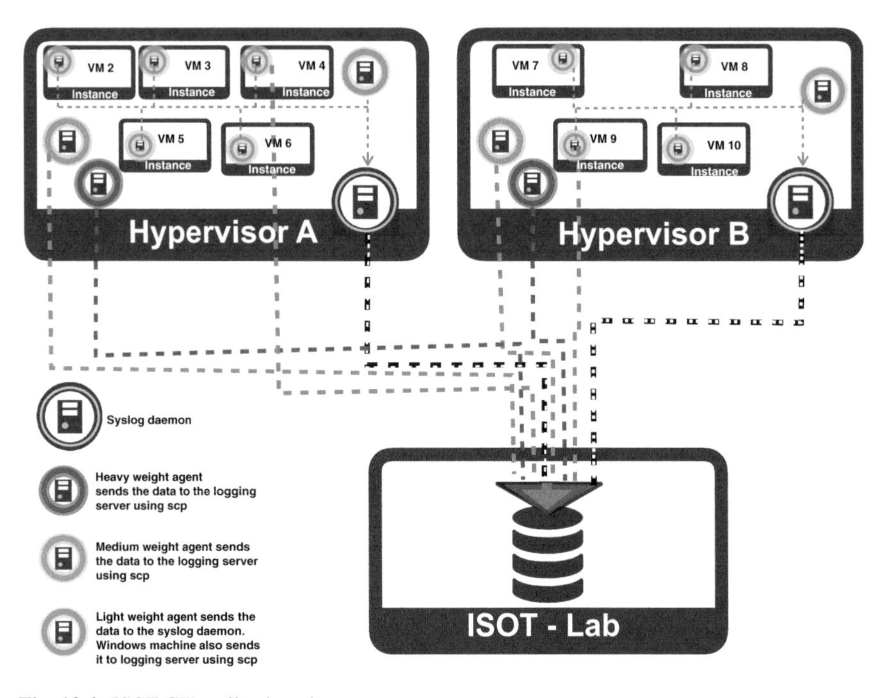

Fig. 13.4 ISOT-CID collection elements

different according to the targeted source, but in general they are very similar as they collect information from the same data sources except for the hypervisor agents that collect extra data and more detailed traces. The main pseudo code for the collector agents is shown in Algorithm.

Algorithm 1 Light weight agents

```
send anything printed to stdout and stderr to the syslog
WhileTrue get information about all the processes
get the CPU and disk utilization report
get virtual memory statistics
get information about network connections
get the list of open files
tcpdump the light weight network traffic
sleep 5 s
kill tcpdump
```

When the collector agents are invoked during collection processes, they extract and collect the required data from low level logs in all specified sources. The collected binary logs by the agents are forwarded through a secure channel to a dedicated log repository server in the ISOT-lab for data storage and future processing. Furthermore, the data collected from each source is sorted chronologically on the server. Figure 13.4 briefly describes the collection processes.

The collector agents can be classified and mapped into the three cloud layers (instances, hypervisors and networks) as follows: Instance-based agents, Hypervisor-based agents and Network-based agents into the three cloud stack layers. The goal of each agent is to collect specific type of real time data that is relevant to the environment in which it resides as described next.

- *Host-based agent:* The host-based agent is installed on each VM in ISOT-cloud environment. The agent's responsibility is to collect runtime information from the VM's data sources such as, operating system and network, and forward them to the log server in the ISOT-lab.

- *Hypervisor-based agent:* The hypervisor-based agent runs on each hypervisor in ISOT-cloud environment. Its main responsibility is to collect runtime information from the hypervisor data sources and forward them to the log server.

- *Network-based agent:* The network-based agent also runs on each hypervisor and collects runtime data from the hypervisor network traffic and forward them to the log server.

By spreading the agents between all cloud stack layers we were able to build a rich cloud dataset. We collected different data types by targeting a variety of data sources. These data sources are: system calls, cpu utilization, windows events log, memory dumps and tcp dumps.

ISOT-CID collection was performed over several months for the VM instances, and several days and time slots for the Hypervisors and the Network traffics (more details will be provided when describing the attack scenarios).

13.6 Attack Scenarios

As discussed previously in Sect. 13.5.2, the aim of our cloud environment setup is to obtain a rich cloud dataset by gathering different data types from variety of sources. Moreover, to amplify its richness and make it comprehensive, we performed multiple types of attacks that are ranked as the top threats in cloud computing. All the traces of these attacks were collected from all the previously specified data sources. This is essential as cloud IDS researchers/developers need a cloud dataset that covers different attack scenarios in advance so that they can effectively and productively go through the development and evaluation cycles.

In addition, there are two major categories of users in cloud computing: the insiders who have some sort of access to the cloud infrastructure, and the outsiders who are not. Both of this kind of users are included in our ISOT-CID attack scenarios. Likewise, the ISOT-CID attacks are divided into inside and outside attacks that are performed by an insider and outsider attackers, respectively. The inside attacks were originated from within the ISOT-cloud environment and targeted either the internal cloud infrastructure or the outside world. In our case, we opted for two scenarios:

1. The attacks are performed by a malicious insider who has high privileges (e.g. root access) on the hypervisor nodes in our ISOT-cloud environment.
2. The attacks are originated from an instance inside the ISOT-cloud environment. This instance could be either compromised or used on purpose as a bot to attack other instances in the ISOT-cloud environment or the outside world.

The outside attacks are defined as the attacks that target the ISOT- cloud environment from the outside world. In the outside attacks, in addition to the daily attacks that were performed by stranger attackers, we carried out different masquerades and DoS attack scenarios from outside world targeting ISOT-cloud instances. Attack types and scenarios are discussed next.

13.6.1 Inside Attack Scenarios

Most cloud vendors provide little to no information about their policies regarding recruitment procedures, the level of privileges assigned to their employees, and the legal actions that could be taken against the malicious insiders. This lack of transparency can encourage the insiders, especially employees, to carry out their malicious activities, and to even have escalated privileges that they are not supposed to have. With these privileges and the knowledge about the resources to attack, an employee can easily gain access to confidential data [31]. In most cases, this can even go undetected as the insider activities often bypass the provider firewall or IDS. Moreover, current data protection mechanisms, such as encryption, have failed to prevent the inside data theft attacks [39]. Therefore, it is not surprising that the number of data breaches by malicious insiders have been increasing [4] and malicious insiders are ranked as the 3rd top threat in cloud computing [19].

In the inside attack scenarios we performed, multiple attacks were run by malicious insider, a user having root access to the ISOT-cloud hypervisors. These attacks violate confidentiality, privacy, integrity and protection of cloud user's data. We discuss below the different categories of inside attacks enacted.

13.6.1.1 Revealing User's Credentials

Two different techniques were used to attack the ISOT-cloud environment and reveal user's credentials. The summary of the first attack scenario as follows:

Step 1 Scan the running instances and their associated disks.
Step 2 Extract the names of the instances from their *libvirt.xml* specifications.
Step 3 Focus on a single instance, and locate and mount its disk.
Step 4 Copy the password and shadow files, and the *.ssh* folder.

Step 5 Copy the private key to the attacker workstation.
Step 6 Finally, compromise the ISOT instance using the copied private key.

The second inside attack was performed by a root privileged user to get memory and strace dumps. These dumps were obtained using *gcore* and *strace*, respectively. As shown in [11], once the memory dumps are available, the inside attacker can extract user credential and private information, including passwords in clear text.

13.6.1.2 Extracting User's Confidential Data

Following the same steps as above, once the inside attacker has mounted the instance's disk he can search, copy, delete and change any user's data, similar to copying the *.ssh* folder above.

13.6.1.3 Back Door and Trojan Horse

In this scenario, the inside attacker shutdown the instance using virsh tools, mounted its disk using guestfish tools, and created Trojan horse for *ls* and *ps*. He also added an attractive script called *psb* that runs a malicious code when executed (this is a kind of back door to access the machine later on).

The other inside attack scenarios that we performed were carried out from compromised ISOT-cloud instances. After two instances were compromised, the inside attacker configured one of them to be a stepping stone, which potentially could be part of a botnet. Although this stepping stone was setup by an insider attacker, it could have been launched by the owner of the instance himself.

13.6.1.4 Compromised Instance and Introductory Botnet Activities

These attacks are carried out from an inside attacker who is either aiming at compromising an instance and using it for his malicious activities, or owning an instance that behaved as a normal instance for a while but it then turns out to be part of a botnet.[1] Once the attackers compromise an instance, they start attacking the neighboring instances. The summary of the attack scenario for this step is as follows:

Step 1 After a successful access to an instance VM8 from a previous outside attack (see Sect. 13.6.2), the attacker started to discover the internal network by searching local IP information and scanning the subnet. A few VMs with open ports were then discovered.

[1]In our particular scenario, the botnet command and control (C&C) server is hosted remotely, outside the cloud and only one bot is in the cloud.

Step 2 A file was transferred via *scp* from the attacker machine to the compromised instance VM8. The file was used as a password dictionary.
Step 3 The attacker installed *hydra* on VM8 and launched internal network scans using *nmap*. The machines discovered were then subjected to brute-force attacks using the password dictionary file. One of these attacks was successful against the instance VM5.
Step 4 A quick ssh login into the internal host VM5 from VM8 was logged.
Step 5 Finally, the attacker installed *hping3* on VM8 and ran several failed DoS attempts against VM4.

13.6.2 Outside Attack Scenarios

As we previously emphasized, the cloud is becoming an attractive target to outside data hunters. Having a centralized environment for virtual machines and data is a hacker's dream that is becoming true with the widespread adoption of cloud computing. Many attack attempts are made every single second to access instances and exploit any vulnerabilities within cloud. We can easily see that in our ISOT-CID; outside attackers are performing many different masquerade and DoS attacks attempts. Just within few months of operations, the Compute Canada's West cloud detected 5 compromised instances, most of which had weak passwords that the outsider attackers cracked using brute-force attacks. Moreover, at least one of these instances was compromised within one hour of its deployment.

In the outside attack scenarios, we performed multiple attacks against ISOT-cloud environment to breach the confidentiality, privacy, integrity, protection and availability of cloud user's data by performing masquerade and DoS attacks. Moreover, once the instance is compromised, it is leveraged to attack other instances from within the cloud environment, executing the inside attack scenarios that were outlined above.

13.6.2.1 Masquerade Attack

Masquerade attacks are very challenging in cloud computing as they can escape detection because any related trace can be buried in the massive amounts of data generated from user audits and activities across cloud environments. A masquerader is an insider or outsider attacker who steals the legitimate user credentials and uses them to authenticate as a legal user. This allows the attacker to fool the system and get access to its abundant resources illegally. In ISOT-CID attack scenarios, we performed different masquerade attacks launched from the outside. As an illustration, we outline the steps of one of these masquerade attacks below:

Step 1 Scanning subnet of the VMs using *nmap* scanner.
Step 2 Password dictionary attack against VM8 using *Ncrack* scan leading to credential discovery.

Step 3 SSH login to VM8 using the cracked credentials.
Step 4 Running *ifconfig* to display network IP address info/range, local IP address determined.
Step 5 *Nmap* scan (against internal network) from VM8 to uncover a few hosts with open ports.
Step 6 Transferring a file using *scp* from the attack host to VM8 (to be used as password dictionary).
Step 7 Installing *hydra* on VM8.
Step 8 Internal network scanning using *nmap* to discover 10 IP addresses (4 hosts up).
Step 9 Cracking the password using *hydra* against one of the internal machines VM5.
Step 10 SSH login to VM5 using the cracked password to check the ssh connection.
Step 11 Setting up *hping3* on VM8 after checking that it was not previously installed.
Step 12 Launching *nmap* scan against VM4 to discover a host with 2 open ports.

13.6.2.2 Denial of Service (DoS) Attack

The second attack scenario performed from the outside against ISOT-cloud environment is Denial of service (DoS) attack. In a DoS attack, the attacker overloads the targeted victim's system or service with huge number of requests consuming large amounts of system resources and leaving the system or the service unavailable. Not only that, in a cloud environment DoS attacks can render the whole cloud infrastructure unusable. In fact, two instances running on the same hypervisor are usually sharing the physical resources of the node. If one of them is subject to a DoS attack, the effect of this attack leaks into the second instance. In addition, consuming huge resources leads to an expensive bill that the victim will need to pay. As opposed to the cloud being a target for DoS attacks, many malicious users are leveraging the "unlimited" amount of resources in the cloud to launch massive outside DoS attacks. It is worth mentioning that, in cloud computing paradigm, DoS attacks were marked by the Cloud Security Alliance as the first of seven top threats [6]. The reader should also keep in mind that DoS attacks are the most common attacks on the Internet, and are hard to detect as distinguishing between DoS attack traffic and legitimate system unavailability is an open problem. In ISOT-CID, we designed and launched three types of DoS attacks. We performed the most common DoS attack, which targets the transport layer, and exploits the flaws in TCP three-way handshake procedures, called the TCP SYN attack. We used the open source *hping3* tool [36] to create a SYN flood attack. The first TCP SYN attack was performed on a compromised instance inside ISOT-cloud environment against another virtual machine in another ISOT-cloud zone. The attack flooded the instance with TCP SYN packets consuming its resources, hindering its use, and causing it to crash.

The second attack was performed from an outside machine against an instance inside the ISOT-cloud environment. The last attack was performed from a compromised instance inside ISOT-cloud environment against the logging server in the ISOT-lab. The 1st and 2nd DoS attack scenarios were the last attack steps from the 1st attack scenario, which we explained before. (The full and detailed attack scenarios are available for interested readers along with ISOT-CID documentations). To complete the 1st attack round, we add the following DoS attacks:

Step 1 From VM8, an unsuccessful *syn* flood DoS attack was launched against VM4.

Step 2 Another *syn* flood DoS attack against VM4 from the attacker machine caused VM4 to crash and we had to reboot it.

The 1st attack round for inside and outside attacks are depicted in Fig. 13.5.

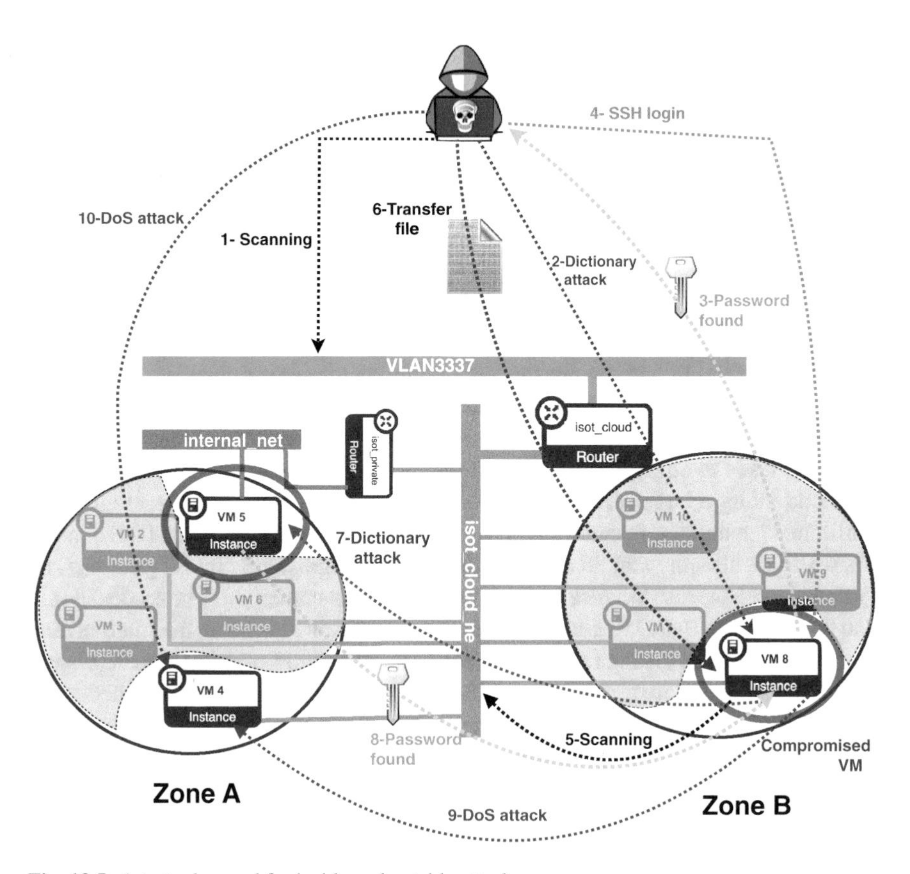

Fig. 13.5 1st attack round for inside and outside attacks

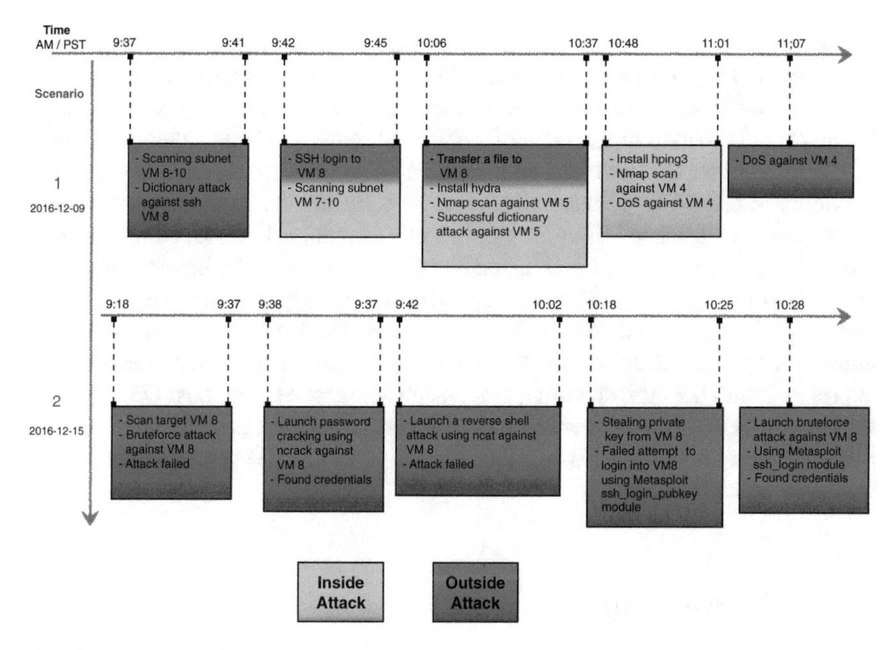

Fig. 13.6 Summary of the first and second attack scenarios for inside and outside attacks

13.6.3 Attack Time and Locations

In designing our ISOT-CID attack scenarios, we chose different attack times and distinct attacker's geographical locations. The attacks were divided into 4 rounds, averaging for 1 hour for each round, and spread over 4 different days. Moreover, the attackers performed and completed the attack scenarios from two different continents, precisely Europe and North America. The data collected during these times were labeled as a malicious traffic and the remaining as a normal traffic. Figure 13.6 summarizes the first and second attack scenarios for inside and outside attacks.

13.7 Collected Data

13.7.1 System Calls

A *system call* is a mechanism by which a user-space application requests a service from the kernel space of the OS [7, 15, 48]. Although every OS (like *nix and Windows) has its own specific system calls that are accessed via dedicated APIs, all of them follow the same bidirectional process in the sense that all requests from a user-level application to the OS are delivered by a system call interface to the kernel before they are executed on hardware. This allows the kernel to run the code on behalf of the user in the privileged space and get the information requested by the user via restricted system calls [7]. Furthermore, each one of these system calls has a unique number that is written in the kernel's system call table to identify them. Examples of system calls for OS (like *nix and WinX) are fork, pipe, read, CreateProcess and ReadFile. In ISOT-CID collection process, we collected the system calls from both hypervisors and instances. They provide useful information, like opened and accessed files, execution time stamps and the count of each system call number executed by the application. This information reflects every activity occurring inside the instances and hypervisors, and the communication between them. In fact, when a guest OS initiates a system call to access the host resources, the hypervisor intercepts and interprets it into the corresponding system call supported by the host OS. By doing so, the hypervisor has full control over the host OS resources like CPU, memory, persistent storage, I/O devices, and the network, and each VM access to any of these resources is accounted for.

13.7.2 Network Traffic

Network traffic is the main mechanism for exchanging information between computers. It is defined as a sequence of packets traveling between two communication endpoints known as source and destination. Each packet consists of a transport layer header, a network layer header, a data-link layer header and a payload representing the actual data exchanged [14]. These network packets can provide rich statistics that is essential for developing network security tools. In a cloud computing infrastructure such as OpenStack, the network traffic travels through different network points. In our case, the outside traffic first arrives via the entry and top-rack switches to dedicated network nodes, called neutron nodes, and then travels to the hypervisor on the compute nodes passing through our internal physical switches. Finally, the hypervisor delivers the traffic to the designated instance. With this OpenStack networking configuration, we have three different network flows: external, internal and local traffic. The external traffic is between the virtual machine and an outside machine (an “instance-outside” traffic). The internal traffic is between the compute node and another compute node (“hypervisor-hypervisor” traffic). The local traffic is between two instances on the same Compute node (“instance-instance” traffic). In ISOT-CID, we were able to collect all of these three different traffics from both hypervisors and instances. By setting the network interface into promiscuous mode, we were able to collect all the network traffic that passes through all the network interfaces on the hypervisors and instances (except for the special network interfaces not involved in our network traffic). The utility TCPdump was used to capture the network traffic [27] on*nix OSs and Netsh on winX OS, and stored in packet capture (pcap) format. We performed two kinds of network traffic data collections: the network traffic without payload on both hypervisors and instances, and the full network traffic on the hypervisors only (the interested readers can find more details in ISOT-CID documentations).

13.7.3 *Memory Dump*

Memory dump is a procedure by which the contents of a physical memory, representing the internal states of the system, are printed out or written to a storage drive. Memory dumps are usually generated when a crash occurred or during debugging [28]. They are very helpful to pin point the source of a crash, to debug system errors and performance issues, to obtain information about the running programs and the overall state of a system, to detect any intrusion or attack, to locate deleted or temporary information, and to carry out malware and forensic investigation.

Unfortunately memory dumps capture sensitive information, like passwords, encryption keys, and the contents of a private document, which can be very harmful if it ends up in the hand's of an attacker [11]. The ISOT-CID contains memory dump files that were collected using *gcore* command on the processes of the instances running of the ISOT-cloud hypervisors. The memory dumps collected resulted in large amount of data as each instance had around 7.5 GB of memory.

13.7.4 *Event Logs*

On any computer system, there is a specific auditing or event log service used to capture any predefined events, such as error reports, system alerts, and diagnostic messages that happen during the daily operation of the system. This event log service usually runs in background and listens to event sent by processes running on the local or remote system [33].

13.7.4.1 Windows Event Logs

In WinX OS, an event is defined as a notification posted to the user and raised by a hardware, software, or system component [10]. The event log service is responsible for logging and organizing these events in WinX OS. There are three types of event logs stored on Windows machine: Security, System and Application event logs. In general, security logs record events related to sensitive components of the system, such as the authentication component, as well as events related to resource usage, such as file operations and process invocations. The system log contains events logged by the system components, for example, driver failures and hardware issues. Finally, the application events can be triggered by any application as long as it implements the event API. The developer can generate events by calling the proper Windows event API. Windows event logs are not only useful in diagnosing problems, but also crucial in monitoring activities on a system. ISOT-CID includes all of these event log files types from the two windows instances that run in different zones and have been involved in the attack scenarios.

13.7.4.2 Unix Logs

Messages, notifications and events generated by *nix subsystems and services are usually logged as text under /var/log in specific files. Each process can virtually log its messages into a dedicated file. For example, sshd logs its authentication events into /var/log/auth.log and kernel messages are sent to /var/log/messages. This of course leads to unmanageable large amounts of log files, especially when dealing with many machines. To avoid that, we installed and configured rsyslog on the *nix machines to collect the logs in a central and secured location on the ISOT log server. Our scripts use the *logger* command to send the collected information to the rsyslog daemons, which in turn forwards them to the ISOT lab logging server.

13.7.5 Resource Utilization

One of the main functions of the operating system is the scheduling of the resources to serve all the running processes [35]. Scheduling CPU is the most important scheduling function of the OS as without it only a single program can use the CPU from start to finish and the rest of processes will starve [47]. When no task is requesting CPU, the processor becomes idle. By summing up the idle times during a given number of CPU ticks, one can deduce the overall CPU utilization. There are two types of CPU utilization, user CPU utilization and system CPU utilization. Both presenting the percentage of time spent performing user and system tasks, respectively. Analyzing the CPU utilization during different states, especially over long periods of time, allows for building a profile for the system. This profile can be enhanced by collecting other statistics, such as disk utilization. By monitoring the usage of resources one can build a baseline profile that characterizes the normal behavior of the system. Any deviation from such a baseline profile can be detected and subjected to further investigation. ISOT-CID provides different information about the usage of resources on all instances and hypervisors. This information was collected using commands like *ps* and *iostat*.

ISOT-CID also contains more information about memory and disk IO in its documentations. Figures 13.7 and 13.8 illustrate some excerpts from ISOT-CID.

```
2016-12-15T08:24:27-08:00 hpisot-dj logIt.sh: procs -----------memory---------- ---swap-- -----io---- -system-- ------cpu-----
2016-12-15T08:24:27-08:00 hpisot-dj logIt.sh:  r  b   swpd   free   inact   active   si   so    bi    bo   in   cs us sy id wa st
2016-12-15T08:24:27-08:00 hpisot-dj logIt.sh:  2  0     0   4750936 394696  2355556   0    0     1    14   29    4  0  1 98  1  0
2016-12-15T08:24:28-08:00 hpisot-dj logIt.sh:  0  0     0   4750944 394692  2355664   0    0     0     0   28   43  0  0 100  0  0
2016-12-15T08:24:28-08:00 hpisot-dj sshd[1713]: Failed password for root from 122.194.229.39 port 13956 ssh2
2016-12-15T08:24:28-08:00 hpisot-dj sshd[1713]: Disconnecting: Too many authentication failures for root from 122.194.229.39 port 13956 ssh2 [preauth]
2016-12-15T08:24:28-08:00 hpisot-dj sshd[1713]: PAM 5 more authentication failures; logname= uid=0 euid=0 tty=ssh ruser= rhost=122.194.229.39  user=root
2016-12-15T08:24:28-08:00 hpisot-dj sshd[1713]: PAM service(sshd) ignoring max retries; 6 > 3
2016-12-15T08:24:29-08:00 hpisot-dj sshd[1715]: Failed password for root from 59.63.166.83 port 21770 ssh2
2016-12-15T08:24:29-08:00 hpisot-dj logIt.sh:  0  0     0   4752240 394688  2354568   0    0     0     0  135   71  0  1 99  0  0
```

```
2016-12-15T08:24:27-08:00 hpisot-dj logIt.sh: MemTotal:        7679804 kB
2016-12-15T08:24:27-08:00 hpisot-dj logIt.sh: MemFree:         4750936 kB
2016-12-15T08:24:27-08:00 hpisot-dj logIt.sh: MemAvailable:    7217908 kB
2016-12-15T08:24:27-08:00 hpisot-dj logIt.sh: Buffers:           87776 kB
2016-12-15T08:24:27-08:00 hpisot-dj logIt.sh: Cached:          2641048 kB
2016-12-15T08:24:27-08:00 hpisot-dj logIt.sh: SwapCached:            0 kB
2016-12-15T08:24:27-08:00 hpisot-dj logIt.sh: Active:          2355556 kB
2016-12-15T08:24:27-08:00 hpisot-dj logIt.sh: Inactive:         394696 kB
2016-12-15T08:24:27-08:00 hpisot-dj logIt.sh: Active(anon):      74600 kB
2016-12-15T08:24:27-08:00 hpisot-dj logIt.sh: Inactive(anon):   110920 kB
```

- Virtual memory statistics

```
2016-12-15T08:24:25-08:00 hpisot-dj logIt.sh: disk- ------------reads------------ ------------writes----------- -----IO------
2016-12-15T08:24:25-08:00 hpisot-dj logIt.sh:        total merged sectors      ms  total merged sectors       ms    cur    sec
2016-12-15T08:24:25-08:00 hpisot-dj logIt.sh: vda    20226    0   2900002   411604 1427026 3998 12380304  158566696  0   7265
2016-12-15T08:24:25-08:00 hpisot-dj logIt.sh: vdb     9780    0    78234    289032 1970862 2622 19754664  116462564  0   2428
2016-12-15T08:24:25-08:00 hpisot-dj sshd[1713]: Failed password for root from 122.194.229.39 port 13956 ssh2
2016-12-15T08:24:26-08:00 hpisot-dj logIt.sh: vda    20226    0   2900002   411604 1427026 3998 12380304  158566696  0   7265
2016-12-15T08:24:26-08:00 hpisot-dj logIt.sh: vdb     9780    0    78234    289032 1970862 2622 19754664  116462564  0   2428
2016-12-15T08:24:26-08:00 hpisot-dj sshd[1715]: Failed password for root from 59.63.166.83 port 21770 ssh2
2016-12-15T08:24:27-08:00 hpisot-dj logIt.sh: vda    20226    0   2900002   411604 1427026 3998 12380304  158566696  0   7265
```

```
2016-12-15T08:24:23-08:00 hpisot-dj logIt.sh: Linux 3.16.0-4-amd64 (hpisot-dj) #01112/15/2016 #011_x86_64_#011(2 CPU)
2016-12-15T08:24:23-08:00 hpisot-dj logIt.sh:
2016-12-15T08:24:23-08:00 hpisot-dj logIt.sh: avg-cpu:  %user   %nice %system %iowait  %steal   %idle
2016-12-15T08:24:23-08:00 hpisot-dj logIt.sh:            0.12    0.00    1.34    0.69    0.06   97.79
2016-12-15T08:24:23-08:00 hpisot-dj logIt.sh:
2016-12-15T08:24:23-08:00 hpisot-dj logIt.sh: Device:            tps    kB_read/s    kB_wrtn/s    kB_read    kB_wrtn
2016-12-15T08:24:23-08:00 hpisot-dj logIt.sh: vda               2.47         2.48        10.57    1450001    6189980
2016-12-15T08:24:23-08:00 hpisot-dj logIt.sh: vdb               3.38         0.07        16.86      39117    9877332
```

- The CPU and disk utilization report

Fig. 13.7 Excerpt from the dataset: sample virtual memory statistics and the CPU and disk utilization for one of the hypervisors

```
2016-12-15T08:24:29-08:00 hpisot-dj logIt.sh: COMMAND     PID TID   USER   FD   TYPE            DEVICE SIZE/OFF  NODE NAME
2016-12-15T08:24:29-08:00 hpisot-dj logIt.sh: systemd       1       root  cwd   DIR             254,1     4096        2 /
2016-12-15T08:24:29-08:00 hpisot-dj logIt.sh: systemd       1       root  rtd   DIR             254,1     4096        2 /
2016-12-15T08:24:29-08:00 hpisot-dj logIt.sh: systemd       1       root  txt   REG             254,1  1309072   107008 /lib/systemd/systemd
2016-12-15T08:24:29-08:00 hpisot-dj logIt.sh: systemd       1       root  mem   REG             254,1    18640   107161 /lib/x86_64-linux-gnu/libattr.so.1.1.0
2016-12-15T08:24:29-08:00 hpisot-dj logIt.sh: systemd       1       root  mem   REG             254,1    14664   109709 /lib/x86_64-linux-gnu/libdl-2.19.so
2016-12-15T08:24:29-08:00 hpisot-dj logIt.sh: systemd       1       root  mem   REG             254,1   448440   106511 /lib/x86_64-linux-gnu/libpcre.so.3.13.1
2016-12-15T08:24:29-08:00 hpisot-dj logIt.sh: systemd       1       root  mem   REG             254,1    31784   109721 /lib/x86_64-linux-gnu/librt-2.19.so
2016-12-15T08:24:29-08:00 hpisot-dj logIt.sh: systemd       1       root  mem   REG             254,1    92888   106513 /lib/x86_64-linux-gnu/libkmod.so.2.2.8
2016-12-15T08:24:29-08:00 hpisot-dj logIt.sh: systemd       1       root  mem   REG             254,1    19016   106545 /lib/x86_64-linux-gnu/libcap.so.2.24
```

- The list of open files

```
2016-12-15T08:24:29-08:00 hpisot-dj logIt.sh: Active UNIX domain sockets (servers and established)
2016-12-15T08:24:29-08:00 hpisot-dj logIt.sh: Proto RefCnt Flags       Type       State         I-Node   PID/Program name    Path
2016-12-15T08:24:29-08:00 hpisot-dj logIt.sh: unix  2      [ ]         DGRAM                    8195     1/init              /run/systemd/notify
2016-12-15T08:24:29-08:00 hpisot-dj logIt.sh: unix  2      [ ACC ]     STREAM     LISTENING     8197     1/init              /run/systemd/private
2016-12-15T08:24:29-08:00 hpisot-dj logIt.sh: unix  2      [ ]         DGRAM                    8714     1/init              /run/systemd/journal/syslog
2016-12-15T08:24:29-08:00 hpisot-dj logIt.sh: unix  2      [ ]         DGRAM                    8210     1/init              /run/systemd/shutdownd
2016-12-15T08:24:29-08:00 hpisot-dj logIt.sh: unix  6      [ ]         DGRAM                    8215     1/init              /run/systemd/journal/dev-log
2016-12-15T08:24:29-08:00 hpisot-dj logIt.sh: unix  2      [ ACC ]     SEQPACKET  LISTENING     8220     1/init              /run/udev/control
2016-12-15T08:24:29-08:00 hpisot-dj logIt.sh: unix  2      [ ACC ]     STREAM     LISTENING     8222     1/init              /run/systemd/journal/stdout
2016-12-15T08:24:29-08:00 hpisot-dj logIt.sh: unix  4      [ ]         DGRAM                    8224     1/init              /run/systemd/journal/socket
```

- Information about network connections

```
2016-12-15T08:24:31-08:00 hpisot-dj logIt.sh: tcpdump: listening on any, link-type LINUX_SLL (Linux cooked), capture size 262144 bytes
2016-12-15T08:24:31-08:00 hpisot-dj logIt.sh: tcpdump: verbose output suppressed, use -v or -vv for full protocol decode
2016-12-15T08:24:31-08:00 hpisot-dj logIt.sh: listening on any, link-type LINUX_SLL (Linux cooked), capture size 262144 bytes
2016-12-15T08:24:31-08:00 hpisot-dj logIt.sh: 08:24:30.345112  In fa:16:3e:3e:7c:39 ethertype IPv4 (0x0800), length 68: 142.104.64.196.514 > 172.16.1.23.42332: Flags [.], ack 14178$
2016-12-15T08:24:31-08:00 hpisot-dj logIt.sh: 08:24:30.345518 Out fa:16:3e:b2:23:fb ethertype IPv4 (0x0800), length 227: 172.16.1.23.42332 > 142.104.64.196.514: Flags [P.], seq 1:1$
2016-12-15T08:24:31-08:00 hpisot-dj logIt.sh: 08:24:30.347131  In fa:16:3e:3e:7c:39 ethertype IPv4 (0x0800), length 68: 142.104.64.196.514 > 172.16.1.23.42332: Flags [.], ack 160, $
2016-12-15T08:24:31-08:00 hpisot-dj logIt.sh: 08:24:30.347352 Out fa:16:3e:b2:23:fb ethertype IPv4 (0x0800), length 224: 172.16.1.23.42332 > 142.104.64.196.514: Flags [P.], seq 160$
2016-12-15T08:24:31-08:00 hpisot-dj logIt.sh: 08:24:30.348828  In fa:16:3e:3e:7c:39 ethertype IPv4 (0x0800), length 68: 142.104.64.196.514 > 172.16.1.23.42332: Flags [.], ack 316, $
2016-12-15T08:24:31-08:00 hpisot-dj logIt.sh: 08:24:30.349260 Out fa:16:3e:b2:23:fb ethertype IPv4 (0x0800), length 227: 172.16.1.23.42332 > 142.104.64.196.514: Flags [P.], seq 316$
2016-12-15T08:24:31-08:00 hpisot-dj logIt.sh: 08:24:30.350689  In fa:16:3e:3e:7c:39 ethertype IPv4 (0x0800), length 68: 142.104.64.196.514 > 172.16.1.23.42332: Flags [.], ack 475, $
2016-12-15T08:24:31-08:00 hpisot-dj logIt.sh: 08:24:30.351129 Out fa:16:3e:b2:23:fb ethertype IPv4 (0x0800), length 228: 172.16.1.23.42332 > 142.104.64.196.514: Flags [P.], seq 475$
```

- Tcpdump the light weight network traffic

Fig. 13.8 Excerpt from the dataset: sample open files, network connections and TCPdump traces for an hypervisor

13.7.6 Excerpt From the Dataset

This section provides snapshots from the dataset that was collected from one hypervisor during one attack scenario. Fig. 13.7 shows collected data about virtual memory statistics and the CPU and disk utilization.

Figure 13.8 provides information about the list of open files, network connections and the collected TCPdump for the lightweight network traffic.

13.8 Conclusion and Future Work

IDSs are by nature inherently pattern recognition systems. The design and evaluation of such systems require appropriate datasets. It is a fact that cloud computing faces several of the same security threats as conventional computing networks and devices. But, cloud computing also involves several new threat vectors due to the specificity of its underlying architecture and operation. This calls for new datasets that would adequately capture the characteristics of the threat landscape faced by cloud computing. ISOT-CID is an initial response toward addressing such need. Although, undoubtedly, there are several limitations in the dataset, we believe that it will prove useful for academia and industry in developing effective cloud intrusion detection systems. Our future work will consist of strengthening and extending the dataset by covering more threat vectors and attack scenarios. The current dataset includes only introductory botnet traces. In the future, we will extend and diversify the cloud traces by involving more instances and different botnet architectures. We are also working, in parallel, in developing new intrusion detection models that will leverage the richness of the dataset for effective protection.

Acknowledgements We would like to express our gratitude and appreciation for Qassim University and the Ministry of Education of the Kingdom of Saudi Arabia for sponsoring this work. Also this gratitude goes to Compute Canada and the University of Victoria, especially Kim Lewall, Ryan Enge and Eric Kolb, for given us this opportunity.

References

1. C.S. Alliance, Sampling of issues we are addressing (2009), https://www.cloudsecurityalliance.org/issues.html#ediscovery. Accessed 3 Feb 2017
2. F.M. Aymerich, G. Fenu, S. Surcis, An approach to a cloud computing network, in *First International Conference on the Applications of Digital Information and Web Technologies, 2008. ICADIWT 2008* (IEEE, New York, 2008), pp. 113–118
3. L.D. Babu, P.V. Krishna, A.M. Zayan, V. Panda, An analysis of security related issues in cloud computing, in *International Conference on Contemporary Computing* (Springer, Berlin, 2011), pp. 180–190

4. W. Baker, M. Goudie, A. Hutton, C. Hylender, J. Niemantsverdriet, C. Novak, D. Ostertag, C. Porter, M. Rosen, B. Sartin, et al., Verizon 2010 Data Breach Investigations Report. Verizon Business (2010)
5. T. Berners-Lee, R. Fielding, L. Masinter, Std66 rfc3986: Uniform resource identifier (uri) generic syntax. ietf full standard rfc (2005)
6. G. Brunette, R. Mogull et al., *Security Guidance for Critical Areas of Focus in Cloud Computing v2. 1* (Cloud Security Alliance, 2009), pp. 1–76
7. I. Burguera, U. Zurutuza, S. Nadjm-Tehrani, Crowdroid: behavior-based malware detection system for android, in *Proceedings of the 1st ACM Workshop on Security and Privacy in Smartphones and Mobile Devices* (ACM, 2011), pp. 15–26
8. R. Buyya, C.S. Yeo, S. Venugopal, Market-oriented cloud computing: vision, hype, and reality for delivering it services as computing utilities, in *10th IEEE International Conference on High Performance Computing and Communications, 2008. HPCC'08* (IEEE, 2008), pp. 5–13
9. N.S. Chauhan, A. Saxena, Energy analysis of security for cloud application, in *2011 Annual IEEE India Conference (INDICON)*, (IEEE, 2011), pp. 1–6
10. M. Corporation, Windows event log service (2011), https://technet.microsoft.com/en-us/library/dd315601(v=ws.10).aspx. Accessed 3 Feb 2017
11. S. Davidoff, *Cleartext Passwords in Linux Memory* (Massachusetts Institute of Technology, Massachusetts, 2008), pp. 1–13
12. I. Drago, M. Mellia, M.M. Munafo, A. Sperotto, R. Sadre, A. Pras, Inside dropbox: understanding personal cloud storage services, in *Proceedings of the 2012 ACM Conference on Internet Measurement Conference* (ACM, 2012), pp. 481–494
13. C. Fry, M. Nystrom, *Security Monitoring: Proven Methods for Incident Detection on Enterprise Networks* (O'Reilly Media, Sebastopol, 2009)
14. F. Fuentes, D.C. Kar, Ethereal versus tcpdump: a comparative study on packet sniffing tools for educational purpose. J. Comput. Sci. Coll. **20**(4), 169–176 (2005)
15. J. Guo, S.L. Johnson, Intrusion detection using system call monitors on a bayesian network, U.S. Patent Application 11/677,059, 21 Feb 2007
16. M. Hamdaqa, L. Tahvildari, Cloud computing uncovered: a research landscape. Adv. Comput. **86**, 41–85 (2012)
17. S. Hanna, A security analysis of cloud computing. Cloud Comput. J. (2009)
18. D. Högberg, An Applied Evaluation and Assessment of Cloud Computing Platforms (2012)
19. D. Hubbard, M. Sutton et al., *Top Threats to Cloud Computing v1. 0* (Cloud Security Alliance, 2010)
20. M. Jensen, J. Schwenk, J.M. Bohli, N. Gruschka, L.L. Iacono, Security prospects through cloud computing by adopting multiple clouds, in *IEEE International Conference on Cloud Computing (CLOUD), 2011* (IEEE, 2011), pp. 565–572
21. P.J. Kaur, S. Kaushal, Security concerns in cloud computing, in *High Performance Architecture and Grid Computing* (Springer, 2011) pp. 103–112
22. H.A. Kholidy, F. Baiardi, Cidd: A cloud intrusion detection dataset for cloud computing and masquerade attacks, in *Ninth International Conference on Information Technology: New Generations (ITNG)* (IEEE, 2012), pp. 397–402
23. R. Kumar, S.P. Lal, A. Sharma (2016), https://github.com/uspscims/cloudsecurity. Accessed 3 Feb 2017
24. R. Kumar, S.P. Lal, A. Sharma, Detecting denial of service attacks in the cloud, in *Dependable, Autonomic and Secure Computing, 14th International Conference on Pervasive Intelligence and Computing, 2nd International Conference on Big Data Intelligence and Computing and Cyber Science and Technology Congress (DASC/PiCom/DataCom/CyberSciTech), 2016 IEEE 14th International Conference* (IEEE, 2016), pp. 309–316
25. H. Lee, J. Kim, Y. Lee, D. Won, Security issues and threats according to the attribute of cloud computing, in *Computer Applications for Security, Control and System Engineering* (Springer, 2012), pp. 101–108
26. H. Li, X. Tian, W. Wei, C. Sun, A deep understanding of cloud computing security, in *Network Computing and Information Security* (Springer, 2012), pp. 98–105

27. J. Liebeherr, M.E. Zarki, *Mastering Networks: An Internet Lab Manual* (Addison-Wesley Longman Publishing Co., Inc., Boston, 2003)
28. A. Mehta, Process and apparatus for collecting a data structure of a memory dump into a logical table, US Patent 5,999,933, 7 December 1999
29. P. Mell, T. Grance, *Effectively and Securely Using the Cloud Computing Paradigm*. NIST, Information Technology Laboratory pp, 2009), pp. 304–311
30. C. Metz, Ddos Attack Rains Down on Amazon Cloud. The Register (2009)
31. C. Modi, D. Patel, B. Borisaniya, A. Patel, M. Rajarajan, A survey on security issues and solutions at different layers of cloud computing. J. Supercomput. **63**(2), 561–592 (2013)
32. M. Moorthy, M. Rajeswari, Virtual host based intrusion detection system for cloud. Int. J. Eng. Technol. 0975–4024 (2013)
33. J.D. Murray, *Windows NT Event Logging* (O'Reilly & Associates, Inc., 1998)
34. C. Rong, S.T. Nguyen, M.G. Jaatun, Beyond lightning: a survey on security challenges in cloud computing. Comput. Electr. Eng. **39**(1), 47–54 (2013)
35. F. Sabrina, C.D. Nguyen, S. Jha, D. Platt, F. Safaei, Processing resource scheduling in programmable networks. Comput. Commun. **28**(6), 676–687 (2005)
36. S. Sanfilippo, Hping3 (8)-Linux Man Page (2005)
37. N. Sénica, C. Teixeira, J.S. Pinto, Cloud computing: a platform of services for services, in *International Conference on ENTERprise Information Systems* (Springer, Berlin, 2011), pp. 91–100
38. M.K. Srinivasan, K. Sarukesi, P. Rodrigues, M.S. Manoj, P. Revathy, State-of-the-art cloud computing security taxonomies: a classification of security challenges in the present cloud computing environment, in *Proceedings of the International Conference on Advances in Computing, Communications and Informatics* (ACM, New York, 2012), pp. 470–476
39. S.J. Stolfo, M.B. Salem, A.D. Keromytis, Fog computing: mitigating insider data theft attacks in the cloud, in *2012 IEEE Symposium on Security and Privacy Workshops (SPW)* (IEEE, New York, 2012), pp. 125–128
40. H. Takabi, J.B. Joshi, G.J. Ahn, Security and privacy challenges in cloud computing environments. IEEE Secur. Priv. **8**(6), 24–31 (2010)
41. H.Y. Tsai, M. Siebenhaar, A. Miede, Y. Huang, R. Steinmetz, Threat as a service?: virtualization's impact on cloud security. IT Prof. **14**(1), 32–37 (2012)
42. Uspscims: (2016), https://github.com/uspscims/cloudsecurity
43. A. Verma, S. Kaushal, Cloud computing security issues and challenges: a survey, in *International Conference on Advances in Computing and Communications* (Springer, Berlin, 2011), pp. 445–454
44. C. Wang, S. Balaouras, L.C.: Q&a: Demystifying cloud security. an empowered report: getting past cloud security fear mongering. Technical report, Forrester Research, Inc (2010)
45. L. Wang, J. Shen, H. Jin, Research on cloud computing security. Telecommun. Sci. **26**(6), 67–70 (2010)
46. M. Waschke, *Cloud Standards: Agreements that Hold Together Clouds* (Apress, New York, 2012)
47. R.K. Yadav, A.K. Mishra, N. Prakash, H. Sharma, An improved round robin scheduling algorithm for CPU scheduling. Int. J. Comput. Sci. Eng. **2**(04), 1064–1066 (2010)
48. R.V. Yampolskiy, V. Govindaraju, Behavioural biometrics: a survey and classification. Int. J. Biom. **1**(1), 81–113 (2008)
49. Z. Yandong, Z. Yongsheng, Cloud computing and cloud security challenges, in *2012 International Symposium on Information Technology in Medicine and Education (ITME)*, vol. 2 (IEEE, New York, 2012), pp. 1084–1088
50. P. You, Y. Peng, W. Liu, S. Xue, Security issues and solutions in cloud computing, in *2012 32nd International Conference on Distributed Computing Systems Workshops (ICDCSW)* (IEEE, New York, 2012), pp. 573–577
51. S. Zhang, S. Zhang, X. Chen, X. Huo, Cloud computing research and development trend, in *Second International Conference on Future Networks, 2010. ICFN'10* (IEEE, New York, 2010), pp. 93–97

52. Q. Zhang, L. Cheng, R. Boutaba, Cloud computing: state-of-the-art and research challenges. J. Int. Serv. Appl. **1**(1), 7–18 (2010)
53. W. Zhao, An initial review of cloud computing services research development, in *2010 International Conference on Multimedia Information Networking and Security (MINES)* (IEEE, New York, 2010), pp. 324–328

Chapter 14
Fog Assisted Cloud Computing in Era of Big Data and Internet-of-Things: Systems, Architectures, and Applications

Rabindra K. Barik, Harishchandra Dubey, Chinmaya Misra, Debanjan Borthakur, Nicholas Constant, Sapana Ashok Sasane, Rakesh K. Lenka, Bhabani Shankar Prasad Mishra, Himansu Das and Kunal Mankodiya

R. K. Barik (✉) · C. Misra · B. S. P. Mishra · H. Das
KIIT University, Bhubaneswar, Odisha, India
e-mail: rabindra.mnnit@gmail.com

C. Misra
e-mail: cmisra@yahoo.com

B. S. P. Mishra
e-mail: mishra.bsp@gmail.com

H. Das
e-mail: das.himansu2007@gmail.com

H. Dubey
Center for Robust Speech Systems, University of Texas at Dallas, Dallas, Texas TX-75080, USA
e-mail: harishchandra.dubey@utdallas.edu

D. Borthakur · N. Constant · K. Mankodiya
Wearable Biosensing Lab, University of Rhode Island, Kingston, Rhode Island RI-02881, USA
e-mail: dborthakur@my.uri.edu

N. Constant
e-mail: kabuki4774@gmail.com

K. Mankodiya
e-mail: kunalm@uri.edu

S. A. Sasane
Savitribai Phule Pune University, Pune, India
e-mail: sapanasasane@yahoo.com

R. K. Lenka
IIIT Bhubaneswar, Bhubaneswar, India
e-mail: rakeshkumar@iiit-bh.ac.in

B. S. P. Mishra et al. (eds.), *Cloud Computing for Optimization: Foundations, Applications, and Challenges*, Studies in Big Data 39,
https://doi.org/10.1007/978-3-319-73676-1_14

Abstract This book chapter discusses the concept of edge-assisted cloud computing and its relation to the emerging domain of "Fog-of-things (FoT)". Such systems employ low-power embedded computers to provide local computation close to clients or cloud. The discussed architectures cover applications in medical, healthcare, wellness and fitness monitoring, geo-information processing, mineral resource management, etc. Cloud computing can get assistance by transferring some of the processing and decision making to the edge either close to client layer or cloud backend. Fog of Things refers to an amalgamation of multiple fog nodes that could communicate with each other with the Internet of Things. The clouds act as the final destination for heavy-weight processing, long-term storage and analysis. We propose application-specific architectures GeoFog and Fog2Fog that are flexible and user-orientated. The fog devices act as intermediate intelligent nodes in such systems where these could decide if further processing is required or not. The preliminary data analysis, signal filtering, data cleaning, feature extraction could be implemented on edge computer leading to a reduction of computational load in the cloud. In several practical cases, such as tele healthcare of patients with Parkinson's disease, edge computing may decide not to proceed for data transmission to cloud (Barik et al., in 5th IEEE Global Conference on Signal and Information Processing 2017, IEEE, 2017) [4]. Towards the end of this research paper, we cover the idea of translating machine learning such as clustering, decoding deep neural network models etc. on fog devices that could lead to scalable inferences. Fog2Fog communication is discussed with respect to analytical models for power savings. The book chapter concludes by interesting case studies on real world situations and practical data. Future pointers to research directions, challenges and strategies to manage these are discussed as well. We summarize case studies employing proposed architectures in various application areas. The use of edge devices for processing offloads the cloud leading to an enhanced efficiency and performance.

Keywords Big data · Edge computing · Fog computing · Cyber-physical systems · Cloud computing · Smart health · Geoinformation systems · Information systems

14.1 Introduction

Cloud GIS framework facilitated sharing and exchange of geospatial data belonging to various stakeholders. It created an environment that enabled wide variety of users to retrieve, access and disseminate geospatial data along with associated meta data in a secure manner [31]. It has the functionality for decision making, storage for various data types, fetching data and maps as per user's demand, querying, superimposing and analyzing the data and generating final reports (maps) to administrators and planners[63]. CloudGIS framework has leveraged for environmental monitoring, natural resource management, healthcare, land use and urban planning, watershed, marine and coastal management [32]. There are numerous emerging applications of

CloudGIS framework. CloudGIS has the ability to integrate and analyze heterogeneous thematic layers along with their attribute information to create and visualize alternative planning scenarios. User friendly CloudGIS has made Geospatial Information Systems (GISs) a preferred platform for planning in global, regional, national and local level using various analyzes and modeling functionalities. CloudGIS integrates common geospatial database operations such as query formation, statistical computations and overlay analysis with unique visualization functionalities [32]. These features distinguish CloudGIS from other spatial decision support systems. It is a widely used tool in public and private sector for explaining events, predicting outcomes and designing strategies [70].

Geospatial data contains informative temporal and geospatial distributions [48]. In traditional setup of CloudGIS framework, we send the data to cloud server where these are processed and analyzed. This scheme takes large processing time and require high Internet bandwidth. Fog computing overcomes this problem by providing local computation near the clients or cloud. Fog computing enhances the CloudGIS by reducing latency at increased throughput. Fog devices such as Intel Edition and Raspberry Pi etc. provides low-power gateway that can increase throughput and reduces latency near the edge of geospatial clients. In addition, it reduces the cloud storage for geospatial big data. Also, the required transmission power needed to send the data to cloud is reduced as now we send the analysis results to cloud rather than data. This leads to improvement in overall efficiency. Fog devices can act as a gateway between clients such as mobile phones [7, 12].

The increasing use of wearables in smart telehealth system led to generation of huge medical big data. Cloud and fog services leverage these data for assisting clinical procedures. IoT Healthcare has been benefited from this large pool of generated data. It suggests that the use of low-resource machine learning on Fog devices which kept close to wearables for smart telehealth. For traditional telecare systems, the signal processing and machine learning modules are deployed in cloud that process physiological data. This book chapter also presents a Fog architecture that relied on unsupervised machine learning big data analysis for discovering patterns in physiological data.

So both geospatial and medical data was processed at the edge using of fog devices and finally has stored at the cloud layer. The present book chapter has made the following contributions:

- It gives the detail concepts and architectural framework of about the edge, cloud and fog computing;
- It presents the big data concept in the field of geospatial and medical applications;
- It discusses about management of services, optimization, security and privacy in fog computing as well as edge computing environment;
- There are two case studies i.e. malaria disease management and telemonitoring of patients with Parkinson's diseases have been elaborated with the use of different fog assisted cloud architecture;

14.2 Edge Assisted Cloud Computing

14.2.1 Cloud and Edge

Data are gradually produced and processed at the edge of the network. Same works have been done before in micro data center and cloudlet. Cloud backend does not efficiently handle the processing of big data that is generated at the edge. Edge computing allows for more edge devices or sensors to interface with the cloud in larger scale [34, 62]. Since the cloud computing environments are not efficient for edge data due to throughput and latency need. The data has to be transferred to cloud and processing being done there. The analysis is later to be communicated to the edges. In light of these facts, we need another layer of computational nodes referred as Fog computing to tackle local processing and inference at reduced latency and increases throughput [55].

In the edge computing concept, the things are both data producers and/or consumers. Both play important roles. At the network edge, the nodes perform computational tasks as well as request service from the cloud. Edge computers store data temporarily, perform computations and in this way translate some of the analysis from cloud to edge node. In addition, it reduced latency, transmission power and increase throughput. The design of edge nodes and their arrangement is decided by the application requirements such as remote telecare of patients with Parkinson's disease [23, 28, 52]. The edge computing paradigm manages the computations near sensor nodes such as wearables, smart watches [21, 64]. Edge computing offers several advantages over cloud processing. It assists the cloud backend, in a way that it is complementary analysis happens at edges and cloud such as preliminary diagnosis of diseases at edge and long-term decision making in cloud backend.

For example, a smart mobile is the edge between the cloud layer and body things, a gateway in a smart home is the edge between the cloud and home things, a micro data center and a cloudlet is the edge between the cloud and a mobile device. From these point of view, the edge computing is the exchangeable with fog computing. But edge is focused more towards the things side where fog computing is mainly focusing on the infrastructure side [3, 57, 62].

14.2.2 Cloud Computing

Cloud facilitates on-demand resources and facilities over the web. It possesses adequate storage and computational infrastructure for data analysis and visualization. Cloud computing provided a transition from desktop to cloud servers. Various web processing architectures have created in an open environment with shared assets [70]. It was facilitated by four distinct types of service model, i.e., infrastructure as a service (IaaS), platform as a service (PaaS), software as a service (SaaS) and database as a service (DaaS). These four types of service models are depicted in

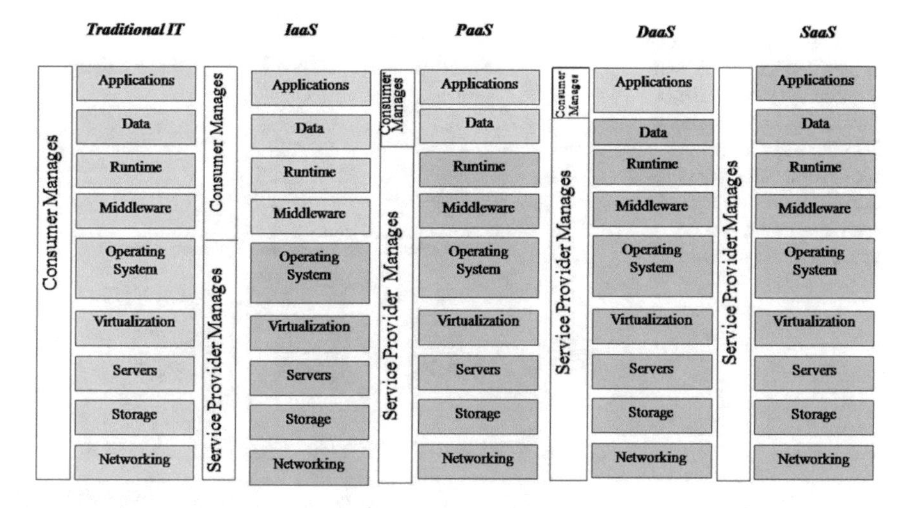

Fig. 14.1 The characteristics of cloud computing service models

Fig. 14.1. IaaS model includes virtual machines (VM) and storage area for storing of data. IaaS balances the computational load in cloud computing. Users could access the software and/or install those through VM. Users have access to hard system vai IaaS model. This way the end users have online access to hardware platform as per their needs [14]. PaaS ensures availability of programming models via web. It has possibility of execution of these program for addressing the user's need. It ensures that web apps could be used without installation or download of any software [9, 44]. SaaS service model allows software deployment in a way that users could run applications without installing those on their system. However, this model is limited to a set of services [1, 47]. A more flexible model, DaaS provide its users access to a database without software installation or hardware setup or configuration. All of the administrative and maintenance tasks are handled by the service providers. Customers can have control over the database depending on the service providers [15, 42]. All the above discussed models were used for various application areas. Particularly SaaS and DaaS model are dedicated for geospatial applications. Analysis and storage of geospatial data requires a robust framework. Such system works on top of PaaS service model in CloudGIS framework. It is a robust platform using that organizations interrelate with technologies, tools and expertise to nurture deeds for producing, handling and using geographical statistics and data.

It deploys a unique instance, multi tenant design that permits more than one client to contribute assets without disrupting each other. This integrated hosted service method helps installing patches and application advancements for users' transparency. Another characteristic is embrace of geospatial web services and as an established architectural methodology in engineering [46]. Many cloud platforms uncover the applications statistics and functionalities via geospatial web services [30]. These permit clients to query or update different types of cloud services and applications.

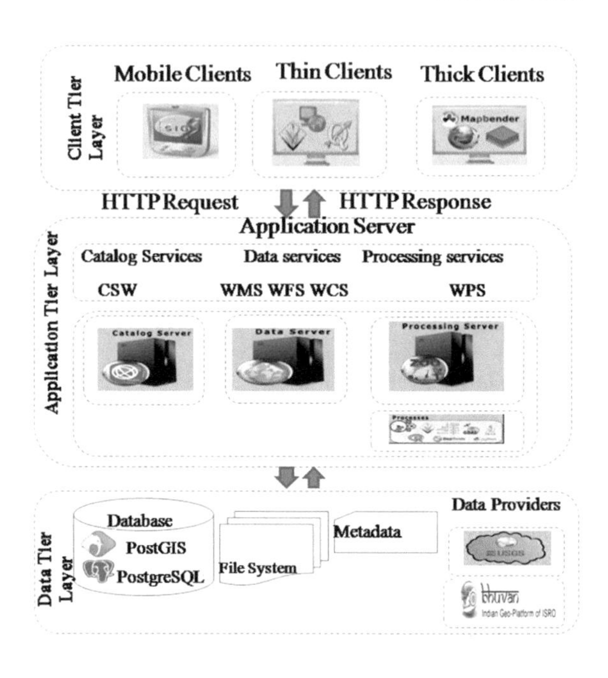

Fig. 14.2 Systems view of CloudGIS framework with three categories of clients (thick, thin and mobile). It has client-tier layer, application-tier layer divided into different web services and data-tier layer with geospatial and meta-data storage [29]

It is a tool that assimilates software cloud with enterprise SOA infrastructure [58]. Figure 14.2 shows the system architecture for CloudGIS framework [29].

The client tier layer consists of three types of clients i.e. thin, thick and mobile clients. These three types of clients have the main function to visualization functionality for geospatial information. Mobile clients are users operating through mobile devices. The users those are working on web browsers are defined to be thin clients. In thin clients, users do not require any additional software for the operation. Thick clients are the users processing or visualizing the geospatial data in standalone system where it requires additional software for full phase operation [38]. The application tier comprises the main geospatial services executed by servers. It is an intermediate amongst the different clients and providers. Dedicated server for each application is operated for services such as Web Map Service (WMS), Web Coverage Service (WCS), Web Feature Service (WFS), Web Catalog Service (CSW) and Web Processing Service (WPS) /emphetc. [58]. Dedicated application server is responsible for requests to and from clients. In addition, services include three types of applications namely catalog, data servers and processing servers. Catalog servers are employed to search the meta information regarding the stored spatial data. It is an important components for controlling geospatial information in cloud. The catalog service has implementation of a standard publish-find-bind model. This model is defined by OGC web service architecture. Data server deals with the WMS, WCS and WFS [17]. Processing server offers geospatial processes that allow different clients to smear in WPS standard geospatial data [66]. Detail explanation of every process done by client request, forward the desire processing service with input of several factors, specifies and provides definite region in leaping box and feedbacks with composite standards. Data tier layer has spatial data long with related information. System utilizes the layer to store, recover, manipulate and update the geospatial data for further analysis. Data can be stored in different open source DBMS packages, simple file system or international organizations (e.g., Bhuvan, USGS, Open Street Map, Google) [67]. The system architecture of CloudGIS shows that geospatial data are one of the key components in data layer for the handling geospatial analysis [61]. The widespread use of cloud for geospatial data storage and analysis lead to generation of large amount of data from different areas [69]. This gives rise to the geospatial big data discussed in next section.

14.2.3 Fog Computing

Cloud computing paradigm has the limitation as most of the cloud data centers are geographically centralized. These data centers are not located near the proximity of users or devices. Consequently, latency-sensitive and real time computation service requests by the distant cloud data centers has often suffer large network congestion, round trip delay and degraded service quality. Edge computing is an emerging technology for resolving these issues [53]. Edge computing provide the computation nodes near the edge, i.e., close to the clients. It enables the desired data processing at the edge network that consists of edge devices, end devices and edge server [37]. Devices are the smart objects, mobile phones whereas routers, bridges, wireless access points are the employed as edge servers. These components function co-operatively for supporting the capability of edge computation [62]. This paradigm ensures fast response to computational demands and after computation, the analysis report could be transferred to cloud backend. Edge computing is not associated with any kind of cloud services model and communicate more often with end devices. Various combination of edge and cloud computing lead to emergence of several computational paradigms such as mobile cloud computing (MCC), mobile edge computing (MEC) [57], Fog computing. These are considered the budding extension of edge and cloud computing as shown in Fig. 14.3 [56]. Cisco proposed Fog computing in 2012 [20] as a paradigm that decentralizes the computations towards users so that quality of service be improved [59]. Fog computing do not require computational resources from cloud data centers [60]. In this way, data storage and computation are brought closer to the users leading to reduced latencies as compared to communication overheads with remote cloud servers [71]. It refers to a computing paradigm that uses interface kept close to the devices that acquire data. It introduces the facility of local processing leading to reduction in data size, lower latency, high throughput and thus power efficient systems. Fog computing has successfully applied in smart cities [35] and healthcare [28].

Fog devices are embedded computers such as Intel Edison and Raspberry Pi that acts a gateway between cloud and mobile clients [5]. From the above discussions, we can see that it requires an efficient, reliable and scalable Fog computing based GIS framework for sharing and analysis of geospatial and medical big data across

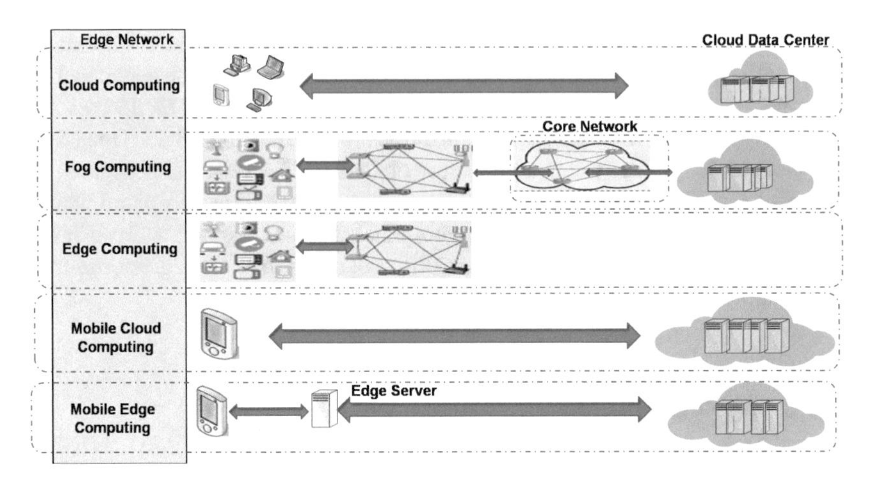

Fig. 14.3 Different schemes of mobile cloud computing, edge computing, Fog and cloud computing [53]

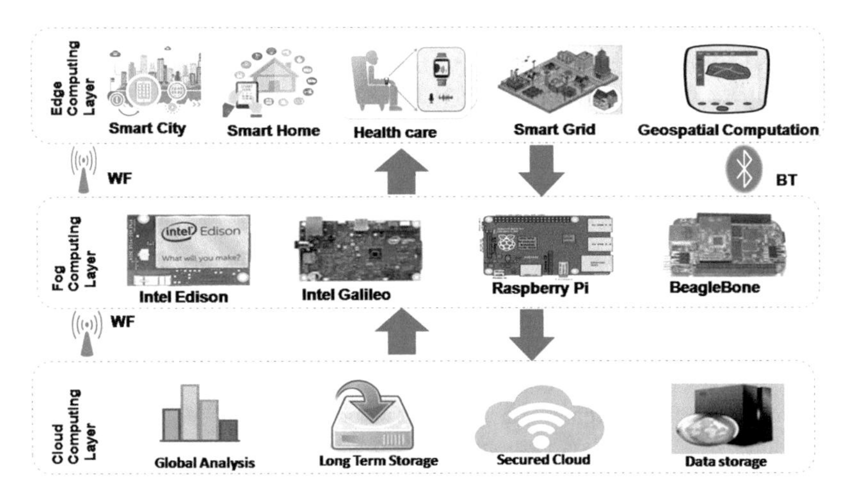

Fig. 14.4 Fog computing is as an intermediate layer between edge computing layer and cloud layer. The Fog computer layer has enhanced the efficiency by providing computing near the edge computing devices. This framework is very much useful for geospatial application, healthcare, smart city, smart grid and smart home etc. [23]

the web. Fog computing is a novel idea that helps to increase throughput and reduce latency at the edge of the client with respect to cloud computing environment (See Fig. 14.4).

14.2.4 Geospatial and Medical Big Data

As Big data from geospatial and medical domains is increasing in terms of variety, size it demands new architectures for efficient analysis. It includes large data sets whose sizes is beyond the ability of commonly used software tools. It requires special tools to acquire, curate, manage, and process big data in realistic time frames [2, 26]. Big data is generated in multiple forms. Most of the big data is semi-structured, quasi-structured or unstructured, that makes the processing challenging. Big data if processed in an efficient manner helps in finding useful analytics related to disease prevention, crime or businesses etc. The data size is increasing at an exponential rate as a result of increased use of wearable and mobile devices, radio-frequency identification (RFID) readers, and remote sensing. Geospatial data has always big data with the combination of remote sensing, GIS and GPS data [48]. In these days, big data analytics for geospatial systems is receiving attention. Geospatial big data usually refer to datasets beyond the processing capability of traditional infrastructures [51].

Geospatial data are categorized into raster, vector and graph data. Raster include geospatial images obtained from satellites, security cameras and aerial vehicles. The raster data are utilized by government agencies for analysis, prediction and decision making. Feature extraction, change detection and pattern mining are some examples of analyzing raster data. Vector consist of points, lines and polygons features. For example, in Google maps, the various temples, bus stops and churches are marked with points data whereas lines and polygons correspond to the road networks. Spatial correction, pattern analysis and hot spot detection is usually done using vector data. Graph appears in form of road networks. Here, an edge represents a road segment and a node denotes an intersection or a landmark.

Due to diversity of health related ailments and variety of treatments and outcomes in health sector, there are numerous health care data have generated. That gives rise to the concept of medical big data. Electronic health records, clinical registries, biometric data, patient-reported data, medical imaging and different administrative claim record are the main sources for medical big data. Medical big data have several typical features that are different from big data from other disciplines. These data are often hard to access and investigators in the medical area are aware of risks related to data privacy and access [49].

Big data poses some challenges for data analysis with cloud computing. Reliability, manageability and low cost are the key factors that make cloud computing useful for data processing. However, the security and privacy are the main concerns for processing sensitive data. Particularly in health geo-informatics applications with sensitive data, we require secure processing [36]. For minimizing privacy and security threats, the data should be utilized as per the user's context for limited amount of data access within the model. After processing the data should be transferred to the next level for final processing of data analysis. That will benefit the data security and privacy.

14.3 Performance Criterion

14.3.1 Management of Services

In consideration of management service at edge and fog network, it has argued that there are different kinds of features such as isolation, extensibility, differentiation and reliability should assure. Every services in edge and fog network, it has to be execute with proper priority of services. For example, in health care system, heart failure detection has the higher order priority as compared with other services [37, 64].

14.3.2 Optimization

Edge computing leads to improved latency, power efficient, at high throughput and reasonable quality of service. These attributes make edge computing an attractive alternative for efficient analysis of big data. Initially, businesses would require investment to augment cloud with edge computing and optimize it for a given application. The data collection, processing and transfer could be optimized for a given application. While some general purpose architectures can be built, the optimization for each application have to be done separately. It is important to note that in era of edge and fog computing, we assist the cloud computing by employing edge computers. We do not intend to replace cloud by edge or Fog computers as cloud is still required for long-term contextual storage and analysis [21, 57]. Latency is an important aspects of cloud based systems. For some applications such as remote monitoring of patients, it is more important than other. For improved latency, the data analysis should be performed To decrease the latency, the workload should better to finish in the nearest layer which has enough calculation ability to the things at the edge, cloud and fog network. From the point of view with respect to latency, high bandwidth reduces the transmission time, particularly for large data set. It has to be correctly decide the workload allocation in each and every layer, it needs to consider the bandwidth usage information and computation capability in the layers to keep away from delay and competition [37, 62].

14.4 Case Study I: Malaria Disease Management

14.4.1 Health GIS

For medical solutions, sharing patient's private data, preparing it further analysis have been challenging partly due to heterogeneous nature of data and also their geographical distributions. It lead to confluence of geo-information system and healthcare solutions into health GIS paradigm. Occurrence and spread of diseases such as Malaria, Chikungunya is correlated to geographical boundaries and related to time related factors. So, during few months of an year and for certain regions of a state or country these are more important than others. In past, Cloud computing has facilitated a means of data collection, process, monitoring and managements of disease information through web interface [65]. However, interoperability, integrating various heterogeneous data remains challenges in cloud-based health GIS systems. The problem can be tackled in different ways. With the emergence of fog computing, it is possible to have dedicate fog nodes are handling heterogeneity, scalability and interoperability issues. The cloud would still be the final destination of metadata and/or processing results [33].

From the different literature reviews, it is summarized that, it requires an efficient, reliable and scalable fog computing based framework for enhanced analysis of geospatial big data. We proposed the FogGIS framework and associated methods such as compression techniques, overlay analysis, energy saving strategies, scalability issues [50], various time analysis and comparative study. We considered the geospatial data of malaria vector borne disease positive maps of Maharastra, India from 2011 to 2014 for performing a case study.

14.4.2 Malaria Vector Borne Disease Positive Maps of Maharastra, India

Maharashtra is a state in the western region of India. It is also the second most populous state and third largest by area. It is bordered by the Arabian sea to the west, Karnataka to the south Gujarat and the Union territory of Dadra and Nagar Haveli to the northwest, Telangana to the southeast, Madhya Pradesh to the north, Chhattisgarh to the east and Goa to the southwest. It spans over an area of 307,731 km2 or 9.84% of the total geographical area of India. In total, it has 41000 villages and 378 urban cities. It has one of the highest levels of urbanization among all Indian states.

The secondary health data positive cases of malaria and number of death due to malaria are collected from the National Vector Borne Disease Control Program (NVBDCP), New Delhi. Climatic data includes all the surface parameters like temperature, rainfall, humidity, wind speed etc. are collected from the National Data Centre (NDC) and India Meteorological Department (IMD), Pune, India.

The inputs of positive cases and the deaths (number of persons) were fed in Quantum GIS software and region wise maps with district boundaries were generated. It includes incidence of malaria with the interval of 2011–2014 to see the trend and pattern of the incidence of Malaria in Maharashtra and finally find out the trend with the help of linear regression equation: $y = a + bx$ where b value shows the rate of change per decade thus the trend of malaria from 2011 2014 generated in the form of Map. The creation of geospatial database are significant and tedious assignment where efficacy in system development and implementation. Integrated geo-health database creation include stages such as inputs of data on geo-health data and related non-geo-health attributes data, its authentication by connecting with same set of data. Geospatial database delivers a platform in that organizations interrelate with technologies to nurture actions for spending, handling and generating geo-health data. The development of geo-health database supports in various administrative and political levels through these decision-making functions. Quantum GIS 2.14.3 is the OS GIS software selected to examine the competences w.r.t. creation of geospatial database. The procedure model of geo-health database creation is recurring or frequent in nature and each operation improves the study and strategy steps through assessment and testing of a complete component. In complete component, Quantum GIS has set up an malaria geo-health database by the help of political map of India. QGIS is also used for integrated geo-health database creation. After geo-health database designed, the next step is to propose the model for better analysis of geo-health data.

14.4.3 Proposed Architecture

This section describes various components of the proposed *GeoFog4Health* framework and discusses the methods implemented in it. It has discussed the hardware, software and methods used for compression of geospatial big data. It has been employed Intel Edison and Raspberry Pi as fog computing device in proposed *GeoFog4Health* framework [5]. Intel Edison is powered by a rechargeable lithium battery and contains dual-core, dual-threaded 500MHz Intel Atom CPU along with a 100 MHz Intel Quark micro controller. It possesses 1 GB memory with 4 GB flash storage and supports IEEE 802.11 a,b,g,n standards. It connects to WIFI and has been used UbiLinux operating system for running compression utilities. Raspberry Pi B Platforms have been used. Raspberry Pi has consisted of a 900 MHz 32-bit quad-core ARM Cortex-A7 CPU with 1 GB RAM. For WiFI connectivity in Raspberry Pi, it has been used WIFI dongle of Realtek RTL8188CUS chip set. In the proposed framework, it has been used both Intel Edison and Raspberry Pi in every fog node for better efficiency in time analysis that will be discuss in later section.

Figure 14.5 has shown the proposed *GeoFog4Health* framework, it has categorized into four layers as cloud layer, fog layer, intermediate fog Layer and client-tier layer. The cloud layer is mainly focused on overall storage and analysis of geo-health data. In cloud layer, it has been implemented to add GeoSpark [72] for real time geospatial big data processing on the top of the Hadoop Ecosystem. The fog layer works as middle tier between client-tier layer and intermediate fog Layer. It has been experimentally validated that the fog layer is characterized by low power consumption, reduced storage requirement and overlay analysis capabilities. In fog layer, all fog nodes have been developed with Intel Edison and Raspberry Pi processor for processing of geo-health data. The additional intermediate fog layer has been added between fog layer and cloud layer for load overhead in fog layer. Thus, intermediate fog layer has been used for refinement of processing and storing of geo-health data in cloud storage area. In client-tier, the categories of users have been further divided

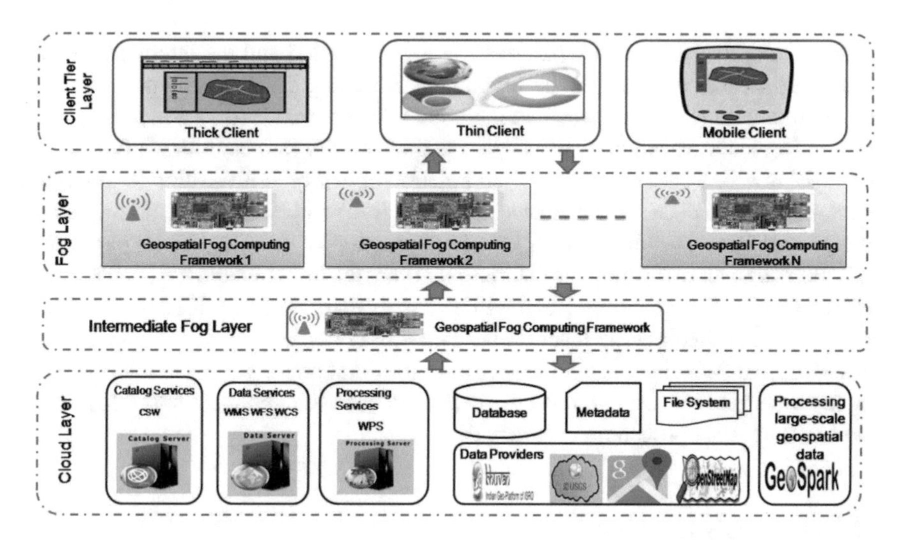

Fig. 14.5 Conceptual diagram of the proposed *GeoFog4Health* architecture with four layers (client-tier layer, fog layer, intermediate fog layer and cloud layer [6])

into thick clients, thin clients and mobile clients environment respectively. Processing and analysis of geo-health data can be possible within these three environments. In the proposed framework, the processing of every fog node, it is found that lots of energy has been realized and it should be properly managed. It also experiments the different overlay analysis and lossless compression techniques in *GeoFog4Health* framework.

14.4.4 Energy Efficiency

In this section, the analytical model has introduced for the energy saving management of intermediate fog layer in *GeoFog4Health*. The proposed framework has investigated the energy saving management using finite buffer batch service buffering system with change over time and multiple vacations. It has been studied that the overall message delay in the uplink channel and performance of mean number of fog node data packets in the buffer, buffering delay and probability of blocking in the fog layer. Lots of energy has been required for handling heavy traffic of fog node data from fog and intermediate fog layer. With vacation mode operation, intermediate fog layer node does not listen to the node of fog layer continuously but it alternates the active state and the vacation state. It has considered a finite buffer batch service buffering system with multiple vacation and changeover time.

Let, it has been assumed a and b as the threshold values of activating the intermediate fog layer service and service capacity, respectively. Whenever the intermediate fog layer node finished all its work, it goes to vacation, an internal timer that is exponentially distributed with parameter θ is then started and the intermediate fog layer node awakes to check the buffer content of the fog layer. When upon awaking the intermediate fog layer finds that there are still less than $j(0 \le j \le a-2)$ data frames, it goes to vacation again. If the number of data frames in the buffer of the fog layer is $a-1$ either at a service completion epoch or at a vacation termination point, the intermediate fog layer service will wait for some more time that is called changeover time. The change over time is exponentially distributed with parameter γ. If there is an arrival during the change over time, the intermediate fog layer service will start immediately, otherwise, it will go for a vacation period. If after a vacation period, the intermediate fog layer finds a non-empty buffer, it serves all data frames present at that point and also all new data frames that arrive while the intermediate fog layer service is working, until the buffer becomes empty again at the fog layer end and the whole procedure is repeated.

14.4.4.1 Analytical Model

It has been considered a Markov chain with the state space $\{(i, j)|0 \le i \le N, j = 0, 1 \bigcup (a-1, 2)\}$ where i gives the buffer size and j represents the state of the server. The process is in the state $(i, 0)$ if there are i data frames waiting in the buffer and the server is in sleep mode. It is in state $(i, 1)$ if there are i data frames waiting in the base station buffer and the mobile station service unit is busy and it is in state $(a-1, 2)$ if there are $a-1$ data frames in the buffer and the server is waiting in the system. Using probabilistic argument at steady state, we obtain the following system of equations

$$\beta P_{0,0} = \alpha P_{0,1}, \tag{14.1}$$

$$\beta P_{i,0} = \beta P_{i-1,0} + \alpha P_{i,1}, \;\; 1 \le i \le a-2, \tag{14.2}$$

$$(\beta+\theta) P_{a-1,0} = \beta P_{a-2,0} + \gamma P_{a-1,2}, \tag{14.3}$$

$$(\beta+\theta) P_{i,0} = \beta P_{i-1,0}, \;\; a \le i \le N-1, \tag{14.4}$$

$$\theta P_{N,0} = \beta P_{N-1,0}, \tag{14.5}$$

$$(\beta+\alpha) P_{0,1} = \beta P_{a-1,2} + \alpha \sum_{s=a}^{b} P_{s,1} + \theta \sum_{s=a}^{b} P_{s,0}, \tag{14.6}$$

$$(\beta+\alpha) P_{i,1} = \beta P_{i-1,1} + \theta P_{i+b,0} + \alpha P_{i+b,1}, \;\; 1 \le i \le N-b, \tag{14.7}$$

Using normalization condition $\sum_{i=0}^{N} P_{i,0} + \sum_{i=0}^{N} P_{i,1} + P_{a-1,2} = 1$ we recursively solved the equations.

14.4.4.2 Performance Measures

The state probabilities of the incoming job request at arrival times are known, we can find out various performance measuring parameters like average number of job requests in the buffer L_q, average time spending in the buffer W_q and the probability of blocking (PBL). They are given by $L_q = \sum_{i=1}^{N} i P_{i,0} + \sum_{i=1}^{N} i P_{i,1} + (a-1)P_{a-1,2}$. The probability of blocking is given by $PBL = P_{N,0} + P_{N,1}$. The average time spending in the buffer using Little's rule is $W_q = L_q/\beta'$, where $\beta' = \beta(1 - PBL)$ is the effective arrival rate.

14.4.5 *Results and Discussions*

14.4.5.1 Data Compression and Overlay Analysis

In the present research, it has been used a well known popular compression algorithms for reduction of the malaria geo-health data size. The concept of compression in geo-health data have been used in several areas like in network and mobile GIS framework for minimization of their data size [16, 43, 73]. Present study, it has been translated the various compression techniques from mobile platform to the proposed framework [55]. After completion of compressed resultant geospatial data at fog layer, it later transmits to the cloud layer. The cloud layer has stored the compressed data or decompressed the data before processing, analyzing and visualizing the resultant one. Presently, we are using lossless compression techniques such as .rar, .gzip, .zip, Various lossless compression techniques has been applied at fog Layer and the result has shown in Table 14.1. In this section, data analysis particularly overlay analysis is performed for malaria vector borne disease positive maps of Maharastra, India. Overlay analysis is one of the important data analysis in that, it has superimpose various geospatial data in a common platform for better analysis of raster and vector geospatial data. It has been found that 2 number of shape files related to malaria information mapping are overlaying with google satellite layer. In the present study, it has been used the malaria death mapping data of Maharastra from 2011–2014; has been processing in *GeoFog4Health*. The overlay analysis of various vector data and raster data of particular area has been performed. Initially, the developed datasets have

Table 14.1 Result of compression in proposed framework using malaria geo-health data

Geo-health data	Original data size (MB)	.rar compressed size (MB)	.gzip compressed size (MB)	.zip compressed size (MB)
India boundary	2.96	2.5	2.2	1.2
Death mapping	0.98	0.55	0.41	0.32

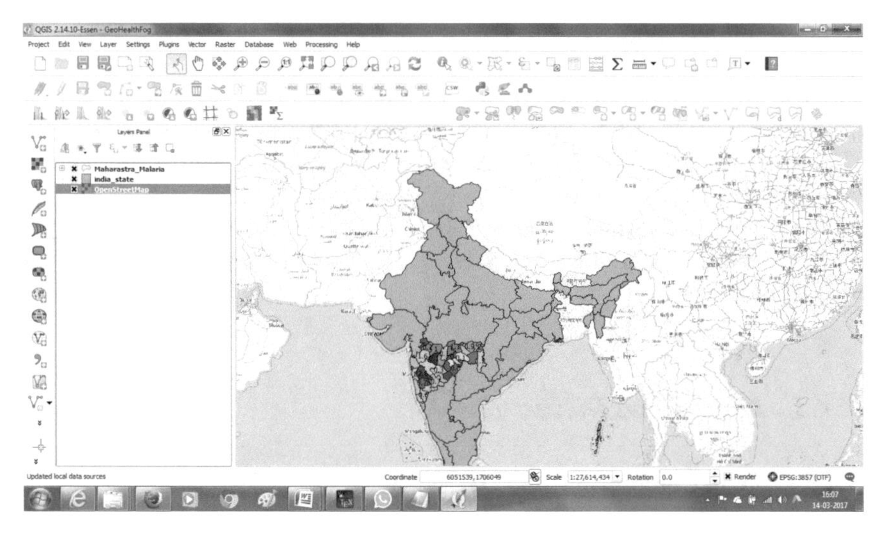

Fig. 14.6 Integrated geo-health database of Malaria

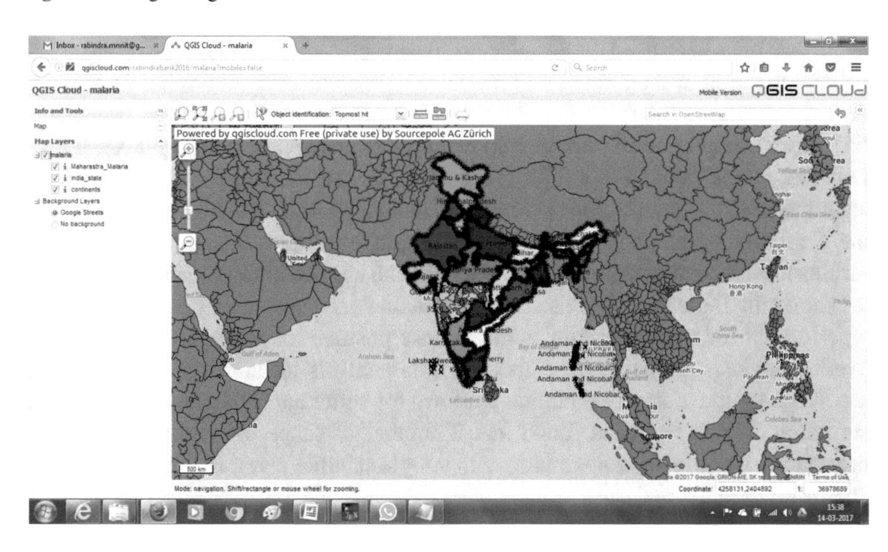

Fig. 14.7 Overlay operation on thin client environment in qgis cloud [39]

been opened with Quantum GIS; desktop based GIS analysis tools, and performed some join operations [8].

The desired overlay operation has been done with standalone application, are known as thick client operation that has been shown in Fig. 14.6. In Fig. 14.6, it has been visualized with OpenStreet maps and the two shape files are invoked in one platform on Quantum GIS desktop environment. In Quantum GIS, plugin named as QGISCloud has been installed. The said plugin has the capability of storing various

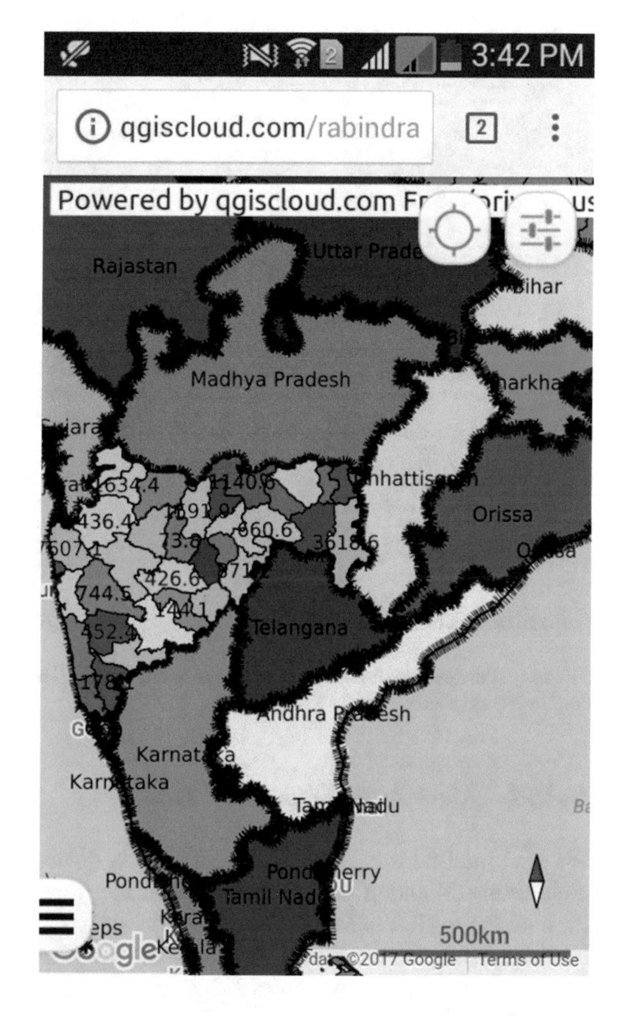

Fig. 14.8 Overlay operation on mobile client environment in qgis cloud [40]

raster and vector data set in cloud database for further overlay analysis.After storing in cloud database, it also generates the mobile and thin client link for visualization of both vector and raster data set. Figures 14.7 and 14.8 shows the overlay analysis on thin and mobile client respectively. It observes that the overlay analysis is a useful technique for visualization of geo-health data. In *GeoFog4Health*, it has been used Raspberry Pi in every Fog Node for better efficiency in time analysis. In the proposed framework, during the processing of every fog node consumes so many energy and cost. Thus, it is found that lots of energy has been realized and it should be properly management. So the next section describes the better strategy for energy efficiency and management in *GeoFog4Health*.

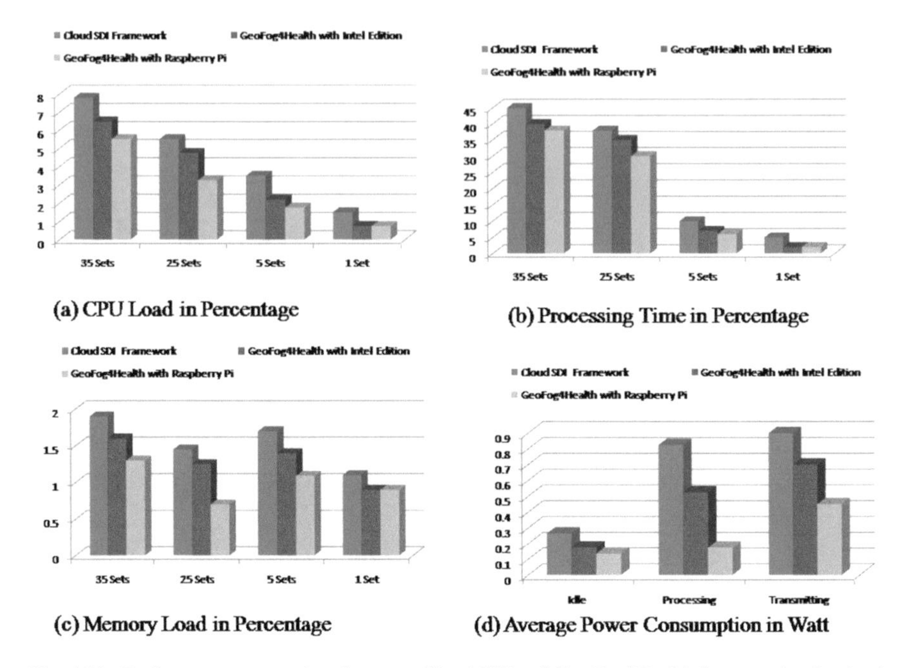

Fig. 14.9 Performance comparison between Cloud GIS and *GeoFog4Health* framework using intel edition and raspberry Pi

14.4.5.2 Analysis of Computation Time

We used Intel Edition and Raspberry Pi as fog device in proposed *GeoFog4Health* architecture. Running time of Intel Edison is greater time than that of Raspberry Pi. Intel Edison has been produced the processing time of order of NLog(N) where N defines the size of dataset. It has found that Raspberry Pi has completed the desire process almost two times faster than that of Intel Edison with scale up to larger than 125 or more data sets. The main network has been designed in each framework between the client-tier layer and the cloud layer. It is assumed that the mean arrival rate of transmitted data would be once per minute assuming that the fog node can place in the locations where only a small number of devices in that area exist. The average waiting time for each and every fog node has been calculated by Little's Law [41]. It has been used malaria positive information geospatial data for the different test of bench-marking experiment. It has also calculated the average load of memory, CPU, processing time in percentage and average power consumption in watt. Figure 14.9 shows the various performance comparison between cloud GIS and *GeoFog4Health* framework using Intel Edition and Raspberry Pi processor. From the comparison analysis, it has been shown that while running one set at a instant of time, the average waiting time for Cloud GIS framework is 189:45 s, the average waiting time for *GeoFog4Health* framework with Intel Edition processor is 73:57 s where as with Raspberry Pi processor has around 10:20 seconds. It has also experimented that the service rate with Raspberry Pi is one third of Intel Edition in *GeoFog4Health* framework. It is found that the proposed framework with Raspberry Pi has been consuming 199 mW = s where as *GeoFog4Health* framework with Intel Edition has 522 mW = s when both these frameworks are in active state.

14.4.5.3 Comparative Analysis of Cloud GIS and Proposed Architecture

Both Cloud GIS and *GeoFog4Health* framework have specific meaning for a service range with in the cloud computing environment and client-tiers that provide the mutual benefit to each other and interdependent services that leads to the greater storage capacity, control and communication possible anyplace with in the specified range [18]. Table 14.2 outlines the comparison characteristics of Cloud GIS and *FogGIS* framework [6].

Table 14.2 Comparison of Cloud GIS and FogGIS architecture

Characteristics	Cloud GIS	FogGIS
Bandwidth requirements and internet connectivity	It requires clients to have network connectivity through entire realm of services, bandwidth required depends on total amount of geospatial data being processed	It operates autonomously with or without Internet, the data and/or results are synced later when Internet become available. Thus, it is more flexible scheme.
Size	At cloud layer, processing has done with large amount of geospatial data at a time and each typically contains tens of thousands of integrated servers	At fog layer, a fog node in each location can be small or as required to meet another fog node for customer or client demands.
Operation	It operates in facilities and environments selected by the specific domain with well trained technical experts	In FogGIS framework the environments primarily decided by the customer's requirements.
Deployment	It requires highly sophisticated and suitable strategically planning for deployment	It requires minimal planning for deployment but challenges is to connect with one fog node to other intermediate fog node.
Server locations	It requires centralized server in a small number of big data centers distributed environment	It often requires distributed servers in many locations and over large geographical areas, closer to users along with fog-to-fog range or cloud-to-thing range. Distributed fog nodes and systems has been controlled either in centralized or distributed manners depending upon the clients/fog node.

14.5 Case Study II: Telemonitoring of Patients with Parkinson's Disease

14.5.1 Telemonitoring of Patients with Parkinson's Disease

Telehealth monitoring is very effective for the speech language pathology, and smart devices can be effective in such situations. Several signs indicate the relationship of dysarthria, speech prosody and acoustic features. Patients with Parkinson's disease are always accompanied by dysarthria. Characterized by the monotony of speech, variable rate, reduced stress, imprecise consonants and breathy and harsh voice. It suggested that excessive F0 variation and range in speakers with severe dysarthria exist. Another important acoustic feature for dysarthria is the amplitude of the speech uttered by the patients with Parkinson's disease. It also mentioned about reduced vocal intensity in hypokinetic dysarthria in Parkinson disease. For assisting to the patients, it presents a Fog Computing architecture, SmartFog (see Fig. 14.10) that relied on unsupervised clustering for discovering patterns in pathological speech data obtained from patients with Parkinson's disease(PD). The patients with PD use smartwatch while performing speech exercises at home. The speech data were routed into the Fog computer via a nearby tablet/smart phone [24].

14.5.2 Proposed Architecture: SmartFog

The Fog computer extracts loudness and fundamental frequency features for quantifying pathological speech. The speech features were normalized and processed with k-means clustering. When we see an abnormal change in features, results are uploaded to the cloud. In other situations data is only processed locally. In this way, Fog device could perform "smart" decision on when to upload the data to cloud computing layer for storing. It developed two prototype using Intel Edison and Raspberry Pi. Both of the prototypes were used for comparative analysis on computation time. Both systems were tested on real world pathological speech data from telemonitoring of patients with Parkinson's disease. The increasing use of wearables in smart

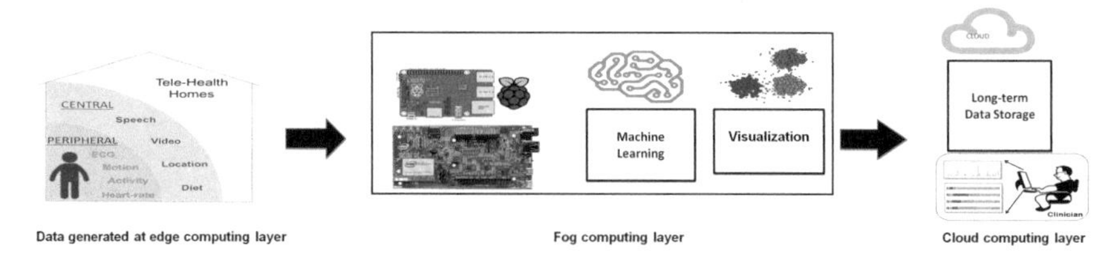

Fig. 14.10 Proposed SmartFog architecture for enhanced analytics in wearable internet of medical things. It is developed and evaluated for telehealth application

telehealth system led to generation of huge medical big data [23, 25, 49]. The telehealth services leverage these data for assisting clinical procedures. It also suggested to use of low-resource machine learning on Fog devices kept close to the wearable for smart telehealth [13, 22, 27].

That is why in the present book chapter, it discussed the development and implementation of a smartfog framework. This developed framework helps mitigate the amount of data processed at the cloud layer. It is also playing as the role of orchestrator in the process of data acquisition, conditioning, analysis and short-term storage. Finally, it compared the results of benchmarking experiments to those of realistic payloads to explore limitations in scale. Edge computing refers to the enabling technologies which allow computation to be performed at the edge of the network. For downstream data and upstream data, it utilized the cloud services and IoT services simultaneously. In this case, it defined edge as any computing and network resources along with the path between cloud data centers and data sources.

14.5.3 Fog-Based Machine Learning

14.5.3.1 Clinical Feature Extraction

Feature engineering is the initial step in any machine learning analysis. It is the process of proper selection of data metric to input as features into a machine learning algorithm. In K-means clustering analysis, the selection of features that are capable of capturing the variability of the data are essential for the algorithm to find the groups based on similarity. The subjects were patients with Parkinson's disease and the features chosen were the average fundamental frequency (F0) and Average amplitude of the speech utterance. Speech data from the patients with Parkinson's disease were collected. For analysis 164 speech samples were considered.These samples comprised of sound files with utterances as a short /a/, a long /a/, a normal then high pitched /a/ ,a normal then low pitched /a/ and phrases. The feature extraction is done with the help of Praat scripting language [11]. For pitch, the algorithm performs an acoustic periodicity detection on the basis of an accurate auto correlation method. For calculating the intensity the values in the sound are first squared, then convolved with a Gaussian analysis window. The intensity is calculated in decibels.

14.5.3.2 K-Means Clustering

K-means clustering is a type of unsupervised learning, that is used for exploratory data analysis of no labeled data [10]. K-means is a method of vector quantization and is quite extensively used in data mining. The main goal of this algorithm is to find groups in the data, the number of groups represented by the variable K. This algorithm works iteratively to assign each data point to one of K groups based on the features that are provided. The input to the algorithm are the features and the value of K. K centroids are initially randomly selected, then the algorithm iterates until convergence. This algorithm aims to minimize the squared error function J. where chosen distance measure is Euclidean distance between the data point and cluster center. Feature Engineering is an essential part in this algorithm. Authors in [45], uses optimized K-means, that clusters the statistical properties such as variance of the probability density functions of the clusters extracted features. In [54] the authors have used clustering on database containing feature vectors extracted from Malay digits utterances. The features extracted in [54] were the Mel Frequency Cepstral Coefficients (MFCC). In our work we have chosen average fundamental frequency and average intensity as features extracted from the speech files for applying K-means clustering.

14.5.4 *Results and Discussions*

14.5.4.1 K-Means Clustering Plot

For our analysis we have chosen speakers with 164 speech samples with utterances that are a short /a/, a long /a/, a normal then high pitched /a/,a normal then low pitched /a/ and phrases. The features chosen are average fundamental frequency and intensity. Feature extraction is done using praat [11] an acoustic analysis software and using praat scripts that uses standard algorithms to extract pitch and intensity mentioned in the discussion above. The results are shown in the the form of plots.The k-means clustering analysis is done on python programming language. The plots below show the Clusters of the speech data samples used in the analysis.Different colors represent different mutually exclusive groups. The analysis is done with 2, 3 and 4 number of clusters, i.e. the value of k chosen as 2 and 3 and 4 respectively. Figure 14.11a shows the K-means clustering plot for two clusters shown with different colors.The python script is run on Raspberry Pi and Intel Edison to generate the results. Figure 14.11b displays the k-means cluster plot for 4 clusters designated with four different colors in a 3D plot.Each observation belongs to the cluster with the nearest mean in k-means clustering.We have used k-means for feature learning performed in the fog device.

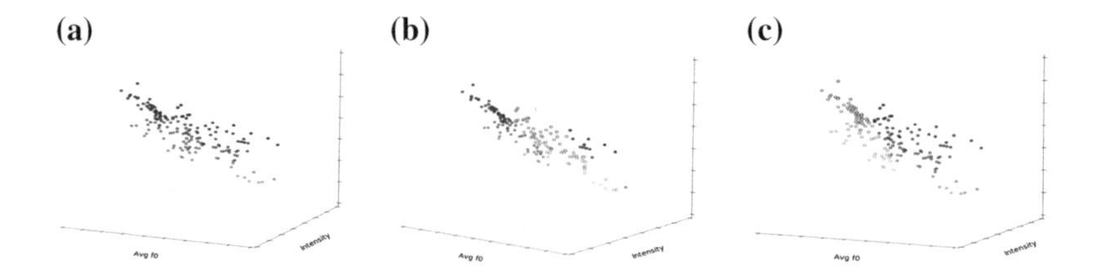

Fig. 14.11 K-means clustering plot

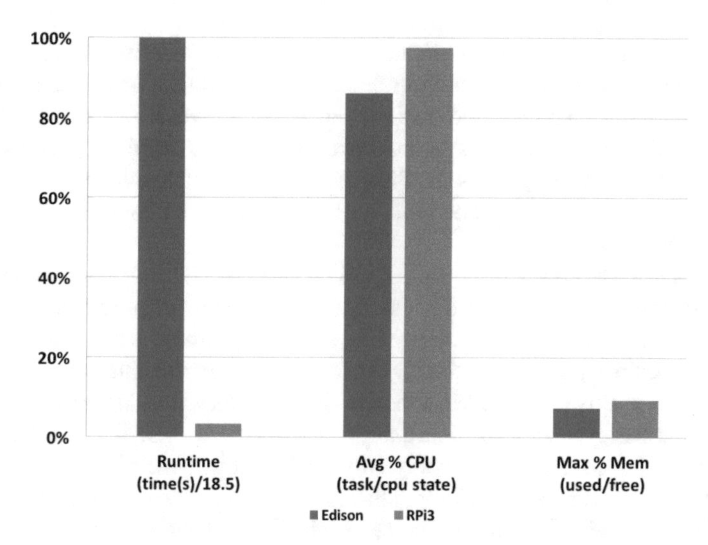

Fig. 14.12 A comparison of intel edison and raspberry pi

Figure 14.11c shows the k-means clustering plot for three clusters with different colors in 3D.

14.5.4.2 Performance Comparison

The Raspberry Pi provides a low-cost computing terminal. The Edison is a deeply embedded IoT computing module. There is a difference of processor speed and power consumption in Edison and Raspberry Pi. The Machine Learning algorithms were run on both of the devices and their Run time, average CPU usage and Memory usage have been calculated.

The Fig. 14.12 shows comparison of Intel Edison and raspberry Pi fog devices. The ideal system will minimize run time, maximize CPU usage, and use a modest amount of memory. The raspberry Pi either outperformed or matched the Edison in each of these criterion. The raspberry Pi was not capable of generating a graphical output for this type of analysis in a real-time response threshold of 200 ms. How ever, without a need for complex graphics, the raspberry Pi was able to reach the threshold clocking in at 160 ms.

14.6 Conclusions

In this book chapter, we developed and validated GeoFog, Fog2Fog architectures for application-specific case studies. Intel Edison processor and Raspberry Pi were used as Fog processors in fog computing layers. Fog nodes not only reduce storage requirements but also results in efficient transmission at improved throughput and latency. Fog of things is collection of all nodes between client layer and cloud. The edge computing done on fog nodes creates an assistive layer in scalable cloud computing. With increasing use of wearables and internet-connected sensors, enormous amount of data is being generated. Clean and curating such data for extraction of useful features can be easily done on fog nodes. The cloud could be reserved for long-term analysis.

In this study, we proposed and validated a Fog-based GIS framework for enhanced analysis of geo-health data. Fog devices reduced the storage requirements, transmission power leading to overall efficiency. Fog computing enhances the data analysis by increasing the throughput and reducing the latency. Geo-health data of malaria vector borne disease positive maps of Maharashtra state in India was used for case study. We analyzed the energy saving and cost analysis for proposed *GeoFog4Health* architecture. Further, the comparison of computation time showed the efficacy of proposed Fog architecture over Cloud GIS for enhanced analysis of geo-health data. Thus, the fog devices add edge intelligence in geo-health data analysis by introducing local processing within cloud computing environments.

Fog computing emphasizes proximity to end-users unlike cloud computing along with local resource pooling, reduction in latency, better quality of service and better user experiences. This paper relied on Fog computer for low-resource machine learning. As a use case, we employed K-means clustering on clinical speech data obtained from patients with Parkinson's disease (PD). Proposed Smart-Fog architecture can be useful for health problems like speech disorders and clinical speech processing in real time as discussed in this paper.Fog computing reduced the onus of dependence on Cloud services with availability of big data.There will be more aspects of this proposed architecture that can be investigated in future.We can expect Fog architecture to be crucial in shaping the way big data handling and processing happens in near future.

Acknowledgements This material is based upon work supported by the National Science Foundation under Grant No. (#1652538). Any opinions, findings, and conclusions or recommendations expressed in this material are those of the author(s) and do not necessarily reflect the views of the National Science Foundation.

References

1. A. Amiri, Application placement and backup service in computer clustering in software as a service (SaaS) networks. Comput. Oper. Res. **69**, 48–55 (2016)
2. J. Andreu-Perez, C.C. Poon, R.D. Merrifield, S.T. Wong, G.Z. Yang, Big data for health. IEEE J. Biomed. Health Inf. **19**(4), 1193–1208 (2015)
3. R. Barik, H. Dubey, R.K. Lenka, K. Mankodiya, T. Pratik, S. Sharma, Mistgis: Optimizing geospatial data analysis using mist computing. in *International Conference on Computing Analytics and Networking (ICCAN 2017)* (Springer, 2017)

4. R. Barik, H. Dubey, K. Mankodiya, Soa-fog: Secure service-oriented edge computing architecture for smart health big data analytics. in *5th IEEE Global Conference on Signal and Information Processing 2017* (IEEE, 2017), p. 15
5. R.K. Barik, H. Dubey, A.B. Samaddar, R.D. Gupta, P.K. Ray, FogGIS: Fog computing for geospatial big data analytics. arXiv preprint http://arxiv.org/abs/1701.02601arXiv:1701.02601 (2016)
6. R. Barik, H. Dubey, S. Sasane, R.K. Lenka, C. Misra, N. Simha, K. Mankodiya, Fog computing-based enhanced geohealth big data analysis. in *2017 International Conference on Intelligent Computing and Control, I2C2* (IEEE, 2017)
7. R. Barik, R.K. Lenka, H. Dubey, N.R. Simha, K. Mankodiya, Fog computing based SDI framework for mineral resources information infrastructure management in india. in *2017 International Conference on Intelligent Computing and Control, I2C2* (IEEE, 2017)
8. R. Barik, A. Samaddar, R. Gupta, Investigations into the efficacy of open source GIS software. Map World Forum (2009)
9. S. Bera, S. Misra, J.J. Rodrigues, Cloud computing applications for smart grid: A survey. IEEE Trans. Parallel Distribut. Syst. **26**(5), 1477–1494 (2015)
10. C.M. Bishop, *Neural Networks for Pattern Recognition* (Oxford university press, Oxford, 1995)
11. P. Boersma, D. Weenink, *Praat-a System for Doing Phonetics by Computer [Computer Software]* (Institute of Phonetic Sciences, University of Amsterdam, The Netherlands, 2003)
12. F. Bonomi, R. Milito, J. Zhu, S. Addepalli, Fog computing and its role in the internet of things. in *Proceedings of the First Edition of the MCC Workshop on Mobile Cloud Computing* (ACM, 2012), p. 13–16
13. D. Borthakur, H. Dubey, N. Constant, L. Mahler, K. Mankodiya, Smart fog: Fog computing framework for unsupervised clustering analytics in wearable internet of things. in *5th IEEE Global Conference on Signal and Information Processing 2017* (IEEE, 2017), p. 15
14. A. Botta, W. De Donato, V. Persico, A. Pescape, Integration of cloud computing and internet of things: a survey. Future Gener. Comput. Syst. **56**, 684–700 (2016)
15. H.T. Chang, T.H. Lin, A database as a service for the healthcare system to store physiological signal data. PloS one **11**(12), e0168935 (2016)
16. F. Chen, H. Ren, Comparison of vector data compression algorithms in mobile gis. in *2010 3rd IEEE International Conference on Computer Science and Information Technology (ICCSIT)*, vol. 1, (IEEE, 2010), p. 613–617
17. Z. Chen, N. Chen, C. Yang, L. Di, Cloud computing enabled web processing service for earth observation data processing. IEEE J. Sel. Top. Appl. Earth Obs. Remote Sens. **5**(6), 1637–1649 (2012)
18. M. Chiang, T. Zhang, Fog and iot: An overview of research opportunities. IEEE Internet Things J. **3**(6), 854–864 (2016)
19. N. Constant, D. Borthakur, M. Abtahi, H. Dubey, K. Mankodiya, Fog-assisted wIoT: A smart fog gateway for end-to-end analytics in wearable internet of things. arXiv preprint arXiv:1701.08680 (2017)
20. A.V. Dastjerdi, H. Gupta, R.N. Calheiros, S.K. Ghosh, R. Buyya, Fog computing: Principles, architectures, and applications. arXiv preprint arXiv:1601.02752 (2016)
21. S. Dey, A. Mukherjee, Robotic slam: a review from fog computing and mobile edge computing perspective. in *Adjunct Proceedings of the 13th International Conference on Mobile and Ubiquitous Systems: Computing Networking and Services* (ACM, 2016), p. 153–158
22. H. Dubey, N. Constant, K. Mankodiya, RESPIRE: A spectral kurtosis-based method to extract respiration rate from wearable ppg signals. in *2nd IEEE/ACM International Conference on Connected Health: Applications, Systems and Engineering Technologies (CHASE)* (IEEE, Philadelphia, USA, 2017)
23. H. Dubey, N. Constant, A. Monteiro, M. Abtahi, D. Borthakur, L. Mahler, Y. Sun, Q. Yang, K. Mankodiya, Fog computing in medical internet-of-things: Architecture, implementation, and applications. in *Handbook of Large-Scale Distributed Computing in Smart Healthcare* (Springer International Publishing AG, 2017)

24. H. Dubey, J.C. Goldberg, K. Mankodiya, L. Mahler, A multi-smartwatch system for assessing speech characteristics of people with dysarthria in group settings. in *2014 IEEE 16th International Conference on e-Health Networking, Applications and Services (Healthcom)* (IEEE, 2015)
25. H. Dubey, R. Kumaresan, K. Mankodiya, Harmonic sum-based method for heart rate estimation using ppg signals affected with motion artifacts. J. Ambient Intell. Hum. Comput. (2016)
26. H. Dubey, M.R. Mehl, K. Mankodiya, BigEAR: Inferring the ambient and emotional correlates from smartphone-based acoustic big data. in *IEEE International Workshop on Big Data Analytics for Smart and Connected Health* (IEEE, Washington DC, USA, 2016)
27. H. Dubey, A. Monteiro, L. Mahler, U. Akbar, Y. Sun, Q. Yang, K. Mankodiya, FogCare: fog-assisted internet of things for smart telemedicine. Future Gener. Comput. Syst. (2016)
28. H. Dubey, J. Yang, N. Constant, A.M. Amiri, Q. Yang, K. Makodiya, Fog data: Enhancing telehealth big data through fog computing. in *Proceedings of the ASE BigData and SocialInformatics 2015* (ACM, 2015), p. 14
29. K. Evangelidis, K. Ntouros, S. Makridis, C. Papatheodorou, Geospatial services in the cloud. Comput. Geosci. **63**, 116–122 (2014)
30. S. Fang, Y. Zhu, L. Xu, J. Zhang, P. Zhou, K. Luo, J. Yang, An integrated system for land resources supervision based on the iot and cloud computing. Enterprise Inf. Syst. **11**(1), 105–121 (2017)
31. J. Georis-Creuseveau, C. Claramunt, F. Gourmelon, A modelling framework for the study of spatial data infrastructures applied to coastal management and planning. Int. J. Geogr. Inf. Sci. **31**(1), 122–138 (2017)
32. G. Giuliani, P. Lacroix, Y. Guigoz, R. Roncella, L. Bigagli, M. Santoro, P. Mazzetti, S. Nativi, N. Ray, A. Lehmann, Bringing GEOSS services into practice: A capacity building resource on spatial data infrastructures (SDI). Trans. GIS **21**, 811–824 (2016)
33. C. Granell, O.B. Fernandez, L. Daz, Geospatial information infrastructures to address spatial needs in health: collaboration, challenges and opportunities. Future Gener. Comput. Syst. **31**, 213–222 (2014)
34. N. Gupta, R.K. Lenka, R.K. Barik, H. Dubey, Fair: A hadoop-based hybrid model for faculty information retrieval system. in *2017 International Conference on Intelligent Computing and Control (I2C217), IEEE, June 23–24, 2017* (IEEE, Coimbatore, India, 2017), p. 16
35. G.P. Hancke, G.P. Hancke Jr. et al., The role of advanced sensing in smart cities. Sensors **13**(1), 393–425 (2012)
36. L. He, P. Yue, L. Di, M. Zhang, L. Hu, Adding geospatial data provenance into SDIa service-oriented approach. IEEE J. Sel. Top. Appl. Earth Obs. Remote Sens. **8**(2), 926–936 (2015)
37. T. Higashino, Edge computing for cooperative real-time controls using geospatial big data. in *Smart Sensors and Systems* (Springer, 2017), p. 441–466
38. http://boundlessgeo.com/products/opengeo-suite/. Accessed 27th Jan 2017
39. http://qgiscloud.com/rabindrabarik2016/malaria?mobile=false. Accessed 27th Jan 2017
40. http://qgiscloud.com/rabindrabarik2016/malaria?mobile=true. Accessed 27th Jan 2017
41. https://www.isixsigma.com/dictionary/littles-law/. Accessed 12th Jan 2017
42. A. Jain, N. Mahajan, Introduction to database as a service. in *The Cloud DBA-Oracle* (Springer, 2017), p. 11–22
43. H. Ji, Y. Wang, The research on the compression algorithms for vector data. in *International Conference on Multimedia Technology (ICMT), 2010* (IEEE, 2010), p. 14
44. B. Joshi, B. Joshi, K. Rani, Mitigating data segregation and privacy issues in cloud computing. in *Proceedings of International Conference on Communication and Networks* (Springer, 2017), p. 175–182
45. H.A. Kadhim, L. Woo, S. Dlay, Novel algorithm for speech segregation by optimized kmeans of statistical properties of clustered features. in *2015 IEEE International Conference on Progress in Informatics and Computing (PIC)* (IEEE, 2015), p. 286–291
46. Z. Khan, D. Ludlow, R. McClatchey, A. Anjum, An architecture for integrated intelligence in urban management using cloud computing. J. Cloud Comput. Adv. Syst. Appl. **1**(1), 1 (2012)

47. S.H. Kim, S.Y. Jang, K.H. Yang, Analysis of the determinants of software-as-a-service adoption in small businesses: Risks, benefits, and organizational and environmental factors. J. Small Bus. Manag. (2016)
48. J.G. Lee, M. Kang, Geospatial big data: challenges and opportunities. Big Data Res. **2**(2), 74–81 (2015)
49. C.H. Lee, H.J. Yoon, Medical big data: promise and challenges. Kidney Res. Clin. Pract. **36**(1), 3 (2017)
50. R.K. Lenka, R.K. Barik, N. Gupta, S.M. Ali, A. Rath, H. Dubey, Comparative analysis of spatialhadoop and geospark for geospatial big data analytics. in *2nd International Conference on Contemporary Computing and Informatics (IC3I 2016)* (IEEE, 2016)
51. Y. Ma, H. Wu, L. Wang, B. Huang, R. Ranjan, A. Zomaya, W. Jie, Remote sensing big data computing: challenges and opportunities. Future Gener. Comput. Syst. **51**, 47–60 (2015)
52. L. Mahler, H. Dubey, C. Goldberg, K. Mankodiya, Use of smartwatch technology for people with dysarthria. in *In the Proceedings of the Motor Speech Conference* (Madonna Rehabilitation Hospital, 2016)
53. R. Mahmud, R. Buyya, Fog computing: A taxonomy, survey and future directions. arXiv preprint http://arxiv.org/abs/1611.05539arXiv:1611.05539 (2016)
54. S. Majeed, H. Husain, S. Samad, A. Hussain, Hierarchical k-means algorithm applied on isolated malay digit speech recognition. Int. Proc. Comput. Sci. Inf. Technol. **34**, 33–37 (2012)
55. A. Monteiro, H. Dubey, L. Mahler, Q. Yang, K. Mankodiya, Fit: A fog computing device for speech tele-treatments. in *2nd IEEE International Conference on Smart Computing (SMART-COMP 2016)* (IEEE, At Missouri, USA, 2016)
56. A. Munir, P. Kansakar, S.U. Khan, Ifciot: integrated fog cloud iot architectural paradigm for future internet of things. arXiv preprint http://arxiv.org/abs/1701.08474arXiv:1701.08474 (2017)
57. S. Nunna, K. Ganesan, Mobile edge computing. in *Health 4.0: How Virtualization and Big Data are Revolutionizing Healthcare* (Springer, 2017), p. 187–203
58. S.S. Patra, R. Barik, Dynamic dedicated server allocation for service oriented multi-agent data intensive architecture in biomedical and geospatial cloud. in *Cloud Technology: Concepts, Methodologies, Tools, and Applications* (IGI Global, 2015), p. 2262–2273
59. S. Sareen, S.K. Gupta, S.K. Sood, An intelligent and secure system for predicting and preventing zika virus outbreak using fog computing. Enterprise Inf. Syst. 121 (2017)
60. S. Sarkar, S. Chatterjee, S. Misra, Assessment of the suitability of fog computing in the context of internet of things. IEEE Trans. Cloud Comput. (2015)
61. B. Schaffer, B. Baranski, T. Foerster, Towards spatial data infrastructures in the clouds. in *Geospatial Thinking* (Springer, 2010), p. 399–418
62. W. Shi, J. Cao, Q. Zhang, Y. Li, L. Xu, Edge computing: vision and challenges. IEEE Internet Things J. **3**(5), 637–646 (2016)
63. J. Smith, W. Mackaness, A. Kealy, I. Williamson, Spatial data infrastructure requirements for mobile location based journey planning. Trans. GIS **8**(1), 23–44 (2004)
64. X. Sun, N. Ansari, EdgeIoT: mobile edge computing for the internet of things. IEEE Commun. Mag. **54**(12), 22–29 (2016)
65. B. Vanmeulebrouk, U. Rivett, A. Ricketts, M. Loudon, Open source gis for hiv/aids management. Int. J. Health Geogr. **7**(1), 53 (2008)
66. X. Wang, H. Zhang, J. Zhao, Q. Lin, Y. Zhou, J. Li, An interactive web-based analysis framework for remote sensing cloud computing. ISPRS Ann. Photogramm. Remote Sens. Spat. Inf. Sci. **4**, W2 (2015)
67. B. Wu, X. Wu, J. Huang, Geospatial data services within cloud computing environment. in *2010 International Conference on Audio Language and Image Processing (ICALIP)* (IEEE, 2010), p. 1577–1584
68. C.P. Yang, Geospatial cloud computing and big data (2017). https://doi.org/10.1016/j.compenvurbsys.2016.05.001
69. C. Yang, R. Raskin, M. Goodchild, M. Gahegan, Geospatial cyberinfrastructure: past, present and future. Comput. Environ. Urban Syst. **34**(4), 264–277 (2010)

70. C. Yang, Q. Huang, Z. Li, K. Liu, F. Hu, Big data and cloud computing: innovation opportunities and challenges. Int. J. Digit. Earth **10**(1), 13–53 (2017)
71. S. Yi, C. Li, Q. Li, A survey of fog computing: concepts, applications and issues. in *Proceedings of the 2015 Workshop on Mobile Big Data* (ACM, 2015), p. 37–42
72. J. Yu, J. Wu, M. Sarwat, Geospark: A cluster computing framework for processing largescale spatial data. in *Proceedings of the 23rd SIGSPATIAL International Conference on Advances in Geographic Information Systems* (ACM, 2015), p. 70
73. H. Zhu, C.P. Yang, Data compression for network GIS. in *Encyclopedia of GIS* (Springer, 2008), p. 209–213

Chapter 15
Secure Smart Vehicle Cloud Computing System for Smart Cities

Trupil Limbasiya and Debasis Das

Abstract We acquire or provide most of the services with the help of the Internet in the fast-growing world. We should deploy various kinds of systems globally so that other users can practice the same effortlessly, and use storage framework to provide conveniences world-wide effectively. Nowadays, we practice vehicular communication technology to exchange diversified data for varied intentions, which helps vehicle operators in diversified manners such as traffic awareness, weather conditions, road assistance, automatic toll payment system, etc. In this chapter, we converse about data transmission through vehicular ad-hoc networks, cloud computing in the vehicular technology. After that, we survey of different schemes related to secure routing and geo-location details of vehicles. We illustrate concerning security demands, possible attacks, and challenges in the vehicular cloud computing (VCC) architecture. Conclusively, we suggest a new identification scheme to get access of the VCC system from the user end, which can be secured against varied attacks. Moreover, we do analysis of the suggested system to determine security worthiness and measure total required time to execute the phases.

Keywords Attack · Authentication · Cloud computing · VANET

The original version of this chapter was revised: Missed out author corrections have been incorporated. The erratum to this chapter is available at
https://doi.org/10.1007/978-3-319-73676-1_18

T. Limbasiya
BITS Pilani - K.K. Birla Goa Campus, Goa, India
e-mail: limbasiyatrupil@gmail.com

D. Das (✉)
K. K. Birla Goa Campus, Goa, India
e-mail: deba16@gmail.com

B. S. P. Mishra et al. (eds.), *Cloud Computing for Optimization: Foundations, Applications, and Challenges*, Studies in Big Data 39,
https://doi.org/10.1007/978-3-319-73676-1_15

15.1 Introduction

We use different advanced applications to make easy and better experience of available services with the help of network systems in the modern society. For this purpose, we should employ high-configured hardware for designing concepts, software for various operations, and network with respect to the transmission channel in the development of advanced systems. We should arrange these requirements in such a way so that it can run efficiently in different environments. Otherwise, there is wastage of various infrastructures (buildings, hardware devices, network components, and software computations). Customers ask to provide a network system, which can accomplish the communication functionality among several computing devices for a limited time. This kind of network arrangement is known as the ad-hoc network, which is flexible, infrastructure-less, for a specific purpose, within a specified range, for mobile devices. It can be worthwhile to military applications, rescue, vehicular systems, home networking, finding surroundings, and much more. A wireless ad hoc network (WANET) is a distributed kind of complicated radio framework. This type of network is structured for a short duration of time because it does not depend on a pre-planned infrastructure, and each node co-operates in a direction by forwarding packets to other nodes. Therefore, the selection of an upcoming forwarding node is periodically focused on the network link up. A mobile ad hoc network (MANET) is a subcategory of WANET, especially for portable appliances. It is helpful to deliver significant data-grams through single hop or multi-hop communication systems. It can be applicable to personal area networks, civilian environment, emergency circumstances. However, there are some concerns such as packet loss, regularly disconnections, limited power resources, fixed communication bandwidth, varied link capacity, broadcast nature, mobility, frequent range updating.

We do have different famous applications of the VANETs, which help us in various circumstances. To better enhancement of ad-hoc facility in the field of vehicular technology, scientists introduced the concept of the vehicular ad-hoc network (VANET), which is focused on the vehicular technology to fulfill available opportunities in the fast-growing world. In the current decade, there is an attention on VANET, and it is a collection of passing vehicular data in a radio network that implements the information communication technology (ICT) to accommodate state-of-the-art services regarding the vehicle traffic and position based upon the vehicle cloud management. Presently, VANET gained notable attention due to the opportunity of investing novel and agreeable explications in the fields of current status, moving details, traffic response, street safety, and Intelligent Transportation Systems (ITS).

Figure 15.1 shows the general structure of vehicular cloud computing connections of various kinds of information sharing links. Vehicles are connected with other vehicles as well as infrastructures to carry out significant packets at appropriate ends within a reasonable period to make out it affective in the vehicular cloud technology infrastructure.

Cloud computing is an enumerating archetype, where many computing devices are linked in the public or private interfaces, to contrive scalable foundation for

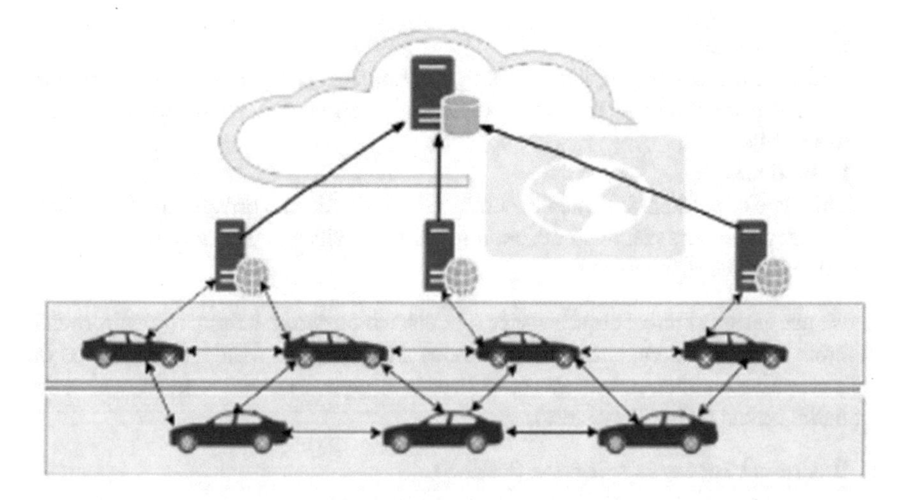

Fig. 15.1 The vehicular cloud computing framework for smart society

administrative intention, data and file storage in a dynamic manner. Some of the typical advantages of cloud computing are flexibility, reduced cost, increased storage, expeditious deployment, green computing, rapid data availability, updated software accessibility, advanced versatility, etc. Clouds carry out enormous interests for both individuals and businesses. Clouds promote in financial profits, outsourcing mechanisms, everywhere anytime accessibility, equipment sharing, on-demand scalability, and co-operation adaptability. Clouds depreciate between the demand for user engagement by masking technical specifications such as permissions, software upgrades, and maintenance from its consumers. Additionally, clouds could extend better security protections against various attacks over public working environments. However, there are some drawbacks such as data recovery, data protection, management strength, data availability in certain circumstances. There are some impressions (e.g. parallel data computation, extraordinary need of responsive applications, virtualized IT infrastructure, large desktop applications, etc.), which motivate us towards cloud usage in the fast-growing technology world. There are three types of cloud namely private, public, and hybrid.

1. **Public Cloud**
   Public clouds are controlled by third bodies; they deliver the higher economies of scale to clients, as the infrastructure expenses are shared between various levels of customers. They deliver services to every customer at reduced cost, pay-as-you-go to plan, etc. All consumers distribute the equivalent infrastructure equipments with certain configurations, security features, and availability variances. These are maintained by the cloud service providers. One of the advantages of a public cloud is that it may be larger than an enterprise cloud.

2. **Private Cloud**
   Private clouds are formulated for a single enterprise. The primary intent is to work out data protection obstacles and extend higher command, which are not present in a public cloud.
3. **Hybrid Cloud**
   This type of cloud is designed to combine both public and private clouds at a time so that customers can avail different kinds of services in a secure manner along with effective management.

We use various kind of cloud service models, which can be helpful to avail/provide different facilities to customers in numerous ways. They are categorized based on their effective characteristics. We have listed six types of cloud models, which are available on the technology world.

1. **Business Process as a Service (BPaaS)**
   A primary objective of BPaaS is to contrive services to other companies for business collaboration. It connects business process management along with more than one features to a cloud deployment system. One industry can provide data processes to other companies for specific reasons.
2. **Communications as a Service (CaaS)**
   A communication is the most imperative segment of any device or human life in the today's digital world. To fulfill requirements of consumers, CaaS permits to interact with others as a portion of the communication system. Skype is an example of CaaS.
3. **Database as a Service (DBaaS)**
   It is a platform to accomplish database-related functionality. There are different cloud providers, which offer database services to various levels of customers to provide/avail their services. Some third party industries avail this type of facility for numerous intentions.
4. **Infrastructure as a Service (IaaS)**
   It is a facility for outsourcing the infrastructure of a network in the interest to the enterprise customers regarding numerous intentions. The infrastructure is managed on the Internet and by the third party industries for computing various operations in a legitimate fashion. There are different IaaS providers, which help to the organizations in availing related conveniences to provide a sufficient level of duties within a reasonable time without any incidents.
5. **Platform as a Service (PaaS)**
   A main purpose of PaaS is to allow functionality of any applications at any machines without installing certain programs. A user can access an application through any computing system, which will redirect to a server and users can avail services of the particular application directly. Before this, a company has to deploy an application on the cloud server else customers cannot acquire the specified needs. Online programming languages and development tools are examples of a PaaS, which are available to the customers through the cloud scenario.

6. **Software as a Service (SaaS)**
   It is on-demand software applications, which can be accessed from a consumer via the Internet. Most of the web browser focused services are instances of SaaS. We can consider Hotmail, Facebook, LinkedIn, etc. as an example of SaaS.

We use the concept of wireless technology from last four and half decades, with few omissions. Researchers came up with an emergent research arena vehicular technology from the past one and half decade, which effects in multiple computerized policies for different intentions. The thought of cloud computing lighted from the understanding of computing infrastructure establishment physically at one place and other organizations may obtain their usefulness (to hire the infrastructure to run their systems) from several places at a time with varied benefits. This dominant plan has been recommended because of a high-speed Internet, relatively low-cost, advances in parallel and shared computing and distributed databases, virtualization, etc. One of the key interests of cloud computing is scalable accessibility to computing devices and information technology (IT) services [1].

There are several vehicular cloud computing services such as accident warnings, vehicle maintenance, intelligent parking controlling, road situation awareness, safety applications, traffic management, etc. The vehicular cloud computing (VCC) is an innovative model, which provides a facility to vehicles for interacting and collaborating to compute the data, understand insight into the environment, generate the outcomes and more commonly distribute resources. Eltoweissy et al. [2] presented the vehicular cloud idea, which supports the on-board devices in participating vehicles. Sometimes, vehicle users park their vehicles for long durations, and others are stuck in overcrowded transportation and progress moderately by modifying their location in the wireless networks. Ultimately, our vehicles consume vital time in the street and may face wavering positions periodically. In this circumstance, the vehicles will support local authorities for fixing traffic occurrences within a limited period, which is not achievable by the municipal traffic administrative system centers simply due to the shortage of enough computational supports. We consider that the vehicles may be proficient of determining difficulties in many circumstances that may need an countless time for a converged system.

Chapter organization: In Sect. 15.2, we present a brief discussion on different vehicular protocols of vehicle data transmission, secure routing, geo-positing of vehicles, and vehicular cloud computing to address related attack vulnerabilities. In Sect. 15.3, we explain about security needs, possible attacks, and challenges in the vehicular cloud computing schemes. In Sect. 15.4, we suggest a verification scheme to get access to the VCC system legitimately. In Sect. 15.5, we do security analysis of the suggested model to identify the security level against multiple attacks. Finally, we conclude our chapter in Sect. 15.6. References are at the end.

15.2 Literature Survey

The theory of co-operative collision avoidance (CCA) is performed using the routing layer as well as medium access control (mac) to classify collision avoidance in the vehicle emergency circumstances. Authors [3] practiced this kind of outline for highway traffic protection in the V2V technology. A warning message distribution can be preceded rapidly in case of V2V communication rather than V2I transmission. An optical wireless communication V2V transmission scheme [4] operates a specific CMOS image sensor for acknowledging to light energy fluctuations and an output circuit with the help of an LED transmitter and a camera receiver.

Now, we discuss regarding the secure routing protocols in the vehicular technology world. There are different kind of routing etiquettes such as topology-based, broadcast-based, geography-based, cluster-based, and traffic-aware. These algorithms work on their own characteristics to achieve better performance in the specific circumstances.

Topology-based algorithms transmit or forward data-grams using existing transmission connections. There are some standard etiquettes such as AODV, PRAODV, PRAODV-M, ADOV+PGB, DSR, PFQ-AODV [5–9]. Authors practiced various scenarios to execute their thoughts. Scientists [10] designed a new trust model to reduce trust bias and to maximize the routing administration. They conducted a comparable study of their recommended routing etiquette against the non-trust based and bayesian trust-based routing etiquettes. The effects illustrated that the proposed protocol was capable to deal with selfish operations and was flexible against trust-based attacks. Moreover, it can effectively trade off packet overhead and delay for a meaningful gain as the performance ratio.

An etiquette [11] was recommended to transfer the data packets through the quickest path amongst the verified vehicles. Transferring the data in a very associated path with a limited connection is a dilemma, which magnifies the system achievement and decides the authorized vehicles in this system. The experiment outcomes have been reviewed for better secure path routing protocol, anchor-based street, greedy traffic-aware routing, path length, average end-to-end delay, traffic-aware routing, and network gap encounter.

Researchers [12] came up with a new etiquette, which enables high-resolution robust localization based on the vehicle sites. Communication systems related to the vehicle work in the wireless networks for better functionality implementation in the scheme. But, we should take care of some constraints such as cost-effective operations, information alteration, correct position, false location, etc. Scientists [13] proposed SeRLOC skeleton focused on a one-way hash function and symmetric key, which can resist against remarkable preservation vulnerabilities (sybil attack and wormhole attack) in different circumstances. There are some drawbacks such as advanced keys distribution, public and private keys generation, execution speed, maintenance of keys, storage, protection of the private key, etc. A new ring signature policy has been implemented to succeed over multiple concerns. In this method, three users (the supervisors, the deployers, and the clients) play an important role in carrying out a complete model. Ring signature algorithm fulfils anonymity, decentralization, integrity, originality, and privacy [14].

Yan et al. [15] suggested secure vehicle position frameworks, which enable security peculiarities (availability, confidentiality, and integrity). They discussed different paths to make simpler integrity, to improve the existence, and to offer cryptographic schemes. Sun et al. [16] advised an identity-based vehicular data defending the scheme without using certificates to preserve the privacy of the vehicle users in certain states. They also examined other security demands (authenticity, confidentiality, non-repudiation, and information reliability) in the proposed model. This arrangement has been performed to defeat data location disclose offenses.

Wu et al. [17] suggested a framework focused on position of vehicles. The mixture of ad-hoc networks and RSU infrastructures comprise present probabilistic RSU amendment etiquette practices effective localization to predict the number and position of nearest RSUs. A routing switch framework is used to ensure quality of service (QoS) in different network connectivity and deployment adumbrations. The execution of hybrid routing system is assessed under both practical and really test based experiments. They also discussed an administration of opportunistic routing, practicing a carry-and-forward scheme to work out the forwarding disruption difficulty in VANETs. Scientists [18] found two types of vehicle management mapping attacks and suggested a new-vehicle management scheme to preserve privacy related to a present position on the vehicle. They also described the scenario for the adversary to realize the practical outcome of the proposed model.

Cloud computing frameworks facilitate organizations to commute cost of outsourcing computations on-interest. There may be concerns related to user or data authentication, information confirmation, data privacy, data availability, etc. However, there are significant benefits (cost savings, reliability, less environmental impact), which lead to the acceptance of cloud computing in the highly expanding technology world. Hence, most of the industries have accepted cloud services in their day-to-day businesses. Authors [19] introduced a trusted cloud computing platform, which enables IaaS contributors to present secure services during the initialization stage.

We should consider technical concerns such that browser-based authentication at cloud end, future browser enhancements, cloud malware injection, metadata spoofing, flooding, etc. [20]. Wang et al. [21] recommended an auditing system to obtain the privacy of public cloud database, which combines the theory of the homomorphic authenticator focused on asymmetric key along with random masking. This model mainly enables two demands, which are as the followings: (1) the cloud database audit process can be implemented without the local data and extra computing resource requirements by the third party auditor. (2) A system would not be affected (in terms of additional exposures about private details) during the execution of cloud data audit procedures. A third party auditor can conduct numerous auditing engagements concurrently by using the bilinear aggregate signature method.

Scientists [22] proposed a privacy-preserving public auditing etiquette, which enables public data auditing, no additional impediment to the cloud server, quick execution. As we know, users can ask for the stored data (at the cloud end) anytime due to different goals and hence, there is always a requirement of parallel computing. The batch auditing scheme can satisfy it, and it has been achieved in the proposed protocol.

A new threshold reputation-based scheme [23] is proposed, and it is focused upon the population dynamics efficiently to counter the node compromise attacks, stimulate the cooperation with intermediate nodes, increase vehicular nodes attention and understand the truth of maintaining the probability of conveying data-grams for credits. This etiquette is intended to determine the dilemma of withstanding layer-adding attack. The vehicle privacy is quite preserved from both (the transportation manager and cloud) ends. Ultimately, formal security proofs and the extensive simulations have been explained the effectiveness in their proposed system in opposing the complicated attacks and potency in different parameters, e.g. high delivery ratio, special reliability, and moderate average delay in cloud-assisted vehicular delay tolerant networks.

15.3 Security Discussion in the Vehicular Cloud Computing

The vehicular cloud computing is a vast system and is not an easy architecture to handle simply. However, it is highly helpful for different types of vehicle-related services easily in the today's technology enabled universe. This kind of system communicates in the public environment during their packet transmission stages and vehicles distribute significant information with other vehicles or infrastructure units for the specific intention. Hence, the data comes in the picture and data regarding care should be taken by the concerned authority else there are opportunities to attackers for disturbing various natures of connected systems efficiently. Consequently, we explain different security needs, security challenges, and security attacks in the vehicular cloud computing. Figure 15.2 describes various needs, attacks, and challenges in terms of vehicular cloud security in the present world.

15.3.1 Security Requirements

We know that a system has to fulfill security requirements else a system or users may be exposed, and an attacker can avail advantages on the same, which leads to serious trouble for various appliances and/or humans throughout the technology world. We illustrate essential needs of security (authentication, accountability, availability, confidentiality, integrity, privacy) into the real-time systems. Figure 15.2 indicates a security taxonomy (needs, attacks, and challenges) of the vehicular cloud computing.

In our day to day life, we use vehicle communication sessions for numerous intentions as mentioned earlier. Here, we consider an example of the toll plaza and toll plazas are mostly found on national highways and bridges and we spend an ample amount of time over a queue for paying a tax there. Consequently, there is a wastage

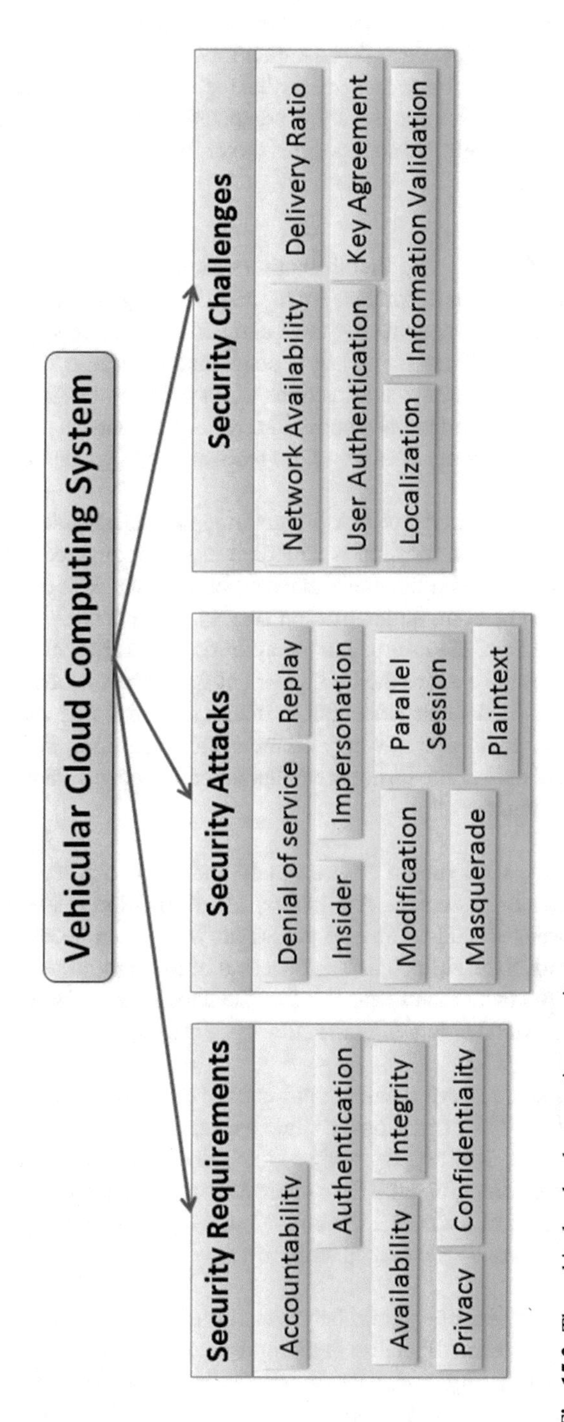

Fig. 15.2 The vehicular cloud computing security taxonomy

of money by using unnecessary fuel. Furthermore, the mobility of vehicles gets interrupted by this method, which takes a higher travel period, more consumption of fuel and pollution level gets increased near to that region, etc. The methods frequently used by enterprises and technology-enabled organizations are the automatic payment systems, which provide multiple superiorities rather than traditional methodology in the vehicular systems.

1. **Accountability**
   Every person who works with an information system should have specific responsibilities for information assurance. An individual is engaged with the particular information security management system and he/she is responsible for the same. Person's duties can be measurable by a concerned authority. The security goal that generates the requirement for actions of an entity to be traced uniquely to that entity, and it supports non-repudiation, restriction, fault isolation, intrusion detection and prevention, and after-action recovery and legal action.
2. **Authentication**
   Authenticity is the property of holding genuine and being able to be validated as well as trusted. Authentication is a method in which the credentials provided by users are checked with the credentials stored in a database (either on a local operating system or at an authentication server). If credentials match, the process is completed and the user is granted authorization for access. Vehicles have been identified as the basis of their unique identity number, which is directly connected to their individual bank account. A matter related to financing comes from the picture and hence, people worry on the same. Therefore, there is a need of proper level authentication during communication between vehicle and payment collection arrangement.
3. **Availability**
   It means that a system should ensure timely and reliable access of facility and use of resources to customers. Availability of information refers to ensure that authorized parties should be able to access the data when needed. Information only has value if the right people can access it at the right time. This feature of the automatic payment system ensures that the details and information about the vehicle will be available to affiliated authority in case of any need or verification.
4. **Confidentiality**
   A system should preserve details about authorized clients, access, and owned information. Data has extraordinary value especially in today's world. There is always a requirement of cryptography so that only authorized device/person can understand the meaningful data. Here, there is a need to compute details concerning bank accounts of users. All the confidential data provided by the customer should be taken care by the related authority.
5. **Integrity**
   A certain level of security should be present in the information computation to restrict unauthorized modification or disruption of the communicated data in the public medium. Integrity is related to the data into the system/human. Information only has value if and only if it is correct at the required moment. The system should

include similar functionality, which ensures the customers that the information will not be leaked and will not be manipulated by any unauthorized person(s).

6. **Privacy**
   Privacy refers to information shared with visiting sites, how that information is used, what information has been shared with whom, or if that information is used to track users or not. A term often used to describe an individual's anonymity and how safe they feel at a location, about personal activities, regarding private details, etc. Most of the present web-related services share their clients' data with others. But, Systems should provide a certain level of assurance related to sharing users' confidential data and privacy.

15.3.2 Security Attacks

An attack is an interference in a legitimate facility, that can affect the vital information/data from the customers and/or system elements. The attack is a spiteful exercise, which can be performed to collapse the software and/or hardware segments of an arrangement. It can influence pertinent data of either a server or a client or both. In this manner, private data to the server or customer can be divulged or familiar to others. There are numerous attacks, which can disturb vehicle operators or vehicle cloud computing framework.

1. **Denial of Service**
   Denial of service (DoS) is a one type of attack, which can be practiced by a moderate knowledgeable people to disturb various levels of computing facility, which can be availed by different system users. The main intention of DoS attack is to keep busy computing servers with fake service requests by malicious attackers. Hence, authorized users may not gain particular services at appropriate time and then, an user might not obtain needed facility during emergency circumstance(s). For example, a vehicle user is interested in knowing the traffic status of precise road network, but he/she may not receive proper packets containing fruitful data within a limited time. Then, an user may lose his/her valuable time.
2. **Impersonation**
   An attacker can impersonate a legal road side unit (RSU) infrastructure to other vehicle users if he/she has some information from previous conversations as being long time same range sharing vehicle user. The vehicle customer regularly updates its stored information when the RSUs do not update data frequently, therefore, resynchronization can be transpired. Hence, the vehicular cloud storage is not updated as per current status. If the RSU is not protected, there is an opportunity of a server impersonation; if a vehicle operator is not secured, then a user impersonation attack can happen in the VCC system. Thus, an attacker can acquire the stored private data. A primary conception behind this attack is to target upcoming vehicle transactions (regarding significant information) to avail a facility without the knowledge of valid vehicle users and/or RSUs. This type of

attack also acts in silence mode; therefore, it is not easy to capture the identity of a malicious vehicle customer.

3. **Insider**
   An insider attack is a one kind of malicious activity implemented on numerous networks or computer systems by illegal objects or legal objects those are already available within the same range of the vehicle(s). An attacker can gain the access to restricted information or confidential data when this attack is performed in a vehicle span or VCC system. An insider attack is typical in any computer system or network because anyone might or might not be the truthful in advance technology world. An authorized officer can misuse related user credentials for different objectives.
4. **Modification**
   A modification is hard to identify in the VCC. The main objective behind a modification attack is to alter transferred data. A User/RSU sends a packet to other RSUs/users; at that moment, an attacker tracks packets, transforms the message information, and retransmits the changed information to the RSU/user. Thus, the RSU will not receive correct packet data. In other words, the attacker keeps recording of earlier vehicle communication transactions. After words, when the certified vehicle users access a system/network, the attacker changes communication requests from the users and re-sends reformed packets to RSUs/VCC. Thus, legal users will not be capable to contact the system/network.
5. **Masquerade**
   A masquerade attack is one kind of attack that practices an invalid identity of the vehicle user. Such an attack can be prosecuted over an insecure verification processing. Masquerade attacks are performed by utilizing known pass-codes and login credentials of original vehicle users. A masquerade attack can depend upon the prominence of the users. When the communication medium is not dominant in verification schemes, an attacker can proceed for this type of attack to identify significant information effortlessly that is passed in the public communicating channel.
6. **Parallel Session**
   When more than one login request from different vehicle customers are received concurrently for the same users/RSUs, at that moment, the occurrence of a parallel session attack transpiring will arise. Such a parallel session attack is practiced to similar communication transactions, methods and tasks. In these advance technology systems, computations are frequently executed in an identical way due to limited time availability. Therefore, a parallel session attack can be performed in the vast technology world easily. A simple objective of the attacker is a disturbance during significant vehicle transactions.
7. **Plaintext**
   An attacker will choose some portion of normal text. A plaintext is concomitant in public-key cryptosystem especially. A well-known plaintext attack is linear cryptanalysis against block and stream ciphers. Its principal objective is to obtain data and to reduce performance ratio of the VCC system. The plaintext attack is performed by an attacker if the relevant information is transferred in a simple form.

In other words, secure cryptographic functions are not in use during the authentication mathematical calculations. Therefore, an adversary can be succeeded to get meaning of the transmitted texts easily.

8. **Replay**
   A replay attack is practiced by an attacker to stop/delay the transmitted packets through a public communication channel so that a request (to get some service) made by a legitimate vehicle operator may not be available to the recipient end to proceed with further step(s). There is a situation, where a delay of significant information may disturb the system/human at the extreme level. Consider that a vehicle is at the toll-plaza (enabled with the automatic toll payment system) and the system is unable to receive the request for payment from the vehicle operator end due to some attack, then there is no use of the automatic toll payment system, which requires more time to complete the payment process.
9. **Session Key Disclosure**
   This type of attack can be applied to the system when an attacker has knowledge of required credentials to compute a session key for establishing a transaction session. There is another situation, where an attacker can calculate needed parameters to enumerate a session key for creating a transaction circumstance for further step(s). Assume that, some vehicle user (acting as an adversary) has computed some other user's session key to do transaction for different facility. This kind of situation can happen to the case of same secret key has been practiced in the system from the server end. Hence, he/she can avail different level of services behalf of a targeted vehicle operator without his/her knowledge.

15.3.3 Security Challenges

Here, we explain different present challenges during the vehicular cloud computing system. Vehicle nodes share their present position after every specified interval. Vehicle operators may get issues regarding denial of service, user authentication, data verification at the receiver end.

1. **Network Availability**
   A vehicle user is interested in getting some information (e.g. petrol bank) within the particular range of kilometers. However, all vehicles have limited network range. Due to this, he/she must request to other vehicle users for assisting on the same. Other vehicle customers might reply to the requester; a vehicle user will receive more responses from different vehicle operators. At the same moment, a vehicle (original sender) may fail to fulfill normal operations to be alive in the network due to high traffic of ripostes. There may be a concern regarding high mobility because vehicles are always in moving situation. Hence, a vehicle may fall into the dead circumstance sometimes due to lack of required significant information. At the end, a vehicle user cannot avail needed information in a reasonable period.

2. **User Authentication**
   We consider the similar scenario (Sect. 15.3.1.1) to clarify a need for user authentication in the vehicular networks. A vehicle operator requested to provide information about nearby public facilities and various vehicle users may reply with different information based upon their knowledge. A vehicle (original sender) has to verify the source of the data because a user may have received information from numerous vehicles. There is a possibility that a vehicle user may act as an adversary in the present network. Attackers have an opportunity to distribute incorrect data among several vehicle consumers. There is another point for a requirement on the user authentication system in the vehicular ad-hoc networks. Consider that there is a network of some vehicle operators presently and another vehicle (not belongs to a current network) is going to join a present network. At the same time, existing vehicle users must verify that a new member is authorized to communicate to other group members or not. Because, members of a present network may share various information with their group members in the future, and a new member can also distribute it in a malicious manner, which leads to various problems in the future. Hence, there is a demand of user authentication in the system of vehicular ad-hoc networks.
3. **Information Validation**
   Here, vehicles share various kind of facts with other vehicle users in case of demand from a precise user(s). However, they can distribute required information through a common communication channel. The adversary has also access to a public transmission medium, and he/she can modify a message (which is sent by a legitimate user to an original sender as a reply) or create a bogus response based on available information of normal vehicle consumers. Vehicles are located at varied places and there is a probability of dissimilar information from many users additionally even though they are legal users. A vehicle (original sender) must confirm the originality and correctness of the received information from numerous vehicles.
4. **Delivery Ratio**
   A delivery of packets to the particular recipient party should be within a reasonable time else a sender cannot have access for requested facility at appropriate time and is not able to perform certain step(s) to accomplish other procedures. Consider that, a vehicle user is interested in informing the situation regarding weather as well as road of particular place to future (few hours) visiting vehicle users so that they can use alternative route to reach the destination without any concerns. Here, the concept of multi-hop communication comes into the picture. Due to some reason(s), a packet is delivered to future (few hours) visiting vehicle users after reaching very close to that place. Hence, data should be available at appropriate recipient within a reasonable time else it is meaningless.
5. **Key Agreement**
   The key agreement is a concept, which is practiced to provide mutual verification at the both ends (sender and receiver). Generally, we believe that a server (which is used to authenticate users) is always genuine. However, there are some conditions in which an attacker can create fake server and tries to establish

communication sessions with original users on the system. In this circumstance, a legal user may be trapped of an adversary and transfer important credentials to an attacker due to considering a legitimate server communication. By using the theory of key management, the sender verifies to the recipient, and the receiver authenticates the sender. In the vehicular technology, we use infra-structureless network as per characteristics of ad-hoc networks. Consequently, there is a requirement of mutual authentication between the receiver(s) and the sender(s). To provide mutual authentication, both (sender and receiver) have to compute different required operations as per designed scheme and have to establish a same session key within a reasonable time. This session key is valid up to a limited duration. It is not an easy deal to provide mutual authentication facility to vehicle operators during their communication sessions in the vehicular ad-hoc network.

6. **Localization**
   Vehicles are in moving situation generally so that we should consider dynamic nature of the vehicles in the system. Thus, there is a need to know the current position about a specific vehicle at different periods to provide better facility to users. Consider that a vehicle user is traveling from one station to other stations, and he/she is interested in knowing about his/her availability time to the destination or sharing his/her present location detail with family members due to awareness of locality. We can assume another circumstance of the medical vehicles (carrying a patient from home to the hospital). Here, staff members (present at hospital) are interested to know current position or reaching time of a patient traveling through the medical vehicle so that they can arrange medical components accordingly in the emergency case.

15.4 Proposed Model for Vehicle User Identification

In this section, we suggest an authentication scheme to verify legitimacy of a vehicle user. If a receiver verifies a vehicle user correctly, then both (sender and recipient) can communicate with each other to share significant information in the vehicular ad-hoc networks. In this framework, we design three phases (initialization, login and authentication) systematically, which have been practiced in the most of identification model generally. Table 15.1 shows used different notations, which can be helpful to understand the recommended design in the vehicular technology. Table 15.2 states an initialization phase, Table 15.3 dictates a login phase and Table 15.4 presents an authentication phase to understand a suggested identification model clearly.

15.4.1 Initialization Phase

This procedure is a one-time exercise for V_i, which should be completed. The initialization phase includes the following steps:

Table 15.1 Notations used in the model

Symbol	Description
V_i	A vehicle user
V_{is}	A sender vehicle user
V_{ir}	A receiver vehicle user
VID_i	An identity of $V_{is/ir}$
PWD_i	A password of V_i
SC_i	A smart-chip for VID_i
SCR_{is}	A smart-chip reader of V_{is}
SCR_{ir}	A smart-chip reader of V_{ir}
RC	A registration point for V_i
$List_{X_{VID_i}}$	A list of personal keys of different VID_i
T_1, T_2	Timestamps
δT	A predefined maximum time to generate SK
ΔT	The permitted delay in transmission
SK	A session key
$h(\cdot)$	A one-way hash function
$\|\|$	The concatenation operation
$\oplus$	The exclusive-or operation (ExOR)

Table 15.2 Initialization phase of the proposed scheme

Vehicle (V_i)		Registration Center (RC)
Chooses VID_i, PWD_i and r_u		
Computes $RPW_i = h(r_u \parallel PWD_i)$		
	$\xrightarrow[\text{Secure channel}]{\{VID_i, RPW_i\}}$	
		Chooses X_{VID_i}
		Computes...
		$A_i = RPW_i \oplus h(X_{VID_i})$
		$B_i = A_i \oplus VID_i$
		$SC_i = \{B_i, h(\cdot)\}$
	$\xleftarrow[\text{Secure channel}]{\{SC_i, List_{X_{VID_i}}\}}$	
Computes $C_i = VID_i \oplus h(r_u \|\| PWD_i)$		
Updates SC_i with C_i		

Table 15.3 Login phase of the proposed scheme

Sender (V_i)		Smart-chip Reader (SCR_{is})	
Inserts VID_i and PWD_i			
	$\xrightarrow[\text{Secure channel}]{\{VID_i, PWD_i\}}$		
		Retrieves B_i, C_i from SC_i	
		Computes...	
		$D_i = B_i \oplus C_i \oplus (VID_i \| \| T_1)$	
			$\xrightarrow[\text{Open channel}]{\{VID_i, D_i, T_1\}}$

Table 15.4 Authentication phase of the proposed scheme

Smart-chip Reader (SCR_{is})		Smart-chip Reader (SCR_{ir})
	$\xrightarrow[\text{Open channel}]{\{VID_i, D_i, T_1\}}$	
		Checks $T_2 - T_1 \leq \Delta T$
		Computes...
		$h'(X_{VID_i}) = (VID_i \| \| T_1) \oplus D_i$
		$h'(X_{VID_i}) \stackrel{?}{=} h(X_{VID_i})$
$SK_S \stackrel{?}{=} SK_R$		

1. V_i chooses VID_i as per vehicle identity number allocated by the respected authority and selects PWD_i as well as r_u (a random number) freely.
2. V_i computes $RPW_i = h(r_u||PWD_i)$.
3. V_i sends VID_i, RPW_i to RC.
4. After receiving it, RC generates a personal key (X_{VID_i}) for VID_i and computes $A_i = RPW_i \oplus h(X_{VID_i})$ plus $B_i = A_i \oplus VID_i$.
5. RC saves credentials (B_i, $h(\cdot)$) into SC_i and installs it along with $List_{X_{VID_i}}$ into a particular vehicle.
6. V_i calculates $C_i = VID_i \oplus h(r_u||PWD_i)$ and updates SC_i with C_i and RPW_i.

15.4.2 Login Phase

This phase is performed between V_i and SCR_{is}. They perform the following steps to get login into the system as well as to proceed further in order to communicate with other vehicles:

1. V_i enters VID_i and PWD_i.
2. SCR_{is} extracts B_i, C_i from SC_i. Then, it calculates $D_i = VID_i \oplus B_i \oplus C_i \oplus T_1$ and transmits $\{VID_i, D_i, T_1\}$ to the receiver.

15.4.3 Authentication Phase

An authentication phase is practiced to check the sender (SCR_{is}) before starting a communication at the recipient end. The receiver (SCR_{ir}) performs subsequent steps as the followings:

1. SCR_{ir} receives VID_i, D_i, T_1 and checks a validity of a verification request by performing a timestamp validation. If it is valid, then proceeds to next step else it drops a request immediately.
2. SCR_{ir} computes $h^{'}(X_{VID_i}) = VID_i \oplus D_i$ and matches with $h(X_{VID_i})$. If both are same, then the receiver starts a communication session. Otherwise V_i terminates a request directly.
3. SCR_{ir} computes a session key, $SK_R = VID_i \oplus D_i \oplus T_1 \oplus T_3$. SCR_{is} computes a session key, $SK_S = VID_i \oplus D_i \oplus T_1 \oplus T_3$. Where, $T_3 = \Delta T + \delta T$.

15.5 Security Analysis

We do analysis of the proposed model in terms of security after considering certain suppositions, and these assumptions have been practiced by multiple researchers [24, 25] in order to measure the security level of different authentication schemes.

1. **Impersonation Attack**
   An impersonation attack can be applied if an attacker ($\mathscr{A}$) has some significant data of earlier packet transmissions as well as basic information (an identity) about a targeted person. Finally, $\mathscr{A}$ can impersonate either receivers or senders if he/she can manipulate these data and is able to put into effect at receiver/sender ends, then we can consider that there is an opportunity to apply an impersonation attack into the scheme. SCR_{is} transfers some parameters (VID_i, D_i, T_1) to SCR_{ir} for verification. However, $\mathscr{A}$ does not have knowledge of r_u, A_i and PWD_i. As a result, $\mathscr{A}$ cannot manipulate any data of a packet request. Therefore, the suggested model is secured against an impersonation attack.
2. **Insider Attack**
   It is an attack, which can be applied by an internal user (he/she belongs to the same organization.) of the system and internal person acts as an attacker ($\mathscr{A}$). So, he/she knows an identity of a user. At that point, he/she does not have familiarity with some credentials (PWD_i, r_u). In this case, s/he should guess one parameter at a time, which is a guessed password ($PWD_i^{'}$). But, $\mathscr{A}$ does not know r_u correctly, and it is difficult to apply an attack with two guessed parameters within a polynomial period. In this way, the proposed scheme can withstand against an insider attack.
3. **Modification Attack**
   In order to apply a modification attack, an attacker should able to update the transmitted credentials (from SCR_{is} to SCR_{ir}) according to his/her wish and should

be succeeded to pass the verification test. $\mathscr{A}$ cannot able to modify the transmitted request by using VID_i, D_i and T_1 in the advised system. To update D_i, $\mathscr{A}$ needs B_i, C_i, RPW_i, which are not accessible to $\mathscr{A}$ easily in a reasonable duration. As a consequence, the recommended mechanism is protected against a modification attack.

4. **Masquerade Attack**
   In this attack, an attacker uses an identity of other users to obtain unauthorized/authorized access to the system. For this action, he/she should be aware of PWD_i and should get access of other user's vehicle physically. We assume that $\mathscr{A}$ makes use of guessed password (PWD_i') but, it is difficult to get physical access to a vehicle without knowledge of an original vehicle owner for an attacker. Now, $\mathscr{A}$ cannot proceed further to execute certain steps in order to be successful. Consequently, the suggested system is secured against a masquerade attack.
5. **Parallel Session Attack**
   A parallel session attack is applied to the system to get rights of authorized user when an original user tries to obtain services of the system for some intention(s). At that time, an attacker ($\mathscr{A}$) intercepts a message request ($\{VID_i, D_i, T_1\}$) and redirects it towards its end. After that, $\mathscr{A}$ updates credential(s) and forwards to allow permission for getting different kind of services illegally. However, a verification test will not be passed at SCR_{ir} end if $\mathscr{A}$ has done any modification, and a session will be terminated automatically. If a session key is not computed, then there is no chance to establish a communication session in the proposed method. Therefore, the advised system can withstand against a parallel session attack.
6. **Plaintext Attack**
   SCR_{is} does not transmit any value in the simple form so that, it is not useful up to some extends, and we use the concept of time-stamp. Here, it is difficult to retrieve any credential without having proper knowledge of other parameters. In order to understand D_i, $\mathscr{A}$ should have knowledge of B_i and C_i but $\mathscr{A}$ does not know these credentials. As a result, a plaintext attack is not feasible in the proposed model.
7. **Replay Attack**
   We practice the concept of time-stamp in the system in order to keep secure communication. If $\mathscr{A}$ tries to change T_1, then he/she requires some amount of time to process this operation. Thus, $\mathscr{A}$ will fail at the recipient end to clear the verification process ($T_2 - T_1 \leq \Delta T$). SCR_{ir} terminates a session straightaway if a time-stamp test fails. In this manner, the suggested scheme can withstand against a replay attack.

15.6 Conclusion

We describe concerning various type of wireless ad-hoc networks and their numerous applications in the smart city concept. Afterwords, we deliberate for the cloud computing systems along with their various level of applications. We discuss different schemes regarding V2V communication, vehicular cloud computing, secure routing, geo-graphic position of vehicles in the vehicular ad-hoc networks. Then, we explain about essential security requirements, possible security attacks, various security challenges in the vehicular technology. Finally, we design a scheme, which is useful to identify the vehicle user in establishing a communication session to share remarkable information. Furthermore, we discuss the security analysis of the same scheme to recognize the security level.

Acknowledgements We are grateful to the anonymous reviewers for their noteworthy time and effort in suggesting their exceptional views on the chapter in the review process. In addition, we are thankful to editors for the handling of this book. This work is partially supported by Early Career Research Award from Science and Engineering Research Board (SERB), Department of Science and Technology(DST), Govt. of India, New Delhi, India (Project Number: ECR/2015/000256).

References

1. S. Olariu, I. Khalil, M. Abuelela, Taking VANET to the clouds. Int. J. Pervasive Comput. Commun. **7**(1), 7–21 (2011)
2. M. Eltoweissy, S. Olariu, M. Younis, Towards autonomous vehicular clouds, in *International Conference on Ad Hoc Networks* (Springer, Berlin, Heidelberg, 2010), pp. 1–16
3. S. Biswas, R. Tatchikou, F. Dion, Vehicle-to-vehicle wireless communication protocols for enhancing highway traffic safety. IEEE Commun. Mag. **44**(1), 74–82 (2006)
4. I. Takai, T. Harada, M. Andoh, K. Yasutomi, K. Kagawa, S. Kawahito, Optical vehicle-to-vehicle communication system using LED transmitter and camera receiver. IEEE Photonics J. **6**(5), 1–14 (2014)
5. C. Perkins, E. Belding-Royer, S. Das, Ad hoc on-demand distance vector (AODV) routing (No. RFC 3561) (2003)
6. V. Namboodiri, M. Agarwal, L. Gao, A study on the feasibility of mobile gateways for vehicular ad-hoc networks, in *Proceedings of the 1st ACM International Workshop on Vehicular Ad Hoc Networks* (2004), pp. 66–75
7. V. Naumov, R. Baumann, T. Gross, An evaluation of inter-vehicle ad hoc networks based on realistic vehicular traces, in *Proceedings of the 7th ACM International Symposium on Mobile Ad Hoc Networking and Computing* (2006), pp. 108–119
8. S.K. Dhurandher, S. Misra, M.S. Obaidat, M. Gupta, K. Diwakar, P. Gupta, Efficient angular routing protocol for inter-vehicular communication in vehicular ad hoc networks. IET Commun. **4**(7), 826–836 (2010)
9. C. Wu, S. Ohzahata, T. Kato, Flexible, portable, and practicable solution for routing in VANETs: a fuzzy constraint Q-learning approach. IEEE Trans. Veh. Technol. **62**(9), 4251–4263 (2013)
10. R. Chen, F. Bao, M. Chang, J.H. Cho, Dynamic trust management for delay tolerant networks and its application to secure routing. IEEE Trans. Parallel Distrib. Syst. **25**(5), 1200–1210 (2014)
11. S.K. Bhoi, P.M. Khilar, SIR: a secure and intelligent routing protocol for vehicular ad hoc network. IET Netw. **4**(3), 185–194 (2014)
12. S. Capkun, J.P. Hubaux, Secure positioning in wireless networks. IEEE J. Sel. Areas Commun. **24**(2), 221–232 (2006)
13. X. Lin, X. Sun, P.H. Ho, X. Shen, GSIS: a secure and privacy-preserving protocol for vehicular communications. IEEE Trans. Veh. Technol. **56**(6), 3442–3456 (2007)

14. C. Zhang, R. Lu, X. Lin, P.H. Ho, X. Shen, An efficient identity-based batch verification scheme for vehicular sensor networks, in *The 27th Conference on Computer Communications, INFOCOM 2008* (IEEE, 2008), pp. 246–250
15. Gongjun Yan, Stephan Olariu, Michele C. Weigle, Providing location security in vehicular Ad Hoc networks. IEEE Wirel. Commun. **16**(6), 48–53 (2009)
16. J. Sun, C. Zhang, Y. Zhang, Y. Fang, An identity-based security system for user privacy in vehicular ad hoc networks. IEEE Trans. Parallel Distrib. Syst. **21**(9), 1227–1239 (2010)
17. D. Wu, Y. Zhang, L. Bao, A.C. Regan, Location-based crowdsourcing for vehicular communication in hybrid networks. IEEE Trans. Intell. Transp. Syst. **14**(2), 837–846 (2013)
18. J. Kang, R. Yu, X. Huang, M. Jonsson, H. Bogucka, S. Gjessing, Y. Zhang, Location privacy attacks and defenses in cloud-enabled internet of vehicles. IEEE Wirel. Commun. **23**(5), 52–59 (2016)
19. N. Santos, K.P. Gummadi, R. Rodrigues, Towards trusted cloud computing. HotCloud **9**(9), 3 (2009)
20. M. Jensen, J. Schwenk, N. Gruschka, L.L. Iacono, On technical security issues in cloud computing, in *IEEE International Conference on Cloud Computing, 2009. CLOUD'09* (2009), pp. 109–116
21. C. Wang, Q. Wang, K. Ren, W. Lou, Privacy-preserving public auditing for data storage security in cloud computing, in *Proceedings of the IEEE INFOCOM* (2010), pp. 1–9
22. C. Wang, S.S. Chow, Q. Wang, K. Ren, W. Lou, Privacy-preserving public auditing for secure cloud storage. IEEE Trans. Comput. **62**(2), 362–375 (2013)
23. J. Zhou, X. Dong, Z. Cao, A.V. Vasilakos, Secure and privacy preserving protocol for cloud-based vehicular DTNs. IEEE Trans. Inf. Forensics Secur. **10**(6), 1299–1314 (2015)
24. R. Madhusudhan, R.C. Mittal, Dynamic ID-based remote user password authentication schemes using smart cards: a review. J. Netw. Comput. Appl. **35**(4), 1235–1248 (2012)
25. T. Limbasiya, N. Doshi, An analytical study of biometric based remote user authentication schemes using smart cards. Comput. Electr. Eng. **59**, 305–321 (2017)

Chapter 16
Video Transcoding Services in Cloud Computing Environment

Sampa Sahoo, Bibhudatta Sahoo and Ashok Kumar Turuk

Abstract Nowadays, online video consumption is an outstanding source of infotainment. Current social media era allows people to communicate with others around the world via Facebook, LinkedIn, YouTube and other platforms by sharing/sending photos, videos over the Internet. The proliferation of viewing platforms, file formats, and streaming technologies generate the need for video transcoding. The transcoding process ensures that video content can be consumed from any networks and devices, but it is a time-consuming, computation-intensive method and requires high storage capacity. The rise of video distribution and consumption makes the video service providers face unpredictable CAPEX and OPEX, for delivering more videos across multi-screens and networks. A cloud-based transcoding is used to overcome the limitations with on-premise video transcoding. The virtually unlimited resources of the cloud transcoding solution allow video service providers to pay as they use today, with the assurance of providing online support to handle unpredictable needs with lower cost. This chapter is designed to discuss various techniques related to cloud-based transcoding system. Various sections in this chapter also present the cloud-based video transcoding architecture, and performance metrics used to quantify cloud transcoding system.

16.1 Introduction

The Internet is now an important part of entertainment media, i.e., a user can watch a video of their choice or watch live events or matches through the Internet. The volume of media assets is increasing rapidly due to the growth of on-line viewing, social

S. Sahoo (✉) · B. Sahoo · A. K. Turuk
NIT Rourkela, Rourkela 769008, India
e-mail: sampaa2004@gmail.com

B. Sahoo
e-mail: bibhudatta.sahoo@gmail.com

A. K. Turuk
e-mail: akturuk@gmail.com

B. S. P. Mishra et al. (eds.), *Cloud Computing for Optimization: Foundations, Applications, and Challenges*, Studies in Big Data 39,
https://doi.org/10.1007/978-3-319-73676-1_16

media and mobile outlets. Current social media era allows people to communicate with others around the world via Facebook, LinkedIn, YouTube and other platforms by sharing/sending photos, videos over the internet. The variation in video quality, file sizes, and compression codecs makes the job of media professionals critical to maintaining it. Growth in other technologies like internet connectivity, increase in bandwidth put additional pressure. More recently, a significant new data modality has emerged due to unstructured data from video and images. A plethora of videos generated by digital devices demands attention in the current Big Data market. The video uploading rate of most popular online video archiving systems YouTube is around 100 video per minute. In recent time, one of the research challenges is analysis and processing of video data to support anytime anywhere viewing irrespective of devices, networks, etc. [1]. Video service providers use broadcast-quality video streaming services to reach local and worldwide audiences irrespective of networks and devices. Streaming is the process of fragmenting the video files into smaller pieces and delivered it to the destination. Streaming use buffering technology to collect several packets before the file being played. For example, Imagine a glass filled with water with a hole at the bottom, then there is a constant stream of water drainage as long as there is enough water in the glass. The streaming technology applies to both live streams and progressive downloads for audio and video on demand. Streaming can be done at the streaming server or by renting streaming service provided by streaming service providers who can host the video on the cloud. Streaming service can be rented on an hourly or monthly basis, i.e., the user needs to pay only for the resources consumed.

Cloud computing is used to provide ubiquitous, on-demand network access such that user can access computing resources anywhere and at any time through the Internet. It is built on the base of distributed computing, grid computing, and virtualization, which delivers the on-demand computing resources over the Internet on a pay-for-use basis. Users can access the resources without considering the installation details and use it as required with paying only for the used units. The cloud service model can be Software as a service (SaaS), Platform as a service (PaaS), Infrastructure as a service (IaaS). SaaS is a subscription model to access software running on Service Providers servers. Few examples of SaaS applications are Google Apps, Box, Dropbox. PaaS provides a computing platform for development and deployment of web applications without buying and maintaining software and infrastructure required for it. Examples of PaaS include Google App Engine, Microsoft Azure Services, Force.com, etc. Infrastructure resources like storage, network, and memory are delivered as on-demand service in the IaaS cloud service model. Examples of IaaS providers are Amazon Web Services, Rackspace, etc. [2–6]. A virtualization technique used by cloud computing allows splitting of coarse-grained physical resources into fine-grained virtual resources to be allocated on-demand [7]. The cloud-based video delivery architecture allows storage and distribution of single, high quality, high bitrate video files in multiple formats without the expense of purchasing and maintaining own infrastructure [8]. In Video-on-Demand (VoD) cloud, the content provider rent resources (e.g. storage and bandwidth) from the cloud provider. The resources can be rescaled based on fluctuating demands of the user to satisfy some

Quality of Service (QoS) (e.g. video start-up delay). There are several advantages of moving video streaming services to the cloud and are listed as follows:

(i) The company converts upfront capital expenditure (CAPEX) to operating expense (OPEX). Cloud eliminates the massive capital investment for on-premise hardware (e.g. servers, storage arrays, networks) and software. It also puts an end to investment for continuously expanding and upgrading on-premise infrastructures.
(ii) In the cloud, a user pays as they go for processing intensive services such as encoding (transcoding), asset management and storage of video streaming. The payment can be paid on a transaction basis, monthly subscription or as an annual fee. For example, a file-based video content can be transcoded on an hourly, pay-as-you-go basis in the Amazon Web Services (AWS) marketplace.
(iii) To build a massively scalable encoding on-premise platform with support for latest devices and players is costly and not trivial. But, the infrastructure needed for this can be upgraded by the cloud provider without the knowledge of user very easily and with less effort.
(iv) A user can work from anywhere in the world at any time.
(v) The time-consuming uploads, downloads or inefficient bandwidth use is eliminated and thus making cloud time-efficient.
(vi) The cloud offers flexibility and scalability. On-premise scaling-up/down of resources is not hassle free sometimes. As resources can't be added on the fly content providers, need to start the whole process of buying and maintaining the new resources. Whereas in the cloud, the addition of new resources is simple and quick as the user only need to change its requirement details. The cloud service provider will accordingly either reduce or add resources for the user and charge for the same.
(vii) Whether Video delivery is on-demand or live streaming cloud ensures high quality and stability.

Different video streaming services used in practice are storage, transcoding, content delivery. This study mainly focuses on video transcoding service. The Cisco white paper [9] discusses why and how the cloud can support video delivery to multiscreen, i.e., more devices. London Olympic, 2012 is considered as a milestone that takes traditional viewing to a new level, i.e., shifting towards connected devices like tablets, smartphones, etc. for anytime anywhere viewing. The challenge lies in creating an effective multiscreen offer with the consideration of a different combination of devices, networks, service platforms, etc. The cloud architecture used for both homogeneous and heterogeneous environments could reduce the potential cost from 36 percent to 13 percent compared to traditional video architecture. One of the ways to manage the cost of multiscreen access is to allow a temporary bandwidth increase for premium consumers, who are ready to pay extra for the better experience. The paradigm shift will also help video service providers to reduce CAPEX/OPEX as well as coping with the growth of online video industry. According to statistics presented in [10], 462 million active Internet users are there in India, which is 34.8% of the whole population and 13.5% of the world Internet users. Despite a significant percentage of Internet users, the average connection speed in India is 3.6 Mbps, the lowest in the Asia-Pacific region. The average peak connection speed is also the lowest in India with 26.1 Mbps. In Asia-Pacific region, South Korea has the highest average connection speed, i.e., 27 Mbps, and average peak connection speed in Singapore is the highest with 157.3 Mbps [11]. From the report we can see that country-wise, there is variation in Internet speed. So, a single format of a video for all will not be sufficient. The advancement of mobile devices, tablets, PC support multiple formats, and it adds additional challenges to the video service providers. So, there is a need of video format conversion that will satisfy the demand for the various user devices and network. The format conversion process is known as transcoding.

An online video consumed by a user on multiple screens, such as digital TVs, smartphones, and tablets, need to be conveyed in a device suitable format. Video content providers require many formats of a single video file to provide service to users with varying need. It is practically impossible to prepare a video in all formats as it requires large content storages. There is also continuous development in the field of coding and encoding technology, codecs, etc. So, there is a need for a solution that will convert a video into the required format with less effort and cost [12, 13]. The conversion of video from one digital format to another format termed as transcoding. Video transcoding method helps video content providers to generate all the possible formats of a single video [14]. To provide such a transcoding capability, video content providers need enormous computing powers and storage space. Video service providers want their videos to look good and playable irrespective of devices or platforms. The proliferation of video distribution and consumption makes the video service providers to face unpredictable CAPEX and OPEX, to deliver more videos across multi-screens. Video transcoding solution requires enormous computing power and must deal with issues related to cost, scalability, video quality, delivery flexibility, and ubiquity. Cloud computing has emerged as a front-runner to give a solution to time-consuming and resource-aware transcoding process.

Video transcoding was initially employed to reduce the video file size, but now the priority has changed, i.e., transcoding is not only used to reduce the file size but also make the video viewable across platforms, enable HTTP streaming and adaptive bitrate delivery. Transcoding may results degradation in video quality. So, it is desirable to start with a high-quality mezzanine file and carefully do the transcoding based on specific target formats and content types. Transcoding a large, high-definition video to a diverse set of screen sizes, bit rates, and quality requires a lot of time, computation, storage capacity. To overcome the difficulty associated with the transcoding process content providers are using the cloud services. The cumbersome transcoding is simple, take less time and pocket-friendly in the cloud as compared to in-house process. A user only needs to specify its requirements and subscribe the services provided by the cloud which is only a single click away. The rest of the task, i.e., resource allocation and time-consuming transcoding process will be performed in the background. Finally, the user is charged only for the resources consumed without much overhead. Before discussing transcoding in the cloud further first, we present various terms used in video transcoding in next Sect. 16.1.1.

16.1.1 Terms in Video Transcoding

A video consists of video sequences where each video sequence consists of Group of Pictures (GOPs). Each GOP has several video frames. Usually, transcoding is performed at GOP level or frame level. Few more terms related to video transcoding are

(i) Codec: The method used by a system to determine the amount of change between frames is called a codec. Codec stands for Compressor-Decompressor, and it either encode or decodes the video.
(ii) Bitrate (Data rate): It is the amount of data that is used for each second of video (Kbps, Mbps, etc.). Bit rate can be constant (CBR), i.e., the same amount of data every second or variable (VBR), i.e., the amount of data is adjusted depending on changes between frames.
(iii) Resolution: Resolution is the actual size of the video (1 frame) measured in pixels. For example 1920×1080 resolution = 2,073,600 pixels. Let each pixel uses 24 RGB color bit, then the size of one frame is $2{,}073{,}600 \times 24 = 0.25$ MB (1 MB = 8388608bit).
(iv) Frame rate: Number of frames shown every second is known as frame rate. Popular frame rates are 24 fps, 30 fps, 50/60 fps, etc. If 1 frame size is 0.25 MB then bandwidth (data rate) requirement of a video @60 fps is 15 MB/s, whereas @24 fps is 6 MB/s.

From the above calculation, we can see that even a video with few frames need a significant amount of bandwidth. For a movie or long duration video, it will be even more. A video is also demanding a substantial amount of storage. Different ways to deal with massive storage and bandwidth requirements are: buy and use infrastructure or convert video into a format (transcode) that will consume less storage and bandwidth. As transcoding is time-consuming and computationally intensive, cloud-based video conversion is preferable to reduce a provider's expense.

16.2 Video Transcoding Process

Transcoding (Encoding) is the process of changing the input video file from one format to another for video delivery to different programs and devices without losing originality [15]. Transcoding is commonly used as an umbrella term that covers some digital media tasks, such as transrating, trans-sizing [16]. Transcoding is typically the change of codec, bitrate or resolution. The change of bitrates, i.e., 5 Mbps to 3 Mbps or 1.8 Mbps, etc. are known as transrating, and change of resolution (video frame size), i.e., 1080p–720p is referred as trans-sizing. Transcoding involves following two steps: first decoding is done to convert input video file to uncompressed format and after that re-encoding is done to generate data format supported by the user's device [17]. Screen size 1080p racking up the pixel dimensions to 1920×1080 for

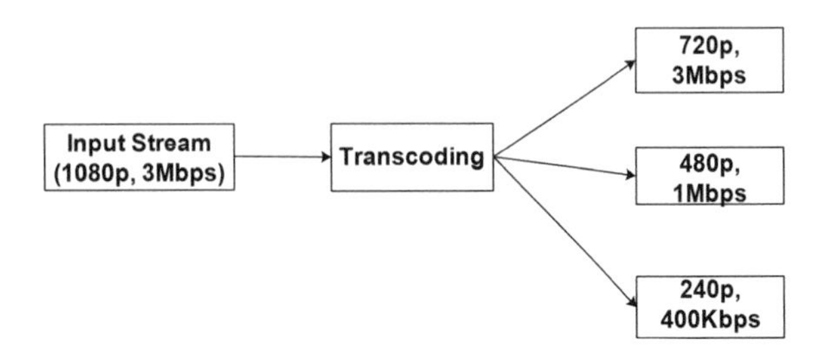

Fig. 16.1 Transcoding Process

full high definition. Here p is for progressive scanning, meaning each line is drawn in sequence upon screen refresh. 3 Mbps represents bit rate. Figure 16.1 shows that how transcoding will generate different formats. The encoded video file can be delivered on-demand or live. The video file can be transferred entirely before playing it (downloading) or stream to the user device. Video content delivery depends on the distance between user location and media server containing the requested file. If the distance is less content delivery is fast, but if distance is more user experience choppiness, loading lag, poor quality. The advantage of transcoding in the cloud is lower cost, virtually unlimited scalability, and elasticity to counter peak demand in real-time. The cloud transcoding solution allows video service providers to pay-as-you-use, with the assurance of providing online support to handle unpredictable needs [18, 19]. Video transcoding service in cloud uses popular cloud service model that include Software as a Service (SaaS), Platform as a Service (PaaS), Infrastructure as a Service (IaaS), and hybrid model [20].

Video service providers want their videos to look good and playback irrespective of devices or platforms. The proliferation of video distribution and consumption makes the video service providers to face unpredictable CAPEX and OPEX, to deliver more videos across multi-screens (smartphones, PCs, TVs, tablets, etc.) and network (2/3/4G, mobile, broadcast, etc.). Encoding is not just compression; it means having to choose and accelerating, not declining as the number of renditions are needed to support the diversity of user devices and networks. The encoding solutions, whether in-house infrastructure, the third party need to deal with cost, scalability, video quality, delivery flexibility, and ubiquity issues. One of the solution to solve these problems is cloud-based transcoding. The advantage of transcoding in the cloud is lower cost, virtually unlimited scalability, and elasticity to counter peak demand in real-time. The cloud transcoding solution allows video service providers to pay as they use today, with the assurance of providing online support to handle unpredictable needs [21]. Operators, service providers, and content providers see the benefits of using standard servers in the cloud and want to move away from special appliances or dedicated hardware. Pushing video to the cloud, in real-time requires a high-speed, highly available network. The selection of cloud for any application depends on following primary requirements: bandwidth, storage, cost, security, and accessibility. The total bandwidth required for a video stream varies depending on the number of frames or images being captured per second as well as the quality of the images being captured. The availability and affordability of bandwidth are not consistent from city to city, country to country. Bandwidth and storage requirement calculation is an essential step in the planning or design of any cloud-based video delivery system.

16.3 Cloud-Based Video Transcoding Architecture

The system architecture of the cloud-based video transcoding service adopted from [13, 19, 22, 23] is shown in Fig. 16.2. The architecture consists of many components like a streaming service provider, a video repository, splitter, transcode manager, transcoding VMs (Servers), video merger and caching storage. The *streaming service providers* like YouTube, Netflix accepts user's request and checks if required video is present in *video repository* or not. If the video is present in its desired format, then starts streaming the video. However, if the coveted video is in another format than the one requested by the user, online transcoding is done using cloud resources. The service provider charged according to the amount of resource reserved in clouds. For online transcoding first, the video is split into several segments, or chunks by *video splitter*. The video segments are mapped to transcoding VMs by the transcode manager to be transcoded independently. The video segments can be of equal size, an equal number of frames, equal number of GOPs, equal-size with an odd number of intraframes or different size with an equal number of intraframes [24, 25].

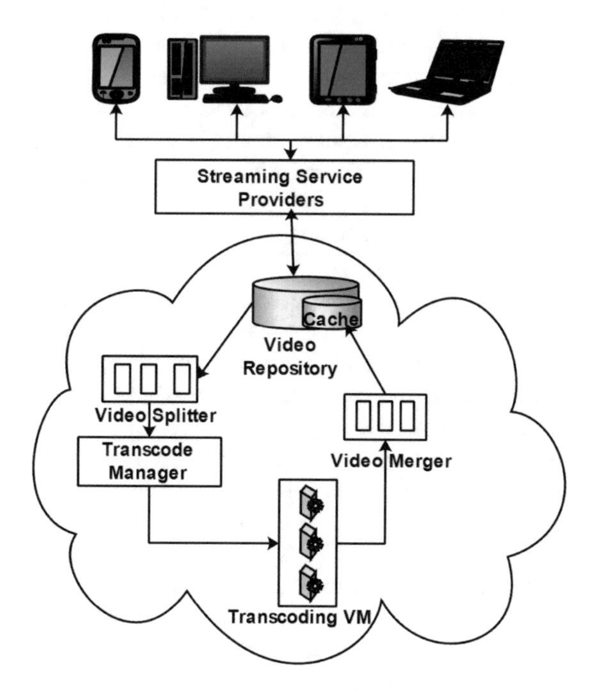

Fig. 16.2 Cloud-based Transcoding Architecture

The *transcode manager* dispatches the transcoding jobs (video segments) to appropriate VMs. The goal of the manager's mapping of jobs to VMs is to satisfy some user related QoS like minimum startup delay, cost, etc. It is the responsibility of the manager to scale the capacity of the transcoding service up and down by adding and removing VM instances based on user demands.

The *transcoding VM* is used to transcode the source videos into targeted videos with desired video specification concerning format, resolution, quality, etc. with certain QoS constraints. Each transcoding VM is capable of processing one or more simultaneous transcoding task. There are two possibilities to do transcoding inside a VM. First, all segments of a video are transcoded in a single VM and second different parts of a video in different VMs simultaneously. Since the first approach is centralized user may have to wait for a video segment if it is not transcoded at the time of the request arrival. But it eliminates the overhead of maintaining different segments from the different VMs. The advantage of the second strategy is that always there is something to serve the user, as several segments are transcoded at a time. This method suffers if VM does not supply the asked segment at the time of the request.

Video Merger is used to place all the video streams in the right order to create the resulting transcoded stream. After the transcoding operation, a copy of the transcoded video is stored in the video repository to avoid repetition of the video transcoding process for further request of a video.

From the literature study, we found that video access rate follows a long tail distribution, i.e., only a few videos (i.e., Popular) are frequently accessed, whereas the user rarely streams, many other videos. All the possible forms of popular and frequently accessed videos are stored in cache storage. The unpopular video requested by the user is transcoded online and served to the user. Along with this essential component, some researchers used prediction module for transcoding on the fly to reduce waste of bandwidth and terminal power [23, 26, 27]. Xiaofei et al. use prefetching and social network aware transcoding [28]. Adnan et al. and Zixia et al. proposed stream-based and video proxy based transcoding architecture respectively [22, 29]. The overall process must be completed before the deadline, i.e., the delivery time of the video frame.

Figure 16.3 shows the overall working of a cloud-transcoding process. When a user request arrives at the streaming service provider the video repository invoked to check whether the video is present in required form or not. If the video is in inquired form, then it is directly served to the user. If required form is missing in the video repository, then online transcoding is implemented to produce demanded form. The first step of the online transcoding process is to break the video into several segments for easy processing. Video splitter is used to divide original video into several video chunks. These video chunks are forwarded to one of the transcoding VMs. The transcoding VMs convert this chunks into requested format. In the end, video merger is used to arrange the transcoded video chunks of a video and store them in the repository so that new request of the same video format can be directly served. As soon as all the operations are over, the streaming service provider start streaming the requested video.

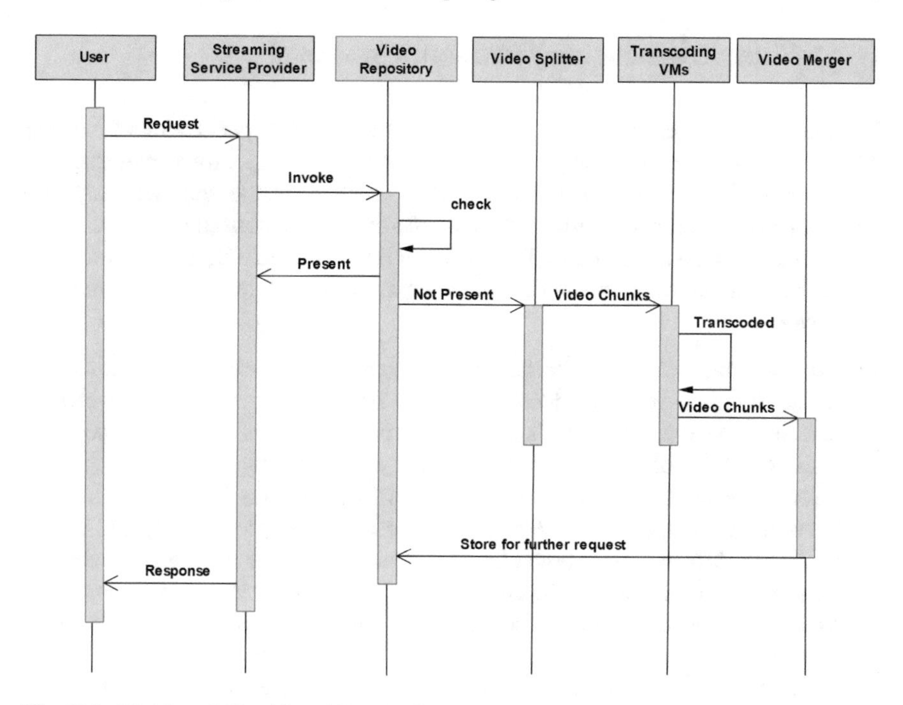

Fig. 16.3 Working of Cloud-based Transcoding Process

Table 16.1 Time in different phases of a transcoding process

Time	Definition
Reach Time (RT)	Time for a request to reach from user to the streaming server
Check Time (CT)	Time spent by the streaming server to check video repository for required video format
Split Time (ST)	Time to split the original video into several video chunks
Transcode Time (TT)	Time to transcode video segments into requested format
Merging Time (MT)	Time for arranging and storing transcoded segments of a video
Response Time (RT)	Time to give response to user by the streaming service provider

Table 16.1 shows time for different phases of online video transcoding process, starting from user request reach to a streaming service provider to response sent by the service provider. Let D be the deadline of a frame to be delivered then total time taken to transcode a frame must be less than or equal to D as shown in Eq. 16.1.

$$RT + CT + ST + TT + MT + RT \leq D \tag{16.1}$$

16.4 Video Transcoding Techniques in Cloud

The cumbersome transcoding is simple, take less time and pocket-friendly in the cloud as compared to in-house process. A user only needs to specify its requirements and subscribe the services provided by the cloud which is only a single click away. The rest of the task, i.e., resource allocation and time-consuming transcoding process will be performed in the background. Finally, the user is charged only for the resources consumed without much overhead. Following are different transcoding techniques in the cloud.

A. **Pre-transcoding**: One of the ways to deal with the video transcoding problem is pre-stored multi-version videos. This process stores various transcoded versions of each video to serve all user's requests and different types of client devices. The main drawback of this approach is the usage of large storage and computation resources. Even though pre-transcoding is a widely used method in VoD industry, but it induces a high cost. The increase in cost is due to provisioning and upgradations of fast-growing transcoding infrastructure, storage overhead to maintain all versions of all videos. It becomes cost prohibitive for small and medium size streaming service providers. The explosive growth of video streaming demands on a broad diversity of the client devices makes it ineffective [13, 30].
B. **On-demand Transcoding**: On-demand/real-time/online video transcoding performs conversion only when a particular format is not present in the storage. This method reduces the expenses of storage as it eliminates the need to store all versions. The on-demand resource reservation in the cloud makes the video transcoding process simpler and less costly. A leading content delivery network provider Akamai uses on-demand transcoding. Along with all the advantages, some challenges associated with this method are over/under-provisioning of cloud resources. The reserved resources must be future ready. Transcoder performs on-demand transcoding only for the segment that is not present in storage but requested by the user [30]. A QoS and cost aware on-demand transcoding of video streams using cloud resources presented in [13] with the aim of minimizing incurred cost of stream providers. Both on-demand and pre-transcoding are done to reduce the transcoding overhead. Pre-transcoding of video based on popularity based prediction is done into specific formats and resolution when the transcoding request falls (e.g. at night). A user is given several choices if its required format is not in local storage. If a user disagrees with the options, on-demand transcoding is done to generate user-specified form [26]. In [19] QoS-aware online video transcoding in the cloud is presented. Akamai advocates transcoding of multiple files in parallel for a fast turnaround time. Akamai the leading content delivery network (CDN) service provider of media and software distribution, uses stream packaging feature to detect the set of stream formats required by the user device and do the formatting and packaging of video streams at the network edge servers on-demand. The computation at edge server eliminates the shortcomings of pre-transcoding and centralized server transcoding process

as it reduces additional storage cost and content management overhead [31]. Transcoding improves throughput (bandwidth) while preserving quality.

C. **Partial Transcoding**: In partial transcoding, some segments of a video are converted to other formats and stored in cloud storage. Based on the user viewing pattern rest segments of a video is transcoded on-demand. The purpose of doing so is to reduce operational cost, i.e., transcoding cost plus storage cost. Another approach is to store multiple versions of a popular segment and only one highest quality version for unpopular segments. If a user request for a segment that is not in the cloud storage, do the transcoding in real-time. Authors in [32] proposed partial transcoding scheme for content management in media cloud. Initially, a partial subset of video contents in different bitrates is stored in local storage that can be directly consumed. If the user specified format is not present in local storage, then online transcoding is done. The purpose of this approach is to reduce long-run overall cost. In [30] partial transcoding proposed based on user viewing pattern.

D. **In-network Transcoding**: Here, transcoding service can be placed on nodes inside the network, i.e., the introduction of a transcoding service into a network infrastructure. But, it requires routers that support this service processing. Further service placement of routers requires redesigning of the underlying network architecture. Thus, the practical implementation of this is not applicable to the existing network architecture [30].

E. **Bit-rate Reduction Transcoding**: A bitrate is the number of bits processed per unit time i.e. usually bits per second. The data rate for a video file is the bitrate. Example: bit rate of a standard definition DVD is 6 megabits per second (mbps) whereas video for phones is in kilobits per second (kbps). As the high rate video demands high network bandwidth, the video stream bit rate is reduced to ensure smooth streaming. This process is also known as transrating. Video segmentation is used to perform bit rate reduction transcoding in a distributed computing environment. The distributed transcoder do transcoding of different segments parallelly on several machines to speed-up transcoding process [25]. Cloud computing is an extension of distributed computing so the parallel bit-rate reduction transcoder can be implemented with or without any modification. This paper [13] introduces bit-rate, spatial and temporal resolution reduction transcoding techniques.

   Video data at 1920×1080 pixels, 60 frames per second means original frame size is 1920×1080 pixels and each pixel is sampled 60 times per second. Spatial resolution ascertains information related to each frame and temporal resolution defines the change between frames represented by frames per second. Example 1080 HD is a case of spatial resolution or containing more pixels, but 720 HD is a case of temporal resolution or containing more frames per second. Based on the reduction in the temporal or spatial domain following type of transcoding are possible.

F. **Spatial-Resolution Reduction Transcoding**: The spatial resolution indicates the encoded dimension size of a video. In spatial resolution reduction transcoding macro-blocks of an original video are removed or combined without sacrificing the video quality. Spatial resolution reduction transcoding produces an output video with a smaller resolution, but with same frame rate than the original video. In [24] spatial resolution reduction transcoding implemented in multi-core distributed system. Since the spatial resolution reduction method reduces the resolution, the transcoder has to deal with less number of bits. Since a virtual machine (VM) can have multiple cores, this process can be extended to the cloud environment.

G. **Temporal-Resolution Reduction Transcoding**: In a temporal resolution, transcoding frames are dropped to support frame rate of the client device and reduce the required transcoding time. To reduce the transcoding time a temporal resolution reduction transcoding is used for the cloud environment in [22]. Here the transcoder drops a small proportion of video frames from a video segment of continuous video content delivery to the user.

H. **Scalable Transcoding**: Scalable coding consists of a base layer (minimum bit-rate) and several enhanced layers (gradually increase the bit rate). Depending on the link capability of the user device one or more enhanced layer delivered along with a base layer. There is always something to play (i.e., base layer). In [28] an adaptive video streaming is proposed that can adjust the streaming flow with scalable video coding technique based on the feedback of link quality. This paper also discusses the advantage of this method like an efficient use of bandwidth. The video encoding strategy presented in Zencoder a cloud-based video encoding service, white paper [33] maintain a base format (e.g. MP4) that is playable on a broad range of devices. Then decide the number and type of renditions. For example, short duration (e.g. 1 min) videos can be easily downloadable, so only a few versions are sufficient. But long duration videos like movies cant be downloadable in a single go so need extra attention. As the user always expect a high-quality video, many renditions are required that can be used according to available network bandwidth and the user device.

Video transcoding can be performed in the client device or on-premise architecture or in any third party architecture like a cloud. The compute intensive and time-consuming transcoding job suffers from low processing power and energy sources of the client device (e.g. smart phone). So, it is not feasible to perform transcoding on client devices [13]. The in-house architecture suffers from scalability issues. Let there is infrastructure for a particular rendition sets. After sometimes a new set of renditions is required. To satisfy the new demand, the process of buying hardware and installing software will be time-consuming, cumbersome and costly. This process may also suffer from over/under-provisioning of resources which will have an adverse impact on the companys investment. Apart from economic inefficiency, there is also wastage of resources as most of the time servers are in an idle state [13, 19, 27]. Forecasting transcoding capacity is difficult due to the proliferation of delivery options, the unpredictability of the volume of video assets, a significant lead time required to acquire and prepare the necessary infrastructure [18]. To overcome the difficulties of in-house infrastructure to deal with the flash crowd, over/under-provisioning of resources, video service providers can look into cloud technology. The scalable cloud architecture allows on-demand resource reservation with less maintenance and cost. The compute intensive and time-consuming video transcoding can use on-demand cloud resources to reduce the expenses of video service providers [13, 19].

16.5 Performance Metrics

Performance metrics determined the benefit/lack of system designs. These are the metrics used to evaluate the working of a system in all conditions (favourable or unfavourable). The following section discusses various performance metrics used by researchers to assess their cloud transcoding system and listed in Table 16.2.

(i) Encoding Delay (Transcoding Time): It is the total time used to convert a video from one format to another.
(ii) Prefetching Delay: This is the time used to pre-fetch video segments for a user based on his/her social network service activity.
(iii) Watching (Click-to-play Delay): This is a time to wait for a user from clicking a video link to the first streaming arrival.
(iv) Transcoding Rate/Speed: It is the number of transcoding operation completed per unit time.
(v) Number of streams with Jitter (Transcoding Jitter): It is the measure of streams with jitter where jitter is the time difference in stream inter-arrival time.
(vi) Number of dropped frames: It is the count of frames being dropped to avoid deadline violation.
(vii) Queue waiting time: Waiting time of a stream in the queue to be encoded is known as queue waiting time.
(viii) Launch Latency: It is the time between the opening of a VM and its become ready to provide service.
(ix) End-to-End Delay: The end-to-end delay is the time between streaming provider starts to deliver streaming media to the time a user device starts playing it.
(x) Utilization Rate: It is defined as the ratio between used transcoding capacities to the total transcoding capacity of the system. Lower utilization rate indicates the larger idle time of the transcoding system.
(xi) Average Response Time: It is the difference between the time of transcoding request to the time of first transcoding response of the system.
(xii) Operational Cost: The overall operational cost is the combination of storage, computing and transcoding cost. The storage and computing cost is adopted from Amazon S3 and Amazon EC2 On-Demand instances respectively [32].
(xiii) VM Provisioning Cost (Cost): It is the cost to acquire resources in the cloud for transcoding process. To reduce the VM provisioning cost, the number of servers required to transcoding process should also be minimized [23].

Table 16.2 Performance metrics

Sl.No.	Reference no.	Performance metrics	Environment	Transcoding tool
1.	[27] [2015]	Cost	Cloud	Own Set-up
2.	[22] [2013]	Transcoding rate, Number of streams with Jitter, Queue waiting time	Cloud	Python with simpy framework
3.	[34] [2016]	Launch latency, End-to-End Delay, Utilization Rate, Average Response Time	Cloud	ffmpeg and Hi-cloud with Windows server 2008 operating system and Amazon EC2
4.	[32] [2015]	Operational Cost (storage cost + computing cost + transcoding cost)	Cloud	Own setup
5.	[29] [2011]	Transcoding Speed, Transcoding Jitters	Cloud	Own set-up
6.	[23] [2013]	VM Provisioning Cost	Cloud	Python with Simpy framework
7.	[26] [2012]	Average CPU Utilization, Cache Hit Rate, Data Transfer Rate	Cloud	ffmpeg
8.	[13] [2016]	Average startup delay, Average deadline miss-rate, Cost	Cloud	Cloudsim
9.	[28] [2013]	Pre-fetching delay, Watching delay, Encoding delay	Cloud	U-cloud server offered by Korean Telecom and JAVA based server application
10.	[19] [2017]	Transcoding time, Targeted Video Chunk Size	Cloud	Python, ffmpeg
11.	[35] [2014]	Energy	Cloud	Own set-up
12.	[30] [2017]	Storage and Transcoding Cost, Average Delay (startup)	Cloud	Own set-up

(xiv) Average CPU Utilization: It measures how efficiently CPU is utilized for transcoding process.

(xv) Cache Hit Rate: If the requested transcoded video is present in cache storage it is a hit. The cache hit rate indicates the ratio between a hit and total transcoded video request.

(xvi) Data Transfer Rate: Data Transfer Rate represents the transfer speed of the transcoded videos from the cloud to the user with the aim of saving user's energy consumption in retrieving the requested videos [26].

(xvii) Average Startup Delay: It is the delay occurred at the beginning of a video stream and represents the delay in receiving the first transcoded video stream [13].

(xviii) Average Deadline Miss-Rate: This delay occurs when a transcoding task misses its presentation deadline during streaming of a video [13].

(xix) Targeted Video Chunk Size: The average video chunk size influences the cloud resources, i.e., the number of CPU cores, so it should be carefully selected [19].

The study shows that transcoding time, i.e., the time taken by the system to transcode a video from one format to another format is an important metric. Other parameters used by researchers are transcoding rate/speed, i.e., the number of video transcoding per time unit. Some researchers used end-to-end delay (time difference between service requests in response), cost, etc. Operational cost, i.e., the cost used to acquire cloud resources such as storage, transcoding is also used by several researchers.

16.6 Conclusion

Initially, video transcoding performed to reduce file size, but now the perception has changed. Now transcoding is considered as a facility that provides options to choose. The options are picked to make the video viewable across platforms, devices, and networks. The cloud transcoding solution allows video service providers to pay as they use, with the assurance of providing online support to handle unpredictable needs, i.e., flash crowd. This paper discusses various aspects of performing a transcoding operation in cloud starting from why to shift the transcoding process to the cloud, to general architecture and metrics used to evaluate a cloud transcoding system. The system model designed for cloud-based video transcoding must consider the constraint, i.e., the online transcoding must be completed before the deadline of the frame. Future studies can be carried out to design and simulate a cloud-based transcoding model such that it will meet the constraint with less overhead. The overhead reduction may be concerning cost, scheduling, and provisioning of cloud resources, energy, response time, transcoding time, etc.

References

1. https://content.pivotal.io/blog/large-scale-video-analytics-on-hadoop
2. S. Sahoo, S. Nawaz, S.K. Mishra, B. Sahoo, Execution of real time task on cloud environment, in *2015 Annual IEEE India Conference* (INDICON) (New Delhi, 2015), pp. 1–5, https://doi.org/10.1109/INDICON.2015.7443778
3. S.K. Mishra, R. Deswal, S. Sahoo, B. Sahoo, Improving energy consumption in cloud, in *2015 Annual IEEE India Conference* (INDICON, New Delhi, 2015), pp. 1–6, https://doi.org/10.1109/INDICON.2015.7443710
4. S.K. Mishra, B.Sahoo, K.S. Sahoo, S.K. Jena, Metaheuristic approaches to task consolidation problem in the cloud, in *Resource Management and Efficiency in Cloud Computing Environments* (IGI Global, 2017), pp. 168–189, https://doi.org/10.4018/978-1-5225-1721-4.ch007
5. S. Sahoo, B. Sahoo, A.K. Turuk, S.K. Mishra, Real time task execution in cloud using mapreduce framework, in *Resource Management and Efficiency in Cloud Computing Environments* (IGI Global, 2017), pp. 190–209, https://doi.org/10.4018/978-1-5225-1721-4.ch008
6. S.K. Mishra, P.P. Parida, S. Sahoo, B. Sahoo, S.K. Jena, Improving energy usage in cloud computing using DVFS, in *International Conference on Advanced Computing and Intelligent Engineering (ICACIE)* (2016), http://hdl.handle.net/2080/2598
7. https://www.ericsson.com/res/thecompany/docs/publications/ericssonreview/2010/cloudcomputing.pdf
8. Kontron Whitepaper, Video optimization in the cloud (2014)
9. http://www.cisco.com/c/dam/en_us/about/ac79/docs/sp/Streaming_Under_the_Clouds.pdf
10. http://www.internetlivestats.com/internet-users/india/
11. https://www.akamai.com/uk/en/our-thinking/state-of-the-internet-report/state-of-the-internet-connectivity-visualization.jsp
12. S. Ko, S. Park, H. Han, Design analysis for real-time video transcoding on cloud systems, in *Proceedings of the 28th Annual ACM Symposium on Applied Computing* (SAC '13) (ACM, New York, 2013), pp. 1610–1615, https://doi.org/10.1145/2480362.2480663
13. X. Li, M.A. Salehi, M. Bayoumi, R. Buyya, CVSS: A cost-efficient and QoS-aware video streaming using cloud services, in *2016 16th IEEE/ACM International Symposium on Cluster, Cloud and Grid Computing (CCGrid)* (Cartagena, 2016), pp. 106–115, https://doi.org/10.1109/CCGrid.2016.49
14. S. Sahoo, I. Parida, S.K. Mishra, B. Sahoo, A.K. Turuk, Resource allocation for video transcoding in the multimedia cloud, in *International Conference on Advanced Computing, Networking, and Informatics (ICACNI)* (2017), http://hdl.handle.net/2080/2722
15. S. Sahoo, B. Sahoo, A.K. Turuk, An analysis of video transcoding in multi-core cloud environment. in *International Conference on Distributed Computing and Networking (ICDCN)* (2017), http://hdl.handle.net/2080/2643
16. https://www.wowza.com/blog/what-is-transcoding-and-why-its-critical-for-streaming (2015)
17. http://coconut.co/video-transcoding
18. http://download.sorensonmedia.com/PdfDownloads/LowRes/whitepaper.pdf (2011)
19. L. Wei, J. Cai, C.H. Foh, B. He, QoS-aware resource allocation for video transcoding in clouds. IEEE Trans. Circuits Syst. Video Technol. **27**(1), 49–61 (2017), https://doi.org/10.1109/TCSVT.2016.2589621
20. http://www.streamingmedia.com/Articles/Editorial/Featured-Articles/Buyers-Guide-to-Cloud-Based-Video-Encoding-and-Transcoding-2015-102483.aspx
21. http://download.sorensonmedia.com/PdfDownloads/LowRes/whitepaper.pdf (2011)
22. A. Ashraf, F. Jokhio, T. Deneke, S. Lafond, I. Porres, J. Lilius, Stream-based admission control and scheduling for video transcoding in cloud computing, in *2013 13th IEEE/ACM International Symposium on Cluster, Cloud, and Grid Computing, Delft*, (2013), pp. 482–489, https://doi.org/10.1109/CCGrid.2013.21

23. F. Jokhio, A. Ashraf, S. Lafond, I. Porres, J. Lilius, Prediction-based dynamic resource allocation for video transcoding in cloud computing, in *2013 21st Euromicro International Conference on Parallel, Distributed, and Network-Based Processing*, (Belfast, 2013), pp. 254–261, https://doi.org/10.1109/PDP.2013.44
24. F. Jokhio, T. Deneke, S. Lafond, J. Lilius, Analysis of video segmentation for spatial resolution reduction video transcoding, in *2011 International Symposium on Intelligent Signal Processing and Communications Systems (ISPACS)* (Chiang Mai, 2011), pp. 1–6, https://doi.org/10.1109/ISPACS.2011.6146194
25. F. Jokhio, T. Deneke, S. Lafond, J. Lilius, Bit rate reduction video transcoding with distributed computing, in *2012 20th Euromicro International Conference on Parallel, Distributed and Network-based Processing* (Garching, 2012), pp. 206–212, https://doi.org/10.1109/PDP.2012.59
26. Z. Li, Y. Huang, G. Liu, F. Wang, Z. L. Zhang, Y. Dai. Cloud transcoder: bridging the format and resolution gap between internet videos and mobile devices, in *Proceedings of the 22nd international workshop on Network and Operating System Support for Digital Audio and Video (NOSSDAV '12)*, (ACM, New York), pp. 33–38, https://doi.org/10.1145/2229087.2229097
27. A. Alasaad, K. Shafiee, H.M. Behairy, V.C.M. Leung, Innovative schemes for resource allocation in the cloud for media streaming applications. IEEE Trans. Parallel Distrib. Syst. **26**(4), 1021–1033 (2015), https://doi.org/10.1109/TPDS.2014.2316827
28. X. Wang, M. Chen, T.T. Kwon, L. Yang, V.C.M. Leung, AMES-cloud: a framework of adaptive mobile video streaming and efficient social video sharing in the clouds. IEEE Trans. Multimed. **15**(4), 811–820 (2013), https://doi.org/10.1109/TMM.2013.2239630
29. Z. Huang, C. Mei, L.E. Li, T. Woo, CloudStream: delivering high-quality streaming videos through a cloud-based SVC proxy, in *2011 Proceedings IEEE INFOCOM* (Shanghai, 2011), pp. 201–205, https://doi.org/10.1109/INFCOM.2011.5935009
30. H. Zhao, Q. Zheng, W. Zhang, B. Du, H. Li, A segment-based storage and transcoding trade-off strategy for multi-version VoD systems in the cloud. IEEE Trans. Multimed. **19**(1), 149–159 (2017), https://doi.org/10.1109/TMM.2016.2612123
31. https://www.akamai.com/us/en/multimedia/documents/content/streaming-toward-televisions-future-4k-video-white-paper.pdf
32. G. Gao, W. Zhang, Y. Wen, Z. Wang, W. Zhu, Towards cost-efficient video transcoding in media cloud: insights learned from user viewing patterns. IEEE Trans. Multimed. **17**(8), 1286–1296 (2015), https://doi.org/10.1109/TMM.2015.2438713
33. https://www.brightcove.com/en/blog/2013/08/new-whitepaper-architecting-a-video-encoding-strategy-designed-for-growth
34. K.B. Chen, H.Y. Chang, Complexity of cloud-based transcoding platform for scalable and effective video streaming services, in *Multimedia Tools and Applications* (2016), pp. 1–18, https://doi.org/10.1007/s11042-016-3247-z
35. W. Zhang, Y. Wen, H.H. Chen, Toward transcoding as a service: energy-efficient offloading policy for green mobile cloud. IEEE Netw. **28**(6), 67–73 (2014), https://doi.org/10.1109/MNET.2014.6963807

Chapter 17
Vehicular Clouds: A Survey and Future Directions

Aida Ghazizadeh and Stephan Olariu

Abstract Vehicular clouds have become an active area of research with tens of papers written and a large number of documented applications. We feel this is a good moment to summarize the main research trends in the area of vehicular clouds and to map out various research challenges and possible applications. With this in mind, the primary objective of this chapter is to present a survey of the state of the art in vehicular clouds, of the current research topics and future directions. We will take a critical look at the various vehicular cloud models proposed in the literature and their applications.

Keywords Vehicular clouds · Intelligent transportation systems · Cloud computing

17.1 Introduction and Motivation

It is intuitively clear that, in order to be effective, Intelligent Transportation Systems should acknowledge and employ the most recent technologies specially in the realm of networks and communications. Back in the 1980s, visionaries predicted that the ability to network vehicles together would be a game changer, redefining the fashion we use our vehicles and roads and provide safer and more efficient transportation means which eventually leads to the creation of the smart cities of the future [64]. This idea has inspired scientists to extend the concept of Mobile Ad-hoc Networks (MANET) to roads, highways and parking lots and create the new notion of *vehicular networks*. Vehicular Network uses many forms of communication such as Vehicle-to-Vehicle (V2V) and Vehicle-to-Infrastructure (V2I) communications in order to enhance the drivers' awareness of impending traffic conditions. In particular, it has been shown that if the drivers are given advance notification of traffic events, they can

A. Ghazizadeh (✉) · S. Olariu
Department of Computer Science, Old Dominion University, Norfolk, VA, USA
e-mail: aghaziza@cs.odu.edu

B. S. P. Mishra et al. (eds.), *Cloud Computing for Optimization: Foundations, Applications, and Challenges*, Studies in Big Data 39,
https://doi.org/10.1007/978-3-319-73676-1_17

make informed decisions that either prevent congestion from occurring or contribute to mitigating its effects [71, 82, 85, 86].

Not surprisingly, in the past two decades, there has been an increasing interest in vehicular networking and its countless possible applications and services. To support vehicular networking and to promote its integration with Intelligent Transportation Systems, the US Federal Communications Commission (FCC) has allocated 75 MHz of spectrum in the 5.850–5.925 GHz band for the sole benefit of Dedicated Short Range Communications (DSRC) [79].

The research community was quick to notice that the allocated DSRC spectrum was excessive for the current designed transportation applications [64]. The excess available bandwidth can be used for launching third-party applications such as vehicular peer-to-peer communication, video and content delivery, vehicle entertainment, safety, travel and commercial applications.

It has been noticed [68] that under present-day state of the practice, the vehicles on our roads and city streets are mere spectators that witness traffic-related events without being able to participate in the mitigation of their effect. Recently, it has been shown that in such situations the vehicles have the potential to cooperate with various authorities to solve problems that otherwise would either take an inordinate amount of time to solve (e.g. routing traffic around congested areas) or could not be solved for lack for adequate resources [82, 86].

Back in 2010. Abuelela and Olariu [1] called on the vehicular networking community to "take vehicular networks to the clouds" to use the vehicular resources available to improve safety, increase efficiency and help the economy. That call to action has led to an explosion in the number of papers that dealt with various aspects and applications of vehicular cloud computing. Indeed, it is fair to say that in recent years, vehicular clouds have become an active area of research with a large number of documented applications [68, 84].

The concept of vehicular cloud and its possible applications are recent and so are many challenges and obstacles in the realm of research and implementation [89]. The main goal of this chapter is to provide a critical review of recent research in vehicular clouds.

At this point, we feel it is appropriate to give the reader a synopsis of the chapter. Section 17.2 offers a brief review of cloud computing and cloud services that has motivated the vision of vehicular clouds. Next, in Sect. 17.3 we offer a brief introduction to Mobile Clouds and review their main components and characteristics. Section 17.4 introduces the earliest models of vehicular clouds and discusses some of their attributes. Section 17.5 introduces generic system assumptions about VCs including a description of the vehicular model, a quick review of virtualization and virtual machine migration as well as a discussion about the merits of static and dynamic VCs. Next, Sect. 17.6 offers a generic look at a generic VC architecture. Finally, Sect. 17.11 offers concluding remarks and maps out directions for future investigations.

17.2 Cloud Computing

Professor McCarthy, an MIT computer scientist and cognitive scientist, was the first to suggest the idea of *utility computing* in 1961. He envisioned that in future computing power and applications may one day be organized as public utility and could be sold through a utility business model, like water or electricity. *"Computing may someday be organized as a public utility just as the telephone system is a public utility. Each subscriber needs to pay only for the capacity he actually uses, but he has access to all programming languages characteristic of a very large system"* Professor McCarthy said at MIT's centennial celebration in 1961.

Cloud computing (CC) is a catchy metaphor for *utility computing* implemented through the provisioning of various types of *hosted services* over the Internet [11]. Indeed, it has been argued [4] that CC was inspired by the concept of utility computing and was enabled by the availability of infrastructure providers whose surplus computational resources and storage that can be rented out to users. In this context, a user may purchase the amount of compute services they need at the moment. As their IT needs grow and as their services and customer base expand, the users will be in the market for more and more cloud services and more diversified computational and storage resources. This helps users to not worry about over-acquiring of resources and services that might not meet their needs in future and protects them from being obligated to pay for what they don't use. As a result, the underlying business model of CC is the familiar *pay-as-you-go* model of metered services, where a user pays for whatever he/she uses and no more, and where additional demand for service can be met in real time [11, 56, 57].

The term "Cloud" started to gain popularity when Google's CEO Eric Schmidt introduced the term Cloud Computing at an industry conference in 2006. Weeks later, Amazon used the term *Cloud* when launching Amazon Elastic Compute Cloud (EC2). Jumping on the bandwagon, the National Institute of Standards [62] defined CC as *"A model for enabling convenient, on-demand network access to a shared pool of configurable computing resources (e.g., networks, servers, storage, applications, and services) that can be rapidly provisioned and released with minimal management effort or service provider interaction"*.

The past decade has witnessed the amazing success of CC, a paradigm shift adopted by information technology (IT) companies with a large installed infrastructure base that often goes under-utilized [11, 13, 16, 56, 90]. The unmistakable appeal of CC is that it provides scalable access to computing resources and to a multitude of IT services. Not surprisingly, CC and cloud IT services have seen and continue to see a phenomenal adoption rate around the world [63]. In order to achieve almost unbounded scalability, availability and security, CC requires substantial architectural support ranging from virtualization, to server consolidation, to file system support, to memory hierarchy design [16, 19, 56]. With the help of distributed and parallel computing, virtualization and high-speed Internet that is available for affordable prices, users easily can rent softwares and infrastructures required for their business [39, 56].

17.2.1 A Review of Cloud Services

CC has been associated with *hosted services* over the past decade. The idea that IT and specialized services and storage and computational resources are available on demand creates the opportunity for business to invest in what they need at the moment and acquire more services as their business grow. Similarly, software developers can easily test their applications and develop their ideas on platforms that are offered by CC.

There are many benefits and opportunities that are provided by CC, including the followings:

- It provides the illusion that infinite computing and storage resources and services are available therefore it helps users to save time on planning ahead and also reduces any potential risks of acquiring excess services.
- It reduces the upfront financial commitments and unnecessary investments for the users and companies and it helps businesses to begin their plans as small as they want and develop and expand it as the number of their customers grow and as their needs escalate.
- It allows users to purchase computing, storage and IT services and resources on a short-term basis, for example for months, days or hours. It also gives users the ability to release these resources as needed, therefore providing more flexibility and also encouraging conservancy.

There are three basic types of conventional cloud services:
Infrastructure as a Service (IaaS): Here the cloud provider offers its customers computing, network and storage resources. A good example is Amazon Web Services (AWS), where Amazon provides its customers computing resources through its Elastic Compute Cloud (EC2) service and storage service through both Simple Storage Service (S3) and Elastic Book Store (EBS) [2];
Platform as a Service (PaaS): PaaS solutions are development platforms for which the development tool itself is hosted in the cloud and accessed through a browser. With PaaS, developers can build web applications without installing any tools on their computers and then deploy those applications without any specialized systems administration skills. Google AppEngine [35] and Microsoft Azure [58] are good examples of this category;
Software as a Service (SaaS): With SaaS a provider licenses an application to customers as a service on demand, through a subscription, in a "pay-as-you-go" model. This allow customers to use expensive software that their applications require, without the hassle of installing and maintaining that software. GoogleAppEngine and IBM [45] are good examples of this category.

17.3 Mobile Cloud Computing

Mobile phones and other mobile devices have become an integral part of our lives. The daily time used on mobile phones and the average daily Internet usage for mobile devices have increased drastically in the recent years. Nowadays consumers demand sophisticated applications, unlimited storage and computing power, however the mobile devices do not have powerful storage and computational power (e.g., CPU and memory) and they provide limited resources.

CC offers users advantages such as servers, networks, storage, platforms and applications at a low cost. These advantages can be used in mobile devices and allow mobile users to not compromise the service quality and rich experience that they desire. Mobile Cloud Computing (MCC) is an integration of CC in the mobile devices and mobile environments [6]. In MCC, the data processing and storage are moved from mobile devices to powerful centralized computing platforms in the cloud. The MCC users can then access these data and the storage services over the wireless connection based on native thin clients or web browsers.

The Mobile Cloud Computing Forum defined MCC as *"An infrastructure where both the data storage and data processing happen outside of the mobile device. Mobile cloud applications move the computing power and data storage away from mobile phones and into the cloud, bringing applications and MC to not just smartphone users but a much broader range of mobile subscribers"*.

In MCC mobile devices such as smartphones, laptops or personal digital assistants (PDA) are connected to the network through the help of base stations, transceiver stations, access points or satellites. Mobile user's requests and information are delivered to the cloud through the Internet. The mobile network operators can provide authentication, authorization and accounting services for the data that is stored in the databases. Cloud controllers then process the request and information and provide users with the corresponding and requested cloud services. Monitoring and calculating the functions of the system will guarantee that the QoS of the system is maintained until the task is completed.

For a useful survey of MCC the readers are referred to [21] where important applications are discussed (Fig. 17.1).

17.4 Vehicular Clouds – How It All Started

In 2010, inspired by the success and promise of conventional clouds, a number of papers have introduced the concept of a *Vehicular Cloud*, (VC, for short), a non-trivial extension, along several dimensions, of the conventional CC paradigm. VCs were motivated by understanding that the current and future vehicles are equipped with powerful computers, transceivers and sensing devices. These resources are usually underutilized and employing them in a meaningful way has compelling economical and environmental impacts. Arguably, the first researchers that introduced the concept

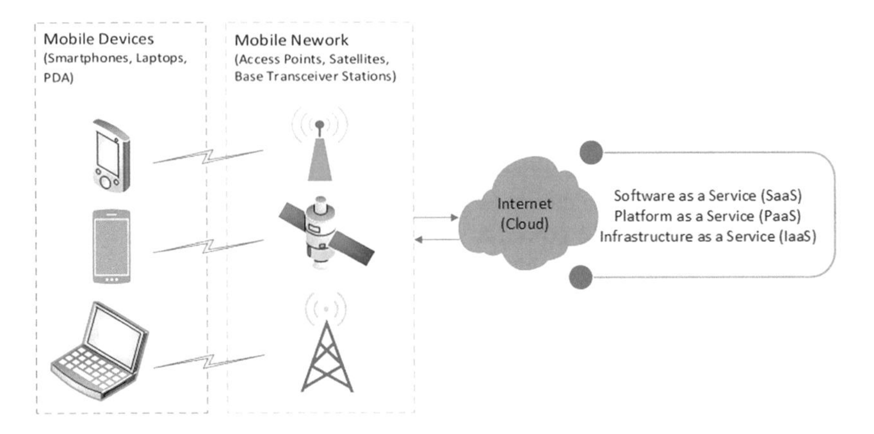

Fig. 17.1 Illustrating a possible MCC architecture

of a VC were Eltoweissy et al. [24], Olariu et al. [66, 67]. Their early papers defined various possible versions of VCs, their various applications and research challenges. Specifically, *autonomous clouds*, a precursor of VCs, were proposed for the first time by Olariu et al. [66] where they also provided an overview of a number of important applications and research challenges. Later, Florin et al. [25] have extended the VC model of Eltoweissy et al. [24] to accommodate the computational needs of military deployments and tactical missions.

The insight that led to VC was that by providing financial opportunities and encouragements, the drivers and owners of vehicles can rent their excess computational and storage resources to customers. This approach is similar to the approach of large companies and corporations that rent out their excess resources in exchange for financial benefits. For example, Arif et al. [3] suggested that in the near future air travelers will park and plug their vehicles in airport long-term parking lots. In return for free parking and other perks, they will allow their vehicles to participate, during their absence, in the airport-run datacenter.

A number of researchers have pointed out that even under current state of the practice, many implementations of VCs are both technologically feasible and economically viable [44, 66, 67, 84]. Given their large array of applications, it is reasonable to expect that, once adopted, VCs will be the next paradigm shift, with a lasting technological and societal impact [68]. It is somewhat surprising that, in spite of a good deal of theoretical papers, no credible implementation attempts of VCs have been reported in the literature. The only notable exceptions are the work of Lu et al. [54] and Florin et al. [26].

One of the fundamental ways in which VCs differ from conventional clouds is the ownership of resources. In VCs, the ownership of the computational resources is scattered over several owners as opposed to a single owner as is the case of conventional clouds run by companies such as Amazon, Google and IBM. A corollary of this is that the resources of the VC are highly dynamic. As vehicles enter the VC, new resources become available while others depart, often unexpectedly, creating a volatile environment where the task of reasoning about job completion becomes very challenging.

Gu et al. [36] published a survey paper where they reviewed key issues in VC mostly concerning its architecture, inherent features, service taxonomy and potential applications. More recently, Whaiduzzaman et al. [84] offered an updated perspective of various research topics in VCs.

Arif et al. [3] were the first to look at the feasibility of datacenters running on top of the vehicles parked at a major airport. Given time-varying arrival and departure rates, they proposed a stochastic model predicting occupancy in the parking lot. Specifically, they provided a closed form for distribution of parking lot occupancy as a function of time, for the variance and number of vehicles. In addition to analytical results, they have obtained a series of empirical results that confirm the accuracy of their analytical derivations and the feasibility of these datacenters.

Further motivation for investigating VCs and their applications was provided by Olariu et al. [68] where it was suggested that VCs are the next paradigm shift, taking vehicular networks to the next level of relevance and innovation. The authors of [68] have also hinted to an entire array of possible applications of VCs. Since the publication of [68], numerous papers have shown how VCs can enhance the solution to many problems of interest in security and privacy, reliability and availability, intelligent transportation systems and the like. Some of these recent papers will be summarized in this chapter.

17.5 Vehicular Clouds – Generic System Assumptions

The objective of this section is to provide an overview of the assumptions about the capabilities of vehicles, virtualization, VM migration and data replication strategies necessary to support the execution of a large variety of user jobs. A high-level description of a possible VC architecture will be presented in Sect. 17.6.

17.5.1 The Vehicle Model

The past decade has seen significant advances and growing interest in vehicular networks and Intelligent Transportation Systems (ITS) with its array of potential applications. The improved availability of wireless Internet in vehicles has encouraged the appearance of innovative applications and services related to traffic management and road traffic safety [64, 87]. At the same time, we have witnessed an unmistakable trend towards development of smart vehicles and making our cities safer and the transportation experience more enjoyable and more comfortable [73].

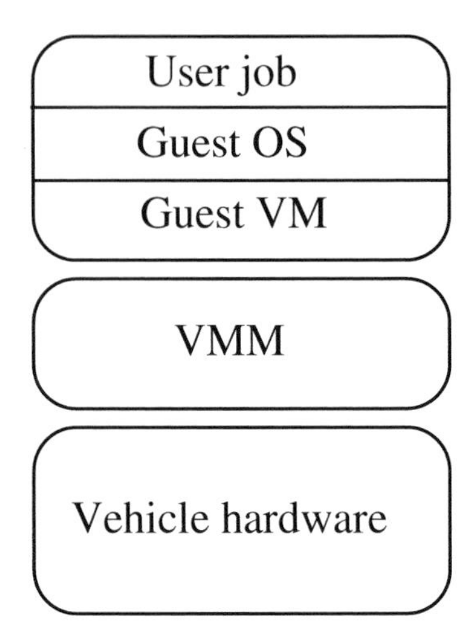

Fig. 17.2 Illustrating a virtualization model for VCs

A typical vehicle is usually equipped with several devices and capabilities such as sensing devices and transceivers, GPS, tracking devices, camera and powerful on-board computers and storage [76].

As will be discussed later in the chapter, we assume that, as illustrated in Fig. 17.2, each vehicle has a virtualizable on-board computer and has been pre-loaded with a suitable Virtual Machine Monitor (VMM). Because of their sophisticated compute capabilities and ample storage, our vehicles are good candidates for servers in a Warehouse-Scale Computer [10].

It is reasonable to assume that in the case of static VCs, the vehicles in the parking lot are plugged into a standard power outlet (to save their battery power) and are provided Ethernet connection to the datacenter.

17.5.2 Virtualization, the Workhorse of VCs

The Virtual Machine (VM) abstraction is one of the pillars of conventional CC [16, 39, 56]. The same concept remains fundamental in VC. While the type of service provided by the VC is largely immaterial, to fix the ideas, in the following discussion we assume that the VC offers IaaS cloud services. It is well known that in IaaS, the users can acquire virtualized computing resources. They are able to choose an operating system and receive a VM with an installed instance of their desired OS. As illustrated in Fig. 17.2, the VC offers the user a virtualized instance of the desired hardware platform and operating system bundled as a VM and guest OS. For fault-tolerance purposes, each user job is assigned to multiple cars. When the VM running the user job on a vehicle terminates execution, the result is uploaded to the datacenter.

As discussed in Sect. 17.5.1, we assume that each vehicle has a *virtualizable* on-board computer similar, but not necessarily identical, to the Intel Itanium, AMD Athlon or Opteron lines of processors. Vehicles are assumed to have been pre-loaded with a VMM in charge of mapping between the guest VM and the host car's resources. Naturally, to increase system performance, VM consolidation is essential [11, 12, 39]. Evaluating VC performance metrics under various VM consolidation disciplines is an interesting and worthwhile research topic and is being addressed by one of the authors of this chapter [89].

However, there are some contexts where the main concern is system dependability and availability only, and not system performance [30]. In those contexts it is acceptable to assume that each vehicle in the VC runs at most one VM.

It is important to distinguish between vehicle residency times in the VC and VM lifetime. For this purpose, it is useful to remember that VMs are associated with user jobs while vehicle residency times reflect the amount of time vehicles spend in the VC. We say that a vehicle leaves *prematurely* when it departs the VC before the VM that it is hosting has terminated execution.

17.5.3 VM Migration and Data Replication

Once a vehicle is ready to leave the VC, a number of actions need to be undertaken. First and foremost, if the vehicle is hosting a guest VM, the VM and all intermediate data stored by the departing vehicle must be *migrated* to an available vehicle in the VC. There are several known strategies for VM migration [9, 48, 49, 72].

In spite of its fundamental importance, somewhat surprisingly, only a few authors addressed VM migration in VCs such as Baron et al. [9], Refaat et al. [72] and Florin et al. [26]. For the purpose of this chapter we are not suggesting any specific migration discipline, although the topic is very important and impacts reliability, system availability and job duration time [89].

Inspired by the data replication strategies adopted by the Google File System (GFS) [33] and the Google record storage system Bigtable [17], in support of reliability, dependability and availability, the datacenter will mandate data replication for multi-stage user applications, such as MapReduce.

As an example, at the end of the Map stage the intermediate result, in the form of a set of key-value pairs will be stored by a number of vehicles in the VC. Likewise, when a VM belonging to a multi-stage user job has to be migrated, it will go (if at all possible) to one of the vehicles that stores relevant intermediate data for that job.

17.5.4 Static and Dynamic VCs

In their seminal papers Eltoweissy et al. [66] and Olariu et al. [67] have argued that in the near future, vehicles have the ability to use their computing resources in a cooperative manner to self-organize into VCs to solve a large number of crucial problems. While stationary and static VC may mimic the behavior of a conventional CC, we should remember that vehicles do not have an absolute static nature and they are involved in several dynamic situations and also unexpected circumstances such as accidents and traffic congestions.

VCs are still in their infancy and likewise the myriad potential applications and novel research challenges facing the VC research community. While a good number of papers were written about various flavors of VCs, there are virtually no studies concerning the practical feasibility of VCs. As already mentioned, a number of recent papers proposed VCs built on top of moving vehicles interconnected by some form of wireless communication fabric; some other authors consider a mix of moving and stationary vehicles. However, as pointed out by Florin et al. [26] their extensive simulations have shown that VCs based on wireless communications (as are, per force, all those inherent to moving vehicles) do not appear to be able to support data-intensive applications effectively. It is our hope that this negative result might serve as a wakeup call to the VC community.

One of the significant research challenges in the realm of VCs is to *identify* conditions under which dynamic VCs can support data-intensive applications. Such applications, with stringent data processing requirements might not be efficiently supported by *ephemeral* VCs, that is to say, VCs where the residency time of vehicles in the cloud is too short for supporting VC setup and VM migration. Similarly, VC implementations with bandwidth-constricted interconnection topologies are not the perfect choice for applications that consume or produce a large amount of data and therefore need to be running for an extended amount of time.

17.6 A High-Level Definition of a VC Architecture

The main goal of this section is to present a bird-eye's view of a possible VC architecture, similar to the one proposed by Florin et al. [26].

Throughout this chapter we deal with a datacenter supported either by a *static VC* built on top of vehicles parked in a adequately large parking lot, or else by a *dynamic* VC involving vehicles in a MVC (see Florin et al. [25]) or, perhaps, moving vehicles in the downtown area of a Smart City where they have full-time connectivity [37, 69, 80]. We assume that a sufficient number of vehicles participate in the VC that it is always possible to find one or more vehicles that can be assigned to an incoming user job.

The participating vehicles share their computational and storage resources to create a VC-based datacenter that then helps companies and corporations to make the most out of their underutilized resourced and potentially save thousands of dollars.

Conceptually, the architecture of the VC is almost identical to the architecture of a conventional cloud [11, 16, 56], with the important difference that the VC is far more dynamic than a conventional cloud. As an illustration, assume that the VC offers only IaaS cloud services. Recall that in IaaS, the users request a hardware platform and specify their preferred OS support. Referring back to Fig. 17.2, the VC offers a virtualized version of an OS based on user preferences and then the assigned jobs can run on the provided VM and guest OS on top of the vehicles. To reduce the rate of unsuccessful jobs and improve reliability, each job is assigned to several vehicles. When the VM running the user job in a specific vehicle finishes execution and the job is done, then the result is uploaded to the datacenter. In this scenario, the datacenter waits for the prescribed number of instances of the user job to terminate and makes a final determination by using, for example, a quorum-based algorithm or voting mechanism [74, 83, 91, 92].

We suspect that the datacenter is equipped with a system that identifies idle vehicles in the VC. In the simplest form, this can be implemented by assigning a status bit to each vehicle in the VC. Upon joining the VC, the status bit is set to available. Upon receiving a job, the datacenter updates the bit to unavailable. For purposes of fault-tolerance, the datacenter is able to assign multiple vehicles to a user job; this assignment utilizes some criteria based on locality. Upon a vehicle departure, the datacenter notifies all other vehicles assigned to the same job.

When a vehicle enters the VC, it will initiate communication with the *Resource Manager* component of the datacenter manager. We assume that the drivers of the vehicle will select an available parking spot at random. The Resource Manager identifies and keeps track of the location of the vehicle. Finally, the *Job Scheduler* identifies and picks the available vehicles for the assignment of jobs.

To get an idea of the type of processing that is going on, consider a user job submitted for execution by the datacenter and refer to Fig. 17.3 that offers a functional view of the datacenter manager module. The requirements of the submitted user job, in terms of the hardware platform that needs to be emulated, the requested OS and the user-specified input data are evaluated by the *Job Admission* daemon. If these requirements can be met by the datacenter, the job is admitted and gets inserted into a queue of jobs awaiting execution. Otherwise, the job is rejected. It would be of interest to mimic the Amazon AWS paradigm offering support for different types of jobs, described by the quality and quantity of system resources they need. To the best of our knowledge, this approach has not been reported on yet.

Once the job makes it to the front of the queue, control passes to the *Virtualization Agent* that bundles the resources specified by the user job into a VM, specific OS, and the input data to the VM. Finally, the Virtualization Agent passes control to the *Job Scheduler* that identifies one vehicle (or a group of vehicles) on which the job is to be executed.

Figure 17.4 presents the logical view of the communication structure of the datacenter. This communication structure is, conceptually, very similar to that of a standard datacenter (see Barroso et al. [11]). The analogy is better understood if one assumes a static VC involving the vehicles in a parking lot. Indeed, one can think of a vehicle in the parking lot as a server in a rack of servers. Accordingly, the parking lot is partitioned logically into clusters of parking spots, regions of clusters and so on. The communication in each cluster is under the control of a switch called the *Cluster Controller*. Similarly, the communication in a region is under the control of a switch called a *Region Controller*. Referring to Fig. 17.4, the parking lot is partitioned into

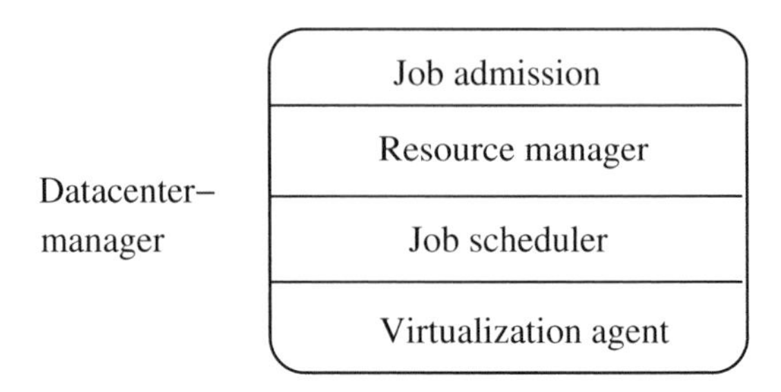

Fig. 17.3 Illustrating the functional view of the datacenter manager assumed in this chapter

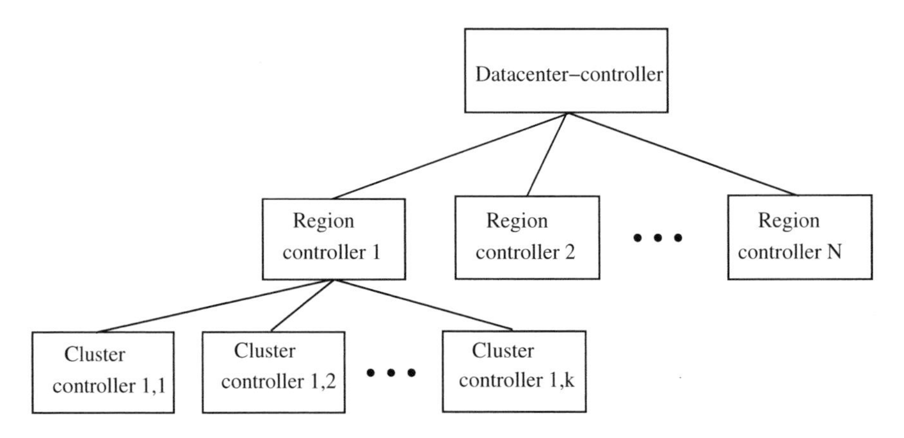

Fig. 17.4 Illustrating the logical view of the datacenter communications

N regions. Each such region consists of k clusters. The datacenter manager discussed above is in control of assigning user jobs to various vehicles in the parking lot.

The datacenter supplies a wired or wireless connection to give each vehicle access to a network fabric connecting each vehicle in the parking lot [3, 60, 61]. Also connected is data storage where data from each job is saved when the job is completed. When a vehicle to which a job was assigned leaves the VC prematurely, the datacenter notifies the other vehicles assigned to the same job. We further assume that, when the need arises, the datacenter can assign, using some suitably lightweight criteria, an available vehicles to the job [11, 56].

17.7 Vehicle Residency Times in the VC, Why Does It Matter?

Although very few authors seem to pay much attention to the residency time of vehicles in the VC, it is of fundamental importance to distinguish between vehicle residency times in the VC and VM lifetime. As mentioned before vehicle residency time shows the amount of time a vehicle spends in the VC. If a vehicle leaves the VC before finishing the job, meaning the VM that is running the job has not terminated, then we can say that the vehicle has left *prematurely*.

The VC literature is full of papers discussing fanciful VC models implemented on top of moving vehicles and supported by ubiquitous and pervasive roadside infrastructure. Those authors do not seem to be concerned with the obvious fact that moving vehicles' residency times in the VC may, indeed, be very short-lived and therefore so is their contribution to the amount of useful work performed. Worse yet, should a vehicle running a VM leave the VC prematurely the amount of work performed by that vehicle may be lost. Unless, of course, special precautions are being taken. Such precautionary measures involve either some flavor of checkpointing [23, 77] or else enforcing some form of redundant job assignment [30–32]. Both approaches have consequences in terms of overhead and impact the mean time to failure [32] and job completion time [28].

While such supporting infrastructure may well be available in the Smart Cities [34, 52], of the future, it does not exist today. This state of affairs is casting a long shadow of doubt on the feasibility of those VC models and is apt to discredit the entire VC concept. Similarly, zero-infrastructure VCs built on top of moving vehicle are utterly unrealistic since the DSRC-mandated V2V transmission range [75], combined with the limited bandwidth inherent to wireless communications, does not permit VM migration and/or data offloading in case vehicles exit the VC.

As a general comment, we note that for tractability reasons, to a first approximation, vehicular residency times in the VC are considered to be independent, exponentially distributed random variables with a common parameter.

17.7.1 A Taxonomy of Residency Times

Given the importance of residency times, it is useful to offer a more general look at VCs in terms of the assumed vehicular residency times in the VC.

- **Long residency times**: At one end of the spectrum, we find VCs characterized by long vehicular residency times. These can be further partitioned depending on whether or not the vehicles are assumed to be parked or mobile. For example, Arif et al. [3] and other researchers have assumed a VC is developed using the vehicles in the *long-term* parking lot of an international airport where residency times are days or maybe even weeks [32]. Intuitively, it is clear that the users of such a parking lot may share their transportation plans with the datacenter at the airport and so the residency times become almost deterministic.

  As another example of a VC characterized by long and stable vehicular residency times consider a plant that operates 24 hours a day, seven days a week. The patrons of the parking lot are working at the plant in eight-hour shifts, that creates a pool of vehicles that can serve as the substructure for a datacenter. We assume that the vehicles in the parking lot are plugged into a standard power outlet and are provided Ethernet connection to the datacenter.

Recently, Florin et al. [25] proposed Military Vehicular Clouds (MVC) an instance of VCs with (relatively) long vehicular residency times. In their vision, a MVC comprises either parked military vehicles or vehicles that move in formation so that, for all practical purposes, the distance between adjacent vehicles remains the same. This controlled mobility regimen applies to military units deployed in support of a tactical mission. The vehicles participating in the MVC have a wireless connection to a Mobile Access Point (MAP) which may be connected to a center of command and control. There are two major differences between MVCs and conventional clouds. First, unlike conventional clouds that are mostly public, MVCs are private, catering to the special needs of the military. Second, because of mobility and of the fact that military vehicles join and leave the MVC in ways that best support the tactical mission at hand, the MVCs are characterized by unpredictable availability of compute resources.

- **Medium residency times**: Some VCs are characterized by shorter vehicle residency times. Such is the case, for example, of VCs build on top of the vehicles in the parking lot of a shopping mall. Observe that the vehicular residency times at the mall are correlated with the average time a standard American spends at the shopping mall in one visit. While vehicle residency times at shopping malls are not directly available, there is a profusion of data concerning shopping pattern statistics – see, for example, [6, 40, 46, 47]. This statistical data indicates that, during normal business hours, when the datacenter services are required, the shopping center parking lot contains a sufficient number of vehicles which makes the task of finding available vehicles, for job assignment, feasible. Also, according to these statistics, the average American shopper spends about *two hours* per visit at the shopping mall. Of course, the exact residency time of any one individual is a random variable and so any form of reasoning here must be probabilistic. One of the challenges in these types of VCs is to identify the types of cloud services that can be offered.

  Yet another instance of a VC characterized by medium vehicular residency times are VCs set up in a Smart City [89] where the vehicles have connectivity through pre-deployed infrastructure provided by the municipality. In such a scenario, whether the vehicle are parked or mobile, as long as they reside within the Smart City (as the vast majority do) they participate in the VC without interruption of service [80].

  The challenge facing the implementation of a datacenter built on medium residency vehicles is to maintain high availability and reliability in the face of the dynamically changing resources.

- **Short residency times**: Finally, one can conceive of instances of VCs involving vehicles with a short residency times. These are vehicles that happen to be in an area that offers free Internet connectivity, such a gas stations or restaurants at highway interchanges, where they can be assigned jobs to process. The vehicles will process their jobs while moving (and not being connected) and will upload their results at the next interchange.

  As in the case of medium residency times, VC involving short-term vehicular residency times are very challenging to deal with and, to the best or our knowledge, no papers in the literature have reported on promoting high system availability and reliability in the face of the dynamically changing resources.

17.8 Services Supported by VCs

In [68] Olariu et al. have introduced three types of services that will be made possible by VCs: Network as a Service (NaaS), Storage as a Service (STaaS), and Cooperation as a Service (CaaS). Since then, other services were proposed. These services will be discussed briefly in Sects. 17.8.1–17.8.6 below.

17.8.1 Network as a Service (NaaS)

It is clear that, at the moment, not every vehicle on the road has Internet connection. Therefore the vehicles that have access to Internet can share their excess Internet with other drivers that request it. Many drivers have limited or unlimited Internet connection through 3G or 4G cellular network and these resources can be under-utilized because, for example, not all drivers are constantly downloading from the Internet. A driver that wants to share the Internet connectivity will advertise it to all the vehicles nearby. Considering the fact that the vehicles are moving on the same direction and with relative speeds, this system can be viewed as a traditional VANET, constructed of a set of static access points and mobile nodes that are moving at low speeds.

For example in Fig. 17.5, vehicle a, and c have 4G and vehicle d has WiFi connectivity through an access point. Vehicles a, d and c broadcast packets and inform other drivers about their intention to share their networks. If any vehicle is interested to rent these Internet services, the requests will be sent to a selected vehicle with a stable Internet connection. There are several factors such as reliability of the network, expected connection time, and speed of the network that should be considered for selecting a candidate.

17.8.2 Storage as a Service (STaaS)

Many vehicles have plenty of on-board storage resources and capabilities, however there are also some vehicles that need extra storage for running complex applications. In addition parked vehicles in the shopping malls or any large or medium-size parking lot can rent out their storage resources to the management of such places. Another example is using this excess storage in backup purpose, peer-to-peer applications and various types of content such as multimedia contents that are larger in size and require more storage support. STaaS in VCs has limitations related to mobility of the

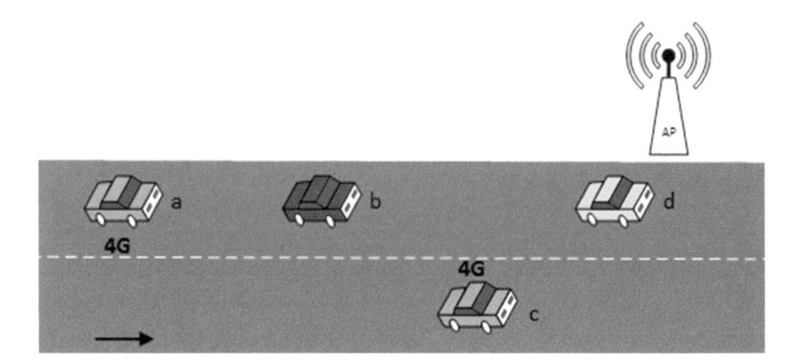

Fig. 17.5 Example of Naas Service in VCs

vehicles. The vehicles are not static forever; hence the users are able to use storage resources for an extended period of time. This limitation is an obstacle against renting storage as a service in VCs.

17.8.3 Cooperation as a Service (CaaS)

Mousannif et al. [59] have introduced Cooperation as a Service, as a new architecture that extends the two types of services in VCs, Network as a Service (NaaS) and Storage as a Service (SaaS). It aims to provide vehicles with a set of services for free, and without the need for increased roadside infrastructure. Vehicular networks can be help in many situations to improve the safety of transportation, decrease traffic congestion, and provide accident warnings, road condition and weather information, parking availability and advertisement [64]. ITS and available 3G and 4G networks can be useful for offering these services. In CaaS, which can be considered as a new form of community service, drivers can obtain a set of services using minimal roadside infrastructure. If no infrastructure is available, these vehicles can take advantage of V2V communication. V2V communications are made possible via DSRC or WAVE (Wireless Access in a Vehicular Environment) [79].

CaaS uses a hybrid publish/subscribe mechanism where the driver (or subscriber) informs the system about his wish to acquire a service and the vehicles/drivers that have subscribed to the same service will cooperate to provide the driver with the necessary information regarding the service, by publishing this information in the network. An example of a cooperative vehicular safety application is the Electronic Brake Warning (EBW) application. When a vehicle breaks sharply, it broadcasts an event-based message, alerting the surrounding vehicles of the braking incident. This information is especially useful in situations that the view of the break light is blocked and it will allow drivers to take necessary precautions. Another example is the Vehicle Stability Warning (VSW) application. In VSW, preceding vehicles alert following vehicles of upcoming potential hazardous conditions such as icy or slippery roads. VSW is an event-based application, similar to EBW. These applications require constant reliable communication between vehicles since they are not effective and reliable if the sent emergency messages are not received.

17.8.4 Computing as a Service (CompaS)

Statistics show that most vehicles are parked for several hours, every day, in parking lots, parking garages or driveways. The computing resources of these vehicles are an untapped and under-utilized and wasted under current state of the practice. The owner of the vehicles can use this opportunity and rent out their on-board capabilities on-demand for hours, days or weeks to customers. For example, in airport long-term parking lots, travelers will plug their vehicles and they will allow users or managements to use the computation power of their vehicles in return for free parking and other benefits. Another example is when vehicles are stuck in traffic jams, they are willing to allow municipal traffic management centers to use their resources and run designed simulations to find a solution for the traffic and alleviate the effects.

17.8.5 Information-as-a-Service (INaaS) and Traffic Information-as-a-Service (TIaaS)

Drivers need different types of information, for safety improvements, advance warnings, news about the sudden crashes or emergency situations. These services are recognized as Information-as-a-Service (INaaS). In [44], Hussain et al. have used a framework of VANET-based clouds namely VuC (VANET using Clouds) and defined another layer, named TIaaS (Traffic Information as a Service). TIaaS provides vehicles on the road with traffic information. Vehicles and drivers share their gathered information about the traffic situations, congestions, road conditions and etc. with the other vehicles and also with the cloud infrastructure. Users that want to get notifications about such information can subscribe to the TIaaS and receive the timely information accordingly.

17.8.6 Entertainment-as-a-Service (ENaaS)

Nowadays travelers seek entertainments to make their trips more enjoyable. Movies, commercials and games can be played on the vehicles screens and make the travel more entertaining and comfortable. These types of services are recognized as Entertainment-as-a-Service (ENaaS).

17.9 Applications of VC

The major objective of this section is to offer an overview of a number of VC applications. Many others, that were discussed in [68] will not be mentioned here.

17.9.1 Security and Privacy in VCs

When users are allowed to share pools of resources in a shared network, security and privacy questions and issues arise. In the past few years, many researchers have suggested solutions for such issues.

The first authors to investigate security and privacy issues in VC were Yan et al. [87] and Yan et al. [88]. They have shown that many of the insecurities found in conventional CC carry over to VCs. In addition, several VC-specific security challenges were identified and also preliminarily solutions were proposed. They have categorized the main targets of attacks into attacks related to confidentiality, integrity and availability. Examples for such attacks are finding the identities of users, personal and sensitive data, code and documents stored on the VC. Such attacks can be done in many ways, for example attackers pretend to be the user that is requesting the service or they can discover a bug or flaw in the system and get access to the sensitive and hidden data that they normally don't have the permission or access to. Yan et al. have argued that the security authentication in VC is challenging due to the mobile nature of the moving vehicles. Specially, authentication of messages with location context is not easy since the location of the vehicles is changing with time. Another vulnerability area in VC is that often because of the legal reasons, the vehicle identity and information is pinned to its owner's identity. However most VC applications use the location information of the vehicle and tracking the location of the vehicle violates the owner's privacy. Pseudonymization has been suggested as a possible solution. In this approach the vehicle's identity is replaced by a pseudonym to protect the driver's privacy.

Huang et al. [43] have proposed a vehicular cloud computing system called PTVC for improving privacy and trust based verification communication. The purpose of this system is to provide a better solution for selection of credible and trustworthy vehicles for forming a VC. The proposed PTVC scheme is composed of a few stages: system setup, privacy-preserving trust-based vehicle selection protocol, privacy-preserving verifiable computing protocol and trust management. In this scheme, a trust authority is responsible for execution and maintenance the the whole system and it generates public and private keys for vehicles and road side units (RSU). When a vehicle wants to join or form a VC, it will try to find the nearest available vehicles with the highest reputation. Participating vehicles that want to transfer data without worrying about leaking privacy, should first encrypt their data, which is done with the help of privacy preserving verifiable computing protocols. Each participating vehicle will receive feedbacks on the performance and participation which then helps in determining the reputation value of the vehicle. The authors have concluded that the security analysis showed that this proposed scheme is practical and effective against various security attacks. This system can identify the untrustworthy vehicles and blocks them from entering the VC.

17.9.2 Disaster Management Using VCs

The importance of emergency response systems and disaster management systems cannot be diminished due to the sudden nature of disasters and the damage, loss and destruction that it brings to human lives and as well as properties. In the past decade we have been seen many disasters such as the Earthquake in Haiti, in 2010 that caused over 200,000 deaths, left two million homeless and three million in need of emergency assistance.

At the time of disaster, VCs plays a very important role in helping with removing people from disastrous and damaged areas and transfer them to safe areas and therefore save many lives and also valuables including information. Raw et al. [70] proposed an Intelligent Vehicular Cloud Disaster Management System (IVCDMS) model and have analyzed the intelligent disaster management through VC network communication modes with the existing hybrid communication protocols. The architecture of this system consists of three interacting service layers named vehicular cloud infrastructure as a service, intelligence layer service and system interface service. The smart interface service helps in transportation of data from one place to another and it acquires data from various sources such as roadside units, smart cell phones, social network etc. The VC infrastructure as a service provides base platform and environment for the IVCDMS. The intelligence layer service provides the necessary computational models, algorithms and simulations; both stochastic and deterministic that then provides the emergency responses strategies by processing the available data from various resources.

Raw et al. [70] have implemented the communication V-P (vehicle-to-pedestrian) protocol which uses cellular phones and wireless network to improve the safety of travelers, pedestrians and drivers. This protocol helps drivers and pedestrians to get informed of one another and have enough time to take proper action and avoid probable accidents and hazards.

17.9.3 VC Data Collection for Traffic Management

VCs can help the ITS community by improving the traffic and accident information sharing. Chaqfeh et al. [18] have proposed an on-demand and pull-based data collection model for better information sharing between vehicles. When a vehicle wants to get information about a specific rout, it will create a route request (RREQ) message that contains details about the destination and the route that the vehicle is interested to get more information about. The requesting vehicle then broadcasts this message to its neighbors. The RREQ message is the re-routed to vehicles in the identified destination. The vehicles at the destination that receive the RREW message then publish their sensor data and gathered traffic information to the network by forming a dynamic VC. Finally a vehicle is selected as the broker to collect the information and transfer and communicate it with the infrastructure. Chaqfeh et al. [18] have showed via simulation that their VC-based approach can effectively support ITS with real-time data, even with not too many participating vehicles and concluded that the participation of small number of vehicles in a dynamic VC is enough for data transferring and data collection. However, these authors have ignored fundamental VC problems including whether VMs will be set up and how system failures can be handled in case vehicles leave the VC unexpectedly.

17.9.4 Sharing Real-Time Visual Traffic Information via VCs

With advances in technology, vehicles are equipped with more sophisticated sensing devices such as cameras and these features give vehicles the ability to act as mobile sensors that provide useful traffic information.

With advances in technology, vehicles are having better quality and improved cameras and sensing devices which can be used to collect traffic and road information.

Users can send or receive accurate and up-to-dated information about the road or traffic conditions by using vehicular cloud services. Kwak et al. [51] have presented an architecture for a possible collaborative traffic image-sharing system which allows drivers in the VC to report and share visual traffic information called NaviTweets. NaviTweets are selected and transformed into an informative snapshot summary of the route. This will provide reliable information about various details of the road condition, traffic, congestion, weather and any data that is useful for drivers and is beneficial for enhancing the safety and speed of the transportation.

Figure 17.6 illustrates an example case for using this system. Assume that Driver A want to get to a destination and does not know which route to take. He has two options, Route 1 and Route 2 and there is not enough available information regarding the safety of the road on the application that he is using. He decides to subscribe to the NaviTweets service to receive other driver's opinions about the road. Driver B that is driving on Route 1 posts a picture, via the VC, about an accident that occurred and resulted in congestion. Driver C that is driving on Route 2 sends a tweet about the slippery bridge and he indicates that there is no traffic on that road. The VC discards the older tweets from previous drivers and keeps the most up-to-date tweets regarding a rout. Driver A then receives a summary of each road and plans his route accordingly.

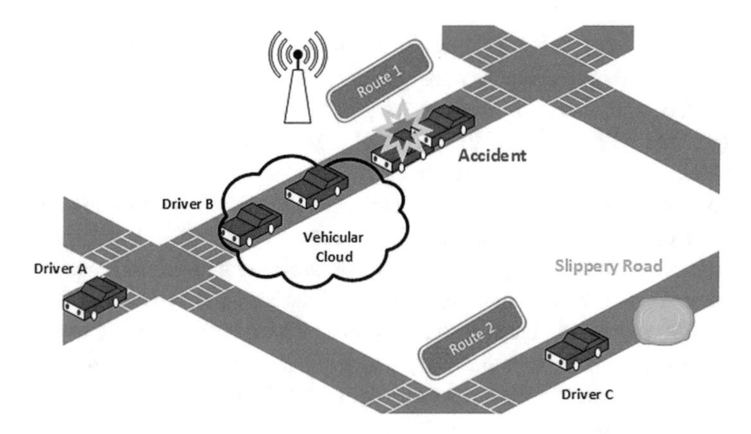

Fig. 17.6 Example scenario of the proposed architecture

17.10 Future Challenges for VCs

The main goal of this section is to point out a few important challenges for the VC.

17.10.1 A Large-Scale Implementation of VCs

As already mentioned in this chapter, to the best of the authors' knowledge, no large-scale implementation of Vcs has been reported in the literature. We are aware of two attempts at addressing this challenge: one is purely simulation-based and was reported in [26]. The other is a recent testbed that we discuss below.

Researchers often desire convenient, cost-effective and reconfigurable testing environments. Developing and testing VC applications is challenging and expensive. One solution to this problem is creating miniature testbeds using robots instead of real vehicles. Recently, this approach was explored by Lu et al. [54] who have attempted to build a VC testbed with different types of mobile robots. Their testbed is constructed from robot vehicles on mini cloud, network, remote cloud and management servers. The VC-bots can be managed and controlled with the help of a user interface (UI) that was built using java. Using this interface, users can see and track vehicles on a map. The interface is a graph-based model that shows straight roads as edges and intersections as nodes. The width of an edge shows the width of the road and the position or coordinate of each node, shows the location of the intersection. Parking lots and open spaces are shown as special nodes connected to the road network. Users can move the robot vehicles from one location to another, using the UI. An advantage of this design is the configurable map and road network.

The testbed contains 19 robot vehicles of four types: VC-trucks, VC-van, VC-sedan, and VC compacts. The types of robot vehicles vary in sensing devices, computing resources, and battery life and an independent WiFi network was used for the purpose of the robot vehicle management. Each robot vehicle is equipped with WiFi interface and also LTE modem, or the purpose of better network connection and in case the robot is out of the range of the WiFi network. In addition, several WiFi routers are deployed as roadside units (RSU).

In this testbed, the cloud resources contain robot vehicles on board mini cloud and remote cloud in the data center. The remote cloud consists of Linux servers managed by OpenStack and provides a cloud management user interface, centralized control and storage. A Kernel-based Virtual Machine (KVM) was installed on the robot vehicles to provide basic virtualization environment. Users can create virtual machines (VM) or import existing VM images into mini cloud, and run their programs and applications on these virtual machines. Live migration is also allowed from a robot vehicle to another.

The testbed of Lu et al. [54] is a work-in-progress and proof-of-concept prototype for VC-bots, an evolving platform for testing vehicular network and VC applications. To the best of our knowledge this is the first time researchers have implemented a testbed for VCs and it is a proof that this area of research should be explored and studied further.

17.10.2 Promoting Reliability and Availability in VCs

It has been pointed out repeatedly [16, 39, 56] that the huge success of CC was due, to a large extent, to the extremely system reliability and availability of datacenters. For example, Amazon and Google are striving to make customer experience more enjoyable by insisting on extremely high system availability under all imaginable conditions [11, 19].

However, as mentioned before, the dynamically changing availability of compute resources due to vehicles joining and leaving the VC unexpectedly leads to a volatile computing environment where promoting system reliability and availability is quite challenging. Yet, it is clear that if VCs are to see an adoption rate and success similar to CCs, reliability and availability issues must be addressed in VCs.

To understand this issue better, assume that a job is assigned to a vehicle currently in the VC. If the vehicle remains in the VC until the job completes, all is well. Difficulties arise when the vehicle leaves the VC before job completion. In this case, unless special precautionary measures are taken, the work is lost and the job must be restarted, taking chances on another vehicle, and so on, until eventually the job is completed. Clearly, losing the entire work of a vehicle, in case of a premature departure, must be mitigated. One possible method is to use checkpointing, a strategy originally proposed in databases [23, 77]. This strategy requires making regular backups of the job's state, and storing them in a separate location. In the event a vehicle leaves before job completion, the job can be restarted, on a new vehicle, using the last backup. While this seems simple and intuitive, checkpointing is not implemented efficiently in VCs. In fact, Ghazizadeh [30] has found that, in VCs, most checkpoints remain unused and introduce a very large overhead.

Alternatively, reliability and availability can be enhanced, as it is often done in conventional CCs, by employing various forms of redundant job assignment strategies [11, 15, 16, 20, 22, 50, 55, 65]. Indeed, the volatility of resources in VCs suggests employing job assignment strategies wherein each user job is assigned to, and executed by, two or more vehicles in the VC [32].

A well-known yardstick for assessing reliability of a system is the *Mean Time to Failure* (MTTF), defined as the expectation of the time until the occurrence of the next system failure [5, 7, 8, 14]. Related to MTTF is the *Mean Time To Repair* (MTTR) that quantifies the expected time until the system is operational again. Barlow and Proschan [8] defined system availability as the long-term probability that the system is operating at a specified time. Thus, the availability of a system can be expressed in terms of MTTF and MTTR as (see also [39], p. 34)

$$AVAIL = \frac{MTTF}{MTTF + MTTR}. \tag{17.1}$$

It is intuitively clear that the longer and more predictable the vehicle residency times in the VC, the easier it is to ensure system reliability and availability. Recently, Ghazizadeh et al. [32] have studied redundancy-based job assignment strategies and have derived analytical expressions for the corresponding MTTF. Similarly Florin et al. [25] have looked at the problem of reasoning about the reliability of MVCs. Building on the results in [32] Florin et al. have developed an analytic model for estimating the job completion time in VCs assuming a redundant job assignment strategy.

Florin et al. [27] has studied the reliability of VCs with *short vehicular residency times*, as is the case of shopping mall parking lots or the short-term parking lot of a large airport. They showed how to enhance system reliability and availability in these types of VCs through redundancy-based job assignment strategies. They offered a theoretical prediction of the corporate MTTF and availability offered by these strategies in terms of the MTTF of individual vehicles and of the MTTR. To the best of the authors' knowledge, this is the first paper that is looking at evaluating the MTTF and availability in VCs built on top of vehicles with a short residency time. Extensive simulations using data from shopping mall statistics [40–42] have confirmed the accuracy of our analytical predictions.

17.10.3 Big Data in VCs

One of the significant research challenges in VCs is to identify conditions under which VCs can support Big Data applications. It is apparent that Big Data applications, with stringent data processing requirements, cannot be supported by ephemeral VCs, where the residency time of vehicles in the cloud is too short for supporting VM setup and migration.

Quite recently Florin et al. [26] have identified sufficient conditions under which Big Data applications can be effectively supported by datacenters built on top of vehicles in a parking lot. This is the first time researchers are looking at evaluating the feasibility of the VC concept and its suitability for supporting Big Data applications. The main findings of [26] are that

- if the residency times of the vehicles are sufficiently long, and
- if the interconnection fabric has a sufficient amount of bandwidth, then Big Data applications can be supported effectively by VCs.

In spite of this result, a lot more work is needed to understand what it takes for VC to be able to support, in a graceful way, data- and processing-intensive applications.

17.10.4 VC Support for Smart Cities

Visionaries depict that the smart cities of the future are fully connected and the future vehicles are equipped with powerful computers capable of deep learning, solving complex problems and and cognitive computing [38, 52, 53, 78]. Such features will allow vehicles to be organized into ubiquitous networks and transfer information as well as find solution for many traffic situations which in return impacts the transportation experience significantly.

The rise of Internet of Things (IoT) and adoption of smart cities create opportunities for creative and efficient management and utilization of the available resources. One of the characteristics of smart cities is the interconnectivity of the city's infrastructure, which allows data to be collected from various human-generated or machine-generated sources. Future vehicles with powerful on-board computers, communication capabilities, and vast sensor arrays are perfect candidates in this hyper-connected environment to utilize into a fluid Vehicular Cloud (VC) capable of performing large-scale computations.

The way we see it, the main challenge for VCs in the context of Smart Cities should be aligned with the 2015–2019 strategic priorities recently spelled out by US-DOT [81]. In order to show relevance of VCs to Smart Cities, we propose to achieve the following objectives:

1. *Enhance urban mobility through information sharing*: VCs should combine detailed knowledge of real-time traffic flow data with stochastic predictions within a given time horizon to help (1) the formation of urban platoons containing vehicles with a similar destination and trajectory; (2) adjust traffic signal timing in order to reduce unnecessary platoon idling at traffic light; and (3) present the driving public with high-quality information that will allow them to reduce their trip time and its variability, eliminate the conditions that lead to congestion or reduce its effect.

2. *Avoid congestion of key transportation corridors through cooperative navigation systems*: Congestion-avoidance techniques that become possible in SC environments will be supplemented by route guidance strategies to reduce unnecessary idling and will limit environmental impact of urban transportation.
3. *Handling non-recurring congestion*: VCs will explore strategies to efficiently dissipate congestion, by a combination of traffic light retiming and route guidance to avoid more traffic buildup in congested areas.
4. *Demonstrate and evaluate VCs*: We will build a small-scale prototype for VCs. The evaluation of the traffic light retiming will be conducted in VCs through simulation. Several traffic simulation systems described in [29] will be evaluated before deciding on the right manner to simulate a platoon-aware system.

17.11 Concluding Remarks and Directions for Future Work

A few years ago, inspired by the success and emergence of CC, researchers have proposed Vehicular Cloud Computing (VCC) which is assumed to be a non-trivial extension of the conventional Cloud Computing (CC) paradigm. VCC was first motivated by understating the fact that the current and future smart vehicles are assembled with powerful computational and storage resources and several transceivers and sensing devices. These resources in vehicles are chronically underutilized and can be rented out to users or shared to solve traffic and safety related problems or used in many different means and applications. In this work, we briefly discussed the conventional Cloud Computing as well as the Mobile Cloud Computing. We provided an overview of the Vehicular Cloud Computing. We studied the details of the architecture and services provided by VCs. In addition we reviewed the recent researches and potential future applications. We also looked at some of the challenges of VCs and discussed possible solutions.

With the rise of Internet of things (IoT) and intelligent transportation systems (ITS) our cities, roads and vehicles are getting smarter and information and communication technologies are improving and becoming more advanced. Such features create great opportunity to not only create a safe, secure and comfortable journey for travelers but also provide the most efficient and least expensive solutions which is the purpose of Vehicular Cloud Computing. A concerted effort among industry, academia, auto-industry and the government is needed for new researches and implementations of VCs. Once adopted VCC will be the next paradigm shift with impressive impacts on economy, environment, safety and transportation.

References

1. M. Abuelela, S. Olariu, Taking VANET to the clouds, in *Proceedings of the ACM MoMM* (Paris, France, 2010)
2. Amazon Inc, Amazon web services (2010), http://aws.amazon.com
3. S. Arif, S. Olariu, J. Wang, G. Yan, W. Yang, I. Khalil, Datacenter at the airport: Reasoning about time-dependent parking lot occupancy. IEEE Trans. Parallel Distrib. Syst. **23**(11), 2067–2080 (2012)
4. M. Armbrust, A. Fox, R. Griffith, A.D. Joseph, R.H. Katz, A. Konwinski, G. Lee, D.A. Patterson, A. Rabkin, I. Stoica, M. Zaharia, Above the clouds: a berkeley view of cloud computing. Technical Report No. UCB/EECS-2009-28 (U.C. Berkeley, 2009)
5. A. Avizienis, J.-C. Laprie, B. Randell, C. Landwehr, Basic concepts and taxonomy of dependable and secure computing. IEEE Trans. Dependable Secur. Comput. **1**(1), 1–23 (2004)
6. K. Bachman, Study; teens still big mall shoppers (2017), http://www.adweek.com/news/advertising-branding/study-teens-still-big-mallshoppers-112487. Accessed 18 Jan 2017
7. R.F. Barlow, F. Proschan, *Mathematical Theory of Reliability* (Wiley, New York, 1967). 3-rd printing
8. R.F. Barlow, F. Proschan, *Statistical Theory of Reliability and Life Testing Probability Models* (Holt, Rinehart and Winston, New York, 1975)
9. B. Baron, M. Campista, P. Spathis, L.H. Costa, M. Dias de Amonim, O.C. Duarte, G. Pujolle, Y. Viniotis, Virtualizing vehicular node resources: feasibility study of virtual machine migration. Veh. Commun. **4**, 39–46 (2016)
10. L.A. Barroso, U. Hölzle, *The Datacenter as a Computer: An Introduction to the Design of Warehouse-Scale Machines* (Morgan & Claypool, San Rafael, California, 2009)
11. L.A. Barroso, J. Clidaras, U. Hölzle, *The Datacenter as a Computer: An Introduction to the Design of Warehouse-Scale Machines*, 2nd edn. (Morgan & Claypool, San Rafael, California, 2013)
12. A. Beloglazov, J. Abawajy, R. Buyya, Energy-efficient resource allocation heuristics for efficient management of datacenters for cloud computing. Future Gener. Comput. Syst. **28**, 755–768 (2012)
13. A. Bento, R. Bento, Cloud computing: a new phase in information technology management. J. Inf. Technol. Manag. **22**(1), 39–46 (2011)
14. B. Bergman, On reliability theory and its applications. Scandinavian J. Stat. **12**, 1–41 (1985)
15. D. Bhagavathi, P.J. Looges, S. Olariu, J.L. Schwing, A fast selection algorithms on meshes with multiple broadcasting. IEEE Trans. Parallel Distrib. Syst. **5**(7), 772–778 (1994)
16. R. Buyya, C. Vecchiola, S. Thamarai Selvi, *Mastering Cloud Computing: Foundations and Applications Programming* (Morgan Kaufman, Elsevier, 2013)
17. F. Chang, J. Dean, S. Ghemawat, W.C. Hsieh, D.A. Wallach, M. Burrows, T. Chandra, A. Fikes, R.E. Gruber, Bigtable: a distributed storage system for structured data, in *Proceedings 7-th USENIX Symposium on Operating Systems Design and Implementation, (OSDI06)* (Seattle, Washington, 2006)
18. M. Chaqfeh, N. Mohamed, I. Jawhar, J. Wu, Vehicular cloud data collection for intelligent transportation systems, in *Smart Cloud Networks & Systems (SCNS)* (IEEE, 2016), pp. 1–6
19. G. DeCandia, D. Hastorun, M. Jampani, G. Kakulapati, A. Lakshman, A. Pilchin, S. Sivasubramanian, P. Vosshall, W. Vogels, Dynamo: amazons highly available key-value store, in *Proceedings of 23-th ACM Symposium on Operating Systems Principles, (SOSP07)* (Stevenson, Washington, 2007)
20. J. Deng, S.-H. Huang, Y. Han, J. Deng, Fault tolerant and reliable computation in cloud computing, in *2010 IEEE GLOBECOM Workshops (GC Wkshps)* (2010), pp. 1601–1605
21. H.T. Dinh, C. Lee, D. Niyato, P. Wang, A survey of mobile cloud computing: architecture, applications, and approaches. Wirel. Commun. Mobile Comput. **13**(18), 1587–1611 (2013)
22. I. Egwutuoha, S. Chen, D. Levy, B. Selic, A fault tolerance framework for high performance computing in cloud, in *2012 12th IEEE/ACM International Symposium on Cluster, Cloud and Grid Computing (CCGrid)* (2012), pp. 709–710

23. R. Elmasri, S.B. Navathe, *Fundamentls of Database System*, 6th edn. (Addison-Wesley, Boston, 2011)
24. M. Eltoweissy, S. Olariu, M. Younis, Towards autonomous vehicular clouds, in *Proceedings of AdHocNets'2010* (Victoria, BC, Canada, 2010)
25. R. Florin, P. Ghazizadeh, A.G. Zadeh, S. Olariu, Enhancing dependability through redundancy in military vehicular clouds, in *Proceedings of IEEE MILCOM'2015* (Tampa, Florida, 2015)
26. R. Florin, S. Abolghasemi, A.G. Zadeh, S. Olariu, Big Data in the parking lot, in *Big Data Management and Procesing*, chapter 21 ed. by K.-C. Li, H. Jiang, A. Zomaya (Taylor and Francis, Boca Raton, Florida, 2017), pp. 425–449
27. R. Florin, A.G. Zadeh, P. Ghazizadeh, S. Olariu, Promoting reliability and availability in vehicular clouds. IEEE Trans. Cloud Comput. submitted (2017)
28. R. Florin, P. Ghazizadeh, A.G. Zadeh, S. El-Tawab, S. Olariu, Reasoning about job completion time in vehicular clouds. IEEE Trans. Intell. Transp. Syst. **18** (2017)
29. D. Gettman, L. Head, Surrogate safety measures from traffic simulation models (2003), http://www.fhwa.dot.gov/publications/research/safety/03050/03050.pd
30. P. Ghazizadeh, Resource allocation in vehicular cloud computing. Ph.D. Thesis, Old Dominion University (2014)
31. P. Ghazizadeh, S. Olariu, A.G. Zadeh, S. El-Tawab, Towards fault-tolerant job assignment in vehicular cloud, in *Proceecings IEEE SCC* (2015), pp. 17–24
32. P. Ghazizadeh, R. Florin, A.G. Zadeh, S. Olariu, Reasoning about the mean time to failure in vehicular clouds. IEEE Trans. Intell. Transp. Syst. **17**(3), 751–761 (2016)
33. A.S. Ghemawat, H. Gobioff, S.-T. Leung, The Google file system, in *Proceedings of 19-th ACM Symposium on Operating Systems Principles, (SOSP03)* (Bolton Landing, NY, 2003), pp. 29–43
34. D.V. Gibson, G. Kozmetsky, R.W.E. Smilor, *The Technopolis Phenomenon: Smart Cities, Fact Systems, Global Networks* (Rowman and Littlefield, 8705 Bollman Place, Savage, MD, 1992)
35. Google, Inc. Google app engine (2010), http://code.google.com/appengine/
36. L. Gu, D. Zeng, S. Guo, Vehicular cloud computing: a survey, in *Proceedings of IEEE Globecom Workshops* (2013), pp. 403–407
37. C. Harrison, I.A. Donnelly, The theory of smart cities, in *Proceedings of the 55th Annual Meeting of the International Society for the Systems Sciences, (ISSS'2011), Hull, U.K.* (2011)
38. D. Hatch, Singapore strives to become 'The Smartest City': Singapore is using data to redefine what it means to be a 21st-century metropolis (2013)
39. J.L. Hennessy, D.A. Patterson, *Computer Architecture a Quantitative Approach* (Morgan Kaufman, Elsevier, 2012)
40. H. Hu, C.R. Jasper, Men and women: a comparison of shopping mall behavor. J. Shopp. Center Res. **11**(1/2), 113–123 (2004)
41. H. Hu, C.R. Jasper, Consumer shopping experience in the mall: Conceptualization and measurement, in *Academy of Marketing Science Annual Conference Proceedings* (Coral Gables, Florida, 2007)
42. H. Hu, C.R. Jasper, A qualitative study of mall shoping behaviors of mature consumers. J. Shopp. Center Res. **14**(1), 17–38 (2007)
43. C. Huang, R. Lu, H. Zhu, H. Hu, X. Lin, PTVC: achieving privacy-preserving trust-based verifiable vehicular cloud computing, in *Proceedings of IEEE Global Communications Conference (GLOBECOM'2016)* (IEEE, 2016), pp. 1–6
44. R. Hussain, F. Abbas, J. Son, H. Oh, TIaaS: secure cloud-assisted traffic information dissemination in vehicular ad hoc networks, in *2013 13th IEEE/ACM International Symposium on Cluster, Cloud and Grid Computing (CCGrid)* (2013), pp. 178–17
45. IBM Inc, IBM smart cloud (2010), http://www.ibm.com/cloud-computing
46. JCDecaux, The mall phenomenon (2017), http://www.jcdecauxna.com/sites/default/files/assets/mall/documents/studies/mallphenomenon.pdf. Accessed 18 Jan 2017
47. JCDecaux, Teen marketplace (2017), http://www.jcdecauxna.com/sites/default/files/assets/mall/documents/studies/TeensFact.pdf. Accessed 18 Jan 2017

48. D. Kapil, E.S. Pilli, R. Joshi, Live virtual machine migration techniques: Survey and research challenges, in *Proceedings of 3rd International IEEE Advance Computing Conference (IACC)* (Ghaziabad, India, 2013)
49. P. Kaur, R.A. Virtual machine migration in cloud computing. Int. J. Grid Distrib. Comput. **8**(5), 337–342 (2015)
50. S. Y. Ko, I. Hoque, B. Cho, I. Gupta, Making cloud intermediate data fault-tolerant, in *Proceedings of the 1st ACM Symposium on Cloud Computing, SoCC '10* (ACM, New York, NY, USA, 2010), pp. 181–192
51. D. Kwak, R. Liu, D. Kim, B. Nath, L. Iftode, Seeing is believing: sharing real-time visual traffic information via vehicular clouds. IEEE Access **4**, 3617–3631 (2016)
52. K. Lakakis, K. Kyriakou, Creating and intelligent transportation system for smart cities: performance evaluation of spatial-temporal algorithms for traffic prediction, in *Proceedings of 14-th International Conference on Environmental Science and Technology* (Rhodes, Greece, 2015)
53. T. Litman, Autonomous vehicle implementation predictions: implications for transport planning. Presented at the 2015 Transportation Research Board Annual Meeting (2015)
54. D. Lu, Z. Li, D. Huang, X. Lu, Y. Deng, A. Chowdhary, B. Li, Vc-bots: a vehicular cloud computing testbed with mobile robots, in *Proceedings of First International Workshop on Internet of Vehicles and Vehicles of Internet* (ACM, 2016), pp. 31–36
55. S. Malik, F. Huet, Adaptive fault tolerance in real time cloud computing, in *2011 IEEE World Congress on Services (SERVICES)* (2011), pp. 280–287
56. D.C. Marinescu, *Cloud Computing, Theory and Applications* (Morgan Kaufman, Elsevier, 2013)
57. D.C. Marinescu, A. Paya, J.P. Morrison, S. Olariu, An approach for scaling cloud resource management. Clust. Comput. **20**(1), 909–924 (2017)
58. Microsoft Corporation, Windows azure (2010), http://www.microsoft.com/windowazure/
59. H. Moussanif, I. Khalil, S. Olariu, Cooperation as a service in VANET: implementation and simulation results. Mob. Inf. Syst. **8**(2), 153–172 (2012)
60. K. Nakano, S. Olariu, Randomized leader election protocols in radio networks with no collision detection, in *International Symposium on Algorithms and Computation, (ISAAC'2000)* (2000), pp. 362–373
61. K. Nakano, S. Olariu, J.L. Schwing, Broadcast-efficient protocols for mobile radio networks. IEEE Trans. Parallel Distrib. Syst. **10**(12), 1276–1289 (1999)
62. National Institute of Standards and Technology (NIST), NIST definition of cloud computing (2009), http://csrc.nist.gov/groups/SNS/cloud-computing/cloud-def-v15.doc
63. National Institute of Standards and Technology (NIST), DRAFT cloud computing synopsis and recommendationsdefinition of cloud computing (2011), http://csrc.nist.gov/publications/drafts/800-146/Draft-NIST-SP800-146.pdf
64. S. Olariu, M.C. Weigle (eds.) *Vehicular Networks: From Theory to Practice* (CRC Press/Taylor & Francis, 2009)
65. S. Olariu, M. Eltoweissy, M. Younis, ANSWER: AutoNomouS netWorked sEnsoR system. J. Parallel Distrib. Comput. **67**, 114–126 (2007)
66. S. Olariu, M. Eltoweissy, M. Younis, Towards autonomous vehicular clouds. ICST Trans. Mob. Commun. Appl. **11**(7–9), 1–11 (2011)
67. S. Olariu, I. Khalil, M. Abuelela, Taking VANET to the clouds. Int. J. Pervasive Comput. Commun. **7**(1), 7–21 (2011)
68. S. Olariu, T. Hristov, G. Yan, The next paradigm shift: from vehicular networks to vehicular clouds, in *Mobile Ad Hoc Networking Cutting Edge Directions*, ed. by S. Basagni, et al. (Wiley and Sons, New York, 2013), pp. 645–700
69. P. Olson, Why Google's Waze is trading user data with local governments (2014), http://www.forbes.com/sites/parmyolson/2014/07/07/why-google-waze-helps-local-governments-track-its-users/
70. R.S. Raw, A. Kumar, A. Kadam, N. Singh, et al., Analysis of message propagation for intelligent disaster management through vehicular cloud network, in *Proceedings of Second International Conference on Information and Communication Technology for Competitive Strategies* (ACM, 2016), p. 46

71. D. Rawat, D. Treeumnuk, D. Popescu, M. Abuelela, S. Olariu, Challenges and perspectives in the implementation of NOTICE architecture for vehicular communications, in *Proceedings of the International Workshop on Mobile Vehicular Networks (MoVeNet)* (Atlanta, GA, 2008)
72. T.K. Refaat, B. Kantarci, H.T. Mouftah, Virtual machine migration and management for vehicular clouds. Veh. Commun. **4**, 47–56 (2016)
73. R. Roess, E. Prassas, W. McShane, *Traffic Engineering*, 4th edn. (Pearson, Boston, 2011)
74. L. F. Sarmenta, Sabotage-tolerance mechanisms for volunteer computing systems. Future Gener. Comput. Syst. **18**(4), 561 – 572 (2002). Best papers from Symposium on Cluster Computing and the Grid (CCGrid2001)
75. SIRIT-Technologies, DSRC technology and the DSRC industry consortium (dic) prototype team. White Paper (2005), http://www.itsdocs.fhwa.dot.gov/research_docs/pdf/45DSRC-white-paper.pdf
76. Texas Transportation Institute, 2012 urban mobility report (2013), http://mobility.tamu.edu/ums/
77. S. Toueg, O. Babaoglu, On the optimum checkpoint selection problem. SIAM J. Comput. **13**(3), 630–649 (1984)
78. A.M. Townsend, *Smart Cities: Big Data, Civic Hackers, and the Quest for a New Utopia* (W. W. Norton, New York, NY, 2013)
79. US Department of Transportation. Standard Specification for Telecommunications and Information Exchange Between Roadside and Vehicle Systems - 5 GHz Band Dedicated Short Range Communications (DSRC) Medium Access Control (MAC) and Physical Layer (PHY) Specifications. ASTM E2213-03 (2003)
80. USDOT, The smart/connected city and its implications for connected transportation (2014)
81. USDOT, 2015–2019 Strategic Plan Intelligent Transportation Systems (ITS) (2015), http://www.its.dot.gov/strategicplan.pdf
82. X. Wang, S. El-Tawab, A. Alhafdhi, M. Almalag, S. Olariu, Toward probabilistic data collection in the NOTICE architecture. IEEE Trans. Intell. Transp. Syst. **17**(12), 3354–3363 (2016)
83. K. Watanabe, M. Fukushi, S. Horiguchi, Expected-credibility-based job scheduling for reliable volunteer computing. IEICE Trans. Inf. Syst. **93**(2), 306–314 (2010)
84. M. Whaiduzzaman, M. Sookhak, A. Gani, R. Buyya, A survey of vehicular cloud computing. J. Netw. Comput. Appl. **40**, 325–344 (2014)
85. G. Yan, S. Olariu, M.C. Weigle, M. Abuelela, SmartParking: A secure and intelligent parking system using NOTICE, in *Proceedings of the International IEEE Conference on Intelligent Transportation Systems* (Beijing, China, 2008), pp. 569–574
86. G. Yan, S. Olariu, D. Popescu, NOTICE: an architecture for the notification of traffic incidents. IEEE Intell. Transp. Syst. Mag. **4**(4), 6–16 (2012)
87. G. Yan, W.D., S. Olariu, M.C. Weigle, Security challenges in vehicular cloud computing. IEEE Trans. Intell. Transp. Syst. **4**(1), 6–16 (2013)
88. G. Yan, S. Olariu, J. Wang, S. Arif, Towards providing scalable and robust privacy in vehicular networks. IEEE Trans. Parallel Distrib. Syst. **25**(7), 1896–1906 (2014)
89. A.G. Zadeh, Virtual machine managent in vehicular clouds. PhD Thesis, Department of Computer Science, Old Dominion University (2018)
90. Q. Zhang, L. Cheng, R. Boutaba, Cloud computing: state-of-the-art and research challenges. J. Internet Serv. Appl. **1**(1), 7–18 (2010)
91. Y.A. Zuev, On the estimation of efficiency of voting procedures. Theory Prob. Appl. **42**(1), 73–81 (1998)
92. Y. Zuev, S. Ivanov, The voting as a way to increase the decision reliability. J. Franklin Inst. **336**(2), 361–378 (1999)

Erratum to: Secure Smart Vehicle Cloud Computing System for Smart Cities

Trupil Limbasiya and Debasis Das

Erratum to:
Chapter 15 in: B. S. P. Mishra et al. (eds.), *Cloud Computing for Optimization: Foundations, Applications, and Challenges*, Studies in Big Data 39, https://doi.org/10.1007/978-3-319-73676-1_15

The original version of the book was inadvertently published without incorporating the author corrections in Chapter 15, which have to be now incorporated. The erratum chapter and the book have been updated with the changes.

The updated online version of this chapter can be found at
https://doi.org/10.1007/978-3-319-73676-1_15

B. S. P. Mishra et al. (eds.), *Cloud Computing for Optimization: Foundations, Applications, and Challenges*, Studies in Big Data 39,
https://doi.org/10.1007/978-3-319-73676-1_18

Printed in the United States
By Bookmasters